CONTROL ENGINEERING

A Series of Reference Books and Textbooks

1. Nonlinear Control of Electric Machinery, *Darren M. Dawson, Jun Hu, and Timothy C. Burg*
2. Computational Intelligence in Control Engineering, *Robert E. King*
3. Quantitative Feedback Theory: Fundamentals and Applications, *Constantine H. Houpis and Steven J. Rasmussen*
4. Self-Learning Control of Finite Markov Chains, *A. S. Poznyak, K. Najim, and E. Gómez-Ramírez*
5. Robust Control and Filtering for Time-Delay Systems, *Magdi S. Mahmoud*
6. Classical Feedback Control: With MATLAB®, *Boris J. Lurie and Paul J. Enright*
7. Optimal Control of Singularly Perturbed Linear Systems and Applications: High-Accuracy Techniques, *Zoran Gajif and Myo-Taeg Lim*
8. Engineering System Dynamics: A Unified Graph-Centered Approach, *Forbes T. Brown*
9. Advanced Process Identification and Control, *Enso Ikonen and Kaddour Najim*
10. Modern Control Engineering, *P. N. Paraskevopoulos*
11. Sliding Mode Control in Engineering, *edited by Wilfrid Perruquetti and Jean-Pierre Barbot*
12. Actuator Saturation Control, *edited by Vikram Kapila and Karolos M. Grigoriadis*
13. Nonlinear Control Systems, *Zoran Vukić, Ljubomir Kuljača, Dali Donlagič, and Sejid Tesnjak*
14. Linear Control System Analysis & Design: Fifth Edition, *John D'Azzo, Constantine H. Houpis and Stuart Sheldon*
15. Robot Manipulator Control: Theory & Practice, Second Edition, *Frank L. Lewis, Darren M. Dawson, and Chaouki Abdallah*
16. Robust Control System Design: Advanced State Space Techniques, Second Edition, *Chia-Chi Tsui*
17. Differentially Flat Systems, *Hebertt Sira-Ramirez and Sunil Kumar Agrawal*

18. Chaos in Automatic Control, *edited by Wilfrid Perruquetti and Jean-Pierre Barbot*
19. Fuzzy Controller Design: Theory and Applications, *Zdenko Kovacic and Stjepan Bogdan*
20. Quantitative Feedback Theory: Fundamentals and Applications, Second Edition, *Constantine H. Houpis, Steven J. Rasmussen, and Mario Garcia-Sanz*

Quantitative Feedback Theory

Fundamentals and Applications

Second Edition

Constantine H. Houpis
Air Force Institute of Technology
Wright-Patterson AFB, Ohio

Steven J. Rasmussen
General Dynamics
Wright-Patterson AFB, Ohio

Mario Garcia-Sanz
Public University of Navarra
Pamplona, Spain

CRC Press
Taylor & Francis Group
Boca Raton London New York

CRC Press is an imprint of the
Taylor & Francis Group, an **informa** business
A TAYLOR & FRANCIS BOOK

CRC Press
Taylor & Francis Group
6000 Broken Sound Parkway NW, Suite 300
Boca Raton, FL 33487-2742

First issued in paperback 2019

ISBN-13: 978-0-8493-3370-5 (hbk)
ISBN-13: 978-0-367-39159-1 (pbk)
Library of Congress Card Number 2005053836

Library of Congress Cataloging-in-Publication Data

Houpis, Constantine H.
Quantitative feedback theory / Constantine H. Houpis, Steven J Rasmussen, Mario Garcia-Sanz.--2nd ed.
p. cm. -- (Control engineering ; 20)
Includes bibliographical references and index.
ISBN 0-8493-3370-9 (alk. paper)
1. Feedback control systems. I. Rasmussen, Steven J. II. Garcia-Sanz, Mario. III. Title.
IV. Control engineering (Taylor & Francis) ; 20.

TJ216.H69 2006
629.8'3--dc22 2005053836

Visit the Taylor & Francis Web site at
http://www.taylorandfrancis.com

and the CRC Press Web site at
http://www.crcpress.com

Preface

The objective of this text is to bridge the gap between the scientific (theoretical) and engineering methods of *Quantitative Feedback Theory* (*QFT*) by applying this multivariable robust control system design technique to real-world problems. Thus, the engineer is at the interface of the real-world with the body of knowledge and theoretical results available in the technical literature. Professor Isaac M. Horowitz, the developer of the QFT technique, has continually stressed the *transparency of QFT*; that is, the ability to visually relate the implementation of the design parameters to the real-world problem, from the onset of the design and throughout the individual design steps. Therefore, it is the purpose of this text to enable and enhance the ability of the engineering student and the practicing engineer to bridge the gap between the scientific and engineering methods. In order to accomplish this goal, the text is void of theorems, corollaries, and/or theoretical lemmas. In other words, this textbook stresses the engineering approach and not the scientific approach – *this is a textbook for engineers.*

Professor Horowitz began developing the *Quantitative Feedback Theory* (*QFT*) in early 1960. Since then great strides have been made in exploiting the full potential of the QFT technique. It is a frequency domain technique for designing a class of control systems for nonlinear plants. The abbreviation *QFT* should not be confused with the physics topic *Quantum Field Theory* (*QFT*). The catalyst that has propelled Horowitz's QFT to the level of being a major robust multiple-input multiple-output (MIMO) control system design method has been the development and availability of viable QFT computer-aided-design (CAD) packages. Through the close collaboration of Professor Horowitz with Professor C.H. Houpis and his graduate students, during the 1980s and the early part of the 90s, successful QFT designs involving structured parametric uncertainty had been completed and published by the Air Force Institute of Technology (AFIT) MS thesis students and faculty. During this period, the first multiple-input single-output (MISO) and MIMO QFT CAD packages were developed at AFIT. Another major accomplishment was the successful implementation and flight test of two QFT designed flight control systems, by Captain S.J. Rasmussen, of the Air Force Wright Laboratory, for the LAMBDA unmanned research vehicle in 1992 and 1993. Also, on April 28, 1995, Dr. Charles Hall of North Carolina State University, announced that four successful flight tests of QFT flight controllers had been accomplished.[53] Based upon these solid accomplishments, an aerospace engineering firm began applying the QFT design method in 1995. Other individuals throughout the world such as Professor E. Eitelberg, Professor E. Boje, Professor S. Jayasuriya, Professor P.O. Gutman, Professor A. Banos, Dr. D.J. Balance, Professor P.S.V. Nataraj, Professor O. Yaniv, Professor Y. Chait *et al* are also applying QFT to design real-world robust control systems. Professor D.S.

Bernstein ably points out the power of the frequency domain analysis, not only for linear systems, but also for nonlinear systems.[76]

In 1986 Professor Houpis published a technical report[41] which was the first attempt to bring under one cover the fundamentals of QFT. The second edition[2] of this Technical Report (TR) brought the material up to the state-of-the-art, and, like the first edition, aimed to provide students and practicing engineers with a document that presented QFT in a unified and logical manner. Refinements based upon the class testing of the first and second editions are incorporated in this text. Much of the material in this text is based upon the numerous articles written by Professor Horowitz, along with his colleagues, and the numerous lectures that he presented at the Air Force Institute of Technology.

In the summer of 1995 Professor Mario Garcia-Sanz, visiting research scientist at the Control System Centre, UMIST (UK), attended a QFT seminar given by Professor Houpis. At that time they started a collaborative effort in furthering the development of QFT and its application to real-world problems. A few months later, on his return to Spain, Professor Garcia-Sanz established the Control Engineering Group at the Public University of Navarra, with QFT as one of the main lines of research and with full encouragement from Professor Horowitz in the field. Since then the group works very closely with industry, especially with the M. Torres Company, CEIT and NASA-JPL. In fact, the group pioneered the application of QFT design techniques to new challenging real-world problems such as the control of large variable-speed multipole synchronous wind turbines, of waste water treatment plants and of multiple spacecraft in formation. Results of this theoretical and applied work are incorporated in this edition.

Both analog and sampled-data (discrete-time) MISO and MIMO feedback control systems are covered in detail. Extensive use is made of the MISO and MIMO QFT CAD packages to assist the reader in understanding and applying the QFT design technique. A PC MIMO QFT CAD, is available with this text, to be used with conjunction of the PC version of MATHEMATICA. This is accomplished by including appropriate examples in each chapter and problems at the end of each chapter in order to reinforce the fundamentals.

The authors have exerted meticulous care with explanations, diagrams, calculations, tables, and symbols. The reader is shown how to make intelligent real-world assumptions based upon mathematics and/or on a sound knowledge of the characteristics and the operating scenario of the plant to be controlled. The text provides a strong, comprehensive, and illuminating account of those elements that have relevance in the analysis and design of robust control systems and in bridging the gap between the scientific and the engineering methods.

Chapters 2 through 7 present the fundamentals of the QFT technique and the associated design procedure for the tracking control problem. This is followed by an extension of the technique to handle external disturbances, the regulator control problem, for a MIMO system. The remaining chapters focus on *bridging the gap between theory and the real world* by presenting *engineering rules* and the factors that are involved, such as simulations, implementation, etc., that are important in implementing a successful robust control system design. Extensive use of the MISO and MIMO QFT CAD packages (see Appendices C-E) is made for the MISO and MIMO examples throughout the text.

Chapter 2 discusses the reasons why feedback is required to achieve the desired system performance. This is followed by presenting an overview of QFT: the design objectives; what is structured parametric uncertainty and its Bode plot and Nichols Chart representations; performance specifications; design overview; and QFT basics. The chapter concludes with an insight into the QFT technique and the benefits of applying this technique.

As Horowitz and his colleagues have shown, an *mxm* MIMO control system can be represented by m^2 *MISO equivalents*. As a result, the QFT technique was initially developed for MISO control systems and was then extended to MIMO control systems. Thus, Chapter 3 presents the fundamentals of the QFT design technique for the analog MISO control system. This is followed by the extension of this technique to MISO discrete-time control systems in Chapter 4.

Chapter 5 begins with an introduction to MIMO plants having structured parametric uncertainty. This is followed by the introduction to the QFT MIMO compensated system formulation and the development of the effective MISO equivalents of a MIMO system. The remaining portion of the chapter adapts the MISO analog QFT design technique of Chapter 3 to the QFT robust design (Methods 1 and 2) of MIMO control systems containing structured parametric uncertainty. Chapters 5 through 9 and Chapter 11 utilize a diagonal **G** compensator/controller matrix whereas Chapter 10 discusses the use of non-diagonal **G** matrix (Method 3).

The QFT Method 1 design technique is discussed in detail in Chapter 6. Aspects such as performance tolerances, sensitivity analysis, simplification of the single-loop structure, high frequency and stability analyses, equilibrium and trade-offs, some universal design features, and the determination of bounds are discussed.

Chapter 7 thoroughly presents the details of Method 2. This method has the advantage of reducing the amount of over-design inherent in Method 1 and is applied when the diagonal dominance condition is not satisfied. Design equations for the *2x2* and the *3x3* MIMO systems are presented with corresponding design guidelines. The Binet-Cauchy formula is applied to determine if a minimum-phase (m.p.) effective plant (det **P**) is achievable.

In Chapter 8 the QFT technique is extended to the design of MIMO control systems with external disturbance inputs; i.e., the regulator control problem. From the state-space equations the corresponding plant and disturbance matrices are derived and the corresponding block diagram representation is shown. Based upon this formulation, the QFT m^2 MISO effective loop equations are derived for the regulator case. This QFT regulator design technique is applied to a real-world design example.

The remaining chapters enhance the emphasis of this text: to bridge the gap. Throughout the preceding chapters the elements that contribute to the *transparency of QFT* have been stressed, where applicable. Based upon these elements, and upon many years of applying the QFT robust control system design technique to many real-world nonlinear problems, *Engineering Rules* are presented in Chapter 9. These rules attempt to bridge the gap between QFT and the real-world problems.

Chapter 10 is a major contribution to this edition which presents the QFT design of non-diagonal compensators/controllers **G** for both tracking and disturbance rejection problems (Method 3). It introduces a non-diagonal **G** decoupling approach based upon a sequential design methodology.

There are a number of factors that contribute to making the decision that a satisfactory control system design has been achieved. Chapter 11 discusses the factors that are involved in a *control system design cycle,* factors that must be kept in mind from the onset of the design process and which bridge the gap between theory and the real world.

Time delay strongly limits the achievable performance of systems and complicates the design of controllers. Traditionally industry has been addressing this problem by the use of a Smith Predictor Controller (SPC). However, the SPC is very sensitive to plant model mismatch, and then, in the presence of plant uncertainty, it may result in a poor performance or an unstable system. Thus, Chapter 12 introduces a method to design a SPC using the QFT technique for systems containing time delay with plant uncertainty in both, the rational part and the time delay.

The concept of *Bridging the Gap*, discussed in Chapter 9, is exemplified in Chapter 13 by the discussion of two real-world QFT designed control systems: the design of a control system for a waste water treatment plant and for a large wind turbine synchronous generator.

Chapter 14 discusses the utilization of a weighting matrix in conjunction with control authority allocation in order to enhance the achievement of the maximum number of the desired system performance specifications. This utilization is especially useful in systems whose plant model matrix is non-square and is nonlinear.

Five new appendices have been added that enhance the technical content of the second edition. Section 2-3.5 is enhanced by Appendix A, *Template Generation.* Appendix B, *Inequalities Bounds Expressions*, provides further insight in the design procedure given in Section 3-16. Two examples of a non-diagonal compensator/controller design to supplement the new Chapter 10 are given in Appendices H (the tracking problem) and I (the disturbance rejection problem). Appendix J, *Elements For Loop Shaping*, provides further insight for the guidelines given in Sections 3-13 and 3-14.

The most effective way of expediting the transfer of QFT knowledge and its corresponding state-of-the-art material is for all authors on this subject to adhere to a standard list of QFT symbols. Thus, this text includes a section entitled *QFT Standard Symbols & Terminology*.

There are many worthwhile robust multivariable control system design techniques available in the technical literature, which are based on both the state-variable approach and on conventional control theory. The applicability of each design technique may be limited to certain classes of design problems. The control engineer must have a sufficiently broad perspective to be able to apply the appropriate technique to the right design problem. Some of the questions that the control engineer must keep in mind (see Chapter 11) in selecting a design method are: (a) can it solve a real-world problem? (b) is the method computationally intensive? (c) can it handle structured parametric uncertainty? (d) which method yields the lowest order compensator or controller? and (e) will the design method result in a control system that can be implemented on the target hardware, etc? For some techniques the designer is assisted by available computer-aided-design (CAD) packages.

This text presents a control system design based on quantitative feedback theory (QFT), which is a very powerful design method when plant parameters vary over a broad range of operating conditions. It incorporates the concept of designing a robust control system that maintains the desired system performance, not only over a prescribed region of plant parameter uncertainty, but also with a degree of control effector failures. The authors believe that this method has proven its applicability to the design of practical MISO and MIMO control systems with low order compensators (controllers) with minimal gain. This textbook provides students of control engineering and the practicing control engineer with a clear, unambiguous, and relevant account of the QFT technique.

The text is arranged so that it can also be used for self-study by the engineer in practice. Included are examples of feedback control systems in various areas of practice (electrical, aeronautical, mechanical, etc.) while maintaining a strong basic QFT text that can be used for study in any of the various branches of engineering. The text has been thoroughly class-tested, thus enhancing its value for classroom and self-study use. The computer-aided-design (CAD) packages of Appendices C through E are available to assist a control engineer in applying the QFT design method. The use of QFT CAD packages are stressed throughout the text. The authors wish to thank John W. Glass of Purdue University for review of the TOTAL-PC CAD software that accompanies this volume.

The authors wish to express their appreciation for the support and encouragement of Professor M. Pachter, at the Air Force Institute of Technology, during the preparation of the 1995 technical report.[41] His wealth of knowledge of the flight control area enhanced its value and his comments with respect to this text are appreciated.

The authors express their thanks to those students who have used this book and to the faculty who have reviewed it for their helpful comments and recommendations. Special recognition is given to Professors E. Eitelberg and E. Boje for their encouragement and support for this second edition. Appreciation is expressed to Dr. J.J. D'Azzo, Professor Emeritus of Electrical Engineering, and Professor Pachter, Air Force Institute of Technology, for the encouragement they have provided in the preparation of this text, as well as to John Glass of Purdue University, who reviewed the TOTAL-PC software.

Special appreciation is expressed to Professor Horowitz for the association and his collaboration with Professor Houpis during the period of 1981 through 1992. The personal relationship with him has been a source of inspiration and deep respect. Appreciation is also expressed to Dr. R.L. Ewing, Air Force Institute of Technology Research associate, for his work on improving and maintaining the TOTAL-PC CAD package.

Acknowledgment of Mr. E. Flinn, U.S. Air Force Wright Aeronautical Laboratories, and his colleagues Mr. J. Morris and Mr. D. Rubertus is made for the support and encouragement in developing and extending the QFT technique, during the 1980s. This support and encouragement was maintained by Mr. M. Davis (retired), Mr. J. Ramage, and Mr. Rubertus (retired) of the Air Force Research Laboratory. Further acknowledgment must be made of the support of Dr. S.N. Sheldon and the support given by the European Office of Aerospace Research and Development of the U.S. Air Force Office of Scientific Research during the latter part of the 20th century.

The thorough review of the 1st edition manuscript by Professor D. McLean of the University of Southampton was of immense value and was greatly appreciated by the authors. Contributions to this second edition were also made by present and former PhD students of Professor Mario Garcia-Sanz at the Control Engineering Group of the Public University of Navarra. Special mention is given to Dr. Xabier Ostolaza, Dr. Juan Carlos Guillen, Dr. Montserrat Gil, Dr. Igor Egana, Dr. Marta Barreras, and Ms. Irene Eguinoa. Their perception and insight have contributed to the clarity and rigor of the presentation.

Constantine H. Houpis
Mario Garcia-Sanz
Steven J. Rasmussen

Contents

QFT Standard Symbols & Terminology

SYMBOL

α_p	The specified peak magnitude of the disturbance response for the MISO system
a.l.	Arbitrarily large
a.s.	Arbitrarily small
$a_{ij} = Lm\ \tau_{ij}$	The desired lower tracking bounds for the MIMO system
$b_{ij} = Lm\ \tau_{ij}$	The desired upper tracking bounds for the MIMO system
a'_{ii}	The desired modified lower tracking bounds for the MIMO system: $a'_{ii} = a_{ii} + \tau_{c_{ii}}$
b'_{ii}	The desired modified upper tracking bounds for the MIMO system $b'_{ii} = b_{ii} - \tau_{c_{ii}}$
$\boldsymbol{B}_D(j\omega_i)$, $\boldsymbol{B}_R(j\omega_i)$, $\boldsymbol{B}_o(j\omega_i)$	The disturbance, tracking, and optimal bounds on $\boldsymbol{L}(j\omega_i)$ for the MISO system
B_h	Ultra high frequency boundary (UHFB) for analog design
B'_h	Ultra high frequency boundary (UHFB) for discrete design
$\mathrm{B_U} = \mathrm{Lm}\mathbf{T}_{\mathrm{R_U}}$	The *Lm* of the desired tracking control ratio for the upper bound of the MISO system

$B_L = \mathrm{Lm}\mathbf{T}_{R_L}$	The *Lm* of the desired tracking control ratio for the lower bound of the MISO system
B_S	Stability bounds for the discrete design
BW	Bandwidth
$\tau_{c_{ij}}$	Allotted portion of the *ij* output due to a cross-coupling effect for a MIMO system
$\delta_D(j\omega_i)$	The (upper) value of *Lm* $T_D(j\omega_i)$ for MISO system
$\delta_{hf}(j\omega_i)$	The dB difference between the augmented bounds of B_U and B_L in the high frequency range for a MISO system
$\delta_R(j\omega_i)$	The dB difference between B_U and B_L for a given ω_i for a MISO system
$\Delta\tau$	The difference between b_{ii} and a_{ii}, i.e., $\Delta\tau = b_{ii} - a_{ii}$
c_{ij}	The interaction or cross-coupling effect of a MIMO system
D	MISO and MIMO system external disturbance input
$\mathcal{D} = \{\boldsymbol{D}\}$	Script cap dee to denote the set of external disturbance inputs for a MIMO system $\mathcal{D} = \{\boldsymbol{D}\}$
F, $\boldsymbol{F} = \{\boldsymbol{f}_{ij}\}$	The prefilter for a MISO system and the *mxm* prefilter matrix for a MIMO system respectively
FOM	Figures of merit (FOM) (see Ref. 1)
G, $\boldsymbol{G} = \{g_{ij}\}$	The compensator or controller for a MISO system and the *mxm* compensator or controller matrix for a MIMO system, respectively. For a diagonal matrix $\boldsymbol{G} = \{g_i\}$
h.f.	High frequency
h.g.	High gain

γ, γ_i	The phase margin angle for the MISO system and for the ith loop of the MIMO system, respectively
γ_{ij}	A function only of the elements of a square plant matrix $\boldsymbol{P}$ (or $\boldsymbol{P}_e$)
k	A running index for sampled-data systems where $k = 0,1,2,...$
kT	The sampled time
J	The number of plant transfer functions for a MISO system or plant matrix for a MIMO system that describes the region of plant parameter uncertainty where $\iota = 1, 2,..., J$ denotes the particular plant case in the region of plant parameter uncertainty
λ	The excess of poles over zeros of a transfer function
L_o, L_{o_i}	The optimal loop transmission function for the MISO system and the ith loop of the MIMO system, respectively
LHP	Left-half-plane
LTI	Linear-time-invariant
MIMO	Multiple-input multiple-output; more than one tracking and/or external disturbance inputs and more than one output
MISO	Multiple-input single-output; a system having one tracking input, one or more external disturbance inputs, and a single output
M_L, M_{L_i}	The specified closed-loop frequency domain overshoot constraint for the MISO system and for the ith loop of a MIMO system, respectively. This overshoot constraint may be dictated by the phase margin angle for the specified loop transmission function
m.p.	Minimum-phase

n.m.p.	Nonminimum-phase
NC	Nichols Chart
ω_b	The symbol for bandwidth frequency of the models for T_{R_U}, T_{R_L} and $\boldsymbol{T} = \{t_{ij}\}$
ω_ϕ, ω_{ϕ_i}	Phase margin frequency for a MISO system and for the ith loop of a MIMO system, respectively
ω_s	Sampling frequency
P	MISO plant with uncertainty
$P_\iota = \{(p_{ij})_\iota\}$	$m x \ell$ MIMO plant matrix where $(p_{ij})_\iota$ is the transfer function relating the ith output to the jth input for plant case ι
$\mathcal{P}$	Script cap pee to denote a set that represents the plant uncertainty for J cases in the region of plant uncertainty, i.e., $\mathcal{P} = \{\boldsymbol{P}_\iota\}$ for a MIMO system
$\boldsymbol{P}_d$	$m_d x m$ MIMO external disturbance matrix
$\boldsymbol{P}_F$	Plant model for a tracking and external disturbance input system which is partitioned to yield $\boldsymbol{P}_e$ and $\boldsymbol{P}_d$
$\boldsymbol{P}_\iota^{-1} = \{(p_{ij}^*)_\iota\}$	The inverted plant matrix for plant case ι where $\ell = m$
$\boldsymbol{P}_{e_\iota} = \boldsymbol{P}_\iota \boldsymbol{W}$	The mxm effective plant matrix when $\boldsymbol{P}_\iota$ is not a square plant matrix and $\boldsymbol{W}$ is an ℓxm weighting or a squaring-down matrix
QFD	Quantitative feedback design based on quantitative feedback theory
$\boldsymbol{Q}_\iota = \{(q_{ij})\}$	An $\ell x \ell$ matrix whose elements are given by $(q_{ij}) = (1/p_{ij}^*)_\iota$

$\mathscr{Q}$	Script cap que to denote a set that represents the plant uncertainty for a MIMO system, i.e., $\mathscr{Q} = \{\boldsymbol{Q}_\iota\}$
R, $\boldsymbol{R} = \{r_i\}$	The tracking input for a MISO system and the tracking input vector for a MIMO system, respectively
RHP	Right-half-plane
SP	Smith Predictor
s	second(s)
T	Sampling time
$\Im \boldsymbol{P}(j\omega_i\}$	Script cap tee in conjunction with $\boldsymbol{P}$ or (q_{ii}) denotes a template, i.e., $\Im \boldsymbol{P}(j\omega_i)$ and $\Im \boldsymbol{Q}(j\omega_i)$ frequency, for a MISO and MIMO plants respectively
T_{R_U}	The desired MISO tracking control ratio that satisfies the specified upper bound FOM
T_{R_L}	The desired MISO tracking control ratio that satisfies the specified lower bound FOM
T_D	The desired MISO disturbance control ratio which satisfies the specified FOM
T_{R_ι}, T_{D_ι}	The MISO tracking and disturbance control ratios for case ι
$T_\iota = \{(t_{ij})_\iota\}$	The mxm MIMO tracking control ratio matrix for plant case ι
$T_{d_\iota} = \{(t_{d_{ij}})_\iota\}$	The mxm MIMO external disturbance control ratio matrix
$\Im_R$	The script cap tee denotes the set that represents the tracking control ratios for J cases, i.e. $\Im_R = \{T_{R_\iota}\}$ for the MISO system and $\Im_{R_i} = \{(\boldsymbol{T}_{R_\iota})_i\}$ for the MIMO system

$\Im_D$ — The script cap tee denotes the set that represents the disturbance control ratios for J cases. i.e., $\Im_D = \{T_{D_\iota}\}$ for the MISO system and $\Im_c = \{(\boldsymbol{T}_{c_{ij}})_\iota\}$ for the MIMO system

τ_{ij} — A set of assigned tolerances on t_{ij} where $a_{ii}^{'}$ and $b_{ii}^{'}$ and $\tau_{c_{ij}}$ are the assigned tolerances for tracking and cross-coupling rejection, respectively

U — The *mx1* controller input vector

UHFB — The ultra high frequency boundary

v, $\boldsymbol{v}$ — The MISO prefilter output and the *mx1* MIMO prefilter output vector, respectively

V — $\lim_{\omega\to\infty} \{Lm\ \boldsymbol{P}_{max} - Lm\ \boldsymbol{P}_{min}\}$ is the dB limiting value for a MISO plant

V_i — $\lim_{\omega\to\infty} \{Lm\ (\boldsymbol{q}_{ii})_{max} - Lm\ (\boldsymbol{q}_{ii})_{min}\}$ is the ith loop template *dB* limiting value for a MIMO plant

$\boldsymbol{W} = \{w_{ij}\}$ — The weighting or squaring-down or mixer matrix

$w' = u + jv$ — *w′-domain variable*; the use of u and v must be interpreted in context

Y, $\boldsymbol{Y} = \{y_{ij}\}$ — The output of a MISO system and the output matrix of a MIMO system, respectively, where $y_{ij} = y_{r_i} + y_{c_{ij}}$

$\boldsymbol{y}$ — The *mx1* plant output vector

y_{r_i} — Is that portion of the ith output due to the ith input

$y_{c_{ij}}$ — Is that portion of the ith output due to c_{ij} (cross-coupling effect or interaction of the other loops)

1

INTRODUCTION

1-1 INTRODUCTION

Automatic control systems permeate life in all advanced societies today. This development has evolved more complex systems in which nonlinearities need to be more effectively addressed in the design process. Such systems act as catalysts in promoting progress and development propelling society into the 21st century. Technological developments have made it possible for high-speed bullet trains; exotic vehicles capable of exploration of other planets and outer space; safe, comfortable and efficient automobiles; sophisticated civilian and military aircrafts; efficient robotic assembly lines; advanced renew energy plants; and efficient environmentally friendly pollution controls for factories. The successful operation of all of these systems depends on the proper functioning of the large number of control systems used in such ventures.

1-2 THE ENGINEERING CONTROL PROBLEM[1]

In general, a control problem can be divided into the following steps:

1. A set of performance specifications is established.
2. The performance specifications establish the control problem.
3. A set of linear differential equations and algebraic equations that describe the physical system is formulated or a system identification technique is applied in order to obtain the plant model transfer functions.
4. A control-theory design approach, aided by available computer-aided-design (CAD) packages or specially written computer programs, involves the following:

   (a) The performance of the basic (original or uncompensated) system is determined by application of one of the available methods of analysis (or a combination of them).

(b) If the performance of the original system does not meet the required specifications a control design method is selected that will improve the system's response.
(c) For plants having structured parameter uncertainty, the Quantitative Feedback Theory (QFT) design technique may be used. Parametric uncertainty is present when parameters of the plant to be controlled vary slowly during its operation or when they are not precisely known, as explained in Chapter 2.

5. Performing a simulation of the designed nonlinear system.
6. Implementing and testing the actual system.

Design of the system to obtain the desired performance is the control problem. The necessary basic equipment is then assembled into a system to perform the desired control function. Although most systems are nonlinear, in many cases the non-linearity is small enough to be neglected, or the limits of operation are small enough to allow a linear analysis to be used. This textbook only considers nonlinear systems having structured parametric uncertainty.

A basic system has the minimum amount of equipment necessary to accomplish the control function. After a control system is synthesized to achieve the desired performance, based upon the nonlinear design technique that is chosen, final adjustments can be made in a simulation, or on the actual system, to take into account the non-linearities that were neglected. A computer is generally used in the design, depending upon the complexity of the system.

The essential aspects of the control system design process are illustrated in Fig. 1.1. The intent of this figure is to give the reader an overview of what is involved in achieving a successful and practical control system design. The aspects of this figure that present the factors that help in *bridging the gap* between theory and the real-world are addressed in the next paragraph. While accomplishing a practical control system design, the designer must keep in mind that the goal of the design process, besides achieving a satisfactory theoretical robust design, is to implement a control system which meets the functional requirements. In other words, during the design process one must keep the real world in mind. For instance, in performing the simulations, one must be able to interpret the results obtained, based upon a knowledge of what can be reasonably expected of the plant that is being controlled. For example, in performing a time simulation of an aircraft's transient response to a pilot's maneuvering command to the flight control system, the *simulation run time* may need to be only 5 *s* since by that time a pilot would have instituted a new command signal. If within this 5 *s* window the performance specifications are satisfied, then it will be deemed that a successful design has been achieved. However, if the performance of interest is the steady-state response, then the simulation run-time must be considerably longer. Another real-world factor is *control authority allocation*, that is, the manner in which the available control power is assigned to the control effectors.

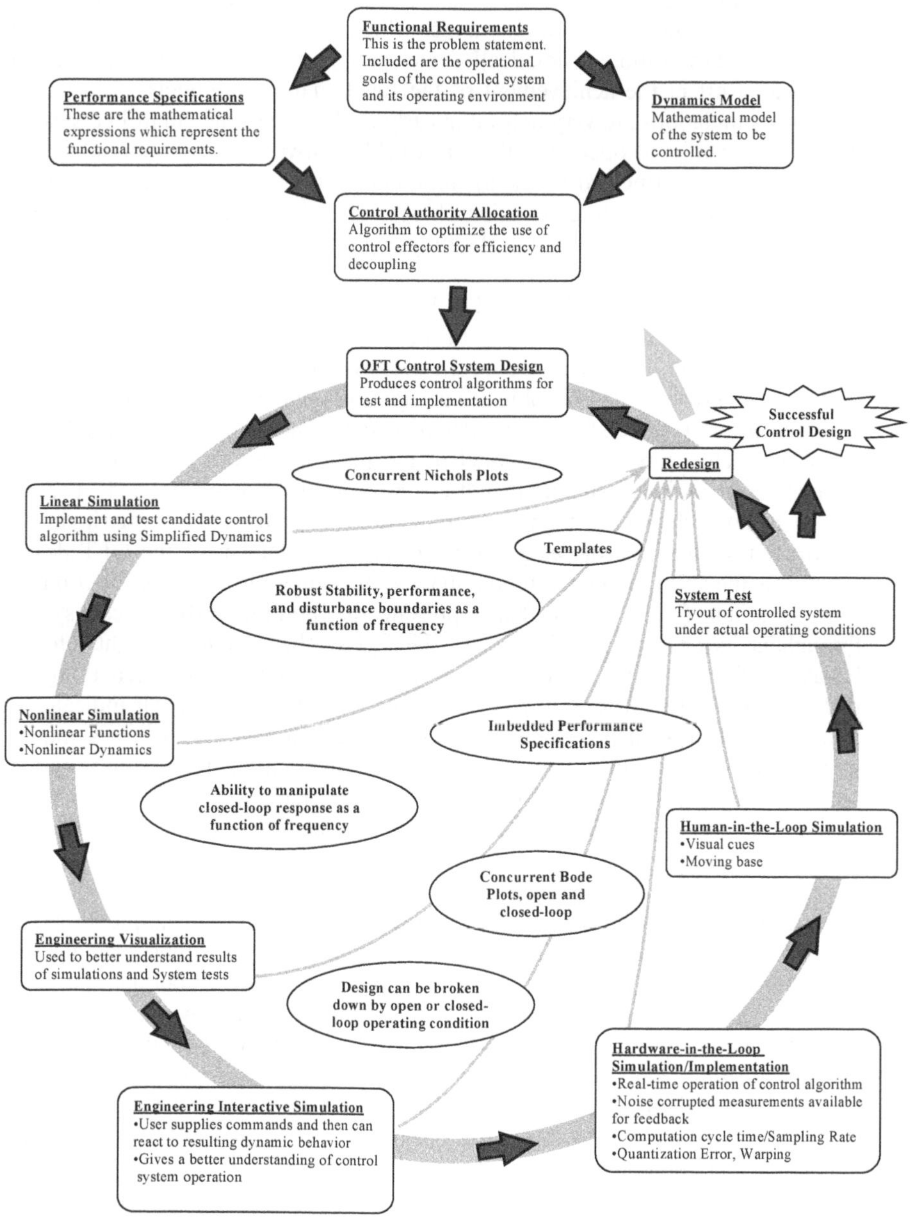

Fig. 1.1 The QFT control system design process: bridging the gap.

This allocation must be based upon a thorough knowledge of the plant that is being controlled and the conditions under which the plant will operate. Position saturation and even more dramatically, rate saturation of the output effectors will significantly affect the achievement of the functional requirements. Linear and nonlinear simulations are very helpful in early evaluation of the controlled system, but if the system is to operate in the real world, hardware-in-the-loop and system tests must be performed to check for un-modeled effects not taken into account during the design and implementation phases. In order to be a successful control system designer, an individual must be fully cognizant of the role corresponding to each aspect illustrated in Fig. 1.1.

Bridging the gap, as illustrated in Fig. 1.1, is enhanced by the transparency of the metrics depicted by the oval items in the interior of the QFT design process. A key element of QFT is embedding the performance specifications, at the onset of the design process. This establishes design goals that enhance and expedite the achievement of a successful design. Another important element is the creation of templates. They represent the model of the plant at various frequencies. The template sizes indicate whether or not a robust design is achievable. If a robust design is not achievable, then the templates can be used as a metric in the reformation of the control design problem. Another element of the QFT design process, is the ability to concurrently analyze frequency responses of the J linear-time invariant (LTI) plants that represent the nonlinear dynamical system throughout its operating environment. This gives the designer insight into the behavior of the system. The designer can use this insight for such things as picking out key frequencies to use during the design process, as an indicator of potential problems such as non-minimum phase behavior, resonance's modes, noise rejection, bandwidth, stability, and as a tool to compare the nonlinear system with the desired performance boundaries. The next element of QFT consist of the design boundaries. During the actual loop shaping process, the designer uses boundaries plotted on the Nichols chart. These boundaries are only guidelines and the designer can exercise engineering judgment to determine if all the boundaries are critical or if some of the boundaries are not important. For example, based on knowledge of the real world system, the designer may determine that meeting performance boundaries below a certain frequency is not important, but it is important to meet the disturbance rejection boundaries below that frequency. Once the initial design has been accomplished, all of the J loop transmission functions can be plotted on a Nichols chart to analyze the results of applying the designed compensator (controller) to the nonlinear system. This gives the designer a first look at any areas of the design that may present problems during simulation and implementation. The last two elements of the QFT design process that help *bridging the gap* is the relation of the controlled system's behavior to the frequency domain design and the operating condition. These relationships enable the designer to better analyze simulation or system test results for problems in the control design. To obtain a successful control design, the controlled system must meet all of the requirements during simulation and system test. If the controlled system fails any of the simulation or system tests, then, using the design elements of QFT, the designer can trace that failure back through the design process and make necessary adjustments to the design. QFT provides many metrics that provide the link between the control design process and real world implementation; this is the *transparency of QFT.*

1-3 QUANTITATIVE FEEDBACK THEORY (QFT)[2-7,14,18-24,26-32,38,52]

QFT is a unified theory that emphasizes the use of feedback for achieving the desired system performance tolerances despite plant uncertainty and plant disturbances. QFT *quantitatively* formulates these two factors in the form of (*a*) the set $\mathfrak{I}_R = \{T_R\}$ of acceptable command or tracking input-output relationships and the set $\mathfrak{I}_D = \{T_D\}$ of acceptable disturbance input-output relationships, and (*b*) a set $\mathcal{P} = \{P\}$ of possible plants which include the uncertainties. The objective is to guarantee that the control ratio $T_R = Y/R$ is a member of $\mathfrak{I}_R$ and $T_D = Y/D$ is a member of $\mathfrak{I}_D$, for all plants P which are contained in $\mathcal{P}$. QFT has been developed for control systems which are both linear and nonlinear, time-invariant and time-varying, continuous and sampled-data, minimum and non-minimum phase, uncertain multiple-input single-output (MISO) and multiple-input multiple-output (MIMO) plants, and for both output and internal variable feedback. It has also been extended to some classes of uncertain distributed systems (in which the plant is described by partial differential equations) and the feedback and specifications are also distributed.

The representation of a MIMO plant with ℓ inputs and m outputs is shown in Fig.1.2. The QFT synthesis technique for highly uncertain linear time-invariant MIMO plants has the following features:

1. The MIMO synthesis problem is converted into a number of single-loop feedback problems in which parameter uncertainty, external disturbances, and performance tolerances are derived from the original MIMO problem. The solutions to these single-loop problems represent a solution to the MIMO plant. It is not necessary to consider the complete system characteristic equation.

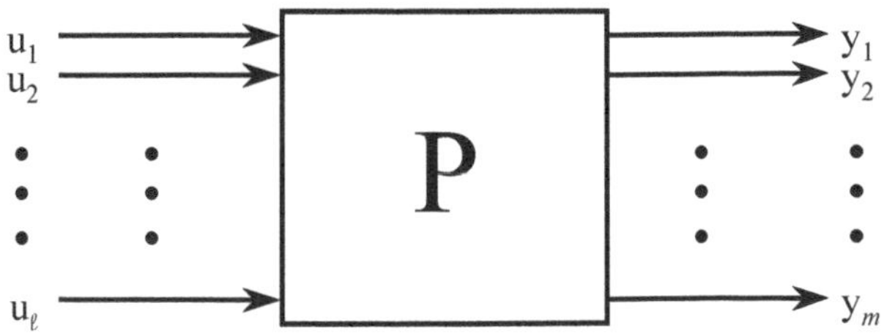

Fig. 1.2 A MIMO plant.

2. The design is tuned to the extent of the uncertainty and the performance tolerances.

This design technique is applicable to the following problem classes:

1. Single-input single-output (SISO) linear-time-invariant (LTI) systems.
2. SISO nonlinear systems. These are rigorously converted to equivalent class 1 systems whose solutions are guaranteed to work for a large problem class.
3. MIMO LTI systems. The performance specifications for each individual closed-loop system transfer function and for all the closed-loop disturbance response functions must be specified.
4. MIMO nonlinear systems. They are rigorously converted to equivalent class 3 systems whose solutions are guaranteed to work for a large class.
5. Distributed systems.
6. Sampled-data systems as well as continuous systems for all of the preceding.
7. Minimum and non-minimum phase systems.
8. Time-delay systems.

QFT is a frequency-domain technique that is applied in this text to MISO systems. Problem classes 3 and 4 are converted into equivalent sets of MISO systems to which the QFT design technique is applied. The objective is to solve the MISO problems, i.e., to find compensation functions which guarantee that the performance tolerances for each MISO problem are satisfied for all P in $\mathcal{P}$. The amount of feedback designed into the system is then tuned to the desired performance sets $\mathfrak{I}_R$ and $\mathfrak{I}_D$ and the given plant uncertainty set $\mathcal{P}$. Also, time-varying and nonlinear uncertain plant sets can be converted into equivalent MISO LTI plant problems to which the MISO frequency-domain technique can be readily applied and where the fundamental tradeoffs are highly visible.

Professor Horowitz's paper *Survey of Quantitative Feedback Theory (QFT)*[52] is an excellent presentation of his robust multivariable control system design technique. The principal properties of QFT and existence theorems are presented along with other important aspects of this technique.

1-4 CONTROL THEORY BACKGROUND

For an understanding of the analog QFT design technique, it is necessary that the reader have a good foundation of fundamental control theory, as covered by the first 14 chapters of Ref. 1. A good understanding of the material in Chapters 1 through 6 and Chapters 11 through 13 of Ref. 13 is required for the study of the discrete-time QFT design technique. Any other comparable textbooks covering similar material (conventional control theory and introductory state-space theory) are satisfactory.

LTI mathematical models that represent the region of structured parametric uncertainty are used in the linear analysis presented in this text. Once a physical system has been described by a set of mathematical equations, they are manipulated to achieve an appropriate mathematical format. When this has been done, the subsequent method of analysis is independent of the nature of the physical system; i.e., it does not matter whether the system is electrical, mechanical, etc. This technique helps the designer to spot similarities based upon previous experience.

1-5 DEFINITIONS AND SYMBOLS

As is the normal procedure in undertaking the study of a new subject matter, one quite often desires to refer to the technical literature for a broadening perspective in enhancing his or her understanding of the material. Thus, a most effective way of enhancing this understanding and expediting the state-of-the-art is for all authors on QFT to adhere to a standard list of definitions and symbols. The section at the beginning of this text entitled *QFT Standard Symbols & Terminology* is consistent with Professor Horowitz's writings.

1-6 QFT APPLICATIONS

The advent of the QFT computer-aided-design (CAD) packages[1,13,15,36] at the Air Force Institute of Technology in the late 80s and early 90s accelerated the application of QFT to many real-world problems. The following sections highlight some of the typical areas of application such as: process control systems, idle speed control for an automotive fuel injected engine, welding control systems, hydraulic actuator control system, wastewater treatment control system, aircraft propulsion control systems, aircraft flight control systems, design of robot controllers, operational amplifier control systems, wind turbine control systems, the design of a position control system utilizing an industrial pneumatic actuator, and the design of an robust control system for flexible manipulators involving articulating flexible structures. Some of these examples are described in the following sections.

1-6.1 QFT AND ROBUST PROCESS CONTROL[9]

Professor P.S.V. Nataraj and his graduate students of The Indian Institute of Technology, Bombay, India have tackled nine problems falling into various process control classes which have been successfully solved using QFT. The specific problems with corresponding problem classes are listed in Table 1.1.

Table 1.1 Process control classes

No.	System	Problem Class
1	Cascade of 5 CSTRs	Lumped, LTI, SISO, mp, Multiple-loop (6-DOF)
2	Distillation col. with side streams	Lumped, LTI, 3x3 MIMO (2-matrix-DOF)
3	Fluidized catalytic cracker	Lumped, LTI, 2x2 MIMO, mp (2-matrix-DOF)
4	High purity distillation	Lumped, LTI, 2x2 MIMO, mp (2-matrix-DOF)
5	Isothermal CSTR	Lumped, nonlinear, SISO (2-DOF)
6	Exothermic CSTR	Lumped, nonlinear, SISO (2-DOF)
7	Exothermic CSTR	Lumped, nonlinear, 2x2 MIMO (2-DOF)
8	Isothermic-fixed	Dist., linear, SISO (2-DOF) bed catalytic reactor
9	Heat equation	Dist., linear, SISO (2-DOF)

Notation: CSTR - Continuous stirred tank reactor(s), DOF - Degree-of-Freedom col - column, mp - minimum-phase, LTI - Linear-time-invariant, Dist - Distributed

1-6.2 IDLE SPEED CONTROL FOR AUTOMOTIVE FUEL INJECTED ENGINE[42,44]

The following is the abstract from the paper entitled "Robust Controller Design and Experimental Verification of I.C. Engine Speed Control" by Dr. M. A. Franchek and G. K. Hamilton, School of Mechanical Engineering, Purdue University.[44]

> Presented in this paper is the robust idle speed control of a Ford 4.6L V-8 fuel injected engine. The goal of this investigation is to design a robust feedback controller that maintains the idle speed within a *150 rpm* tolerance of about *600 rpm* despite a *20 Nm* step torque disturbance delivered by the power steering pump. The controlled input is the by-pass air valve which is subjected to an output saturation constraint. Issues complicating the controller design include the nonlinear nature of the engine dynamics, the induction-to-power delay of the manifold filling dynamics, and the saturation constraint of the by-pass air valve. An experimental verification of the proposed controller, utilizing the nonlinear plant, is included.

The authors' control system is based on controlling an engine such as the one shown in Fig. 1.3. The authors show in their paper that they met all the design objectives and have achieved excellent results.

Fig. 1.3 Fuel injected engine.

1-6.3 WELDING CONTROL SYSTEMS[9]

Mr. A. E. Bently, Process Development and Fabrication Division, Sandia National Laboratories has published the following two papers:

> "Arc Welding Penetration Control Using Quantitative Feedback Theory,"involves a control system that is designed for arc weld penetration control. The feedback signal is obtained by measuring the amount of visible and near-infrared light emitted from the back side of the weld. The system is sensitive enough to use a fiber-optic cable for transmitting the light from the weld to the sensor. This facilitates welding assemblies with limited access to the underside of the weld. Welds of constant penetration have been demonstrated in tests with travel speeds varying from 1.5 to 6 in/min (*0.64-2.54 mm/sec*), and with *200 percent* changes in part thickness. The system also compensates for sharp discontinuities in heat sinking and arc length.

The paper entitled, "Pinch Weld Quality Control Using Quantitative Feedback Theory," involves a system that is based on electrode displacement feedback that greatly improves the quality control of the pinch welding process. A correlation between weld quality and electrode displacement is established for constant force. The QFT designed control system is capable of producing repeatable welds of consistent thickness (and thus consistent quality), with wide variations in weld parameters. This is the first time feedback control had been successfully applied to pinch welding. For both applications Mr. Bently met all the design objectives and achieved excellent results.

1-6.4 CONTROL SYSTEM FOR AN ACTUATOR PLANT[45]

Actuators can be regarded as feedback control systems in their own right. In this application the mechanical feedback link commonly used in actuators is opened and replaced by a sensor and an electronic controller which drives the valve. The latter governs the piston, which is the power element. This control design problem entails the design of a robust controller for an actuator, see Fig. 1.4, that: (1) takes into account the aging of some of the actuator components over its expected lifetime; and (2) the manufacturing tolerances of actuators, such that when an actuator needs to be replaced, the overall control system's robustness is maintained by the replacement. These two factors result in the structured parametric uncertainty of the hydraulic actuator. The resultant QFT design of the actuator's control system achieved the desired degree of robustness.

1-6.5 VISTA F-16 FLIGHT CONTROL SYSTEM (INCLUDING CONFIGURATION VARIATION)[46]

The design of the robust flight control system (FCS) for the VISTA F-16 of Fig. 1.5 was accomplished by an Air Force Institute of Technology student who is an F-16 pilot. He was able to utilize his real-world knowledge of the aircraft and its handling qualities to achieve the desired robust FCS. Traditionally, flight control engineers have taken a conservative, brute force approach to designing a full envelope FCS for an aircraft. First, many design points, which for this design were points representing airspeed vs altitude, within and along the border of the flight envelope plot were selected. Second, individual compensator designs were accomplished for each of these points. Third, smooth transitions between these compensators must be engineered. Making the transitions imperceptible to the pilot is very difficult and time-consuming because each airspeed-altitude design point can be approached from an infinite number of initial conditions. Obviously, if the number of the design points can be reduced, thus reducing the number of transitions required, the design process can be made more efficient, and the resulting FCS is less complex.

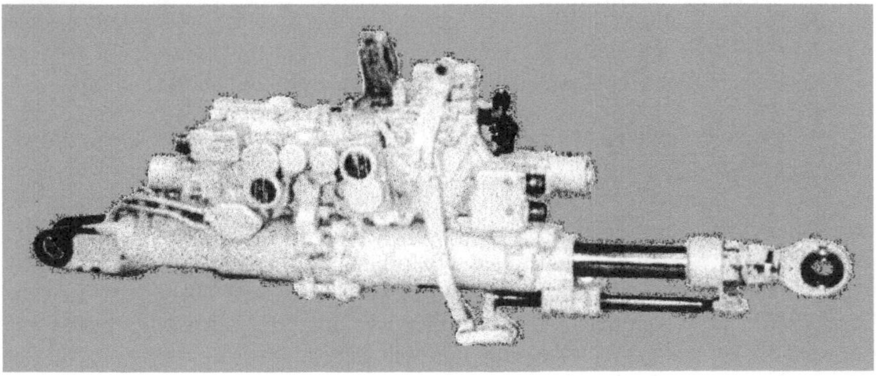

Fig. 1.4 Hydraulic actuator.

Fig. 1.5 VISTA F-16

A way to reduce the number of necessary design points is to apply a robust control design technique to the problem. A compensator synthesized using robust control principles should be able to handle large parts of, if not the whole, flight envelope. Unfortunately, many previous attempts at applying robust control design algorithms to practical, "real-world" problems have been dismal failures.[46] This is because, although the problem is well posed, the resulting compensator is impractical to implement. Either the compensator is of too high order, or its gain is too large to accommodate "real-world" non-linearity. Also, any sensor noise present is accentuated by this gain. The typical reason for these poor results is that the robust design is synthesized in the essentially noiseless world of the digital computer, and then validated on the digital computer through the use of small signal, linear simulation.

A robust control design technique that overcomes the aforementioned pitfalls is the Quantitative Feedback Theory (QFT) design technique. Although a QFT design effort could very easily result in a compensator of high order and of high gain, it does give the designer complete control over the gain and the order of the compensator; hence, QFT is not constrained to produce an impractical compensator. In addition, if a decision is made to decrease or limit the order or gain of a compensator, the performance tradeoffs due to this action can be clearly seen by the designer.

In summary, although excellent FCSs have been designed for aircraft using traditional design methods, the synthesis of those FCS's has been a costly, time-consuming endeavor. Thus, limiting robustness in FCS designs results in a convoluted, complex, full envelope design. QFT offers the ability of incorporating enough robustness to simplify the design process and the resulting FCS, but not so much robustness that the resulting FCS is impractical to implement due to violation of physical limitations imposed by the "real-world" (i.e., actuator saturation or sensor noise amplification). Also, QFT has the feature of utilizing the control system designer's knowledge of the "real-world" characteristics of the plant, etc. during the on-going design process in maximizing the ability to achieve the desired robust system performance. A simulation[47], involving the nonlinear plant, was performed on the Lamars Simulator by the FCS designer - an F-16 pilot. The excellent performance in these simulations demonstrated the viability of a QFT design approach in producing flight worthy aircraft control systems. It illustrated the benefits of designing flight control systems with the QFT robust control system design technique in contrast to the brute force approach of optimizing a flight control system for performance in expected configurations and then scheduling the gains.

1-6.6 DESIGN OF FLIGHT CONTROL LAWS FOR AIRCRAFT WITH FLEXIBLE WINGS USING QUANTITATIVE FEEDBACK THEORY[48]

Aircraft composed of lightweight composite materials are extremely enticing since their structural weight is greatly reduced. However, the control of these aircraft is complicated by the resultant flexibility of the wings. Two avenues of approach are possible: stiffen the wings, thus losing some of the weight reduction benefits, or design the lateral/directional flight control system cognizant of the wing's flexibility. The second approach was used in a QFT design of a lateral/directional robust flight control system for the sub-sonic flight envelope of an F-18 aircraft. The design incorporated a weighting matrix to distribute generalized aileron and rudder commands to the five control surfaces available on the F-18. The additional degree of freedom afforded by the availability of redundant control surfaces and the optimal and coordinated use of all control surfaces allowed for the reduction of the load on the wings, while at the same time meeting military specifications for roll maneuvers. The proposed design incorporated load alleviation concepts to reduce the load on the wings, thus avoiding adverse wing twisting, and met the military specifications, as verified by a nonlinear time simulation.

1-6.7 ROBOT CONTROLLERS[35]

The ultimate objective in robotic arm control research is to provide human arm emulation. Payload invariance is a necessary component of human arm emulation. Model-based controllers require accurate knowledge of payload and drive system dynamics to provide good high speed tracking accuracy. A robust multi- variable control system design technique is required which solves the payload and dynamics uncertainty. Thus, the model-based QFT (MBQFT) design technique is applied which resulted in controllers that are implemented by a series of simple backwards difference equations. MBQFT high speed tracking accuracy was experimentally evaluated on the first three links of the PUMA-500 of Fig. 1.6. This robust design technique increased tracking accuracy by up to a factor of four

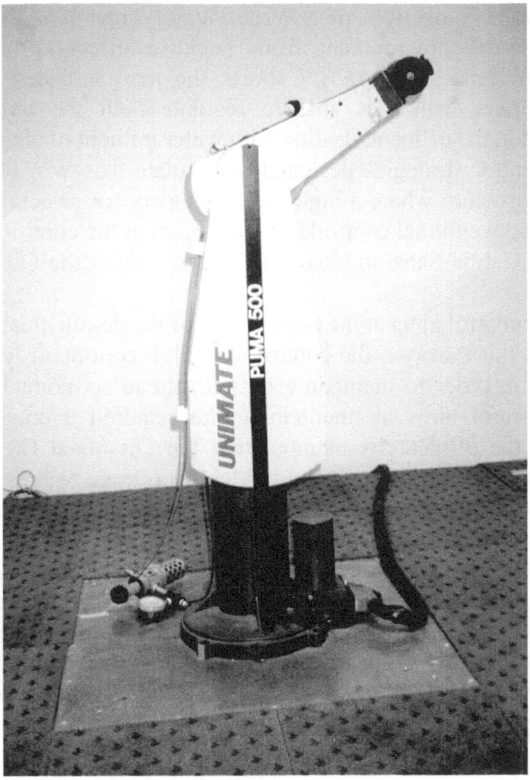

Fig. 1.6 Robot arm.

over the model-based controller performance baseline. The MBQFT tracking performance was robust to both un-modeled drive system dynamics and payload uncertainty. The non-heuristic nature of the MBQFT design and tuning should allow application to a wide range of manipulators. Chapter 10 introduces the design and the experimental results of non-diagonal MIMO QFT controllers for an industrial SCARA robot.[100,137]

1-6.8 OPERATIONAL AMPLIFIERS (OP-AMP)[49]

The QFT robust design methodology was examined, applied, and incorporated into the operational-amplifier (Op-Amp) paradigm, in order to develop an automated design approach for the analog domain. To examine this paradigm, the structured parametric uncertainty involving 741 (BJT) Op-Amp was successfully compensated by the QFT compensator design technique. The QFT robust design methodology provided the insight for the design of a compensated Op-Amp which satisfied the closed-loop performance specifications.

1-6.9 WASTEWATER TREATMENT CONTROL SYSTEM[50,103,104,105,106]

One of the main objectives of a Waste Water Treatment Plant (WWTP) is to protect the water environment from negative effects produced by residual pernicious substances. Figure 1.7 shows the new activated sludge WWTP of Crispijana, Spain, which is able to regulate both the ammonia and nitrate concentration in the effluent, dealing with water influent of about 5000 m^3/hour.

The control strategies designed to regulate that WWTP were based on a hierarchical structure where a high-level or supervisor selects the set-point of the low-level or conventional controllers. The design of the controllers was carried out by Professor Garcia-Sanz and his PhD students using the Quantitative Feedback Theory (QFT).

Nitrate control aims at the optimal use of the de-nitrification potential at any moment. For this purpose, the control algorithm continuously adapts an internal recycle flow in order to maintain a desired nitrate set-point in the anoxic zone. Ammonia control aims at maintaining the required average concentration of ammonia in the effluent by manipulating the Dissolved Oxygen set-point that commands several air flow turbines. Mobile average values of some variables were also introduced in order to eliminate the perturbations associated with the daily 24-hours profiles.

The controllers were designed and verified using long-time dynamic simulations based on a multivariable and nonlinear mathematical model (IWA nº 1) previously calibrated with real data measured in the full-scale WWTP during 12 months. The results obtained in the regulation of the pilot plant show a tighter control of the effluent nitrogen compounds and a significant reduction of the dissolved oxygen demand and in energy demand, rejecting the plant disturbances and insuring robust stability. Chapter 14 describes this example in detail.

Fig. 1.7 Wastewater Treatment Plant of Crispijana, Spain. (Courtesy of AMVISA).

1-6.10 LARGE WIND TURBINE CONTROL SYSTEMS [133,165,174]

Large Wind Turbines (WT) present a very complex multi-objective control problem that combine critical reliability issues with non-linear optimization matters. Advanced QFT robust control strategies have been thoroughly applied by Professor García-Sanz in the design, development and control of the new real Wind Turbines of 1500 and 1650 kW, made by M.Torres company (Fig. 1.8). The WTs are a variable speed, pitch controlled, multi-pole synchronous generator with two controlled IGBT's electrical power converters connected to the stator. The main dimensions are about 72 m of rotor diameter (blades of 36 m) and 65 m of tower.

The principal targets of the more than 20 loops of the control system cover aspects such as the improvement of the maximum power efficiency for every wind speed, the attenuation of the transient mechanical loads and fatigue stresses, the reduction of the electrical harmonics and flicker, and the robustness against parameters variation with a redundant fault tolerance system. In addition some critical problems arise in the design of the WT control system, such as the difficulty to work safely with random and extreme gusts, the complexity introduced by the strongly nonlinear, multivariable and time variable mathematical model and the impossibility to have a direct measurement of the wind speed experienced by the turbine, because of the high uncertainty in the anemometer measurement and the strong influence of the blades movement. These sets of motivations obliged the control engineer to get involved in the design of every dynamic element of the wind turbine from the very beginning of the project, and

Fig. 1.8 Multipole Variable Speed Wind Turbine, 1650 kW. (Courtesy of M.Torres)

to combine advanced control strategies such as QFT robust control techniques, adaptive schemes, multivariable methodology and predictive elements. The actual tests that were carried out in several Wind Turbines, for more than three years, with the proposed QFT control methodologies showed very good behavior of the WT, either in low, medium or high winds and even with the extreme 30 m/s case. Chapter 13 describes this example in detail.

1-6.11 QFT DESIGN OF A PI CONTROLLER WITH DYNAMIC PRESSURE FEEDBACK FOR POSITIONING A PNEUMATIC ACTUATOR[138]

Quantitative feedback theory (QFT) is applied towards the design of a simple and effective position controller for a typical low-cost industrial pneumatic actuator with a 5-port three-way control valve that is subject to disturbing forces. A schematic of the pneumatic servo-actuator under consideration is shown in Fig. 1.9. A simple fixed-gain proportional-integral control law with dynamic pressure feedback is synthesized to guarantee the satisfaction of a priori specified closed-loop performance requirements, including robust stability, tracking performance and disturbance attenuation, despite the presence of non-linearities and parametric uncertainty in the pneumatic functions. Figure 1.10a illustrates the closed-loop block diagram with dynamic pressure feedback. A three degree-of-freedom control structure compatible with the QFT design methodology is shown in Fig. 1.10b. A novel outer-inner design approach is proposed to avoid the synthesis of an unnecessarily complex outer loop controller. The merits of the inner loop feedback are examined from the perspective of system responses to step changes in the reference position and step changes in the disturbing force. Simulation results show clearly that the inner loop feedback improves the closed-loop Fig. 1.10 Block diagram of a pneumatic actuator: (a) closed-loop with disturbance response by eliminating oscillation and reducing the overshoot. The main contribution of this design is the presentation of a systematic approach to the design of position controllers for pneumatic servos with dynamic pressure feedback, within the framework of QFT.

1-6.12 ROBUST CONTROL OF AN ARTICULATING FLEXIBLE STRUCTURE USING MIMO QFT[139]

Articulating flexible structures, such as flexible manipulators, present a difficult closed-loop control problem. In such servo systems, the coupling of the rigid and flexible modes and the non-minimum phase dynamics severely limit system stability and performance. The difficulties in controlling these structures is exacerbated by the denumerable infinite number of flexible modes and associated difficulties in developing accurate dynamic models for controller design. These system features, coupled with the multiple system inputs and outputs to be controlled, make the employment of a robust multivariable control system a practical requirement. Reference 140 examines the multi-input multi-output

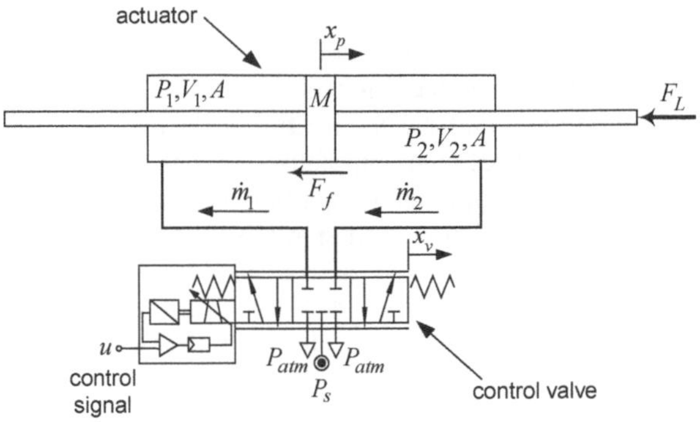

Fig. 1.9 Schematic of a typical valve controlled pneumatic actuator. (Karpenko, M.and N. Sepehri, "QFT Design of a PI Controller with Dynamic Pressure Feedback for Positioning a Pneumatic Actuator," Proc. of the American Control Conference, Boston, MA, pp 5084-5089, June 2004 Courtesy of © 2004 IEEE)

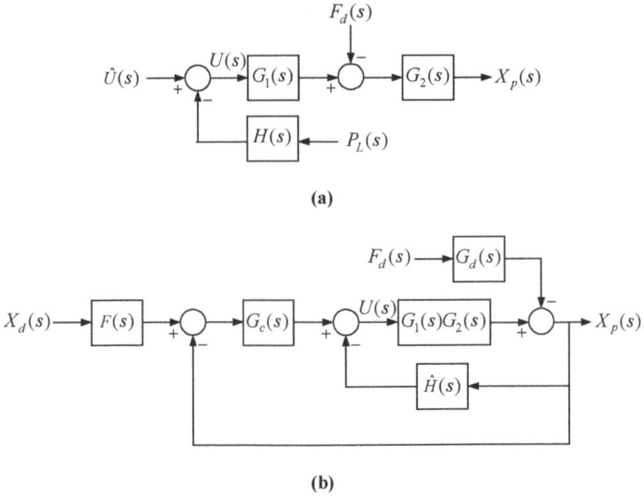

Fig. 1.10 Block diagram of pneumatic actuator: (a) open-loop; (b) closed-loop with dynamic pressure feedback. Three degree-of-freedom QFT structure. (Karpenko, M. and N. Sepehri, "QFT Design of a PI Controller with Dynamic Pressure Feedback for Positioning a Pneumatic Actuator," Proc. of the American Control Conference, Boston, MA, pp 5084-5089, June 2004 Courtesy of © 2004 IEEE)

(MIMO) QFT based control of an articulating flexible structure and presents an enhancement of the theoretical basis for the MIMO QFT design methodologies.

The control problem under consideration is the active vibration control of an articulating single-link flexible manipulator. This is facilitated by an actuation scheme comprised of a combination of spatially discrete actuation, in the form of a DC motor to perform articulation, and spatially distributed actuation, in the form of a piezoelectric transducer for active vibration control. A QFT based multivariable control system is designed that is robust to significant structured and unstructured uncertainty is developed. The QFT design methodology provides a transparent design process, facilitating the trade-off between fast rigid body motion and the regulation of the induced flexible body motion. The non-conservative treatment of the mixed structured and unstructured uncertainty over the cross-over frequency range(s) also permits the synthesis of low "cost of feedback" controllers. Simulations and experimental results jointly confirm the performance and robustness of the control system.

In the process of developing and experimentally validating the QFT based control system, shortcomings in the theoretical basis for the MIMO QFT design methodologies are addressed. Robust stability theorems are developed for the two main MIMO QFT design methodologies, namely the sequential and non-sequential MIMO QFT design methodologies. The theorems complement and extend the existing theoretical basis for the MIMO QFT design methodologies. The dissertation results expose salient features of the MIMO QFT design methodologies and provide connections to other multivariable design methodologies.

1-6.13 COORDINATED QFT BASED CONTROL OF FORMATION FLYING SPACECRAFT[155]

Formation Flying (FF) of multiple spacecraft poses significant research issues for future NASA missions. Due to the limitations of the launch vehicle for access to space and the ability to phase the optical elements over long baselines, separated spacecraft formation flying is becoming a viable means to enable the detection and imaging of Earth type planets with micro-arc-second resolution. Several NASA missions, with high priority science objectives that exploit formation flying technology, in the next two decades include Terrestrial Planet Finder, Stellar and Planet Imager, and Life Finder missions (see Fig. 1.11).

Spacecrafts in formation is mainly a load-sharing control problem when the they try to collaboratively control the relative distances and angles among them. Typically FF approaches avoid the load-sharing problem by moving only one spacecraft at a time (e.g. leader/follower, cyclic architectures, etc.). Moving all the satellites at the same time has additional challenges, as interactive loops and stability problems. In fact non-collaborative controllers in every spacecraft can only be applied with reduced bandwidth objectives to preserve stability.

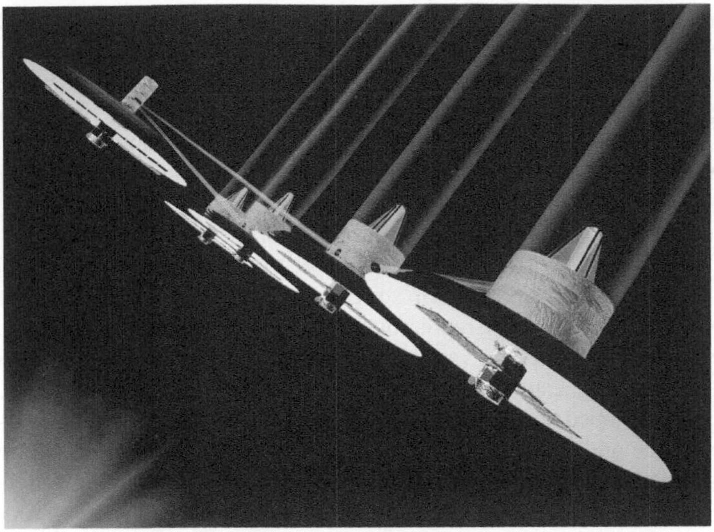

Fig. 1.11 Control of multiple spacecraft in formation. (Artist rendering courtesy of NASA/JPL-Caltech).

At NASA-JPL, Professor Garcia-Sanz and Dr. Hadaegh[155] studied the collaborative control problem of multiple spacecraft in formation with no ground intervention. The work focused on the theory needed to design autonomous and collaborative control strategies to govern the relative distances among satellites, and sharing the load according to frequency specifications. The problem was solved by combining Load-Sharing control theories[156] and the Quantitative Feedback Theory, both in the frequency domain in this design effort.

A coordinate load-sharing control structure for spacecraft in formation and a methodology to deal with their dynamic models with slow time-varying and uncertain parameters were the main objectives covered for this work.

The methodology was applied to three different FF problems: 3DOF deep space, 3DOF low Earth circular orbit and 3DOF low Earth Keplerian orbit. Uncertainty in spacecraft fuel masses and uncertainty in the radius orbit were also considered.

1-7 OUTLINE OF TEXT

The text is essentially divided into two parts. The first part, consisting of Chapters 2 through 8 and Chapter 10, present the fundamentals of the QFT robust control system designs technique for the tracking and regulator control problems and for MISO and MIMO systems. The second part consists of Chapters 9 and 11 through 14. Chapters 9 and 11 focus on *bridging the gap between theory and the real world.* This is accomplished by presenting *Engineering Rules (E.R.)* that contribute to the *transparency* of the design process and the factors that are involved, such as simulation, implementation, etc., that are important in achieving a successful and an implemental control system designs. Chapter 12 presents the QFT design for time-delay systems involving the use of the Smith Predictor. Two challenging real-world QFT examples are presented in Chapter 13: the control systems of a waste water treatment plant and of a large wind turbine synchronous generator. The final chapter, Chapter 14, presents a MIMO QFT control system design using weighting matrices and control authority allocation.

The first part of the text deals with presenting an overview of QFT: the design objectives; explaining what structured parametric uncertainty is and its Bode plot and Nichols chart representations; performance specifications; design overview; and QFT basics. A very important characteristic of *mxm* MIMO control systems is that they can be represented *by* m^2 *MISO equivalent systems.* This property is developed and is followed by a presentation of the fundamentals of the QFT technique and the associated design procedure for analog and discrete-time systems. The applications, given in the second part, include tracking and regulator control problems for MISO analog and discrete-time systems including time-delay systems. Also, this part of the text is devoted to the extension of the QFT technique to MIMO tracking and regulator control systems, both for diagonal and non-diagonal compensators/controllers, and including weighting matrices and control authority allocation.

The book uses the QFT CAD packages (see Appendices. C-E) for the examples contained in the text. These CAD packages have been developed to expedite the analysis and design of robust control systems containing structured parametric uncertainty. A PC MIMO QFT CAD has been developed that can be used in conjunction with the PC version of MATHEMATICA.

Ten appendices enhance and complete the text. Appendices A, B, and J go in depth into the key concepts of QFT; the templates, bounds, and loop shaping, respectively. Appendices F through I present some examples of the main techniques introduced in the text for MISO and MIMO non-diagonal compensators/controllers.

In closing this introductory chapter it is important to stress that feedback control engineers are essentially "system engineers," i.e., people whose primary concern is with the design and synthesis of an overall practical robust control system. To an extent depending on their own background and experience, they rely on, and work closely with, engineers in the various recognized branches of engineering to furnish them with the operating scenario of the plant and the transfer functions, plant matrices, and/or system equations of various portions of a control system.

The following design policy includes factors that are worthy of consideration in the control system design problem:

1. Use proven design methods.
2. Select the system design which has the minimum complexity.

3. Use minimum specifications or requirements that yield a satisfactory system response. Compare the cost with the performance and select the fully justified system implementation.
4. Perform a complete and adequate simulation and testing of the system.

2

INTRODUCTION TO QFT

2-1 QUANTITATIVE FEEDBACK THEORY

This chapter is devoted to presenting an overview of QFT in order to enhance the understanding and appreciation of the power of the QFT technique which is presented in the following chapters. First presented is a discussion of why feedback is needed to achieve performance goals. This is followed by a QFT overview of design objectives, an explanation of structured parametric uncertainty, a design overview, and QFT basics and design. The last few sections are devoted to the insight provided by the QFT technique and to the benefits of QFT.

2-2 WHY FEEDBACK?[51,52]

To answer the question "Why do you need QFT?" consider first the following system. The plant $P(s)$ responds to the input $r(t)$ [$R(s)$] with the output $y(t)$ [$Y(s)$] in the presence of disturbances $d_1(t)$ [$D_1(s)$] and $d_2(t)$ [$D_2(s)$] (see Fig. 2.1). If it is desired to achieve a specified system transfer function $T(s)$ [$= Y(s)/R(s)$] then it is necessary to insert a pre-filter, whose transfer function is $T(s)/P(s)$, as shown in Fig. 2.2. This compensated system produces the desired output as long as the

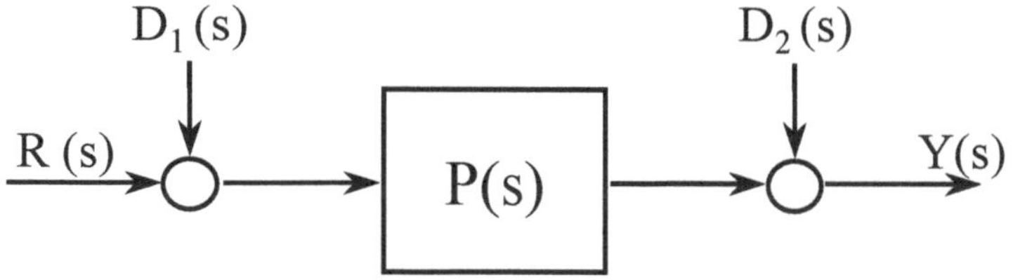

Fig. 2.1 An open-loop system (basic plant).

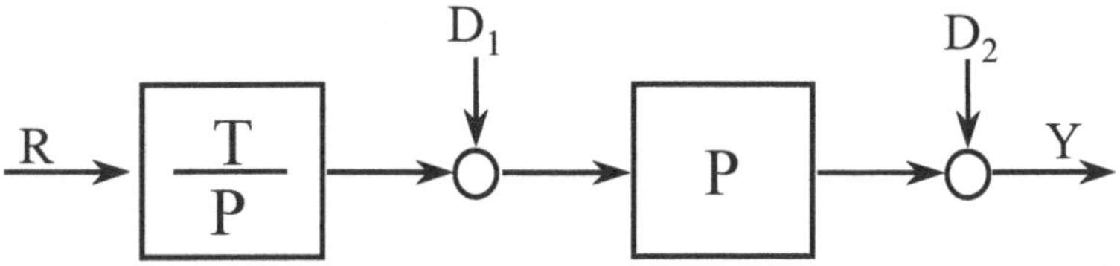

Fig. 2.2 A compensated open-loop system.

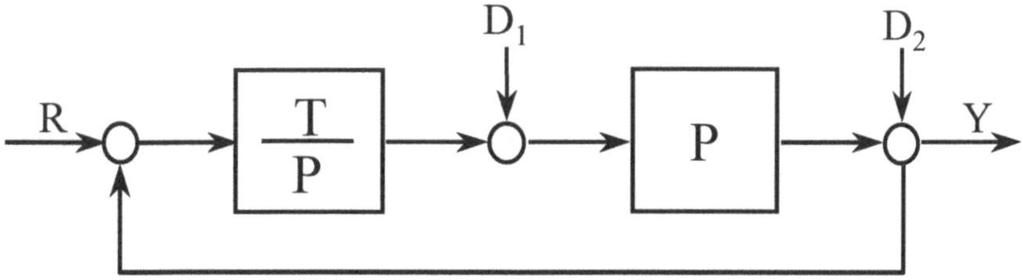

Fig. 2.3 A closed-loop system.

plant does not change, there is no plant uncertainty, and there are no disturbances. This type of system is sensitive to changes in the plant (or uncertainty in the plant), and the disturbances are reflected directly into the output. Thus, it is necessary to feed back the output in order to reduce the output sensitivity to parameter variation and attenuate the effect of disturbances on the plant output, see Fig. 2.3.

In designing a feedback control system, it is desired to utilize a technique that:

a. Addresses at the onset all known plant variations.
b. Incorporates information on the desired output tolerances.
c. Maintains reasonably low loop gain (i.e., reduces the "cost of feed-back").

This last item is important in order to avoid problems associated with high loop gains such as sensor noise amplification, saturation, and high frequency uncertainties.

Consider the control system of Fig. 2.4a, containing a plant uncertainty set, $\mathcal{P}$, that represents a plant with variable parameters. This system has two inputs: $r(t)$ the desired input signal to be tracked and $d(t)$ an external disturbance input signal which is to be attenuated to have minimal effect on $y(t)$. The tracking and disturbance control ratios of Fig. 2.4a, based upon the nominal plant $P_o \in \mathcal{P}$, are, respectively:

$$T_D(s) = \frac{Y(s)}{D(s)} = P_o(s) \tag{2.1}$$

$$T_R(s) = \frac{Y(s)}{R(s)} = P_o(s) \tag{2.2}$$

The sensitivity functions[1] of the open-loop uncompensated system of Fig. 2.4a for the two cases: $Y_R(s)|_{d(t)=0}$ and $Y_D(s)|_{r(t)=0}$ are identical; i.e.:

$$S_{P_o(s)}^{Y_R(s)}(s) = S_{P_o(s)}^{Y_D(s)} = 1 \tag{2.3}$$

For the compensated system of Fig. 2.4b the tracking and disturbance control ratios are, respectively:

$$T_R = \frac{GP_o}{1 + GP_o} = \frac{L_o}{1 + L_o} \tag{2.4}$$

$$T_D = \frac{P_o}{1 + GP_o} = \frac{P_o}{1 + L_o} \tag{2.5}$$

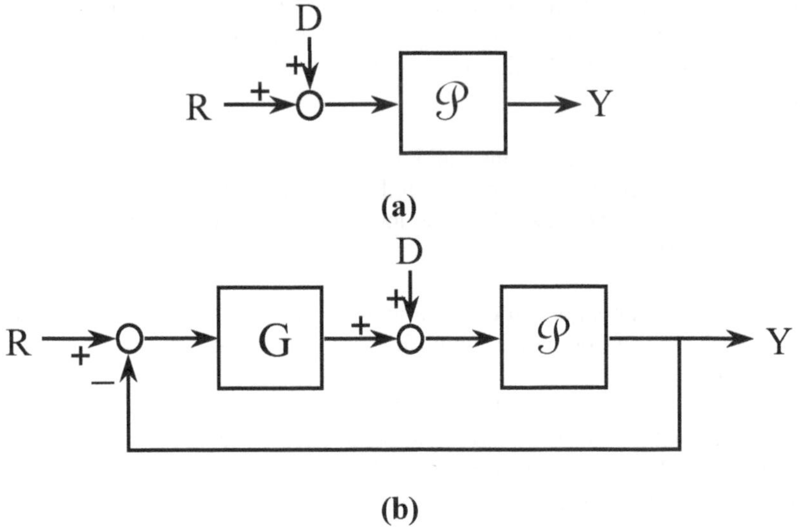

Fig. 2.4 Control systems: (a) open-loop uncompensated sytem; (b) closed-loop compensated system.

where $L_o \equiv GP_o$ is defined as the *nominal loop transmission function*. For the compensated system of Fig. 2.4b, the sensitivity functions for these two cases are also identical; i.e.:

$$S_{P_o(s)}^{Y_R(s)}(s) = S_{P_o(s)}^{Y_D(s)}(s) = \frac{1}{1+GP_o} = \frac{1}{1+L_o} \tag{2.6}$$

Comparing Eq. (2.6) with Eq. (2.3) readily reveals that the effect of changes of the uncertainty set $\mathcal{P}(s)$ upon the output of the closed-loop control system is reduced by the factor $1/[1 + GP_o]$ compared to the open-loop control system. *This reduction is an important reason why feedback systems are used.*

Horowitz has shown[52] that a robust control system design is best achieved by working with L_o and not with the sensitivity function S. The reasons[52] for the choice of L_o are:

1. The sensitivity function is very sensitive to the cost of feedback.
2. A practical optimum design requires working to the limits of the system's performance specifications.
3. The order of the compensator (controller) G can be minimized by incorporating[1,41] the nominal plant P_o into L_o.

The reader is referred to the QFT literature for a more detailed analysis of the sensitivity function with respect to sensor noise and plant parameter uncertainty on system performance.[3,6,40,54]

2-3 QFT OVERVIEW[2,3]

2-3.1 QFT DESIGN OBJECTIVE

Design and implement robust control for a system with structured parametric uncertainty that satisfies the desired performance specifications.

2-3.2 STRUCTURED PARAMETRIC UNCERTAINTY: A BASIC EXPLANATION

2-3.2.1 A Simple Example

To illustrate "*What is Parametric Uncertainty?*" consider the undergraduate laboratory experiment that involves hooking up the d-c shunt motor of Fig. 2.5. Further, consider that the students entered the laboratory on a cold January Monday morning to perform this experiment. The weekend room temperature was set at 50° F but was reset to 70° F when the students entered the room. The students hurriedly hooked-up the motor and set the field rheostat to yield a speed $\omega = 1200$ r.p.m. Upon accomplishing this phase of the experiment they took a one hour break in order to allow the room to reach the desired temperature of 70°. Upon their return they found that the speed of the motor was now 1250 r.p.m. with no adjustments to the applied voltage or of the field rheostat R_f. *Why the change in speed?*

- Due to the heating of the d-c shunt field by the field current i_f and the environmental conditions, the value of R_f increased.
- Which in-turn decreased the value of i_f and in-turn resulted in the increase in speed since speed is inversely proportional to i_f, assuming V_f is constant.
- Therefore, during the operation of the motor, the parameter R_f can vary anywhere within the range $R_{fmin} \le R_f \le R_{fmax}$ due to the variable environmental temperature and field current.
- As a consequence, there is *uncertainty* as to what the actual value of the parameter R_f will be at the instant a command is given to the system.

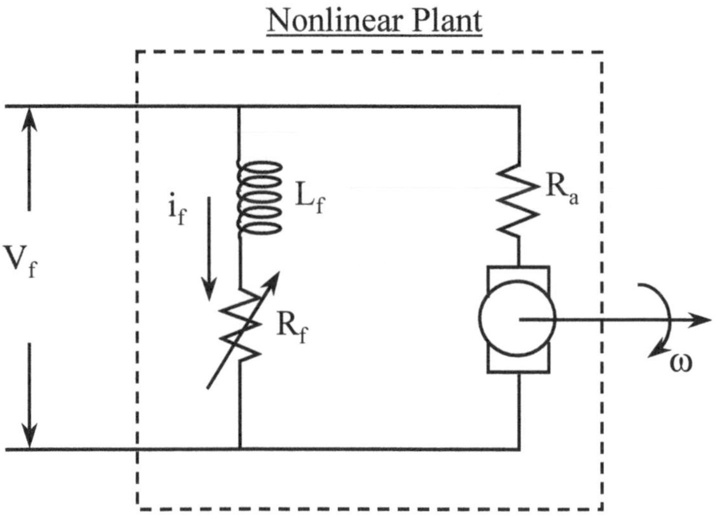

Fig. 2.5 d-c shunt motor.

- Thus, the *parametric uncertainty is structured* because the range of the variation of R_f is known and its effects on the relationship between V_f and ω can be modeled.

2-3.2.2 A Simple Mathematical Description

The transfer function of a d-c servo motor, utilized as a position control device, is:

$$P_\iota(s) = \frac{\Theta_m(s)}{V_f(s)} = \frac{Ka}{s(s+a)} \tag{2.7}$$

where the parameters K and a vary, due to the operating scenario, over the following range: $K \in (K_{min}, K_{max})$ and $a \in (a_{min}, a_{max})$. Over the region of operation, in a position control system, the plant parameter variations are described by Fig. 2.6. The shaded region in this figure represents the region of *structured parametric uncertainty* (region of plant uncertainty). The motor can be represented by six LTI transfer functions P_ι $(\iota = 1,2,\ldots,J)$ at the points indicated on the figure. The Bode plots for these 6 LTI plants are shown in Fig. 2.7.

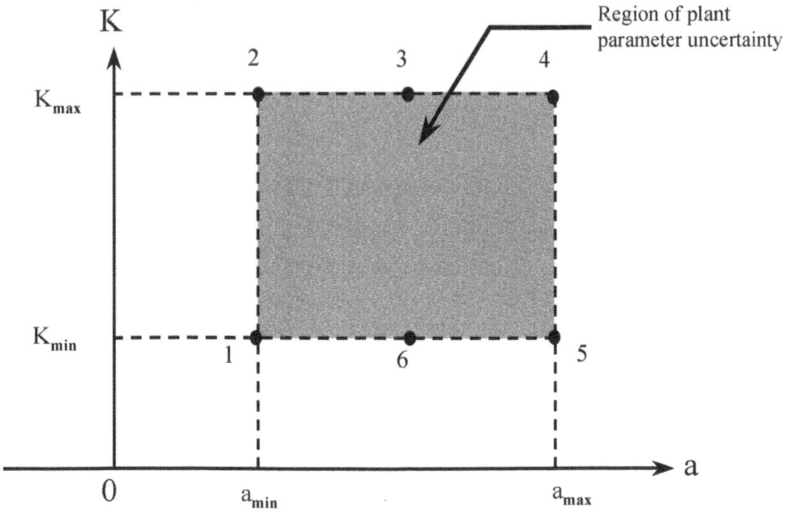

Fig. 2.6 Region of plant parameter uncertainty.

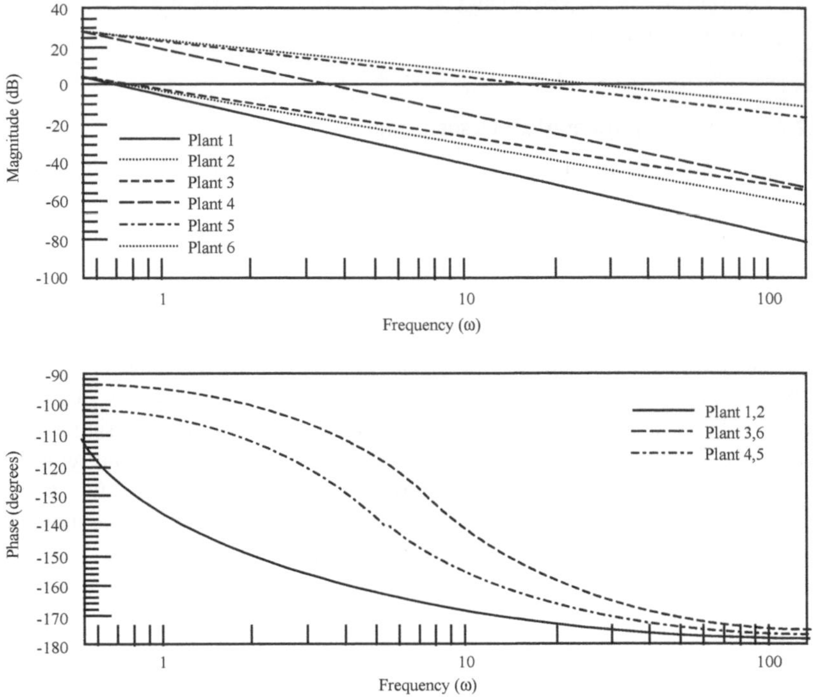

Fig. 2.7 Bode plots of 6 LTI plants: the range of parameter uncertainty.

2-3.3 CONTROL SYSTEM PERFORMANCE SPECIFICATIONS

In many control systems the output $y(t)$ must lie between specified upper and lower bounds, $y(t)_U$ and $y(t)_L$, respectively, as shown in Fig. 2.8a. The conventional time-domain figures of merit, based upon a *step input signal* $r(t) = R_0\, u_{-1}(t)$, are shown in Fig. 2.8a. They are: M_p, peak overshoot; t_r, rise time; t_p, peak time; and t_s, settling time. Corresponding system performance specifications in the frequency domain are, B_U and B_L, the upper and lower bounds respectively, peak overshoot $Lm\ M_m$, and the frequency bandwidth ω_h which are shown in Fig. 2.8b. Assuming that the control system has negligible sensor noise and sufficient control effort authority, then for a stable *linear-time-invariant* (*LTI*) *minimum-phase* (*m.p.*) plant a *LTI* compensator may be designed to achieve the desired control system performance specifications. The case of *non-minimum-phase* (*n.m.p.*) plant is discussed in Chapter 9.

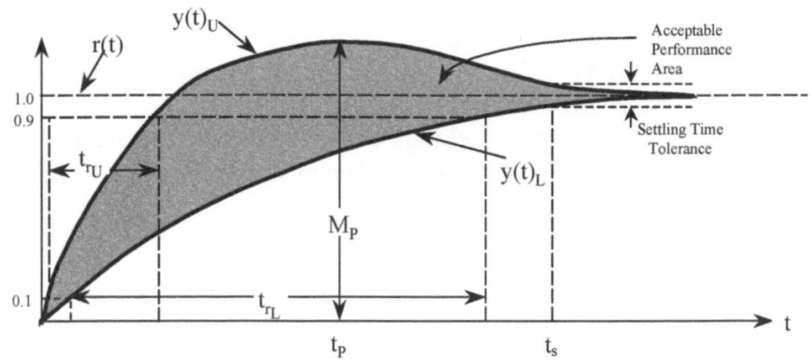

(a)

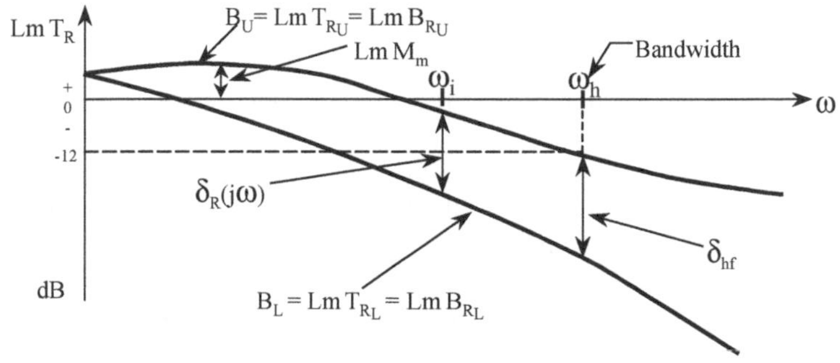

(b)

Fig. 2.8 Desired system performance specifications: (a) time domain response specifications; (b) frequency domain response specifications.

2-3.4 QFT DESIGN OVERVIEW

The QFT design objective is achieved by:

- Representing the characteristics of the plant and the desired system performance specifications in the frequency domain.
- Using these representations to design a compensator (controller).
- Representing the nonlinear plant characteristics by a set of LTI transfer functions that cover the range of structured parametric uncertainty.
- Representing the system performance specifications (see Fig. 2.8) by LTI transfer functions that form the upper B_U and lower B_L boundaries for the design.

- Reducing the effect of parameter uncertainty by shaping the open-loop frequency responses so that the Bode plots of the J closed-loop systems fall between the boundaries B_U and B_L, while simultaneously satisfying all performance specifications.
- Obtaining the stability, tracking, disturbance, and cross-coupling (for MIMO systems) boundaries on the Nichols chart in order to satisfy the performance specifications.

2-3.5 QFT BASICS

Consider the control system of Fig. 2.9, where $G(s)$ is a compensator, $F(s)$ is a pre-filter, and $\mathcal{P}$ is the nonlinear plant. To accomplish a QFT design:

- The nonlinear plant is described by a set of J m.p. LTI plants, i.e., $\mathcal{P} = \{P_\iota(s)\}$ $(\iota = 1,2,..., J)$, which define the structured plant parameter uncertainty. Note: for MIMO systems the elements p_{ij} of the mxm plant matrix $\boldsymbol{P}$ can be n.m.p.,[52] for MISO systems the discussion is restricted to m.p. plants.
- The magnitude variation due to the plant parameter uncertainty, $\delta_P(j\omega_i)$, is depicted by the Bode plots of the LTI plants as shown in Fig. 2.10 for the example of Figs. 2.6 and 2.7.
- J data points (log magnitude and phase angle), for each value of frequency, $\omega = \omega_i$, are plotted on the Nichols chart (NC). A contour is drawn through the data points, for ω_i, that described the boundary of the region that contains all J points. This contour is referred to as a *template*. It represents the region of structured plant parametric uncertainty on the NC and are obtained for specified values of frequency, $\omega = \omega_i$, within the bandwidth (BW) of concern. Six data points (log magnitude and phase angle) for each value of ω_i are obtained, as shown in Fig. 2.11a, for the example shown in Fig. 2.6, to plot the templates, for each value of ω_i, as shown in Fig. 2.11b.

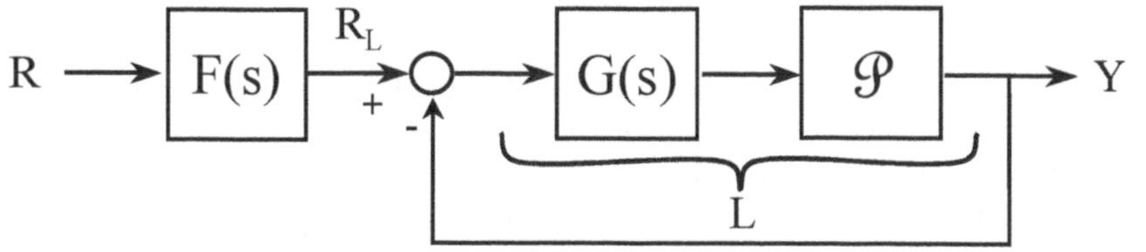

Fig. 2.9 Compensated nonlinear system.

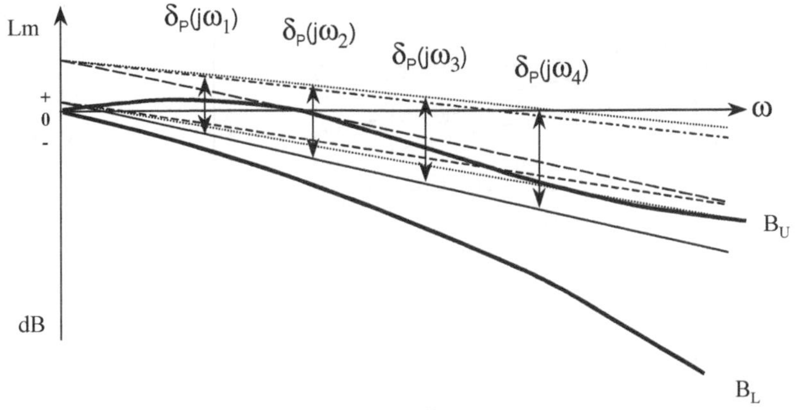

Fig. 2.10 LTI plants.

- The system performance specifications are represented by LTI transfer functions, and their corresponding Bode plots are shown in Fig. 2.10 by the upper and lower bounds B_U and B_L, respectively.

2-3.6 QFT DESIGN

The tracking design objective is to

(a) Synthesize a compensator $G(s)$ of Fig. 2.9 that

- results in satisfying the desired performance specifications of Fig. 2.8,
- results in the closed-loop frequency responses T_{L_ι} shown in Fig. 2.12, and
- results in the $\delta_L(j\omega_i)$ of Fig. 2.12, of the compensated system, being equal to or smaller than $\delta_P(j\omega_i)$ of Fig. 2.10 for the uncompensated system and that it is equal or less than $\delta_R(j\omega_i)$, for each value of ω_i of interest; that is:

$$\delta_L(j\omega_i) \le \delta_R(j\omega_i) \le \delta_P(j\omega_i)$$

(b) Synthesize a pre-filter $F(s)$ of Fig. 2.9 that

- Results in shifting and reshaping the T_{L_ι} responses in order that they lie within the B_U and B_L boundaries in Fig. 2.12 as shown in Fig. 2.13.

Therefore, the QFT robust design technique assures that the desired performance specifications are satisfied over the prescribed region of structured plant parametric uncertainty.

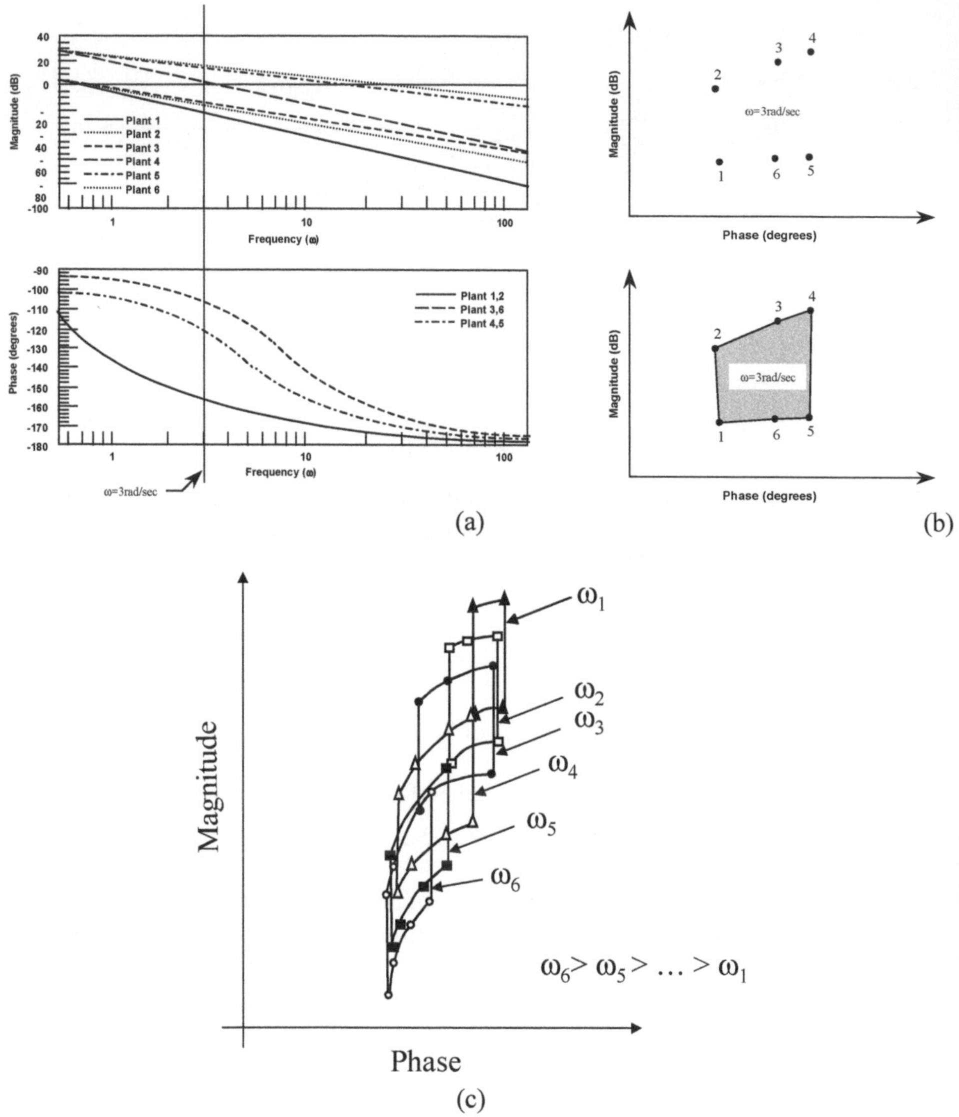

Fig. 2.11 (a) Bode plots of 6 LTI plants; (b) template construction for ω_i = *3 rad/sec*; (c) construction of the Nichols chart plant templates.

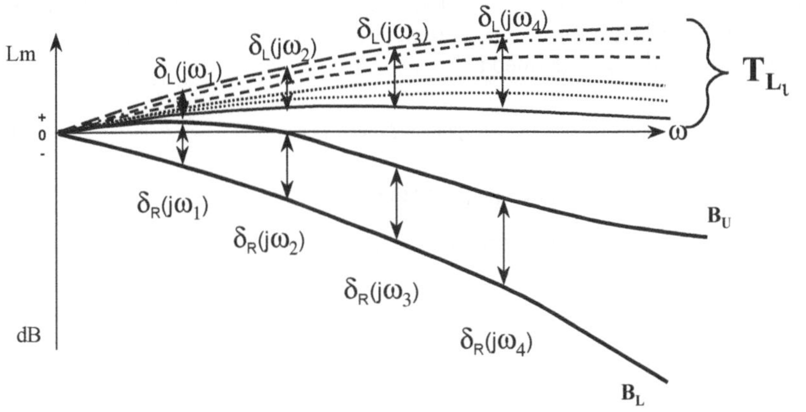

Fig. 2.12 Closed-loop responses: LTI plants with $G(s)$.

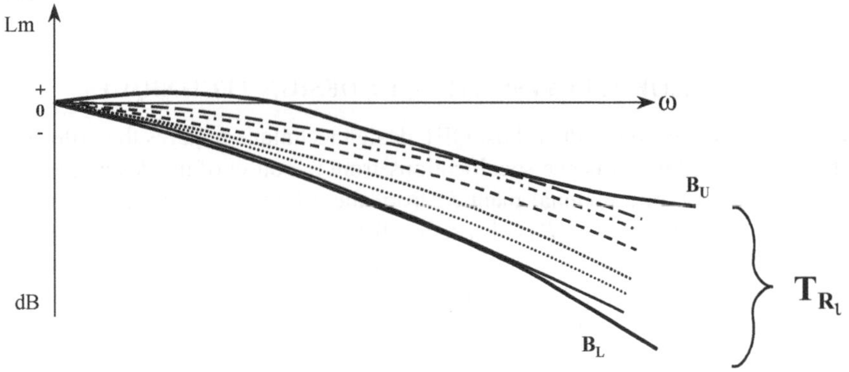

Fig. 2.13 Closed-loop responses: LTI plants with $G(s)$ and F(s).

2-4 INSIGHT TO THE QFT TECHNIQUE

2-4.1 OPEN-LOOP PLANT

Consider the position control system of Fig. 2.9 whose plant transfer function is given by

$$P_\iota(s) = \frac{Ka}{s(s+a)} = \frac{K'}{s(s+a)} \tag{2.8}$$

where $K' = Ka$ *and* $\iota = 1,2,\ldots, J$. The log magnitude variation due to the plant parameter uncertainty, for $J = 6$, is depicted by the Bode plots in Fig. 2.7. The loop transmission $L(s)$ is defined as

$$L_\iota(s) = G(s)P_\iota(s) \tag{2.9}$$

2-4.2 CLOSED-LOOP FORMULATION

The control ratio T_L of the unity-feedback system of Fig. 2.9 is

$$T_{L_\iota} = \frac{Y}{R_L} = \frac{L_\iota}{1 + L_\iota} \tag{2.10}$$

The overall system control ratio T_R is given by:

$$T_{R_\iota}(s) = \frac{F(s)L_\iota(s)}{1 + L_\iota(s)} \tag{2.11}$$

2-4.3 RESULTS OF APPLYING THE QFT DESIGN TECHNIQUE

The proper application of the robust QFT design technique requires the utilization of the prescribed performance specifications from the onset of the design process, and the selection of a nominal plant P_o from the J LTI plants. Once the proper loop shaping of $L_o(s) = G(s)P_o(s)$ is accomplished, a synthesized $G(s)$ is achieved that satisfies the desired performance specifications. The last step of this design process is the synthesis of the pre-filter that ensures that the Bode plots of T_{R_ι} all lie between the upper and lower bounds B_U and B_L.

2-4.4 INSIGHT TO THE USE OF THE NICHOLS CHART (NC) IN THE QFT TECHNIQUE[3]

This section is intended to provide the reader a review of the use of the NC and an insight as to how it applies to the QFT technique.

(1) **Open-Loop Characteristics** – For the nominal plant $\boldsymbol{P}_o(j\omega)$, the nominal loop transmission function is

$$Lm\,\boldsymbol{L}_o = Lm\,\boldsymbol{GP}_o = Lm\,\boldsymbol{G} + Lm\,\boldsymbol{P}_o \tag{2.12}$$

whereas for all other plants, $\boldsymbol{P}(j\omega)$, the loop transmission function is

$$Lm\,\boldsymbol{L} = Lm\,\boldsymbol{GP} = Lm\,\boldsymbol{G} + Lm\,\boldsymbol{P} \tag{2.13}$$

Thus, for $\omega = \omega_i$, the variation $\delta_P(j\omega_i)$ in $Lm\,\mathbf{L}(j\omega_i)$ is given by

$$\delta_P(j\omega_i) = Lm\,\mathbf{L}(j\omega_i) - Lm\,\mathbf{L}_o(j\omega_i) = Lm\,\boldsymbol{P}(j\omega_i) - Lm\,\boldsymbol{P}_o(j\omega_i) \quad \textbf{(2.14)}$$

and its phase angle variation is given by

$$\angle\Delta\boldsymbol{P}(j\omega_i) = \angle\boldsymbol{L} - \angle\boldsymbol{L}_o = (\angle\boldsymbol{G} + \angle\boldsymbol{P}) - (\angle\boldsymbol{G} + \angle\boldsymbol{P}_o) = \angle\mathrm{P} - \angle\boldsymbol{P}_o \quad \textbf{(2.15)}$$

The expression $Lm\,\boldsymbol{P}(j\omega_i) = Lm\,\boldsymbol{P}_o(j\omega_i) + \delta_P(j\omega_i)$, obtained from Eq. (2.14), is substituted into Eq. (2.13) to yield

$$Lm\,\mathbf{L}(j\omega_i) = Lm\boldsymbol{G}(j\omega_i) + Lm\boldsymbol{P}_o(j\omega_i) + \delta_P(j\omega_i) \quad \textbf{(2.16)}$$

(2) **Closed-Loop Characteristics** – The closed-loop system characteristics can be obtained, for a given $\boldsymbol{G}(j\omega)$ and $\boldsymbol{P}_o(j\omega)$, from the plot of $Lm\,\mathbf{L}_o(j\omega)$ vs. $\angle\boldsymbol{L}_o$ shown on the NC in Fig. 2.14[1] Also shown is a plot of a template, $\Im\boldsymbol{P}(j\omega_i)$, whose contour is based upon the data obtained for $\omega = \omega_i$ from Fig. 2.7. This template represents a region of plant parameter uncertainty for ω_i as expressed mathematically by Eqs. (2.15) and (2.16). From the loop transmission plot and its intersections with the *M*- and *α-contours*, the closed-loop frequency response data may be obtained for plotting M_o and α vs. ω. In Fig. 2.15[1] a plot of M_o vs. ω is shown, where

$$\mathrm{M_o}(\mathrm{j}\omega)\angle\alpha(\mathrm{j}\omega) = \frac{\mathbf{Y}(\mathrm{j}\omega)}{\mathbf{R}(\mathrm{j}\omega)} = \frac{\mathbf{L}_o(\mathrm{j}\omega)}{1 + \mathbf{L}_o(\mathrm{j}\omega)} \quad \textbf{(2.17)}$$

(3) **Parametric Variation NC Characteristics** – As an example, consider that for Eq. (2.9) $\boldsymbol{G} = 1\angle 0^\circ$. If point A on the template in Fig. 2.14 represents $Lm\,\boldsymbol{P}_o$ vs. $\angle\boldsymbol{P}_o$, then a variation in P results in:

- a horizontal translation in the angle of P, given by Eq. (2.15)
- and in a vertical translation in the log magnitude value of $\boldsymbol{P}$, given by Eq. (2.14)
- Translations are shown in Fig. 2.14 at points B, C, and D.
- Variation $\delta_P(j\omega_i)$ of the plant, and in turn $\boldsymbol{L}(j\omega_i)$, from the nominal value for $\omega = \omega_i$, over the range of plant parameter variation is described by the template $\Im\boldsymbol{P}(j\omega_i)$ shown in Fig. 2.14
- Consider point A on the template represents the nominal plant $\boldsymbol{P} = P_o(j\omega_i)$.
- Thus, its corresponding closed-loop frequency response is

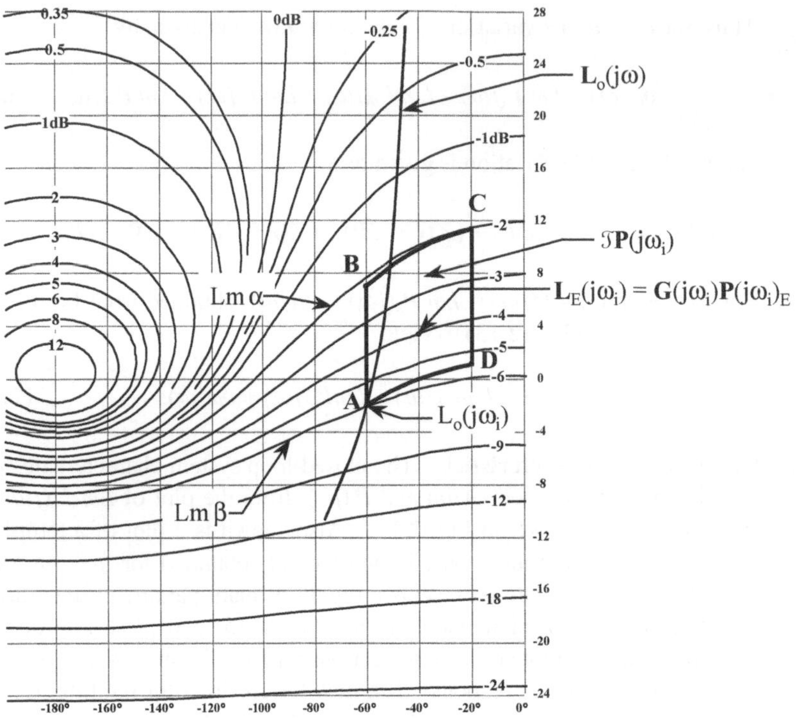

Fig. 2.14 Nominal loop transmission plot with plant parameter area of uncertainty.

$$Lm\ M_A = Lm\ \beta = \text{-6 dB}$$

- For $\boldsymbol{P}(j\omega) = \boldsymbol{P}_i(j\omega)$: point C in Fig. 2.14, its corresponding closed-loop frequency response is

$$Lm\ M_C = Lm\ \alpha = \text{-2 dB}$$

- These values are plotted in Fig. 2.15.
- Note that point A represents the minimum value of $Lm\ \boldsymbol{M}(j\omega_i)$ at $\omega = \omega_i$, i.e.,

$$Lm\ \boldsymbol{T}_{min} = [Lm\ \boldsymbol{M}(j\omega_i)]_{min} = Lm\ \beta$$

- Also, the point C represents the largest value of $Lm\ \boldsymbol{M}(j\omega_i)$ at $\omega = \omega_i$, i.e.,

$$Lm\ \boldsymbol{T}_{max} = [Lm\ \boldsymbol{M}(j\omega_i)]_{max} = Lm\ \alpha$$

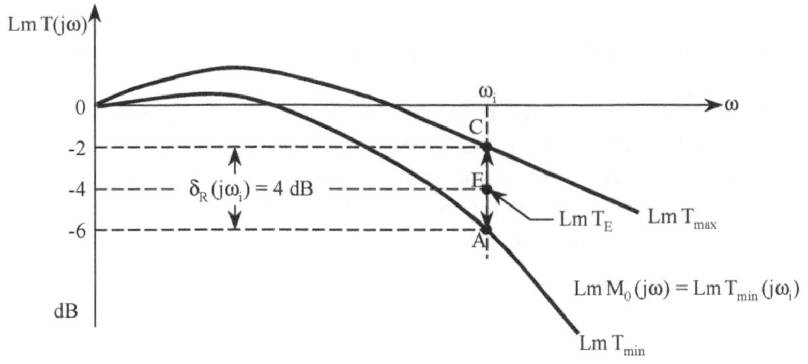

Fig. 2.15 Closed-loop responses obtained from Fig. 2.14.

with a range of plant parameter variation described by the template.

- Thus, the maximum variation in *Lm* ***M***, denoted by $\delta_L(j\omega_i)$, for this example is

$$\delta_L(j\omega_i) = Lm\alpha - Lm\beta = -2 - (-6) = 4 \text{ dB} \tag{2.18}$$

- When $L_o(s)$ is properly synthesized, according to the QFT design technique, then $\delta_L(j\omega_i) \le \delta_R(j\omega_i)$. Shown in Fig. 2.15 is point E, midway between points A and C, that corresponds to point E in Fig. 2.14 which lies within the variation template $\mathfrak{I}\mathbf{P}(j\omega_i)$.
- This procedure is repeated to obtain the maximum variation $\delta_L(j\omega_i)$ for each value of frequency $\omega = \omega_i$ within the desired BW (see Fig. 2.15).
- From this figure it is possible to determine the variation in the control system's figures of merit due to the plant's parameter uncertainty.

This graphical description of the effect of plant parameter uncertainty on the system's performance is the basis of the QFT technique.

2-5 BENEFITS OF QFT

The benefits of the QFT technique may be summarized as follows:

- It results in a robust design which is insensitive to structured plant parameter variation.
- There can be one robust design for the full, operating envelope.
- Design limitations are apparent up front and during the design process.

- The achievable performance specifications can be determined early in the design process.
- If necessary, one can redesign for changes in the specifications quickly with the aid of the QFT CAD package (see Appendix C).
- The structure of the compensator (controller) is determined up front.
- As a consequence of the above benefits, there is less development time for a full envelope design.

2-6 SUMMARY

The purpose of this text is to present the concepts of the QFT technique in such a manner that students and practicing engineers can readily grasp the fundamentals and appreciate its *transparency* in *bridging the gap* between theory and the real world. This is accomplished by limiting the scope of the presentation, for example, to minimum-phase (m.p.) systems. For a more theoretical discussion of QFT, such as existence theorems and non-minimum-phase (n.m.p.) plants, the reader is referred to the literature listed in the reference section of this text. It is highly recommended that the reader, at the conclusion of reading this text, read the excellent paper by Professor I. M. Horowitz entitled *Survey of Quantitative Feedback Theory (QFT)*.[52]

3

THE MISO ANALOG CONTROL SYSTEM

3-1 INTRODUCTION[52]

The multiple-input multiple-output (MIMO) synthesis problem is converted into a number of single-loop feedback problems in which parameter uncertainty, cross-coupling effects, and system performance tolerances are derived from the original MIMO problem. The solutions to these single-loop problems represent a solution to the MIMO plant. It is not necessary to consider the complete system characteristic equation. The design is tuned to the extent of the uncertainty and the performance tolerances. In Chapter 5 the development of a suitable mapping is presented that permits the analysis and synthesis of a MIMO control system having m inputs and m outputs by a set of m^2 equivalent single-loop multiple-input single-output (MISO) control systems.

This chapter builds upon the introduction to QFT that is presented in Chapter 2. It is devoted to presenting an in-depth understanding and appreciation of the power of the QFT technique. This is accomplished by first presenting an introduction of the QFT technique using the frequency-response method as applied to a single-loop MISO system and an overview of the design procedure. Next a discussion of minimum-phase (m.p.) performance specifications is presented. The remaining portion of the chapter is devoted to an in-depth development of the QFT technique as applied to the design of robust single-loop control systems having two inputs, a tracking and an external disturbance input, respectively, and a single output (a MISO system).[1] The in-depth development of the QFT technique as applied to the design of robust MIMO control systems for tracking inputs is presented in Chapters 5 through 7. Chapter 8 presents the analysis and design of a MIMO control system with external disturbances, i.e., the external disturbance rejection problem, and Chapter 10 introduces the synthesis of a fully populated matrix QFT compensator for MIMO systems.

3-2 THE QFT METHOD (SINGLE-LOOP MISO SYSTEM)

The feedback control system in which $\mathcal{P}$ represents the set of transfer functions which describe the region of plant parameter uncertainty, G is the cascade compensator, and F is an input pre-filter transfer function is represented in Fig. 3.1. The output $y(t)$ is required to track the command input $r(t)$ and to reject the external disturbances $d_1(t)$ and $d_2(t)$ and the noise $n(t)$. The compensator G in Fig. 3.1 is to be designed so that the variation of $y(t)$ to the uncertainty in the plant P is within allowable tolerances and the effects of the disturbances $d_1(t)$ and $d_2(t)$ and the noise $n(t)$ on $y(t)$ are acceptably small. Also, the pre-filter properties of $F(s)$ must be designed to the desired tracking by the output $y(t)$ of the input $r(t)$. Since the control system in Fig. 3.1 has two measurable quantities, $r(t)$ and $y(t)$, it is referred to as a two degree-of-freedom (DOF) feedback structure. If the two disturbance inputs are measurable, then it represents a four DOF structure. The actual design is closely related to the extent of the uncertainty and to the narrowness of the performance tolerances. The uncertainty of the plant transfer function is denoted by the set

$$\mathcal{P} = \{P_\iota\} \quad where\ \iota = 1,2,\ldots,J \tag{3.1}$$

and is illustrated by Example 3.1.

Example 3.1 The plant transfer function is

$$P(s) = \frac{K}{s(s+a)} \tag{3.2}$$

where the value of K is in the range [1, 10] and a is in the range [−2, 2]. The region of parameter uncertainty of the plant is illustrated by Fig. 3.2. The uncertainty of the disturbance is denoted by a set of plant disturbances

$$\mathcal{D} \equiv \{D\} \tag{3.3}$$

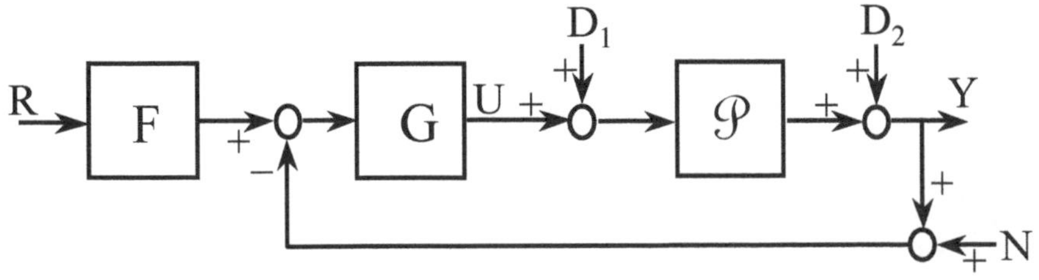

Fig. 3.1 A feedback structure.

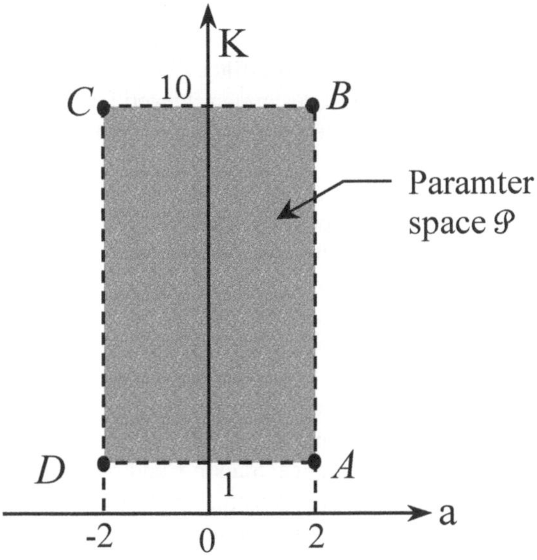

Fig. 3.2 Plant parameter uncertainty.

and the acceptable closed-loop transmittances (control ratios) are denoted by

$$\mathcal{T} \equiv \{T\} \tag{3.4}$$

The design objective is to guarantee that $T_R(s) = Y(s)/R(s)$ and $T_D(s)=Y(s)/D(s)$ are members of the sets of acceptable $\Im_R$ and $\Im_D$, respectively, for all $P \in \mathcal{P}$ and all $D \in \mathcal{D}$. In a feedback system, the principal challenge in the control system design is to relate the system performance specifications to the requirements on the *loop transmission function* $L(s) = G(s)P(s)$ in order to achieve the desired benefits of feedback, i.e., the desired reduction in sensitivity to plant uncertainty and desired disturbance attenuation. The advantage of the frequency domain is that $L(s) = G(s)P(s)$ is simply the multiplication of complex numbers. In the frequency domain it is possible to evaluate $\boldsymbol{L}(j\omega)$ at every ω_i separately, and thus, at each ω_i, the optimal bounds on $\boldsymbol{L}(j\omega_i)$ can be determined.

3-3 DESIGN PROCEDURE OUTLINE

The basic design procedure which is to be followed in applying the QFT robust design technique is outlined in this section. This outline enables the reader to obtain an overall perspective of the QFT technique at the onset. The following sections present a detailed discussion which is intended to establish for the reader a firm understanding of the fundamentals of this technique.

The objective is to design the pre-filter $F(s)$ and the compensator $G(s)$ of Fig. 3.1 so that the specified robust design is achieved for the given region of plant parameter uncertainty. The design procedure to accomplish this objective is as follows:

Step 1: Synthesize the desired tracking model.
Step 2: Synthesize the desired disturbance model.
Step 3: Specify the J linear-time-invariant (LTI) plant models that define the region of plant parameter uncertainty.
Step 4: Obtain plant templates, at specified frequencies, that pictorially describe the region of plant parameter uncertainty on the Nichols chart (NC).
Step 5: Select the nominal plant transfer function $P_o(s)$.
Step 6: Determine the stability contour (U-contour) on the NC.
Steps 7 - 9: Determine the disturbance, tracking, and optimal bounds on the NC.
Step 10: Synthesize the nominal loop transmission function $L_o(s) = G(s)P_o(s)$ that satisfies all the bounds and the stability contour.
Step 11: Based upon Steps 1 through 10, synthesize the pre-filter $F(s)$.
Step 12: Simulate the system in order to obtain the time response data for each of the J plants.

The following sections illustrate this design procedure.

3-4 MINIMUM-PHASE SYSTEM PERFORMANCE SPECIFICATIONS

In order to apply the QFT technique it is necessary to synthesize the desired or model control ratio, based upon the system's desired performance specifications in the time domain. For the minimum-phase (m.p.) LTI MISO system of Fig. 3.1 the control ratios for tracking and for disturbance rejection are, respectively,

$$T_R(s) = \frac{F(s)G(s)P(s)}{1+G(s)P(s)} = \frac{F(s)L(s)}{1+L(s)} = F(s)T(s) \quad with\ d_1(t) = d_2(t) = 0 \qquad \mathbf{(3.5)}$$

$$T_{D_1} = \frac{P(s)}{1+G(s)P(s)} = \frac{P}{1+L} \qquad with\ r(t) = d_2(t) = 0 \qquad \mathbf{(3.6)}$$

$$T_{D_2} = \frac{1}{1+G(s)P(s)} = \frac{1}{1+L} \qquad with\ r(t) = d_1(t) = 0 \qquad \mathbf{(3.7)}$$

Note that for T_{D_l}, the specified maximum value $|y(t_p)| = \alpha_p$, due to $d_l(t) = u_{-1}(t)$, is often used as the disturbance model specification, i.e., maximum $Lm\ T_{D_1} = Lm\ \alpha_p$.

3-4.1 TRACKING MODELS

The QFT technique requires that the desired tracking control ratios be modeled in the frequency domain to satisfy the required gain K_m and the desired time domain performance specifications for a step input. Thus, the system's tracking performance specifications for a simple second-order system are based upon satisfying some or all of the step forcing function figures of merit (FOM) for under-damped (M_p, t_p, t_s, t_r, K_m) and over-damped (t_s, t_r, K_m) responses, respectively. These are graphically depicted in Fig. 3.3. The time responses $y(t)_U$ and $y(t)_L$ in this figure represent the upper and lower bounds, respectively, of the tracking performance specifications; that is, an acceptable response $y(t)$ must lie between these bounds. The Bode plots of the upper bound B_U and lower bound B_L for $Lm\ \boldsymbol{T}_R(j\omega)$ *vs.* ω are shown in Fig. 3.4. Note that for m.p. plants, only the tolerance on $|\boldsymbol{T}_R(j\omega_i)|$ needs to be satisfied for a satisfactory design. For n.m.p. plants, tolerances on $\angle \boldsymbol{T}_R(j\omega_i)$ must also be specified and satisfied in the design process. This text deals only with m.p. plants. The case of n.m.p. plants is discussed in Chapter 8. It should be noted that, for m.p. plants, any desired frequency bandwidth (BW) is achievable whereas for a n.m.p. plants the BW that is achievable is limited.

The modeling of a desired transmittance $T(s)$ is discussed in Ref. 1. It is desirable to synthesize the control ratios corresponding to the upper and lower bounds T_{R_U} and T_{R_L}, respectively, so that $\delta_R(j\omega_i)$ increases as ω_i increases above the 0-dB crossing frequency ω_{cf} (see Fig. 3.4b) of T_{R_U}. This characteristic of

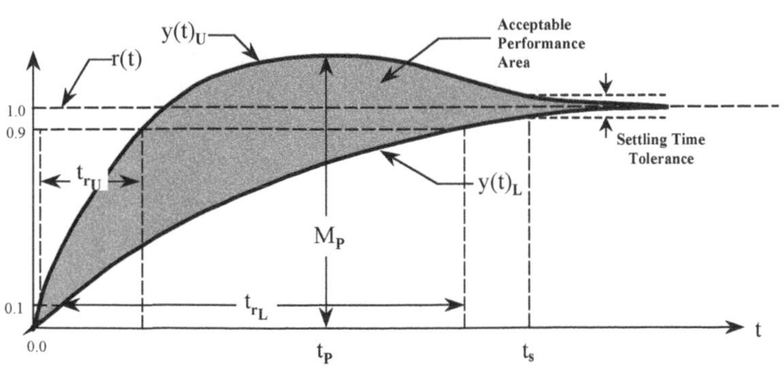

Fig. 3.3 System time domain tracking performance specifications.

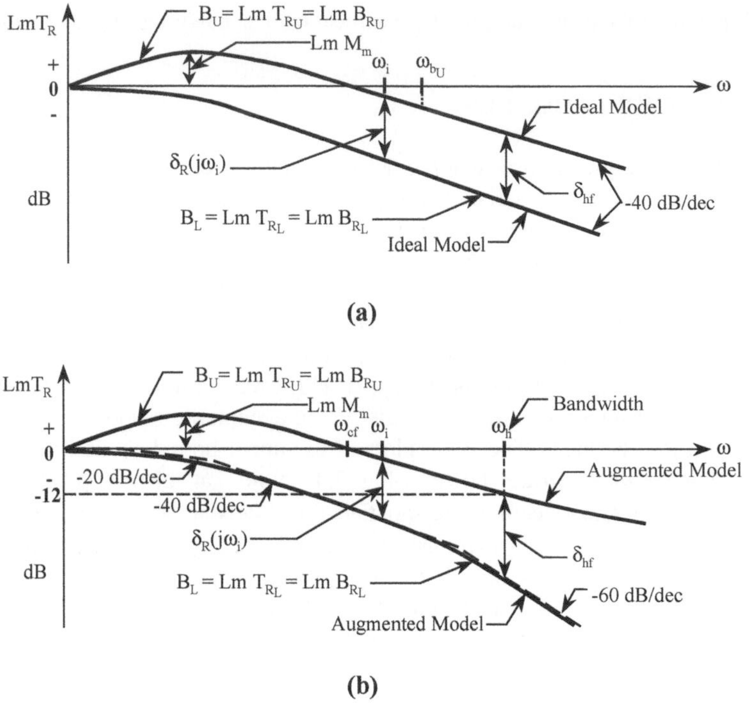

Fig. 3.4 Bode plots of T_R (a) Ideal simple second-order models; (b) The augmented models.

$\delta_R(j\omega_i)$ simplifies the process of synthesizing the loop transmission $L_o(s) = G(s)P_o(s)$ as discussed in Sec. 3-13. To synthesize $L_o(s)$ it is necessary to determine the tracking bounds $\boldsymbol{B}_R(j\omega_i)$ (see Sec. 3-9) which are obtained based upon $\delta_R(j\omega_i)$. This characteristic of $\delta_R(j\omega_i)$ ensures that the tracking bounds $\boldsymbol{B}_R(j\omega_i)$ decrease in magnitude as ω_i increases (see Sec. 3-9).

An approach to the modelling process is to start with a simple second-order model of the desired control ratio T_{R_U} having the form[1]

$$T_{R_U}(s) = \frac{\omega_n^2}{s^2 + 2\zeta\omega_n s + \omega_n^2} = \frac{\omega_n^2}{(s - p_1)(s - p_2)} \tag{3.8}$$

where $\sigma_D = -\zeta\omega_n$, $p_1 = \sigma_D + j\omega_d$, $p_2 = \sigma_D - j\omega_d$, $\omega_n^2 = p_1 p_2$ and $t_s \approx T_s = 4/\zeta\omega_n = 4/|\sigma_D|$ (the desired settling time). The control ratio $T_{R_U}(s)$ of Eq. (3.8) can be represented by an equivalent unity-feedback system so that

$$T_{R_U}(s) = \frac{Y(s)}{R(s)} = \frac{G_{eq}(s)}{1 + G_{eq}(s)} \tag{3.9}$$

where

$$G_{eq}(s) = \frac{\omega_n^2}{s(s + 2\varsigma\omega_n)} \tag{3.10}$$

The gain constant of this equivalent Type *1* transfer function $G_{eq}(s)$ is $K_1 = lim_{s\to 0}[sG_{eq}(s)] = \omega_n/2\zeta$. Equation (3.8) satisfies the requirement that $y(t)_{ss} = R_0 u_{-1}(t)$ for $r(t) = R_0 u_{-1}(\text{t})$. The frequency ω_b for which $|\boldsymbol{T}_{RU}(j\omega_b)| = 0.7071$ is defined as the *system bandwidth frequency* ω_{bU}.

The simplest over-damped model for $T_{RL}(s)$ is of the form

$$T_{R_L}(s) = \frac{Y(s)}{R(s)} = \frac{K}{(s - \sigma_1)(s - \sigma_2)} = \frac{G_{eq}(s)}{1 + G_{eq}(s)} \tag{3.11}$$

where

$$G_{eq}(s) = \frac{\sigma_1\sigma_2}{s[(s - (\sigma_1 + \sigma_2)]}$$

and $K_1 = -\sigma_1\sigma_2/(\sigma_1 + \sigma_2)$. For this system $y(t)_{ss} = R_0$ *for* $r(t) = R_0 u_{-1}(t)$. Selection of the parameters σ_1 and σ_2 is used to meet the specifications for t_s and K_1. The achievement of the desired performance specification is based upon the BW, $0 < \omega < \omega_{h_R}$, which is determined by the intersection of the –12 dB line and the B_U curve in Fig. 3.4b.

Once the ideal models $T_{R_U}(j\omega)$ and $T_{R_L}(j\omega)$ are determined, the time and frequency response plots of Figs. 3.3 and 3.4a, respectively, can then be drawn. Because the models for T_{R_U} and T_{R_L} are both second-order, the high-frequency asymptotes in Fig. 3.4a have the same slope. The high-frequency range (hf) in Fig. 3.4a is defined as $\omega \geq \omega_b$ where ω_b is the model BW frequency of B_U. In addition to achieving the desired characteristic of an increasing magnitude of δ_R of B_U for $\omega_i > \omega_{cf}$, an increasing spread between B_U and B_L is required in the hf range (see Fig. 3.4b), that is,

$$\delta_{hf} = B_U - B_L \tag{3.12}$$

must increase with increasing frequency. This desired increase in δ_R is achieved by changing B_U and B_L, without violating the desired time-response characteristics, by augmenting T_{R_U} with a zero [see Eq. (3.13)] as close to the origin as possible without significantly affecting the time response. This additional zero raises the curve B_U for the frequency range above ω_{cf}. The spread can be further increased by augmenting T_{R_L} with a negative real pole [see Eq. (3.14)] which is as close to the origin as possible but far enough away not to significantly affect the time response. Note that the straight-line Bode plot is shown only for T_{R_L}. This additional pole lowers B_L for this frequency range.

$$T_{R_U}(s) = \frac{(\omega_n^2 / a)(s+a)}{s^2 + 2\varsigma\omega_n s + \omega_n^2} = \frac{(\omega_n^2 / a)(s - z_1)}{(s - \sigma_1)(s - \sigma_2)} \tag{3.13}$$

$$T_{R_L}(s) = \frac{K}{(s+a_1)(s+a_2)(s+a_3)} = \frac{K}{(s-\sigma_1)(s-\sigma_2)(s-\sigma_3)} \tag{3.14}$$

Thus, for these augmented models, the magnitude of $\delta_R(j\omega_i)$ increases as ω_i increases above ω_{cf}.

The manner of achieving a $\delta_R(j\omega)$ that increases with frequency is described below and is illustrated in the example in Sec. 3-17. In order to minimize the iteration process in achieving acceptable models for $T_{R_U}(s)$ and $T_{R_L}(s)$ which have an increasing $\delta_R(j\omega)$, the following procedure may expedite the design process: (a) first synthesize the second-order model of Eq. (3.13) containing the zero at $|z_1| = a \geq \omega_n$ that meets the desired FOM; and (b) then, as a first trial, select all three real poles of Eq. (3.14) to have the value of $|\sigma_3| = a_3 = \omega_n > a_2 = a_1 > |\sigma_D|$. For succeeding trials, if necessary, one or more of these poles are moved right and/or left until the desired specifications are satisfied. As illustrated by the slopes of the straight-line Bode plots in Fig. 3.4b, selecting the value of all three poles in the range specified above insures an increasing δ_R. Other possibilities are as follows: (c) the specified values of t_p and t_s for T_{R_L} may be such that a pair of complex poles and a real pole need to be chosen for the model response. For this situation, the real pole *must* be more dominant than the complex poles. (d) Depending on the performance specifications [see the paragraph below Eq. (3.15)], $T_{R_U}(s)$ may require two real poles and a zero "close" to the origin, i.e., select $|z_1|$ very much less than $|p_1|$ and $|p_2|$ in order to effectively have an under-damped response.

At high frequencies δ_{hf} (see Fig. 3.4b) must be larger than the actual variation in the plant, δ_P. This characteristic is the result of the Bode theorem which states that

$$\int_0^\infty Lm\,\mathbf{S}_P^T\, d\omega = 0$$

Thus, the reduction in sensitivity S_P^T at the lower frequencies must be balanced by an increase in sensitivity at the higher frequencies (see Chapter 8). At some high frequency $\omega_{i \geq} \omega_h$ (see Fig. 3.4b), since $|\boldsymbol{T}_{R_L}| \approx 0$, then

$$\lim_{\omega \to \infty} \delta_R(j\omega_i) \approx B_U - (-\infty) = \infty \text{ dB} \tag{3.15}$$

For the case where $y(t)$, corresponding to T_{R_U}, is to have an allowable "large" overshoot followed by a small tolerable undershoot, a dominant complex-pole pair is not suitable for T_{R_U}. An acceptable overshoot with no undershoot for T_{R_U} can be achieved by T_{R_U} having two real dominant poles $p_1 > p_2$, a dominant real zero ($z_1 > p_1$) "close" to p_1, and a far off pole $p_3 << p_2$. The closeness of the zero dictates the value of M_p. Thus, a designer selects a pole-zero combination to yield the form of the desired time-domain response.

3-4.2 DISTURBANCE REJECTION MODELS[1]

The simplest disturbance control ratio model specification is $|\boldsymbol{T}_D(j\omega)| = |\boldsymbol{Y}(j\omega)/\boldsymbol{D}(j\omega)| < \alpha_p$, a constant, [the desired maximum magnitude of the output based upon a unit-step disturbance input [D of Fig.3.1)]; i.e., for $d_1(t)$: $|y(t_p)| \leq \alpha_p$, and for: $d_2(t)$ $|y(t)| \leq \alpha_p$ for $t \geq t_x$. Thus, the frequency domain disturbance specification is $Lm\ \boldsymbol{T}_D(j\omega) < Lm\ \alpha_p$ over the desired specified BW (see Fig. 3.5). Thus, the disturbance specification is represented by only an upper bound on the NC over the specified

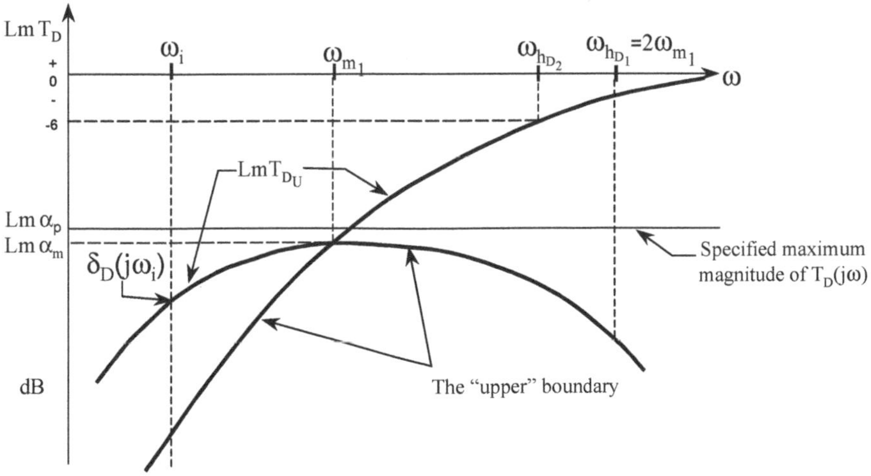

Fig. 3.5 Bode plots of disturbance models for $\boldsymbol{T}_D(j\omega)$.

BW. A detailed discussion on synthesizing disturbance-rejection models for Eqs. (3.6) and (3.7), based upon desired performance specifications, to yield the upper bounds shown in Fig. 3.5 for D_1 and D_2 is given in Ref. 1.

3-5 *J* LTI PLANT MODELS

The simple plant of Eq. (2.7), where $K \in \{1,10\}$ and $a \in \{1,10\}$, is used to illustrate the MISO QFT design procedure. The region of plant parameter uncertainty is illustrated by Fig. 3.6. This region may be described by J LTI plants, where $\iota = 1,2, \ldots J,$ which lie on its boundary. That is, the boundary points 1, 2, 3, 4, 5, and 6 are utilized, as discussed in Chapter 2, to obtain 6 LTI plant models that adequately define the region of plant parameter uncertainty. Note: the numbered points around the contour in Fig. 2.6 are relabeled by letters as shown in Fig. 3.6. This is done in order to simplify the labeling in the figures associated with the discussion in this chapter.

3-6 PLANT TEMPLATES OF $P_\iota(s)$, $\Im P(j\omega_i)$

With $L = GP$, Eq. (3.5) yields

$$Lm\,\boldsymbol{T}_R = Lm\,\boldsymbol{F} + Lm\left[\frac{\boldsymbol{L}}{1+\boldsymbol{L}}\right] = Lm\ F + Lm\ T \tag{3.16}$$

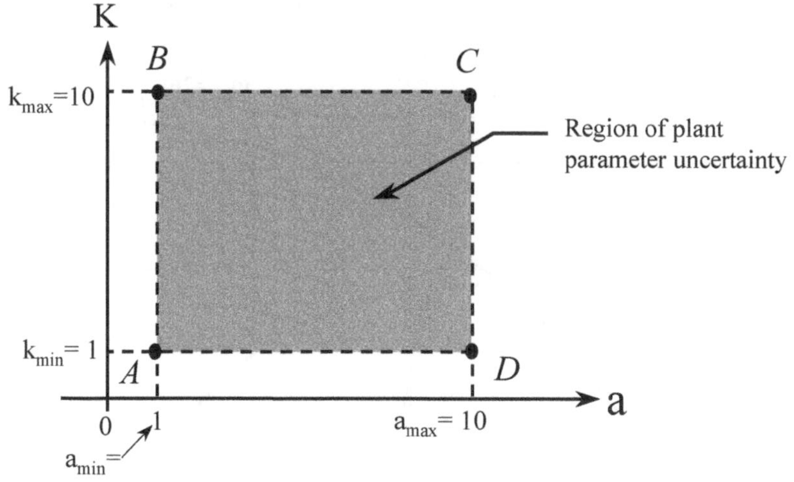

Fig. 3.6 Region of plant parameter uncertainty.

The change in T_R due to the uncertainty in P, since F is LTI, is

$$\Delta(Lm\boldsymbol{T}_R) = Lm\boldsymbol{T}_R - Lm\boldsymbol{F} = Lm\left[\frac{\boldsymbol{L}}{1+\boldsymbol{L}}\right] \tag{3.17}$$

The proper design of $L = L_o$ and F, must restrict this change in T_R so that the actual value of $Lm\ \boldsymbol{T}_R$ always lies between B_U and B_L of Fig. 3.4b. The first step in synthesizing an L_o is to make NC templates which characterize the variation of the plant uncertainty for various values of ω_i over a frequency range $\omega_x \le \omega_i \le \omega_{h_R}$, where $\omega_x < \omega_{cf}$. A guideline for selecting the frequency range for the templates are to select three frequency values below and above the 0-dB crossing frequency ω_{cf}, no less than an octave apart, up to approximately the *-12* dB value of the B_U plot in Fig. 3.4b. In addition, for a Type *0* plant select $\omega_x = 0$ and for Type *1* or higher-order plants select $\omega_x \neq 0$.

To provide more details in obtaining templates, the simple plant of Eq. (2.7) is used whose region of plant uncertainty is depicted in Fig. 3.6. The plant template on the NC can be obtained by mapping the plant parameter uncertainty region. A number of points on the perimeter of Fig. 3.6 are selected, and values of $Lm\ \boldsymbol{P}(j\omega_i)$ and $\angle\boldsymbol{P}(j\omega_i)$ are obtained at each point. These data, for each value of frequency $\omega = \omega_i$, are plotted on a NC as is illustrated in Chapter 2. A curve is drawn through these points and becomes the template $\Im\boldsymbol{P}(j\omega_i)$ at the frequency ω_i. A sufficient number of points must be selected so that the contour of $\Im\boldsymbol{P}(j\omega_i)$ accurately reflects the region of plant uncertainty. In addition to the points *A,B,C,D* marked in Fig. 3.6, it may also be necessary to include additional points on the perimeter.

For the points *A,B,C,* and *D*, at the frequency $\omega = 1$, the data obtained from Eq. (2.7), is, $\boldsymbol{P}_A(j1) = \sqrt{2}/2\angle{-135^\circ}$, $\boldsymbol{P}_B(j1) = 5\sqrt{2}-\angle{-135^\circ}$, $\boldsymbol{P}_C(j1) = 100/\sqrt{101}\ \angle{-95.7^\circ}$ and $\boldsymbol{P}_D(j1) = 10/\sqrt{101}\ \angle{-95.7^\circ}$. These data are plotted on the NC shown in Fig. 3.7. A curve is drawn through the points *A,B,C,D,A* and the shaded area is labeled $\Im\boldsymbol{P}(j1)$. The contour *A,B,C,D* in Fig. 3.7 may be drawn on a plastic sheet (preferably colored) so that a plastic template for $\Im\boldsymbol{P}(j1)$ can be cut and labeled. The templates for other values of ω_i are obtained in a similar manner. A characteristic of these templates is that, starting from a "low value" of ω_i, the templates widen (angular width becomes larger) for increasing values of ω_i. Then, as ω_i takes on larger values and approaches infinity they become narrower (see Fig. 2.11) and eventually approach a straight line of height V dB [see Eq. (3.20)].

For an aircraft, each point *A,B,C,D*, etc. in Fig. 3.7 represents a given flight condition (FC) at $\omega = \omega_i$, i.e., $[\boldsymbol{P}_{A/C}(j\omega_i)]_{FC\iota}$. One of the flight conditions may be identified as the nominal plant P_o.

For the plant of Eq. (2.7) the values $K = a = 1$ represent the lowest point of each of the templates $\Im\boldsymbol{P}(j\omega_i)$ and may be selected as the nominal plant $\boldsymbol{P}_o$ for all

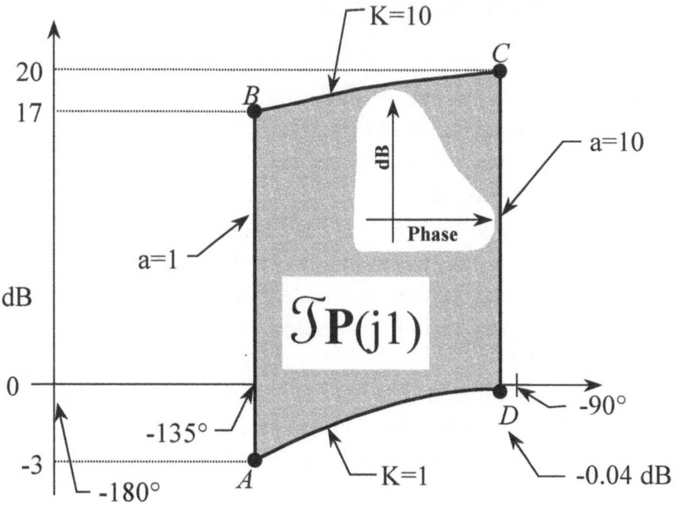

Fig. 3.7 NC characterizing Eq. (2.7) over the region of uncertainty.

frequencies. However, any plant in P can be chosen as the nominal plant[4]. With $\boldsymbol{L} = \boldsymbol{GP}$ and $\boldsymbol{L}_o = \boldsymbol{GP}_o$, as given in Eqs. (2.15) and (2.16):

$$\begin{aligned}\delta_P(j\omega_i) &= Lm\,\boldsymbol{L} - Lm\,\boldsymbol{L}_o \\ &= (Lm\boldsymbol{G} + Lm\boldsymbol{P}) - (Lm\boldsymbol{G} + Lm\boldsymbol{P}_O) \\ &= (Lm\,P - Lm\,P_o) \le \delta_R(j\omega_i)\ dB\end{aligned}$$

and

$$\angle\Delta\boldsymbol{P} = \angle\boldsymbol{P} - \angle\boldsymbol{P}_o$$

Thus, if point A in Fig. 3.7 represents $Lm\ \boldsymbol{P}_o$, a variation in $\boldsymbol{P}$ results in a horizontal translation in the angle of $\boldsymbol{P}$ and a vertical translation in the log magnitude value of $\boldsymbol{P}$. When $\boldsymbol{G}(j\omega)$ represents a specific transfer function, the template of Fig. 3.7 can be converted into a template of $\boldsymbol{L}(j\omega_i)$ by translating it vertically by $Lm\ \boldsymbol{G}(j\omega_i)$ and horizontally by $\angle\boldsymbol{G}(j\omega_i)$. For the template of $\boldsymbol{L}(j\omega_i)$, the values of the M-contours at the intersections with the template are the values of the control ratio $Lm\ \boldsymbol{T}(j\omega_i) = Lm\ [\boldsymbol{L}(j\omega_i)/(1 + \boldsymbol{L}(j\omega_i)]$. The range of values of $\boldsymbol{T}(j\omega_i)$ for the entire range of parameter variation (K and a) can therefore be determined (for complement information about the templates see Appendix A).

3-7 NOMINAL PLANT

While any plant case can be chosen, it is a common practice to select, whenever possible, a nominal plant whose NC point is at the lower left corner of the template at a selected frequency. This nominal plant, defined by a set of parameters with specific values, must be the same for the rest of the frequencies for which templates are to be obtained (see Chapter 9 for further enlightenment).

3-8 *U*-CONTOUR (STABILITY BOUND)

It is well known that $|T(j\omega)| \leq M_L$ establishes a circle in the Nichols Chart (see Fig. 3.8). It also defines the phase margin (PM) and the gain margin (GM). Let $\boldsymbol{L}(j\omega) = \boldsymbol{G}(j\omega)\,\boldsymbol{P}(j\omega j = |\boldsymbol{L}|\,e^{j\varphi}$. In that case the gain margin is $GM = 1/|\quad|$ at the angle $\phi = -180°$. Then, in terms of the M_L circle, GM = $1 + 1/M_L$. Analogously the phase margin is PM = $180° + \phi$ at the angle where ϕ is the phase of $L(j\omega)$ at $|\boldsymbol{L}(j\omega)| = 1$. Hence, in terms of the M_L circle, PM $\geq 180° - 2\cos^{-1}(0.5/M_L)$.

The specifications on system performance in the time domain (see Fig. 3.3) and in the frequency domain (see Fig. 3.4) identify a minimum damping ratio ζ for the dominant roots of the closed-loop system which corresponds to a bound on the value of $M_p \approx M_m$. On the NC this bound on $M_p = M_L$ (see Fig. 3.8) establishes a region which must not be penetrated by the templates and the loop transmission functions $\boldsymbol{L}_\iota(j\omega)$ for all ω. The boundary of this region is referred to as the *universal high-frequency boundary (UHFB)* or *stability bound*, the *U-contour*, because this becomes the dominating constraint on $\boldsymbol{L}(j\omega)$. Therefore, the top portion, *efa*, of the M_L contour becomes part of the *U*-contour. The formation of the *U*-contour is discussed in this section.

For the two cases of disturbance rejection depicted in Fig. 3.1 the control ratios are, respectively, as given in Eqs. (3.6) and (3.7):

$$T_{D_1} = \frac{P}{1+L} \qquad and \qquad T_{D_2} = \frac{1}{1+L}$$

Thus, it is necessary to synthesize an $L_o(s)$ so that the disturbances are properly attenuated. For the present, only one aspect of this disturbance-response problem is considered, namely a constraint is placed on the damping ratio ζ of

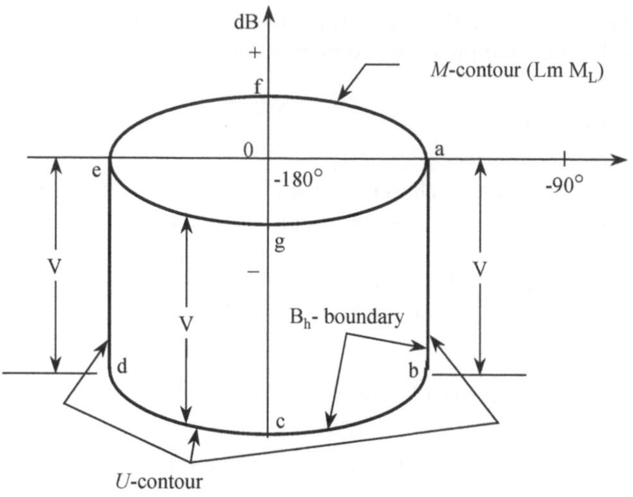

Fig. 3.8 *U*-contour construction (stability contour).

the dominant complex-pole pair of T_D nearest the jω-axis.[13] This damping ratio is related to the peak value of

$$|\boldsymbol{T}(j\omega)| = \left|\frac{\boldsymbol{L}(j\omega)}{1 + \boldsymbol{L}(j\omega)}\right| \qquad \textbf{(3.18)}$$

For example, consider the case for which the dominant complex-pole pair of Eq. (3.18) results in a peak of $Lm\ \boldsymbol{T} = 8$ dB for $\zeta = 0.2$, and 2.7 dB for $\zeta = 0.4$, etc. Although this large peak does not appear in T_R due to the design of the filter $F(s)$, it does affect the response for T_D. If $d(t)$ is very small, a peak of T_D, due to $\zeta = 0.2$, can be "very large," and it may be difficult to achieve the restriction on the peak overshoot α_p of the time response:

$$|c(t)| \le \alpha_p$$

Therefore, it is reasonable to add the requirement

$$|\boldsymbol{T}| = \left|\frac{\boldsymbol{L}}{1 + \boldsymbol{L}}\right| \le M_L \qquad \textbf{(3.19)}$$

where M_L is a constant for all ω and over the whole range of $\mathcal{P}$ parameter values. This results in a constraint on ζ of the dominant complex-pole pair of T_D. This constraint can therefore be translated into a constraint on the maximum value T_{max} of Eq. (3.18). For example, for $Lm\ M_m = 2$ dB, the oval, *agefa*, in Fig. 3.8 corresponds to the *2* dB *M*-contour on the NC. This results in limiting the peak of the disturbance response. A value of M_L can be selected to correspond to the maximum value of T_R. Therefore, the top portion, *efa*, of the *M*-contour on the NC, which corresponds to the value of the selected value of M_L, becomes part of the *U*-contour.

For a large class of problems, as $\omega \rightarrow \infty$, the limiting value of the plant transfer function approaches

$$\lim_{\omega\to\infty} [\boldsymbol{P}(j\omega)] = \frac{K'}{\omega^{\lambda}}$$

where λ represents the excess of poles over zeros of $P(s)$. The plant template, for this problem class, approaches a vertical line of length equal to

$$\begin{aligned} \Delta &\equiv \lim_{\omega\to\infty} [Lm\ \boldsymbol{P}_{\max} - Lm\ \boldsymbol{P}_{\min}] \\ &= Lm\,K'_{max} - Lm\,K'_{min} = \text{V}\ dB \end{aligned} \tag{3.20}$$

If the nominal plant is chosen at $K' = K'_{min}$, then the constraint M_L gives a boundary which approaches the *U*-contour *abcdefa* of Fig. 3.8. (*Note*: For a MIMO plant $\boldsymbol{P} = \{p_{ij}\}$, as $\omega \rightarrow \infty$, the templates may not approach a vertical line if the λ_{ij} are not the same for all p_{ij} elements of the plant matrix. When the λ_{ij} are different, then the widths of the templates are a multiple of 90°.

For the simple plant of Eq. (2.7), where $K \in \{1,10\}$ and $a \in \{1,10\}$ and where $K' = Ka$, applying the limiting condition, $\omega \rightarrow \infty$, to Eq. (3.20) yields

$$\begin{aligned} V = \Delta Lm P = &\lim_{\omega\to\infty} [\{Lm(Ka)_{\max} - Lm(j\omega)^2\} - \\ &\{Lm(Ka)_{\min} - Lm(j\omega)^2\}] \\ = &Lm(Ka)_{\max} - Lm(Ka)_{\min} = Lm100 - Lm1 = 40\ dB \end{aligned} \tag{3.21}$$

For the m.p. plant of Eq. (2.7), where the poles are real, the plant templates have the typical shape of Fig. 3.7.

The high-frequency boundary B_h, the *bcd* portion of the *U*-contour in Fig. 3.8, is obtained by measuring down V dB from the *ega* portion of the *M*-contour as illustrated in this figure. V is determined by Eq. (3.21) which, for this example, is *40* dB. The remaining portions of the *U*-contour, portions *ab* and *de*, not necessarily straight lines, are determined by satisfying the requirement of Eq. (3.19) and $\delta_R(j\omega_i)$. The $\mathfrak{I}\boldsymbol{P}(j\omega_i)$ is used to determine the corresponding tracking bounds $\boldsymbol{B}_R(j\omega_i)$ on the NC in the manner described in Sec. 3-9.

3-9 TRACKING BOUNDS $B_R(j\omega_i)$ ON THE NC

As an introduction to this section, the procedure for adjusting the gain of a unity-feedback system to achieve a desired value of M_m by use of the NC is reviewed. Consider the plot of $Lm\ \mathbf{P}(j\omega)$ *vs.* $\angle\mathbf{P}(j\omega)$ for a plant shown in Fig. 3.9 (the solid curve). With $G(s) = A = 1$ and $F(s) = 1$ in Fig. 3.1, $L = P$. The plot of $Lm\ \mathbf{L}(j\omega)$ *vs.* $\angle\mathbf{L}(j\omega)$ is tangent to the $M = 1$ dB curve with a resonant frequency $\omega_m = 1.1$. If $Lm\ M_m = 2$ dB is specified for $Lm\ \mathbf{T}_R$, the gain A is increased, raising $Lm\ \mathbf{L}(j\omega)$, until it is tangent to the 2-dB M-curve. For this example the curve is raised by $Lm\ A = 4.5$ dB ($G = A = 1.679$) and the resonant frequency is $\omega_m = 2.09$.

Now consider that the plant uncertainty involves only the variation in gain A between the values of 1 and 1.679. It is desired to find a cascade compensator $G(s)$, in Fig. 3.1, such that the specification 1 dB $\leq Lm\ M_m \leq 2$ dB is always maintained for this plant gain variation while the resonant frequency ω_m remains constant. This requires that the loop transmission $\mathbf{L}(j\omega) = \mathbf{G}(j\omega)\mathbf{P}(j\omega)$ be synthesized so that it is tangent to an M-contour in the range of $1 \leq Lm\ M \leq 2$ dB for the entire range of $1 \leq A \leq 1.679$ *and* the resultant resonant frequency satisfies the requirement $\omega_m = 2.09 + \Delta\omega_m$. The manner of achieving this and other time-response specifications is the subject of the remaining portion of this chapter.

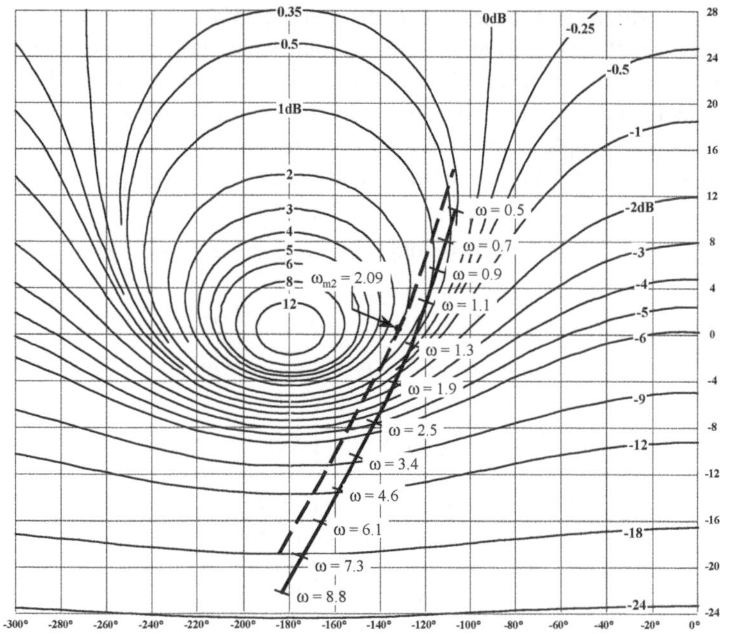

Fig. 3.9 Log magnitude-angle diagram.

It is assumed for Eq. (3.17) that the compensators F and G are fixed (LTI), that is, they have negligible uncertainty. Thus, only the uncertainty in P contributes to the change in T_R given by Eq. (3.17). The solution requires that the actual $\Delta Lm\boldsymbol{T}_R(j\omega_i) \leq \delta_R(j\omega_i)$ dB in Fig.3.4b. Thus, it is necessary to determine the resulting constraint, or bound $\boldsymbol{B}_R(j\omega_i)$, on $\boldsymbol{L}(j\omega_i)$. The procedure is to pick a nominal plant $P_o(s)$ and to derive the bounds on the resulting nominal loop transfer function $L_o(s) = G(s)P_o(s)$.

As an illustration, consider the plot of $Lm\ \boldsymbol{P}(j2)$ *vs.* $\angle\boldsymbol{P}(j2)$ for the plant of Eq. (2.7) (see Fig. 3.6). As shown in Fig. 3.10, the plant's region of uncertainty $\Im\boldsymbol{P}(j2)$ is given by the contour *ABCD*, i.e., $Lm\ \boldsymbol{P}(j2)$ lies on or within the boundary of this contour. The nominal plant transfer function, with $K_o = 1$ and $a_o = 1$, is

$$P_o = \frac{1}{s(s+1)} \tag{3.22}$$

and is represented in Fig. 3.10 by point A for $\omega = 2$ [-13.0 dB, -153.4^o]. Note, once a nominal plant is chosen, it must be used for determining all the bounds $\boldsymbol{B}_R(j\omega_i)$. Since $Lm\ \boldsymbol{L}(j2) = Lm\ \boldsymbol{G}(j2) + Lm\ \boldsymbol{P}(j2)$, then $\Im\boldsymbol{P}(j2)$ is translated on the Nichols chart vertically by the value of $Lm\ \boldsymbol{G}(j2)$ and horizontally by the angle

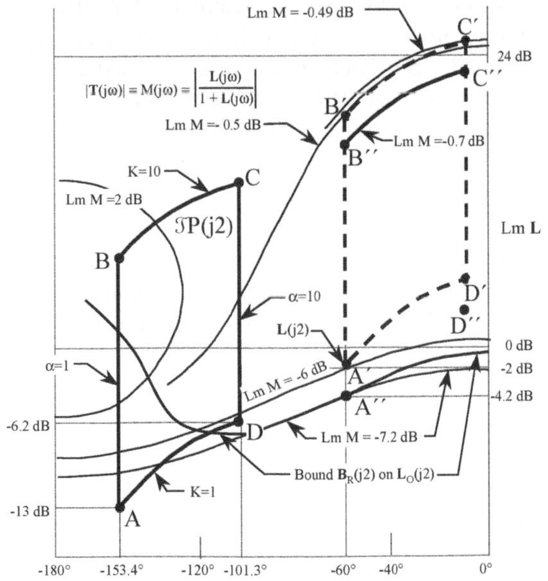

Fig. 3.10 Derivation of bounds $\boldsymbol{B}_R(j\omega_i)$ on $\boldsymbol{L}_o(j\omega)$ for $\omega = 2$.

$\angle G(j2)$. The templates $\Im P(j\omega_i)$ are relocated to find the position of $L_o(j\omega)$ which satisfies the specifications in Fig. 3.4b of $\delta_R(j\omega_i)$ for each value of ω_i. For example, if a trial design of $L(j2)$ requires sliding $\Im P(j2)$ to the position $A'B'C'D'$ in Fig. 3.10, then

$$\left|\text{Lm}\, G(\text{j}2)\right| = \left|\left|\text{Lm}\, L(\text{j}2)\right|_{A'} - \left|\text{Lm}\, P(\text{j}2)\right|_{A}\right| = \left|\left|-2\right| - \left|-13\right|\right| = 11\,\text{dB} \quad \textbf{(3.23)}$$

$$\angle G(\text{j}2) = \angle L(\text{j}2)_{A'} - \angle P(\text{j}2)_{A} = -60^\circ - (-153.4^\circ) = 93.4^\circ \quad \textbf{(3.24)}$$

Using the contours of constant $Lm\ M = Lm\ [L/(1 + L)]$ on the NC in Fig. 3.10, the maximum occurs at point $C'(M = -0.49$ dB$)$ and the minimum at point $A'(M = -6$ dB$)$ so that the maximum change in $Lm\ T$ is, in this case, $(-0.49) - (-6) = 5.51$ dB. If the specifications tolerate a change of 6.5 dB at $\omega = 2$, the above trial position of $Lm\ L_o(j2)$ is well within the permissible tolerance. Lowering the template on the NC to $A''B''C''D''$, where the extreme values of $Lm\ [L/(1 + L)]$ are at $C''(-0.7$ dB) and $A''(-7.2$ dB), yields $Lm\ L(j2)_{C''} - Lm\ L(j2)_{A''} = -0.7 - (-7.2) = 6.5$ dB $= \delta_R(j2)$. Thus, if $\angle L_o(j2) = -60^\circ$, then -4.2 dB is the smallest or minimum value of $Lm\ L_o(j2)$ which satisfies the 6.5 dB specifications for $\delta_R(j\omega_i)$. Any smaller magnitude is satisfactory but represents over design at that frequency. The manipulation of the $\omega = 2$ template, for ease of the design process, is repeated along a new angle (vertical) line, and a corresponding new minimum of $L_o(j2)$ is found. Sufficient points are obtained in this manner to permit drawing a continuous curve of the bound $B_R(j2)$ on $L_o(j2)$, as shown in Fig. 3.10. The above procedure is repeated at other frequencies, resulting in a family of boundaries $B_R(j\omega_i)$ of permissible $L_o(j\omega)$. The procedure for determining the boundaries $B_R(j\omega_i)$ is summarized as follows:

1. From Fig. 3.4b obtain values of $\delta_R(j\omega_i)$ for a range of values of ω_i $(\omega_1, \omega_2, \ldots, \omega_h)$, preferably† an octave apart, over the specified bandwidth. [The selection of ω_h results in the bound $B_R(j\omega_h)$ passing under the U-contour.
2. Place the template $\Im P(j\omega_i)$ on the NC containing the U-contour to determine the bound $B_R(j\omega_i)$ as follows:
   (a) Use major angle divisions of the NC for lining up the $\Im P(j\omega_i)$.
   (b) Select P_o to represent, in general, the lowest point of $\Im P(j\omega_i)$. For the design example shown in this chapter:
   (1) Select point A in Fig. 3.11 as P_o.
   (2) Line up side A-B of $\Im P(j\omega_i)$ on the -90° line, as shown in Fig. 3.11. Move the template up or down until the difference $\Delta Lm\ T_R(j\omega_i)$ between the values of two adjacent M-contours ($Lm\ \alpha$ and $Lm\ \beta$,

† In the low frequency range (desired tracking bandwidth) generally an octave apart will provide a reasonable separation of the bounds on the NC.

respectively) is equal to the value of $\delta_R(j\omega_i)$ obtained from Fig. 3.4b. Thus, in Fig. 3.11, determine the locations of the templates on

(3)

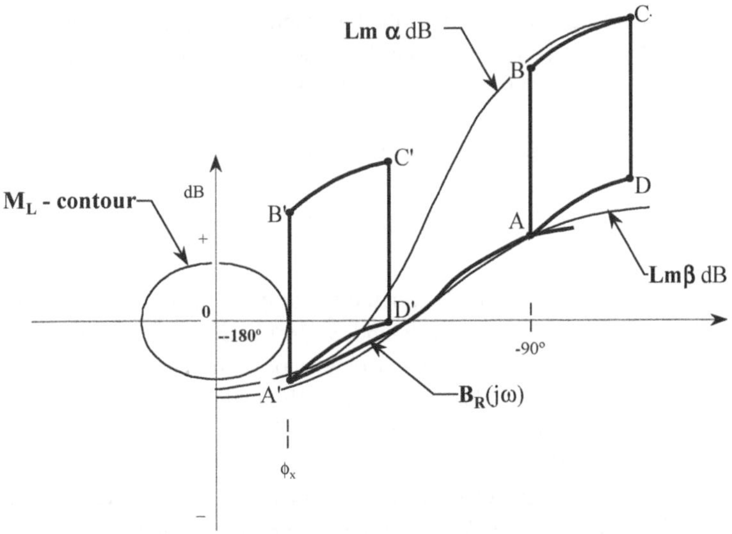

Fig. 3.11 Graphical determination of $\boldsymbol{B}_R(j\omega)$.

the -90° line where

$$\Delta\, Lm\boldsymbol{T}_{\mathrm{R}}(j\omega_i) = Lm\alpha\ -\ Lm\beta\ = \delta_R(j\omega_i) \qquad \textbf{(3.25)}$$

When Eq. (3.25) is satisfied, then point A, on the M-contour = $Lm\ \beta$, lies on the bound $\boldsymbol{B}_R(j\omega_i)$. Mark this point on the NC. For other plants the shape of the template may be such that if point A represents P_o, another point of the template may be the lowest point that satisfies Eq. (3.25). When this equation is satisfied, point A still yields points of $\boldsymbol{B}_R(j\omega_i)$.

3. Repeat step 2 on the lines -100°, -110°, etc., up to -180° or until a point of the template becomes tangent to the M_L-contour. No intersection of the M_L-contour by a template is permissible. For example, in moving the template from the -90° line to the left, the template may eventually become tangent to the M_L-contour at some angle ϕ_x as illustrated in Fig. 3.11. If the template is moved further to the left it will intersect the M_L contour and permit a peak of $\boldsymbol{T}(j\omega)$ greater than M_L. In order to satisfy

the requirement of Eq. (3.19), point A' on the $\phi = \phi_x$ line becomes the left boundary or terminating point for the $\boldsymbol{B}_R(j\omega_i)$ contour and is a point on the U-contour (a point on ab of Fig. 3.8).[3] Draw a curve through all the points to obtain the contour for $\boldsymbol{B}_R(j\omega_i)$. For the plant of this example the U-contour is symmetrical about the -180° axis. Note that obtaining the bounds $\boldsymbol{B}_R(j\omega_i)$ as described in step 2(b)(2) only guarantees that the difference $\delta_L(j\omega_i)$ between the upper bound $Lm\ \boldsymbol{T}_U$ and the lower bound $Lm\ \boldsymbol{T}_L$ for $Lm\ \boldsymbol{T} = Lm\ [\boldsymbol{L}/(1+\boldsymbol{L})]$ will satisfy

$$\delta_L(j\omega_i) \le \delta_R(j\omega_i) = \mathrm{Lm}\boldsymbol{T}_{R_U}(j\omega_i) - \mathrm{Lm}\boldsymbol{T}_{R_L}(j\omega_i)$$
$$= \mathrm{Lm}\boldsymbol{T}_U(j\omega_i) - \mathrm{Lm}\boldsymbol{T}_L(j\omega_i) \tag{3.26}$$

4. Repeat steps 2 and 3 over the range of $\omega_x \le \omega_i \le \omega_h$, generally octaves apart, until the highest bound $\boldsymbol{B}_R(j\omega_i)$ and lowest bound $\boldsymbol{B}_R(j\omega_h)$ on the NC clear the U-contour. For reasonably damped plants ($\zeta > 0.6$), over the entire region of plant uncertainty, the magnitudes of the bounds $\boldsymbol{B}_R(j\omega_i)$ usually decrease as ω increases. Thus, for this type of plant, it is desirable to have $\delta_R(j\omega_i)$ increasing as ω_i increases, as discussed in Sec. 3-4. When this characteristic of $\delta_R(j\omega_i)$ is not observed, it is possible to have $|\boldsymbol{B}_R(j\omega_j)|, > |\boldsymbol{B}_R(j\omega_i)|$ for $\omega_j > \omega_i$. For a plant that is highly under-damped ($\zeta \le 0.6$), over some portion of the region of plant uncertainty, avoid selecting an under-damped nominal plant $P_o(s)$. For this latter situation it is desirable, but not necessary, to synthesize T_{R_U} and T_{R_L} based upon an over-damped $P_o(s)$ model. Note, if γ is specified instead of M_L it dictates the side a-b of the U-contour.

3-10 DISTURBANCE BOUNDS $\boldsymbol{B}_D(j\omega_i)$: CASE 1

Two disturbance inputs are shown in Fig. 3.1. It is assumed that only one disturbance input exists at a time. Both cases are analyzed.

CASE 1 $[d_2(t) = D_0 u_{-1}(t),\ d_1(t) = 0]$

CONTROL RATIO. From Fig. 3.1 the disturbance control ratio for input $d_2(t)$ is

$$T_D(s) = \frac{1}{1+\mathrm{L}} \tag{3.27}$$

Substituting $L = 1/\ell$ into Eq. (3.27) yields

$$T_D(s) = \frac{\ell}{1+\ell} \tag{3.28}$$

this equation has the mathematical format required to use the NC. Over the specified BW it is desired that $|T_D(j\omega)| << 1$, which results in the requirement, from Eq. (3.28), that $|L(j\omega)| >> 1$ (or $|\ell(j\omega)| << 1$), i.e.,

$$\left|T_D(j\omega)\right| \approx \frac{1}{\left|L(j\omega)\right|} = \left|\ell(j\omega)\right| \tag{3.29}$$

DISTURBANCE RESPONSE CHARACTERISTIC. A time-domain tracking response characteristic based upon $r(t) = u_{-1}(t)$ often specifies a maximum allowable peak overshoot M_p. In the frequency domain this specification may be approximated by

$$\left|M_R(j\omega)\right| = \left|T_R(j\omega)\right| = \left|\frac{Y(j\omega)}{R(j\omega)}\right| \le M_m \approx M_p \tag{3.30}$$

The corresponding time- and frequency-domain response characteristics, based upon the step disturbance forcing function $d_2(t) = u_{-1}(t)$, are, respectively,

$$\left|m_D(t)\right| = \left|\frac{y(t)}{d(t)}\right| \le \alpha_p \quad \text{for } t \ge t_x \tag{3.31}$$

and

$$\left|M_D(j\omega)\right| = \left|T_D(j\omega)\right| = \left|\frac{Y(j\omega)}{D(j\omega)}\right| \le \alpha_m \approx \alpha_p \tag{3.32}$$

APPLICATION. Let $L = KL'$ in the *tracking ratio* $T_L = L/(1 + L)$, where K is an unspecified gain, and the specification on the system performance is $M_m = 1.26$ (2 dB). By means of a NC determine the value of K required to achieve this value of M_m and obtain the data to plot $|M(j\omega)|$ *vs.* ω. The plot of $Lm\ L'(j\omega)$ *vs.* $\angle L'(j\omega)$, for $K = 1$, on the NC is tangent to $Lm\ M = 1$ dB contour, resulting in the plot of $Lm\ L(j\omega)$ *vs.* $\angle L(j\omega)$ in Fig. 3.12[1] for $K = 1$. Intersections of $Lm\ L(j\omega)$ with the M-contours provide the data to plot the tracking control ratio $|M(\omega)|$ *vs.* ω.

Now consider the corresponding disturbance control ratio for the same control system. The disturbance transfer function: $T_D = 1/(1 + L) = \ell/(1 + \ell)$ has the desired BW $0 \le \omega \le \omega_2$ for which $|L(j\omega)| >> 1$ and $|\ell(j\omega)| << 1$. Thus Eq. (3.30) applies within the BW region. Table 3.1[1] contains data for two points on the Nichols plot of Fig. 3.12. The plot of $Lm\ \ell(j\omega)$ *vs.* $\angle \ell$ for these 2 points is also shown in this figure. The NC of Fig. 3.12 can be rotated 180° and is redrawn in Fig. 3.13. Since $Lm\ \mathbf{l}(j\omega) = Lm\ [1/\mathbf{L}(j\omega)] = -Lm\ L(j\omega)$, then a negative value of $Lm\ \boldsymbol{\ell}$ yields a positive value for $Lm\ \mathbf{L}$ as shown in Fig. 3.13. Since $L = KL' = 1/\ell$, then

$$\ell(j\omega) = K^{-1}\ell'(j\omega) \qquad \textbf{(3.33)}$$

If $\ell'(j\omega)$ is given and it is required to determine K^{-1} to satisfy Eq. (3.27), then the plot $Lm\ \ell'(j\omega)$ vs. $\angle \ell'(j\omega)$ must be raised or lowered until it is tangent to the Lm α_p-contour ($|T_D|max = \alpha_m$). The amount Δ by which this plot is raised or lowered yields the value of K, i.e., $Lm\ K^{-1} = \Delta$. Note that this is the same procedure used for the tracking example of Fig. 3.12, except that the adjustment in $Lm\ \ell(j\omega)$ is K^{-1}.

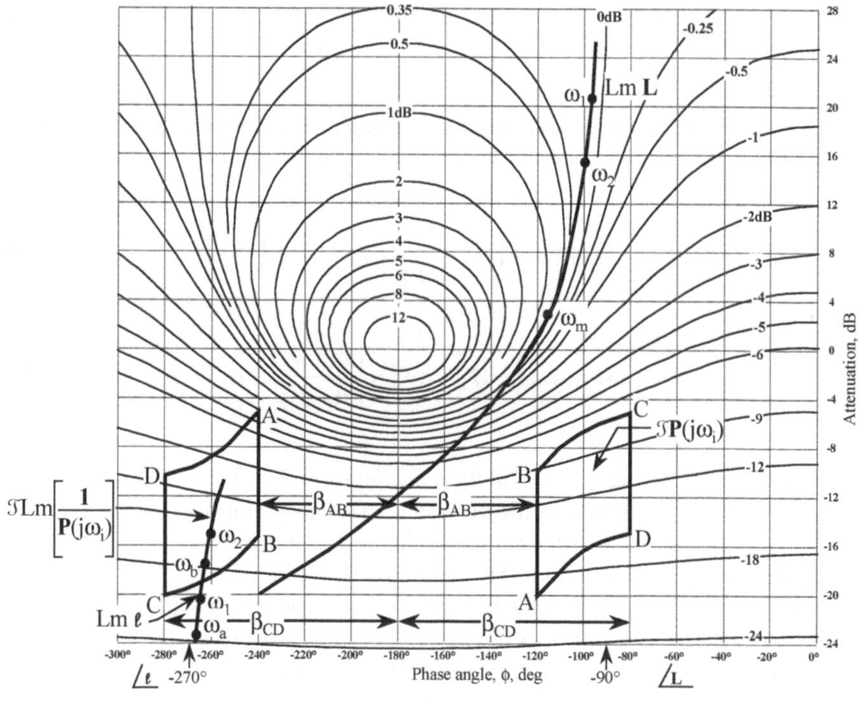

Fig. 3.12 Regular Nichols chart.

TABLE 3.1 Data points for a Nichols Plot

ω	$Lm\,\mathbf{L}$	$\angle\mathbf{L}$	$Lm\,\ell$	$\angle\ell$
ω_1	21	-96°	-21	96° (or –264°)
ω_2	15	-98°	-15	98° (or –262°)

TEMPLATES. For a given plant P having uncertain parameters, consider that its template $\Im\mathbf{P}(j\omega_i)$ for a given ω_i has equal dB differences along it's A-B and C-D boundaries, i.e., for a given $\angle\mathbf{P}(j\omega_i)$,

$$\Delta(Lm\,\mathbf{P}_B - Lm\,\mathbf{P}_A) = \Delta(Lm\,\mathbf{P}_C - Lm\,\mathbf{P}_D) = 10 \text{ dB}$$

This template is arbitrarily set on the NC as shown in Fig. 3.12. Data corresponding to the template location shown in Fig. 3.12 are given in Table 3.2.

TABLE 3.2 Data points for the templates of Fig. 3.12

Points	$\angle\mathbf{P}$	$\angle 1/\mathbf{P}$
A,B	-120°	120° (or –240°)
C,D	-80°	80° (or -280°)

The template of the reciprocal, $Lm\,[1/\mathbf{P}(j\omega_i)]$, is arbitrarily set on NC in Fig. 3.12 for the same frequency as for the template of $Lm\,\mathbf{P}(j\omega_i)$ and for the angles of Table 3.2.[2] Note that the template of $Lm\,[1/\mathbf{P}(j\omega_i)]$ is the same as the template of $Lm\,\mathbf{P}(j\omega_i)$ but is rotated by 180°. Thus, it is located by first reflecting the template of $Lm\,\mathbf{P}(j\omega_i)$ about the -180° axis, "flipping it over" vertically, and then moving it up or down so that it lies between -5 and -20 dB. For the *arbitrary* location of the template of $Lm\,[1/\mathbf{P}(j\omega_i)]$, note that:

(a)

$$\beta_{AB} = 180^o + \angle P = 180^o - 120^o = 60^o$$

$$\beta_{CD} = 180^o + \angle P = 180^o - 80^o = 100^o$$

For $1/\mathbf{P}$, the corresponding angles are

$$\angle(1/\mathbf{P}_{AB}) = -180^o - \beta_{AB} = -180^o - 60^o = -240^o$$

$$\angle(1/\mathbf{P}_{CD}) = -180^o - \beta_{CD} = -180^o - 100^o = -280^o$$

(b) Templates of $Lm\ P(j\omega_i)$ are used for the tracker case $T = L/(1 + L)$ and the templates of $Lm\ [1/P(j\omega_i)]$ are used for the disturbance rejection case of $T_D = 1/(1 + L) = \ell/(1 + \ell)$.

ROTATED NC. Since $Lm\ L$ is desired for the disturbance rejection case, Eq. (3.27), the disturbance boundary $B_D(j\omega_i)$ for $L(j\omega_i) = 1/\ell(j\omega_i)$ is best determined on the rotated NC of Fig. 3.13. Thus, the NC of Fig. 3.12 is rotated clockwise (cw) by 180°, where the rotation of the $Lm\ [1/P(j\omega)]$ template *ABCD* is reflected in Fig. 3.13. The rotated NC is used to determine directly the boundaries $B_D(j\omega_i)$ for $L_D(j\omega_i)$. Point *A* for the simple plant of this design example corresponds to the nominal plant parameters and is the lowest point of the template $\Im P(j\omega_i)$. This point is *again used* to determine the disturbance bounds $B_D(j\omega_i)$. The lowest point of the template must be used to determine the bounds and, in general, may or may not be the point corresponding to the nominal plant parameters.

Based upon Eqs. (3.27) and (3.28) in Fig. 3.5

$$- Lm T_D = Lm[1 + L] \geq - Lm\,\alpha(j\omega_i) > 0 \text{ dB} \tag{3.34}$$

where $\alpha(j\omega_i) < 0$. Since $|L| >> 1$ in the BW, then

$$- Lm\, T_D \cong Lm\, L \geq - Lm\,\alpha(j\omega_i) = - \delta_D(j\omega_i) \tag{3.35}$$

In terms of $L(j\omega_i)$, the constant *M*-contours of the NC can be used to obtain the disturbance performance T_D. This requires the change of sign of the vertical axis in dB and the *M*-contours, as shown in Fig. 3.13.

BOUNDS $B_D(j\omega_i)$. The procedure for determining the boundaries $B_D(j\omega_i)$ is:

1. From Fig. 3.5 the obtain values of $\delta_D(j\omega_i) = Lm\ \alpha_m$ [see Eq. (3.32)] for the same values of frequency as for the tracking boundary $B_R(j\omega_i)$.
2. Select the lowest point of $\Im P(j\omega_i)$ to represent the nominal plant P_o in Fig. 3.13. For the design example used in this chapter select point *A* in Fig. 3.13 as P_o. The same nominal point must be used in obtaining the tracking and disturbance bounds.
3. Use major angle divisions of the NC for lining up the $\Im P(j\omega_i)$. Line up side *A-B* of $\Im P(j\omega_i)$, for example, on the –280° line for ℓ(or the –80° line for L). Move the template up or down until point *A* lies on the *M*-contour that represents $\delta_D(j\omega_i)$. Mark this point on the NC.

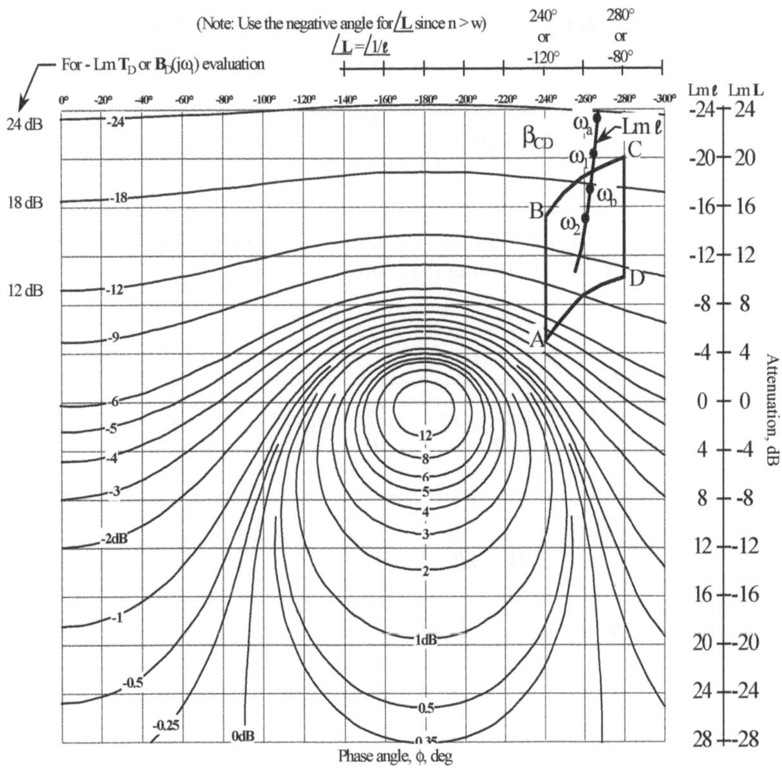

Fig. 3.13 Rotated Nichols chart.

4. Repeat step 3 on the vertical lines for –100°, –120°, etc., up to the –180° line or the *U*-contour. Draw a curve through all the points to obtain the contour for $B_D(j\omega_i)$.
5. Repeat steps 3 and 4 over the desired frequency range $\omega_x \le \omega_i \le \omega_h$.
6. Transcribe these $B_D(j\omega_i)$ onto the NC that contains the bounds $B_R(j\omega_i)$.

Note that when $|L| >> 1$, then, from Eq. (3.34), $|L(j\omega_i)| >> \alpha(j\omega_i)$. Thus the *M*-contour corresponding to $T_D(j\omega_i)$ becomes the boundary $B_D(j\omega_i) = -Lm\ M$ *for* $L(j\omega_i)$. For example, if $\alpha(j\omega_i) = 0.12$, then $L(j\omega_i) = 8.33$ and thus $Lm\ \alpha(j\omega_i) = -18.4$ dB and $Lm\ L(j\omega_i) = 18.4$ dB.

3-11 DISTURBANCE BOUNDS $B_D(j\omega_i)$: CASE 2

CASE 2 $[d_1(t) = D_o u_{-1}(t),\ d_2(t) = 0]$

CONTROL RATIO. From Fig. 3.1 the disturbance control ratio for the input $d_1(t)$ is

$$T_D(j\omega) = \frac{P(j\omega)}{1 + G(j\omega)P(j\omega)} \tag{3.36}$$

Assuming point A of the template represents the nominal plant P_o, Eq. (3.36) is multiplied by P_o/P_o and rearranged as follows:

$$T_D = \frac{P_o}{P_o}\left[\frac{1}{\dfrac{1}{P} + G}\right] = \frac{P_o}{\dfrac{P_o}{P} + GP_o} = \frac{P_o}{\dfrac{P_o}{P} + L_o} = \frac{P_o}{W} \tag{3.37}$$

where

$$W = (P_o/P) + L_o \tag{3.38}$$

Thus, Eq. (3.37) with $Lm\ T_D = \delta_D$ yields

$$Lm\,\mathbf{W} = Lm\,\mathbf{P}_o - \delta_D \tag{3.39}$$

DISTURBANCE RESPONSE CHARACTERISTICS. Based on Eq. (3.30), the time- and frequency-domain response characteristics, for a unit-step disturbance forcing function, are given, respectively, by

$$|M_D(t)| = \left|\frac{y(t_p)}{d(t)}\right| = |y(t_p)| \le \alpha_p \tag{3.40}$$

and

$$|M_D(j\omega)| = |T_D(j\omega)| = \left|\frac{Y(j\omega)}{D(j\omega)}\right| \le \alpha_m \equiv \alpha_p \tag{3.41}$$

where t_p is the peak time.

BOUNDS $B_D(j\omega_i)$. The procedure for determining the boundaries $B_D(j\omega_i)$ is:

1. From Fig. 3.5 obtain the value of $\delta_D(j\omega_i)$ representing the desired model specification $T_D = T_{D_U} = \alpha_p$ for the same values of frequency as for the tracker boundaries $B_R(j\omega_i)$.
2. *Evaluate* in tabular form for each value of ω_i the following items in the order given:

$$Lm\, \boldsymbol{P}_o(j\omega_i) \quad \delta_D(j\omega_i) \quad Lm\, \boldsymbol{W}(j\omega_i) \quad |\boldsymbol{W}(j\omega_i)| \quad \frac{\boldsymbol{P}_o(j\omega_i)}{\boldsymbol{P}_\iota(j\omega_i)}$$

   The ratio $\boldsymbol{P}_o/\boldsymbol{P}_\iota$ is evaluated at each of the four points of Fig. 3.6 for each value of ω_i. It may be necessary to evaluate this ratio at additional points around the perimeter of the *ABCD* contour as shown in Fig. 3.6.
3. Before presenting the procedure for the graphical determination of $\boldsymbol{B}_d(j\omega_i)$, where $\boldsymbol{B}_D = Lm\, \boldsymbol{B}_d$, it is necessary to first review, graphically, the phasor relationship between $\boldsymbol{B}_d$, $\boldsymbol{P}_o/\boldsymbol{P}$, and $\mathbf{W}$ for $\omega = \omega_i$. Equation (3.38), with $\boldsymbol{L}_o$ replaced by its bound $\mathbf{B}_d$, is rearranged to the form

$$\boldsymbol{W} = \frac{\boldsymbol{P}_o}{\boldsymbol{P}} + \boldsymbol{B}_d \quad \rightarrow \quad -\boldsymbol{B}_d = \frac{\boldsymbol{P}_o}{\boldsymbol{P}} + (-\boldsymbol{W}) \tag{3.42}$$

   For arbitrary values of $\boldsymbol{P}_o(j\omega_i)/\boldsymbol{P}(j\omega_i)$ and $\boldsymbol{W}(j\omega_i)$, Fig. 3.14 presents the phasor relationship of Eq. (3.42). Since the values of $\boldsymbol{P}_o(j\omega_i)/\boldsymbol{P}(j\omega_i)$ for $P \in \mathcal{P}$ and $|\boldsymbol{W}(j\omega_i)|$ are known, the following procedure can be used to evaluate $\boldsymbol{B}_d(j\omega_i)$:

   (a) On polar or rectangular graph paper draw $\Im[\boldsymbol{P}_o/\boldsymbol{P}]$ for each ω_i as shown in Fig. 3.15, where point A is the nominal point.

   (b) Based upon Fig. 3.14 and the location of the phasor $-\boldsymbol{W}(j\omega)$, the solution for $-\boldsymbol{B}_d$ is obtained from

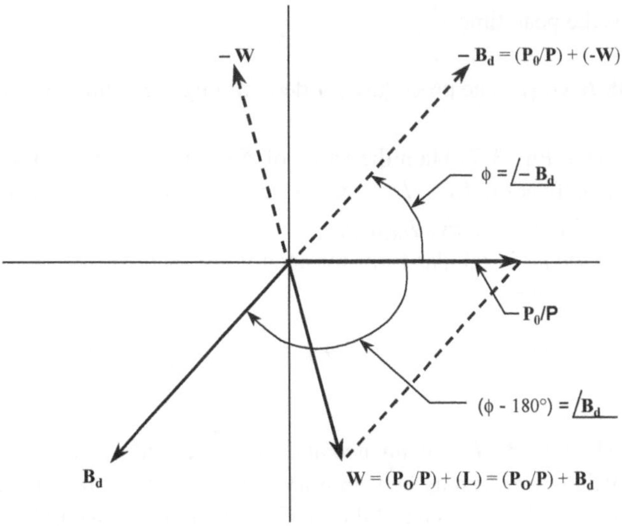

Fig. 3.14 Phasor relationship of Eq. (3.38) with $L_o = B_o$.

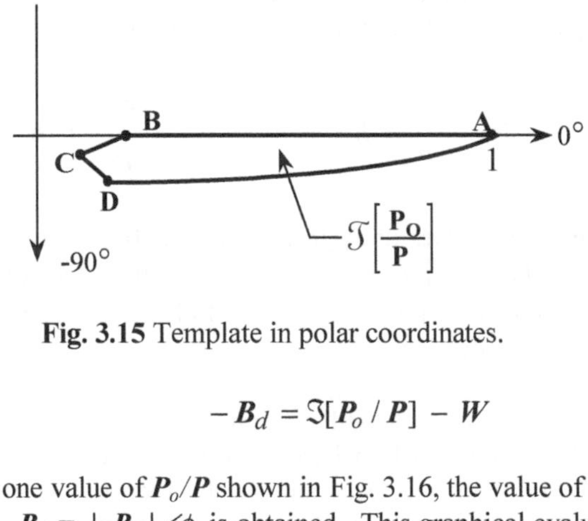

Fig. 3.15 Template in polar coordinates.

$$-B_d = \Im[P_o / P] - W \qquad \textbf{(3.43)}$$

For one value of P_o/P shown in Fig. 3.16, the value of $-W$ is plotted and $-B_d = |-B_d|\angle\phi$ is obtained. This graphical evaluation of $B_d = |B_d|\angle(\phi - 180^o)$ is performed for various points around the perimeter of $\Im[P_o/P]$ in Fig. 3.15. A simple graphical evaluation yields a more restrictive bound (the worst case). Use a compass to mark off arcs with a radius equal to the distance $|W(j\omega_i)|$ at a number of points on the perimeter of $\Im[P_o(j\omega_i)/P(j\omega_i)]$. Draw a curve which is tangent to these arcs to form the first quadrant portion of the Q-contour shown in Fig. 3.16. Depending on the plant type desired for L it may be

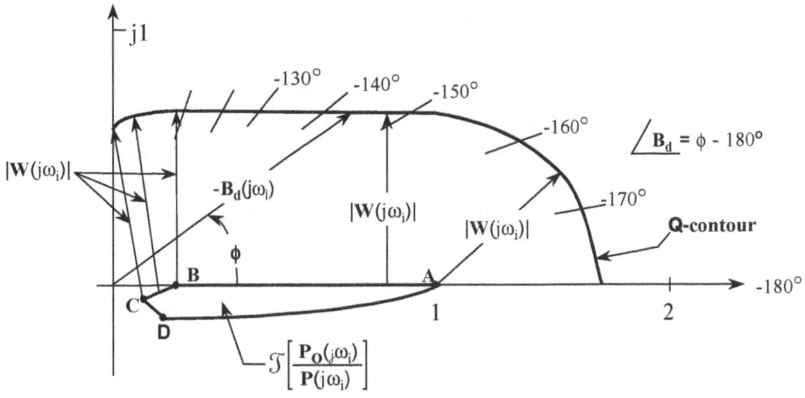

Fig. 3.16 Graphical evaluation of $\boldsymbol{B}_d(j\omega_i)$.

necessary to extend this contour into the second and fourth quadrants.

1. Based upon Eq. (3.43) and Fig. 3.15, the phasor from the origin of Fig. 3.16 to the Q-contour represents $-\boldsymbol{B}_d(j\omega_i)$. This contour includes the plant uncertainty as represented by $\mathfrak{I}[\boldsymbol{P}_o(j\omega_i)/\boldsymbol{P}(j\omega_i)]$. In the frequency range $\omega_x \leq \omega_i \leq \omega_h$, if $|\boldsymbol{W}(j\omega_i)| >> |\boldsymbol{P}_o(j\omega_i)/\boldsymbol{P}(j\omega_i)|$, then the Q-contour is essentially a circle about the origin with radius $|\boldsymbol{W}(j\omega_i)| \equiv |\boldsymbol{B}_d(j\omega_i)|$.
2. Assuming the partial Q-contour of Fig. 3.16 is sufficient, measure from the graph the length $\boldsymbol{B}_d(j\omega_i)$ for every 10° of $\boldsymbol{B}_d(j\omega_i)$ and create Table 3.3 for each value of ω_i.
3. Plot the values of $\boldsymbol{B}_D(j\omega_i)$ from Table 3.3 for each value of ω_i, on the same NC as $\boldsymbol{B}_R(j\omega_i)$.

TABLE 3.3 Data points for the boundary $\boldsymbol{B}_D(j\omega)$

$\angle \boldsymbol{B}_d(j\omega_i)$	$\lvert\boldsymbol{B}_d(j\omega_i)\rvert$	$\boldsymbol{B}_D(j\omega_i) = Lm\, \boldsymbol{B}_d(j\omega_i)$
-180°		
-170°		
-160°		
.		
.		
.		

3-12 THE COMPOSITE BOUNDARY $B_o(j\omega_i)$

The composite bound $B_o(j\omega_i)$ that is used to synthesize the desired loop transmission transfer function $L_o(s)$ is obtained in the manner shown in Fig. 3.17. The composite bound $B_o(j\omega_i)$, for each value of ω_i, is composed of those portions of each respective bound $B_R(j\omega_i)$ and $B_D(j\omega_i)$ that are the most restrictive. For the case shown in Fig. 3.17a the bound $B_o(j\omega_i)$ is composed of those portions of each respective bound $B_R(j\omega_i)$ and $B_D(j\omega_i)$ that have the largest values. For the situation of Fig. 3.17b, the outermost of the two boundaries $B_R(j\omega_i)$ and $B_D(j\omega_i)$ becomes the perimeter of $B_o(j\omega_i)$. The situations of Fig. 3.17 occur when the two bounds have one or more intersections. If there are no intersections, then the bound with the largest value or with the outermost boundary dominates. The synthesized $L_o(j\omega_i)$, for the situation of Fig. 3.17a, must lie on or just above the bound $B_o(j\omega_i)$. For the situation of Fig. 3.17b the synthesized $L_o(j\omega_i)$ *must not lie in the interior* of the $B_o(j\omega_i)$ contour.

3-13 SHAPING OF $L_O(j\omega)$[2,3,5,18]

A realistic definition of optimum[6,55] in an LTI system is the minimization of the high-frequency loop gain K while satisfying the performance bounds. This gain affects the high-frequency response since $\lim_{\omega\to\infty}[L(j\omega)] = K(j\omega)^{-\lambda}$ where λ is the excess of poles over zeros assigned to $L(j\omega)$. Thus, only the gain K has a significant effect on the high-frequency response, and the effect of the other parameter uncertainty is negligible. Also, the importance of minimizing the high-frequency loop gain is to minimize the effect of sensor noise whose spectrum, in general, lies in the high-frequency range (see Chapters 6 and 9). It has been shown that the optimum $L_o(j\omega)$ exists; it lies on the boundary $B_o(j\omega_i)$ at all ω_i, and it is unique.[5,27] Note the bounds can be also calculated using an iterative algorithm that compute them through quadratic inequalities (see Appendix B).

Previous sections describe how tolerances on the closed-loop system frequency response, in combination with plant uncertainty templates, are translated into bounds on a nominal loop transmission function $L(j\omega)$. In Fig. 3.18 the template $\Im P(j\omega_i)$ is located on the corresponding bound $B_o(j\omega_i)$ where point A is on the constant M-curve $Lm\ \beta$, and point C on the constant M-curve $Lm\ \alpha$ such that

$$\delta_R(j\omega) = Lm\ \alpha - Lm\ \beta = 4\text{ dB} \tag{3.44}$$

and where $\alpha = T_{max}$, $\beta = T_{min}$, and $\delta_R(j\omega_i) = 4$ dB.

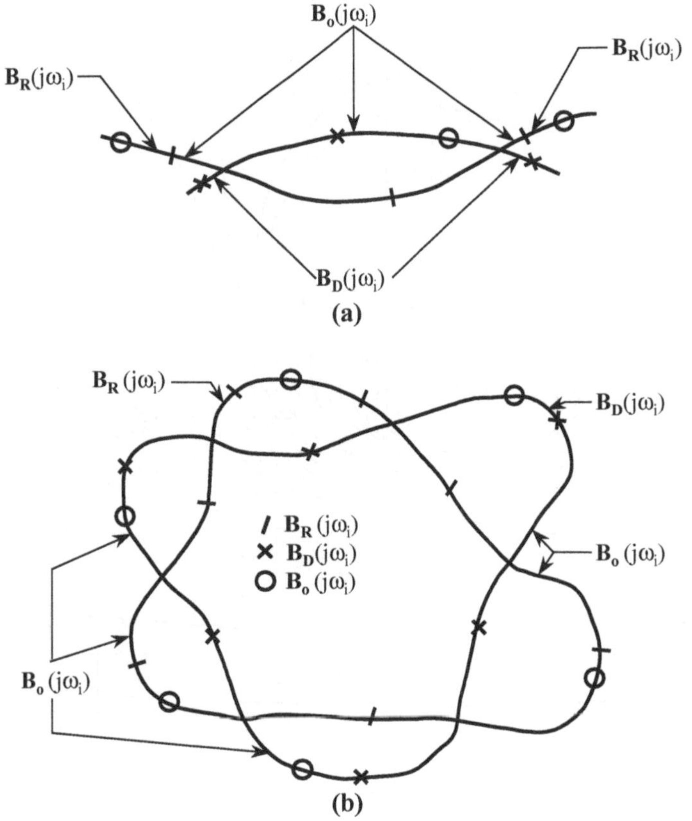

Fig. 3.17 Composite $B_o(j\omega_i)$.

For the set of plant parameters that correspond to point E within $\Im \boldsymbol{P}(j\omega_i)$ and for a synthesized L_o, as shown in Fig. 3.18, the open-loop transfer function is $\boldsymbol{L}_E = \boldsymbol{GP}_E$. The control ratio

$$T_E = \frac{L_E}{1 + L_E} \tag{3.45}$$

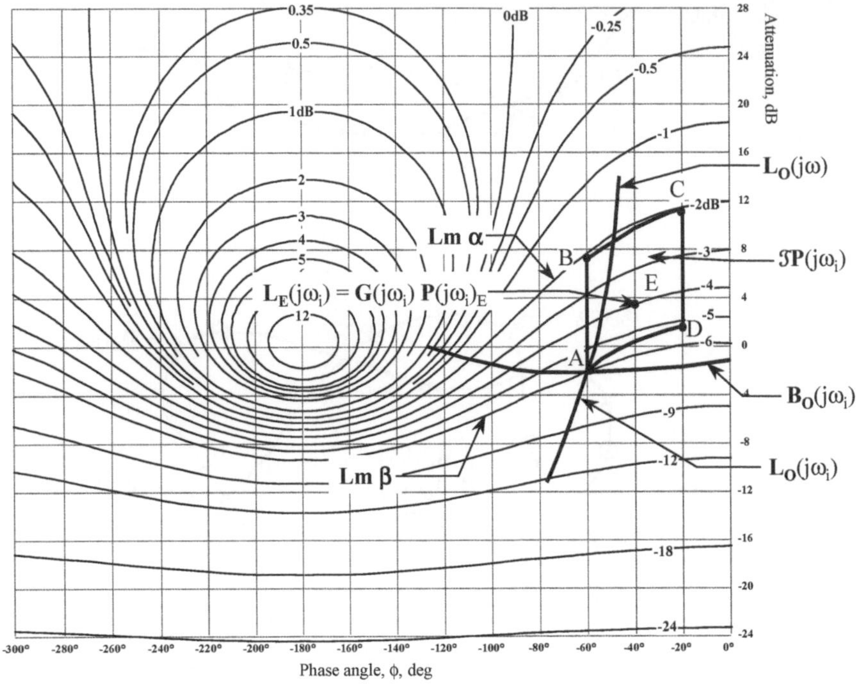

Fig. 3.18 Graphical determination of $Lm\ \mathbf{T}_t(j\omega_i)$ for $P \in \mathcal{P}$.

obtained from the constant M-contours on the NC has the value $Lm\ \mathbf{T}_E = -4$ dB. Thus, any value of $\mathbf{P}$ within $\Im\mathbf{P}(j\omega_i)$ yields a value of $Lm\ \mathbf{T}$ (= $Lm\ \mathbf{T}_t$) of Eq. (3.5) between points A and C in Fig. 3.18. Therefore, any value of $\mathbf{P}$ that lies within the region of uncertainty (see Fig. 3.6 for the example of this chapter) yields a maximum variation in $\mathbf{T}(j\omega_i)$ that satisfies the requirement

$$Lm\,\mathbf{T}_{max} - Lm\,\mathbf{T}_{min} \le \delta_R(j\omega_i) \qquad \mathbf{(3.46)}$$

Thus, proper design of the pre-filter F (see Fig. 3.1) yields a tracking control ratio $\mathbf{T}_R$ that lies between $Lm\ \mathbf{T}_U$ and $Lm\ \mathbf{T}_L$ in Fig. 3.4b.

For the plant of Eq. (2.7), the shaping of $\mathbf{L}_o(j\omega)$ is shown by the dashed curve in Fig. 3.19. A point such as $Lm\ \mathbf{L}_o(j2)$ must be on or above the curve labeled $\mathbf{B}_o(j2)$. Further, in order to satisfy the specifications, $\mathbf{L}_o(j\omega)$ cannot violate the U-contour. In this example a reasonable $\mathbf{L}_o(j\omega)$ closely follows the U-contour up to $\omega = 40$ rad/sec and stays below it above $\omega = 40$ as shown in Fig 3.19. Additional specifications are $\lambda = 4$, i.e., there are 4 poles in excess of zeros, and that it also must be Type 1 (one pole at the origin).

A representative procedure for choosing a rational function $L_o(s)$ which satisfies the above specifications is now described (see also Appendix I). It involves building up the function

$$\mathbf{L}_o(j\omega) = \mathbf{L}_{ok}(j\omega) = \mathbf{P}_o(j\omega)\prod_{k=0}^{w}[K_k \mathbf{G}_k(j\omega)] \qquad \mathbf{(3.47)}$$

where for $k = 0$, $\mathbf{G}_o = 1\angle 0^{\circ}$, and

$$K = \Pi_{k=0}^{w} K_k$$

In order to minimize the order of the compensator a good starting point for "building up" the loop transmission function is to initially assume that $\mathbf{L}_{o0}(j\omega) = \mathbf{P}_o(j\omega)$ as indicated in Eq. (3.47). $\mathbf{L}_o(j\omega)$ is built up term-by-term in order to stay just outside the U-contour in the NC of Fig. 3.19. The first step is to find the $\mathbf{B}_o(j\omega_i)$ which *dominates* $\mathbf{L}_o(j\omega)$. For example, suppose $\mathbf{L}_{o0}(j4) = K_o\mathbf{P}'_o(j4) = 0$ dB $\angle -135^{\circ}$ (point A in Fig. 3.19) and that at $\omega = 1$ the required $Lm\ \mathbf{L}_{o0}(j1)$ is approximately 27 dB. In order for $Lm\ \mathbf{L}_o(j\omega)$ to decrease from 27 to about 0 dB in two octaves, the slope of $\mathbf{L}_o(j\omega)$ must be about -14 dB/octave, with $\angle \mathbf{L}_o < -180^{\circ}$ (see Fig. 3.20). Since the M_L stability contour must be satisfied for all J plants then it is required that $\mathbf{L}_o(j\omega)$ have a phase margin angle γ of 45° over the entire frequency range for which $\mathbf{L}_o(j\omega)$ follows the vertical right-hand side of the U-contour and not just at the 0 dB crossover. Hence $\mathbf{B}_o(j1)$ dominates $\mathbf{L}_o(j\omega)$ more than does $\mathbf{B}_o(j4)$. In the same way it is seen that $\mathbf{B}_o(j1)$ dominates all other $\mathbf{B}_o(j\omega)$ in Fig. 3.19.

The $\mathbf{B}_o(j\omega)$ curves for $\omega < 1$ are not shown in Fig. 3.19 because it is assumed that a slope of -6 dB/octave for $\omega < 1$ suffices (additional values are 33 dB at $\omega = 0.5$, 39 dB at $\omega = 0.25$, etc.). By selecting $L_{o0}(s) = kP'_o(s) = k/[s(s + 1)]$ the first denominator factor of $\mathbf{L}_{o0}$ has a corner frequency at $\omega = 1$ (i.e., a pole at -1) which maintains -135° for $\omega \geq 1$. Thus, the value $Lm\ \mathbf{L}_o(j1)$ on the straight line approximation is selected at 30 dB (to allow for the -3 dB correction at the corner frequency). The function $\mathbf{L}_o(s)$ determined so far is

$$L_{o0}(s) = 31.6 / s(s+1)$$

whose angle $\angle \mathbf{L}_{o0}(j\omega)$ is sketched in Fig. 3.20.

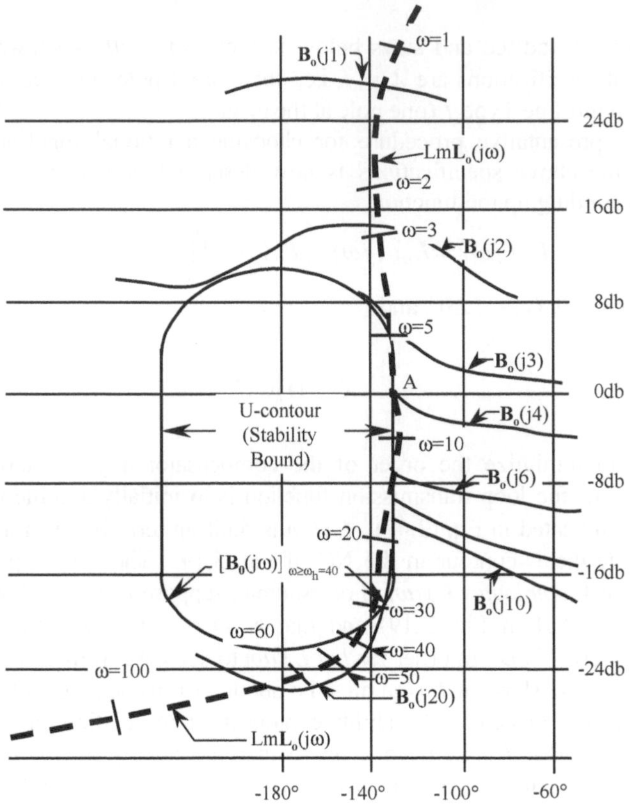

Fig. 3.19 Shaping of $L_o(j\omega)$ on the Nichols chart for the plant of Eq. (2.7).

L_{o0}(jω) violates the -135° bound at $\omega \geq 1.2$, hence a numerator term $(1 + j\omega T_2)$ must be added. At $\omega = 5$, $\angle L_{o0}(j5) = -169^\circ$ (see Fig. 3.20), and therefore a lead angle of 29° is needed at this frequency. Since a second denominator term $(1 + j\omega T_3)$ will be needed, allow an additional 15° for this factor, giving a total of $15^\circ + 29^\circ = 45^\circ$ lead angle required at $\omega = 5$. This is achieved by selecting $T_2 = 0.2$, i.e., a zero at -5. This results in the composite value

$$L_{o1}(s) = \frac{31.6(1 + s/5)}{s(1 + s)}$$

whose phase angle $\angle L_{o1}(j\omega)$ is sketched in Fig. 3.20. In the NC of Fig. 3.19, ω ≥ 5 is the region where the maximum phase *lag* allowed is –135° (i.e., $\angle L_o(j\omega)$ must be $\geq -135^\circ$). At $\omega = 10$, since $\angle L_{o1}(j10) = -112^\circ$, then $135^\circ - 112^\circ = 23^\circ$ more lag angle is needed, and is provided with an additional denominator term $(1 + j\omega T_3)$. However, this is to be followed by an additional numerator term $(1 + j\omega T_4)$

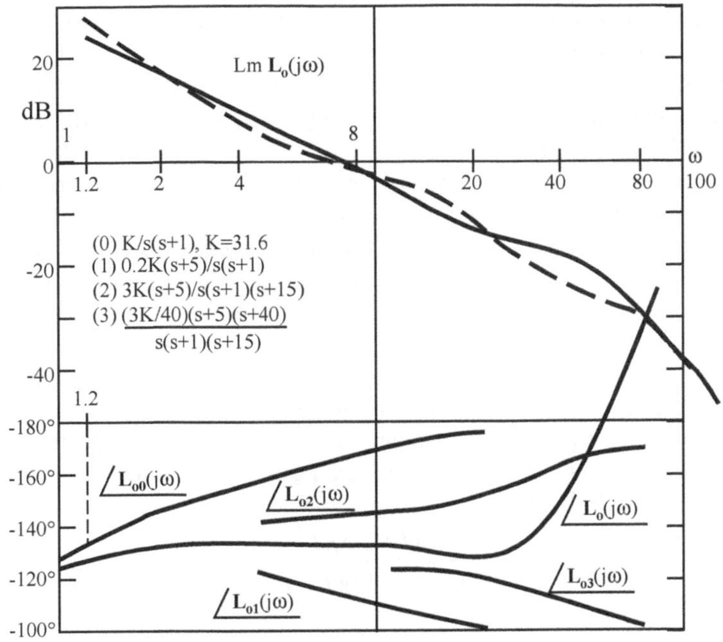

Fig. 3.20 Shaping of $L_o(j\omega)$ on the Bode plot.

so allow about 10° for it, giving $23^\circ + 10^\circ = 33^\circ$ more lag allowable from $(1 + j\omega T_3)$. This requires selecting the corner frequency at $\omega = 15$ (or $T_3 = 1/15$), giving

$$L_{o2}(s) = \frac{31.6(1 + s/5)}{s(1+s)(1+s/15)}$$

A sketch of $\angle L_{o2}(j\omega)$ is shown in Fig. 3.20.

Looking ahead at $\omega = 40$, $Lm\ L_{o2}(j40) = -20$ dB, thus $L_o(j\omega)$ can make its asymptotic left turn *under* the *U*-contour. The plan is to add two more numerator factors, $(1 + jT_4)$ and $(1 + jT_5)$, and finally two complex-pole pairs, in order to have an excess $\lambda = 4$ of poles over zeros and to minimize the *BW*. A zero is assigned at $\omega = 40$ with $T_4 = 1/40$, giving

$$L_{o3}(s) = \frac{31.6(1 + s/5)(1 + s/40)}{s(1+s)(1+s/15)}$$

A sketch of $\angle L_{o3}(j\omega)$ is shown in Fig. 3.20.

In order to achieve (an asymptotic) horizontal segment for *Lm* $\mathbf{L}_o(j\omega)$, before the final *–24* dB/octave slope is achieved, a final zero obtained from $(1 + j\omega T_5)$ is needed. Since the bottom of the *U*-contour (see Fig. 3.19) is at *–22.5* dB, allow a *2* dB safety margin, a *3* dB correction due to $(1 + j\omega T_5)$, and *1.5* dB for the effect of $(1 + j\omega T_4)$, giving a total of $-(22.5 + 2 + 3 + 1.5) = -29$ dB. A damping ratio of $\zeta = 0.5$ is selected for the two complex-pole pairs, so no correction is needed for them. Thus, the final corner frequency for the straight-line curve of *Lm* $\mathbf{L}_{o3}(j\omega)$ is at *-29* dB. Since *Lm* $\mathbf{L}_{o3}(j\omega)$ achieves this at $\omega = 60$, the last corner frequency is at $\omega = 60$, i.e., $T_5 = 1/60$. The resulting phase angle, due to $\mathbf{L}_{o3}$ and $(1 + j\omega/60)$, is *–66*. An angle of *–180°* could be selected at this point, but an additional *15°* margin is allowed (this is a matter of judgment which depends on the problem, which may include the presence of higher-order modes, etc.). This means that *100°* phase lag is permitted: *50°* due to each complex-pole pair ($180° - 66° - 15° \approx 100°$). Thus, a different value of damping ratio, with the appropriate dB correction applied to the log magnitude plot, is chosen, i.e., $\zeta = 0.6$. This locates the corner frequency at $\omega = 100$. Thus

$$L_o(s) = \frac{31.6(1+s/5)(1+s/40)(1+s/60)}{s(1+s)(1+s/15)(1+1.2s/100+s^2/10^4)^2}$$

$$= KP_o(s) \prod_{k=1}^{4} G_k(s) \qquad \textbf{(3.48)}$$

The optimal loop transfer function $\mathbf{L}_o(j\omega)$ is sketched in Fig. 3.19 and is shown by the solid curves in Fig. 3.20. A well designed, i.e., an "economical" $\mathbf{L}_o(j\omega)$ is close to the $\mathbf{B}_o(j\omega)$ boundary at each ω_i. The vertical line at *–140°* in Fig. 3.19 is the dominating vertical boundary for $\mathbf{L}_o(j\omega)$ for $\omega < 5 = \omega_x$, and the right side of the *U*-contour line at *–135°* is the vertical boundary effectively for $\omega_x \approx 5 < \omega < 30 \approx \omega_y$. The final $\mathbf{L}_o(j\omega)$ is good in this respect since it is close to these boundaries. Although there is a "slight" infringement of the *U*-contour, for $\omega > 15$, because of the inherent over design, no reshaping of $\mathbf{L}_o(s)$ is done unless the simulation reveals that the specifications are not met.

There is a tradeoff between the complexity of $L_o(s)$ (the number of its poles and zeros) and its final cutoff corner frequency, which is $\omega = 100$, and the phase margin frequency $\omega_\phi = 7$. There is some phase to spare between $\mathbf{L}_o(j\omega)$ and the boundaries. Use of more poles and zeros in $L_o(s)$ permits this cutoff frequency to be reduced a bit below *100*, but not by much. On the other hand, if it is desired to reduce the number of poles and zeros of $L_o(s)$, then the price in achieving this is a larger cutoff frequency. It is possible to economize significantly by allowing more phase lag in the low-frequency range. If *–180°* is permitted at $\omega = 1$, then a decrease of *Lm* $\mathbf{L}_o(j\omega)$ at a rate of *12* dB/octave can be achieved, e.g., with *Lm* $\mathbf{L}_o(j1) = 25$ dB, then it will be *13* dB at $\omega = 2$ (instead of the present *18* dB). Even with no more savings, this *5*-dB difference allows a cutoff frequency of about $\omega = 70$ instead of *100*.

Figure 3.19 reveals immediately, without any reshaping of $\boldsymbol{L}_o(j\omega)$ required, that reduction (i.e., easing) of the specifications at $\omega = 1$ to about *21* dB (instead of about *26* dB) has the same effect as the above. How badly the specifications are compromised by such easing can easily be checked. The design technique is thus highly "transparent" in revealing the tradeoffs between performance and uncertainty tolerances, complexity of the compensation, stability margins, and the "cost of feedback" in bandwidth.

3-14 GUIDELINES FOR SHAPING $\boldsymbol{L}_o(j\omega)$

Some general guidelines for the shaping of $\boldsymbol{L}_o(j\omega)$ are:

1. For the *beginner* it is best not to use a CAD program (see Fig. D.1 in Appendix D) until step 6. Use the straight-line approximations on the Bode diagram for the log magnitude at the start of the design problem.
2. On the graph paper for the Bode diagram plot the points representing *Lm* $\boldsymbol{B}_o(j\omega_i)$ and the angles corresponding to the right side of the *U*-contour (the desired phase margin angle γ) for the frequency range of $\omega_x \le \omega \le \omega_y$. For the example illustrated in Fig. 3.19 the phase margin angle $\gamma = -45^\circ$ must be maintained for the frequency range $\omega_x = 5 \le \omega \le \omega_y \approx 30$.
3. Do the shaping of *Lm* $\boldsymbol{L}_o(j\omega)$ on the Bode plot using straight-line approximations for *Lm* $\boldsymbol{L}_o(j\omega)$, with the plotted information of step 2, and employ the shaping discussion that follows Eq. (3.48) for the frequency range $\omega < \omega_x$ as guidelines in achieving Eq. (3.48).
4. Use frequencies an octave above and below and a decade above and below a corner frequency for both first- and second-order terms,[13,19] while maintaining the phase-margin angle corresponding to the right side of the *U*-contour in shaping $\boldsymbol{L}_o(j\omega)$.
5. The last two poles that are added to $\boldsymbol{L}_o(j\omega)$ are generally a complex pair (the nominal range is $0.5 < \zeta < 0.7$) which tends to minimize the *BW*.
6. Once $\boldsymbol{L}_o(j\omega)$ has been shaped, determine $F(s)$ (see Sec. 3-15) and then verify that $\boldsymbol{L}_o(j\omega)$ does meet the design objectives by use of a CAD program. If the synthesized $\boldsymbol{L}_o(j\omega)$ yields the desired performance, then the required compensator is given by

$$G(s) = \frac{L_o(s)}{P_o(s)}$$

Specific guidelines for shaping $\boldsymbol{L}_o(j\omega)$ are:

1. An optimum design of $\boldsymbol{L}_o(j\omega)$ requires that $L_o(j\omega_i)$ be on the corresponding bound. In practice, place $\boldsymbol{L}_o(j\omega_i)$ as close as possible to the bound $\boldsymbol{B}_o(j\omega_i)$, but above it, in order to keep the BW of $\boldsymbol{L}_o/[1 + \boldsymbol{L}_o]$ to a minimum.[5]
2. Since exact cancellation of a pole by a zero is rarely possible, any *RHP* poles and/or zeros of $P_o(s)$ should be included in $L_o(s)$. A good starting $L_o(s)$ is $L_{o0}(s) = K_0P_o(s)$. If it is desired that $y(\infty) = 0$ for $d(t) = u_{-1}(t)$, it is necessary to insure that $T_D(s)$ has a zero at the origin. For this situation a possible starting point is $L_{o0}(s) = K_0P_o(s)/s$.
3. If $P(s)$ has an excess of poles over zeros, $\#p's - \#z's = \lambda$, which is denoted *by* e^{λ}, then, in general, the *final form* of $L_o(s)$ must have an excess of poles over zeros of at least $e^{\lambda+\mu}$ where $\mu \geq 1$. If the BW is too large, then increase the value of i. Experience shows that a value of $\lambda + i$ of 3 or more for $L(s)$ yields satisfactory results.[3,5]
4. Generally the BW of $L_o(s)/[1 + L_o(s)]$ is larger than required for an acceptable rise time t_R for the tracking of $r(t)$ by $y(t)$. An acceptable rise time can be achieved by the proper design of the pre-filter $F(s)$ (see Sec. 3-15).
5. Generally it is desirable to first find the bounds $\boldsymbol{B}_D(j\omega_i)$ and then the bounds $\boldsymbol{B}_R(j\omega_i)$. After finding the first $\boldsymbol{B}_R(j\omega_i)$, it may be evident that all or some of the $\boldsymbol{B}_D(j\omega_i)$ are completely dominant compared to $\boldsymbol{B}_R(j\omega_i)$. In that case the $\boldsymbol{B}_D(j\omega_i)$ boundaries are the optimal boundaries, i.e., $\boldsymbol{B}_D(j\omega_i) = \boldsymbol{B}_o(j\omega_i)$.
6. If $\delta_R(j\omega_i)$ is not continuously increasing as ω_i increases, then it will be necessary to utilize complex poles and/or zeros in $G(s)$ in order to achieve an optimal $L_o(s)$. This assumes that the tracking model is not designed to yield this increasing characteristic for $\delta_R(j\omega_i)$.
7. The ability to shape the nominal loop transmission $L_o(s)$ is an art developed by the designer only after much practice and patience.

The success of the compensator design strongly depends on the experience of the designer. The above rules and general guidelines for shaping give a good way to obtain the compensator from the engineering point of view. On the other hand, in the last few years some automatic loop-shaping procedures to help in the QFT compensator design have been developed.[122, 123] Although very often they do not reach the optimum solution and can lose the engineering point of view, they could contribute to show some tracks that help in the compensator design.

3-15 DESIGN OF THE PRE-FILTER $F(s)$[18,55]

Design of a proper $L_o(s)$ guarantees only that the variation in $|\mathbf{T}_R(j\omega)|$, i.e., $\Delta \mathbf{T}_R$, is less than or equal to that allowed. The purpose of the pre-filter is to position *Lm* $\mathbf{T}(j\omega)$ within the frequency domain specifications. For the example of this chapter the magnitude of the frequency response must lie within the bounds B_U and B_L shown in Fig. 3.4b, which are redrawn in Fig. 3.21. A method for determining the bounds on $F(s)$ is as follows: Place the nominal point A of the ω_i plant template on the $\mathbf{L}_o(j\omega_i)$ point of the $\mathbf{L}_o(j\omega)$ curve on the NC (see Fig. 3.22). Traversing the template, determine the maximum $Lm\ \mathbf{T}_{max}$ and minimum $Lm\ \mathbf{T}_{min}$ values of

$$Lm\ \mathbf{T}(j\omega_i) = \frac{\mathbf{L}(j\omega_i)}{1+\mathbf{L}(j\omega_i)} \tag{3.49}$$

obtained from the M-contours. These values are plotted as shown in Fig. 3.21. The tracking control ratio is $T_R = FL/[1 + L]$ and

$$Lm\mathbf{T}_R(j\omega_i) = Lm\mathbf{F}(j\omega_i) + Lm\mathbf{T}(j\omega_i) \tag{3.50}$$

The variations in Eqs. (3.49) and (3.50) are both due to the variation in P; thus

$$\delta_L(j\omega_i) = Lm\mathbf{T}_{max} - Lm\mathbf{T}_{min} \le \delta_R = B_U - B_L \tag{3.51}$$

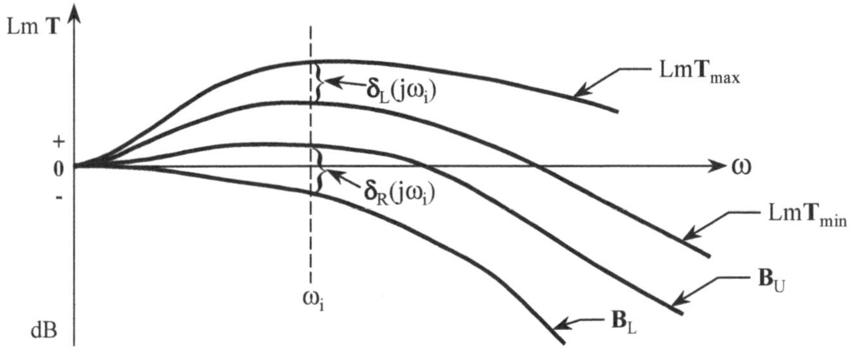

Fig. 3.21 Requirements on $F(s)$.

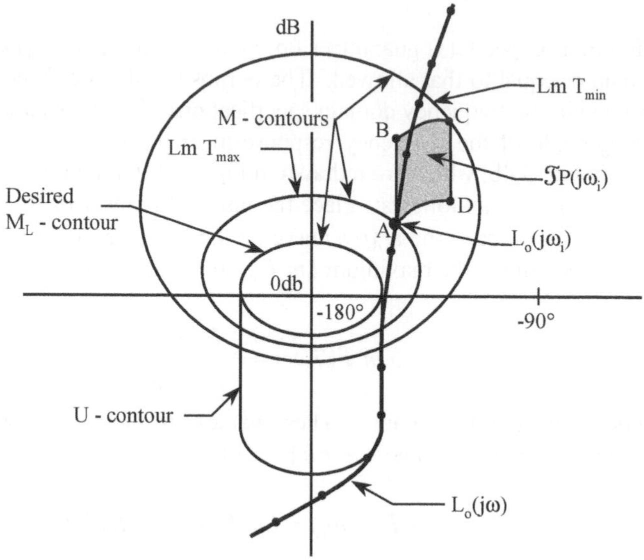

Fig. 3.22 Pre-filter determination.

If values of $\boldsymbol{L}_o(j\omega_i)$, for each value of ω_i, lie exactly on the tracking bounds $\boldsymbol{B}_R(j\omega_i)$, then $\delta_L = \delta_R$. Therefore, based upon Eq. (3.50), it is necessary to determine the range in dB by which $Lm\ \boldsymbol{T}(j\omega_i)$ must be raised or lowered to fit within the bounds of the specifications by use of the pre-filter $\boldsymbol{F}(j\omega_i)$. The process is repeated for each frequency corresponding to the templates used in the design of $\boldsymbol{L}_o(j\omega)$. Therefore, in Fig. 3.23 the difference between the $Lm\ \boldsymbol{T}_{R_U} - Lm\ \boldsymbol{T}_{max}$ and the $Lm\ \boldsymbol{T}_{R_L} - Lm\ \boldsymbol{T}_{min}$ curves yields the requirement for $Lm\ \boldsymbol{F}(j\omega)$, i.e., from Eq. (3.50)

The procedure for designing $F(s)$ is summarized as follows:

1. Use templates in conjunction with the $\boldsymbol{L}_o(j\omega)$ plot on the NC to determine T_{max} and T_{min} for each ω_i. This is done by placing $\Im\boldsymbol{P}(j\omega_i)$ with its nominal point on the point $Lm\ \boldsymbol{L}_o(j\omega_i)$. Then use the M-contours to determine $\boldsymbol{T}_{max}(j\omega_i)$ and $\boldsymbol{T}_{min}(j\omega_i)$ (see Fig. 3.22).
2. Obtain the values of $Lm\ \boldsymbol{T}_{R_U}$ and $Lm\ \boldsymbol{T}_{R_L}$ for various values of ω_i from Fig. 3.4*b*.
3. From the values obtained in steps 1 and 2, plot

$$[Lm\,\boldsymbol{T}_{R_U} - Lm\,\boldsymbol{T}_{\max}] \quad and \quad [Lm\,\boldsymbol{T}_{R_L} - Lm\,\boldsymbol{T}_{\min}] \quad vs \quad \omega \qquad \textbf{(3.52)}$$

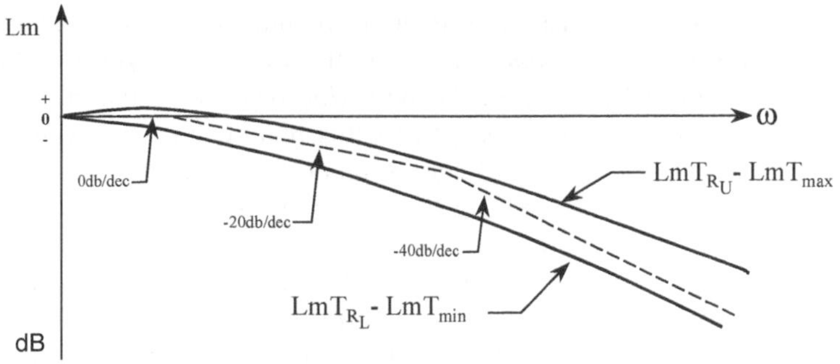

Fig. 3.23 Frequency bounds on the pre-filter $F(s)$.

as shown in Fig. 3.23.

4. Use straight-line approximations to synthesize an $F(s)$ so that $Lm\ \mathbf{F}(j\omega)$ lies within the plots of step 3. For step forcing functions the resulting $F(s)$ must satisfy

$$\lim_{s\to 0}[F(s)] = 1 \qquad \textbf{(3.53)}$$

3-16 BASIC DESIGN PROCEDURE FOR A MISO SYSTEM

The basic concepts of the QFT technique are explained by means of a design example. The system configuration shown in Fig. 3.1 contains four inputs. Parameter uncertainty for the plant of Eq. (2.7) is shown in Fig. 3.6. The first objectives are to track a step input $r(t) = u_{-1}(t)$ with no steady-state error and to satisfy the performance specifications of Fig. 3.3. An additional objective is to attenuate the system response caused by external step disturbance inputs $d_1(t)$ and $d_2(t)$, as described in Sec. 3-4.2. Minimizing the effect of noise is discussed in Secs. 11-3.3 and 11-6. An outline of the basic design procedure for the QFT technique, as applied to a m.p. plant, is as follows:

1. Synthesize the tracking model control ratio $T_R(s)$ in the manner described in Sec. 3-4, based upon the desired tracking specifications (see Figs. 3.3 and 3.4b).
2. Synthesize the disturbance-rejection model control ratios $T_D(s)$ in the manner described in Sec. 3-10, based upon the disturbance-rejection specifications.
3. Obtain templates of $\mathbf{P}(j\omega_i)$ that pictorially describe the plant uncertainty on the Nichols chart for the desired pass-band frequency range.

4. Select a nominal plant from the set of Eq. (3.1) and denote it as $P_o(s)$.
5. Determine the U-contour based upon the specified values of $\delta_R(j\omega_i)$ for tracking, M_L for disturbance rejection, and V for the universal high-frequency boundary (UHFB) B_h in conjunction with steps 6 through 8.
6. Use the data of steps 2 and 3 and the values of $\delta_D(j\omega_i)$ (see Fig. 3.5) to determine the disturbance bound $\boldsymbol{B}_D(j\omega_i)$ on the loop transmission $\boldsymbol{L}_D(j\omega_i) = \boldsymbol{G}(j\omega_i)\boldsymbol{P}(j\omega_i)$. For m.p. systems this requires that the synthesized loop transmission $Lm\ \boldsymbol{L}_D(j\omega_i)$ must be on or above the curve for $Lm\ \boldsymbol{B}_D(j\omega_i)$ on the Nichols Chart (see Fig. 3.19 assuming $\boldsymbol{B}_D = \boldsymbol{B}_o$).
7. Determine the tracking bound $\boldsymbol{B}_R(j\omega_i)$ on the nominal transmission $\boldsymbol{L}_o(j\omega_i) = \boldsymbol{G}(j\omega_i)\boldsymbol{P}_o(j\omega_i)$, using the tracking model (step 1), the templates $\boldsymbol{P}(j\omega_i)$ (step 3), the values of $\delta_R(j\omega_i)$ (see Fig. 3.4*b*), and M_L [see Eq. (3.19)]. For m.p. systems this requires that the synthesized loop transmission satisfy the requirement that $Lm\ \boldsymbol{L}_o(j\omega_i)$ is on or above the curve for $Lm\ \boldsymbol{B}_R(j\omega_i)$ on the Nichols Chart.
8. Plot curves of $Lm\ \boldsymbol{B}_R(j\omega_i)$ versus $\phi_R = \angle \boldsymbol{B}_R(j\omega_i)$ and $Lm\ \boldsymbol{B}_D(j\omega_i)$ versus $\phi_D = \angle \boldsymbol{B}_D(j\omega_i)$ on the same NC. For a given value of ω_i at various values of the angle ϕ, select the value of $Lm\ \boldsymbol{B}_D(j\omega_i)$ or $Lm\ \boldsymbol{B}_R(j\omega_i)$, whichever is the largest value (termed the "worst" or "most severe" boundary). Draw a curve through these points. The resulting plot defines the overall boundary $Lm\ \boldsymbol{B}_o(j\omega_i)$ vs. ϕ. Repeat this procedure for sufficient values of ω_i.
9. Design $\boldsymbol{L}_o(j\omega_i)$ to be as close as possible to the boundary value $\boldsymbol{B}_o(j\omega_i)$ by selecting an appropriate compensator transfer function $\boldsymbol{G}(j\omega)$ (see Appendix I). Synthesize an $\boldsymbol{L}_o(j\omega) = \boldsymbol{G}(j\omega)\boldsymbol{P}_o(j\omega)$ using the $Lm\ \boldsymbol{B}_o(j\omega_i)$ boundaries and U-contour so that $Lm\ \boldsymbol{L}_o(j\omega_i)$ is on or above the curve for $Lm\ \boldsymbol{B}_o(j\omega_i)$ on the Nichols Chart. This procedure achieves the lowest possible value of the loop transmission frequency (phase margin frequency ω_ϕ). Note that $|\boldsymbol{L}_o(j\omega_i)| \geq |\boldsymbol{B}_o(j\omega_i)|$ represents the loop transfer function that satisfies the most severe boundary $\boldsymbol{B}_R$ and $\boldsymbol{B}_D$.
10. Based upon the information available from steps 1 and 9, synthesize an $F(s)$ that results in a $Lm\ \boldsymbol{T}_R$ [Eq. (3.5)] *vs.* ω that lies between $\boldsymbol{B}_U$ and $\boldsymbol{B}_L$ of Fig. 3.4b.
11. Obtain the time-response data for $y(t)$: (*a*) with $d(t) = u_{-1}(t)$ and $r(t) = 0$ and (*b*) with $r(t) = u_{-1}(t)$ and $d(t) = 0$ for sufficient points around the parameter space describing the plant uncertainty (see Fig. 3.6).

Simulation in time domain is always recommended to check the final fulfillment of the initial control specifications. Moreover, if the original specifications are given in time domain, although with the designed compensator the corresponding frequency requirements are accomplished, the final time responses could show some small differences from the initial specifications. That is, because the translation into the frequency domain is good and usually enough of an approximation, but it as an approximation. In fact, it does not exist as a formal and exact translation between these two domains. Thus, if the designer finds such small differences, a second iteration could be useful to improve the controller design.

For the $\boldsymbol{L}_o(j\omega)$ obtained, the plot of $Lm\ \boldsymbol{L}(j\omega)/[1 + \boldsymbol{L}(j\omega)]$ may be larger or smaller than $Lm\ \boldsymbol{T}_{R_U}$ or $Lm\ \boldsymbol{T}_{R_L}$ of Fig. 3.4b, but $\delta_R(j\omega_i)$ is satisfied. By the proper design of the input filter $F(s)$, $Lm\ \boldsymbol{T}_R = Lm\ \boldsymbol{FL}/[1 + \boldsymbol{L}]$ will lie within the bounds of $Lm\ \boldsymbol{T}_{R_U}$ and $Lm\ \boldsymbol{T}_{R_L}$.

In problems with very large uncertainty and in disturbance rejection requiring a very large $|\boldsymbol{L}(j\omega)|$ over a "large" BW, then

$$\left|\frac{\boldsymbol{L}}{1+\boldsymbol{L}}\right| \approx 1 \tag{3.54}$$

and $T \approx F$. For these situations design $F(s)$ in the same manner as for the tracking models of Sec. 3-4 and Ref. 1.

A situation may occur in which it may be impossible to satisfy all the desired performance specifications and a design trade-off decision needs to be made (see Appendix C). For example, the gain that is required to satisfy the dominating $\boldsymbol{B}_D(j\omega_i)$ bounds, i.e., $\boldsymbol{B}_o = \boldsymbol{B}_D$, may be too high resulting in saturation and/or sensor noise effects. The analysis of the equation

$$\boldsymbol{Y}(j\omega) = \left[\frac{\boldsymbol{P}(j\omega)}{1 + \boldsymbol{G}(j\omega)\boldsymbol{P}(j\omega)}\right]\boldsymbol{D}(j\omega)$$

for the condition $|\boldsymbol{G}(j\omega)\boldsymbol{P}(j\omega)| >> 1$, due to the gain in $\boldsymbol{G}(j\omega)$, over the desired BW results in

$$\boldsymbol{Y}(j\omega) \approx \frac{\boldsymbol{D}(j\omega)}{\boldsymbol{G}(j\omega)}$$

Thus, by ignoring the disturbance rejection specification, synthesize the loop transmission function $L_o(s)$ by satisfying the bounds $\boldsymbol{B}_o(j\omega_i) = \boldsymbol{B}_R(j\omega_i)$. The possibility exists that the gain required in $G(s)$ in order to satisfy the tracking performance specifications and the overdesign characteristic of the QFT technique may result in satisfying the performance requirement

$$|\boldsymbol{Y}(j\omega)| = \left|\frac{\boldsymbol{D}(j\omega)}{\boldsymbol{G}(j\omega)}\right| \le \alpha_p$$

The simple plant of Eq. (2.7) is used in the following sections to illustrate the details in applying this QFT design procedure. The CAD package TOTAL-PC[1] (see Appendix D) is used to obtain the data and to execute the design procedures.

3-17 DESIGN EXAMPLE 1

This design example is for the control system of Fig. 3.1 with $r(t) = d_2(t) = u_{-1}(t)$ and $d_1(t) = 0$. The plant transfer function

$$P(s) = \frac{Ka}{s(s+a)} \qquad 1 \le K \le 10 \qquad 1 \le a \le 10 \tag{3.55}$$

as the nominal values: $a = K = 1$.

Step 1. Modeling the tracking control ratio $T_R(s) = Y(s)/R(s)$:

$$T_{R_U} : M_p = 1.2, t_S = 2s$$

$$T_{R_L} : Overdamped\ , t_s = 2s$$

(a) A tracking model for the upper bound, based upon the given desired performance specifications, is tentatively identified by

$$T_{R_U}(s) = \frac{19.753}{s + 2 \pm j3.969} \tag{3.56}$$

A zero is inserted in Eq. (3.56) which does not affect the desired performance specifications but widens δ_{hf} between T_{R_U} and T_{R_L} in the high-frequency range. Thus

$$T_{R_U}(s) = \frac{0.6584(s+30)}{s + 2 \pm j3.969} \tag{3.57}$$

The figures of merit for this transfer function with a step input are $M_p = 1.2078$, $t_R = 0.342$ s, $t_p = 0.766$ s, $t_s = 1.84$ s, $Lm\ M_m = 1.95$ dB, and $\omega_m = 3.4$ rad/sec.

(b) A tracking model (see Sec. 3-4) for the lower bound, based upon the desired performance specifications, is tentatively identified by

$$T_{R_L}(s) = \frac{120}{(s+3)(s+4)(s+10)} \tag{3.58}$$

and is modified with the addition of a pole to yield

$$T_{R_L}(s) = \frac{3520}{(s+4)(s+4)(s+4.4)(s+50)} \tag{3.59}$$

The pole at $s = -50$ is inserted in Eq. (3.59) to widen δ_{hf} further between T_{R_U} and T_{R_L} at high frequencies. The FOM for Eq. (3.59) are: $M_p = 1$, $t_R = 1.02$ s, and $t_s = 1.844$ s. In practice, these synthesized models should yield FOM within *1* percent of the specified values.

(c) Determination of $\delta_R(j\omega_i)$. From the data for the log magnitude plots of Eqs. (3.57) and (3.59), the following values are obtained for $\delta_R(j\omega_i)$:

ω	$\delta_R(j\omega_i)$, dB
0.5	0.851
1	1.01
2	3.749
5	11.538
7	11.735
9	12.194
10	12.594
15	15.224
100	46.81

As seen from the above table, $\delta_R(j\omega)$ has the desired characteristic of an increasing value with increased frequency. Sufficient values of ω up to ω_h should be obtained in order to ensure that this desired characteristic has been achieved.

Step 2. Modeling the disturbance control ratio $T_{D_U}(s) = Y(s)/D_2(s)$.

Since $y(0) = 1$, it is desired that $y(t)$ decay "as fast as possible," i.e., $|y(t)| \leq 0.01$ for $t \geq t_x = 60\ ms$. Let the model disturbance control ratio be of the form

$$T_{D_U} = \frac{Y(s)}{D_2(s)} = \frac{s(s+g)}{(s+g)^2 + h^2} \tag{3.60}$$

For $d_2(t) = u_{-1}(t)$ the output response has the form $y(t) = e^{-gt}\cos ht$. The desired specifications (see Fig. 3.5) are satisfied by choosing

$$T_{D_U}(s) = \frac{s(s+70)}{s + 70 \pm j18} = \frac{1}{1 + L_D} \tag{3.61}$$

For Eq. (3.61) the output decays rapidly so that $|y(t)| \le 0.01$ at $t_x \approx 0.0565$ s.

Steps 3 and 4. Forming the templates of $\boldsymbol{P}(j\omega_i)$.

Analysis of $P(s)$ is based upon the nominal values $K = a = 1$ (point A of Fig. 3.24). The template for each frequency is based on the data listed in Table 3.4 which is computed using Eq. (3.55). The values in Table 3.4 are used to draw the templates $\Im\boldsymbol{P}(j\omega_i)$ shown in Fig. 3.25.

Step 5. Determination of the U-contour.

(a) Determination of V [see Eq. (3.21)] for the plant of Eq. (3.55) yields

$$\lim_{s\to\infty} [\boldsymbol{P}(s)] = \left[\frac{Ka}{s^2}\right] \qquad and$$

$$\Delta Lm[\boldsymbol{P}]_{\max} = Lm[Ka]_{\max} = 40 \text{ dB} \tag{3.62}$$

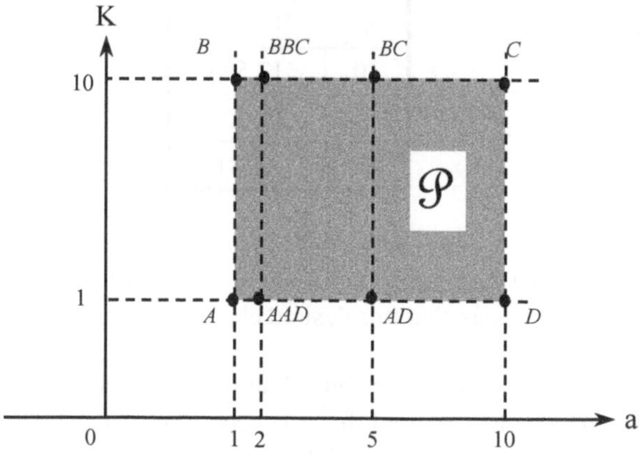

Fig. 3.24 Template points.

TABLE 3.4 Data points for $\Im P(j\omega_i)$

Point	ω_i	**0.5**	**1**	**2**	**5**	**10**	**15**
A	*Lm* **P**	5	-3	-13	-28.1	-40	-47
	$\angle$**P**	-116.6°	-135°	-153.5°	-168.7°	-174.3°	-176.2°
B	*Lm* **P**	25	17	7	-8.1	-20	-27
	$\angle$**P**	-116.6°	-135°	-153.5°	-168.7°	-174.3°	-176.2°
BBC	*Lm* **P**	25.8	19	11	-2.6	-14.2	-21.1
	$\angle$**P**	-104.1°	-116.6°	-135°	-158.2°	-168.7°	-171.4°
BC	*Lm* **P**	26	19.8	13.3	3	-7	-13.5
	$\angle$**P**	-95.7°	-101.3°	-111.8°	-135°	-153.5°	-161.6°
C	*Lm* **P**	26	20	13.8	5	-3	-8.64
	$\angle$**P**	-93°	-95.7°	-101.3°	-116.6°	-13.5°	-146.4°
D	*Lm* **P**	6	-0.04	-6.2	-14.95	-23	-28.6
	$\angle$**P**	-93°	-95.7°	-101.3°	-116.6°	-135°	-146.4°
AD	*Lm* **P**	6	-0.17	6.67	-17	-27	-33.5
	$\angle$**P**	-95.7°	-101.3°	-111.8°	-135°	-153.5°	-116.6°
ADD	*Lm* **P**	5.65	0.97	-9	-22.6	-34.2	-41.1
	$\angle$**P**	-104.1°	-116.6°	-135°	-158.2°	-168.7°	-172.4°

Note: Values of *Lm* are in dB

To determine the value of ω_h, where $\Delta Lm\ \mathbf{P} \approx V = 40$ dB, select various values of ω_i as shown below. For $\omega \approx 100$:

Point	$Lm\ \mathbf{P}(j\omega_i)$
A	−80
B	−60
C	−40
D	−60

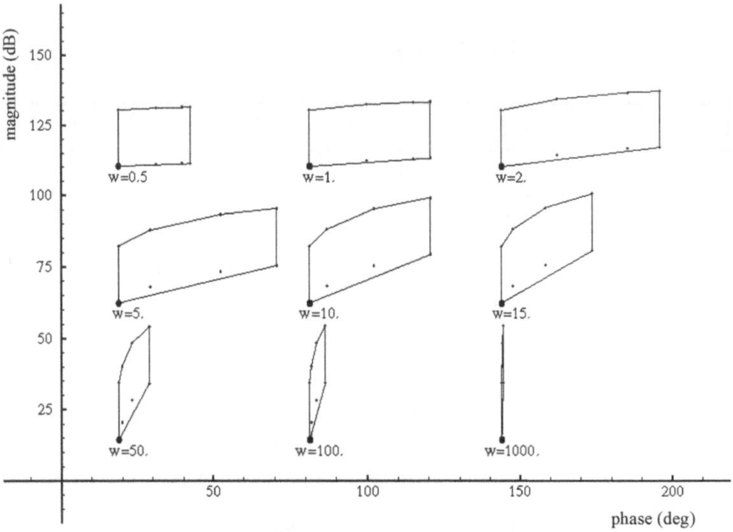

Fig. 3.25 Construction of the plant templates.

From this data it is seen that the maximum value of $\Delta Lm\ \mathbf{P}(j\omega_i)$ occurs between points A and C. For other values of ω_i the maximum values of $\Delta Lm\ \mathbf{P}(j\omega_i)$ are:

Frequency ω_i	*max* $\Delta Lm\ \mathbf{P}(j\omega_i)$ dB
15	38.4
16	38.6
20	39.0
40	39.65

Therefore it can be seen, for this example, that V is achieved essentially at $\omega_h \approx 40$.

(b) Determination of the B_h boundary. Select the M-contour that represents the desired value of M_L for T (see Sec. 3-8). At various points around the lower half of this contour measure down *40* dB and draw the B_h boundary in the manner shown in Fig. 3.8.

Step 6. Determination of bounds $\mathbf{B}_D(j\omega_i)$ for

$$T_{D_U} = \frac{Y}{D} = \frac{1}{1+L} = \frac{\ell}{1+\ell} \tag{3.63}$$

where $\ell = 1/L = 1/PG$ and $Y = DT_{D_U}$. For the disturbance-rejection case:

$$\Delta\, Lm\, \boldsymbol{Y} = \Delta\, Lm\left[\frac{\ell}{1+\ell}\right] = \Delta\, Lm\, \boldsymbol{T}_{D_U} \tag{3.64}$$

Note that there are no lower bounds on the response $y(t)$. Thus, choose $\boldsymbol{L} \geq \boldsymbol{B}_D(j\omega_i)$ to satisfy

$$Lm\left[\frac{\boldsymbol{Y}}{\boldsymbol{D}}\right] = Lm\left[\frac{\ell}{1+\ell}\right] \leq Lm\,[\boldsymbol{T}_{D_U}] \tag{3.65}$$

for each frequency ω_i of $\boldsymbol{T}_D(j\omega_i)$ over the frequency range $0 \leq \omega \leq 15$ rad/sec. The data of Table 3.5 are obtained from Eq. (3.61). Point A of the templates of step 3 yields the maximum value for $\ell = 1/\boldsymbol{L}$. For $0 \leq \omega \leq 15$ the contours of $Lm\, \boldsymbol{T}_D(j\omega_i)$ (see Fig. 3.26) become the contours for $\boldsymbol{B}_D$, i.e., $Lm\, \boldsymbol{T}_D = Lm\, \boldsymbol{B}_D$. For $\omega \geq 20$, the values of $Lm\, \boldsymbol{L}_D$ in Table 3.5, obtained by use of Eq. (3.61), are points of $\boldsymbol{B}_D(j\omega_i)$ on and in the vicinity of the U-contour. For this example, as shown in the next step, the $\boldsymbol{B}_D$ bounds completely dominate over the $\boldsymbol{B}_R$ bounds. Thus, $\boldsymbol{B}_D = \boldsymbol{B}_R$ as shown in Fig. 3.26.

TABLE 3.5 Frequency data for Eq. (3.61)

ω	$Lm\, \boldsymbol{T}_D$, dB	$\angle \boldsymbol{T}_D$,degrees	$Lm\, \boldsymbol{L}_D$
0.5	−43.5	89.65	43.5
1	−37.5	89.3	37.5
2	−31.44	88.6	31.4
5	−23.5	86.4	23.5
10	−17.52	82.85	17.5
15	−14.1	79.31	14.1
20	−11.67	75.81	11.4
50	−4.84	56.8	4.46
100	−1.67	36.17	−2.91
200	−0.453	19.55	−9.05
500	−0.0737	7.99	−17.1
1000	−0.01845	4.007	−21.6
2000	−0.0029954	2.005	−31.0
4000	−0.001154	1.003	−35.2

Step 7. Determination of the bounds $\boldsymbol{B}_R(j\omega_i)$.

By use of the templates, the values of $\delta_R(j\omega_i)$ given in step 2, the M_L contour, and the B_h boundary, the bounds $\boldsymbol{B}_R(j\omega_i)$ and the *U*-contour are determined and are drawn in Fig. 3.27. From this figure it is seen that $\gamma \approx 58^{\circ}$.

Step 8. Determination of the composite bounds $\boldsymbol{B}_o(j\omega_i)$.

An analysis of Figs. 3.26 and 3.27 reveals that the bounds $\boldsymbol{B}_D(j\omega_i)$ for this example all lie above the tracking contours $\boldsymbol{B}_R(j\omega_i)$. Thus, the $\boldsymbol{B}_D$ contours of Fig. 3.26 are more severe and become the $\boldsymbol{B}_o(j\omega_i)$ contours for the overall system of Fig. 3.28.

Step 9. Synthesizing or shaping of $L_o(s)$.

(a) Minimum structure. Since $P(s)$ is Type 1, then $L_0(s)$ must be at least a Type *1* in order to maintain the Type *1* tracking characteristic for $L/(1 + L)$. Thus, the initial $L_o(s)$ has the form

$$L_{o0}(s) = \frac{K_{o1}}{s^m} \tag{3.66}$$

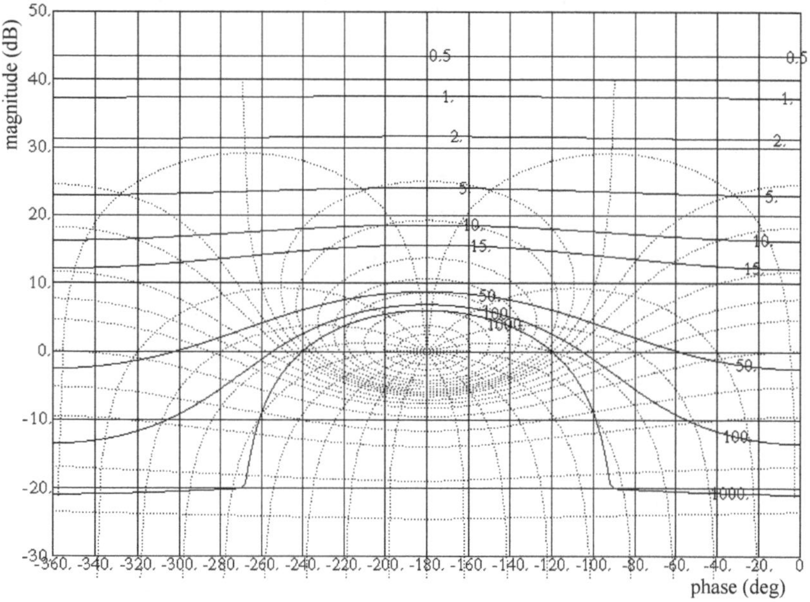

Fig. 3.26 Construction of the $\boldsymbol{B}_D(j\omega)$ boundaries.

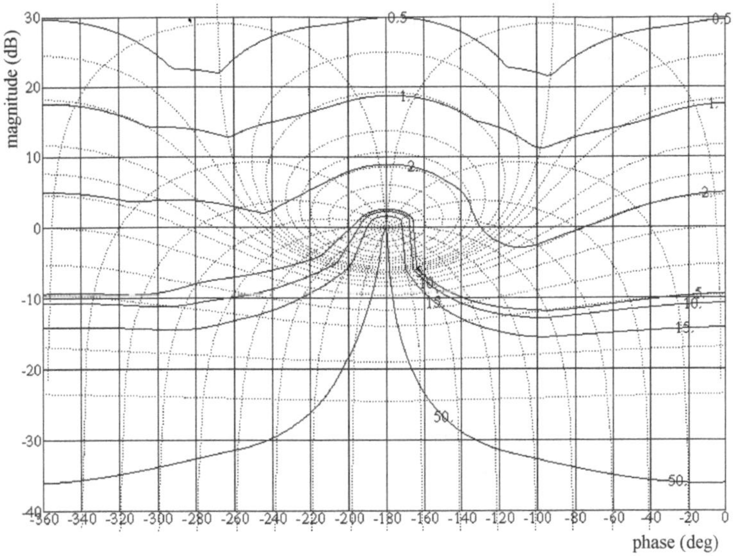

Fig. 3.27 Construction of the $\boldsymbol{B}_R(j\omega_i)$ contours.

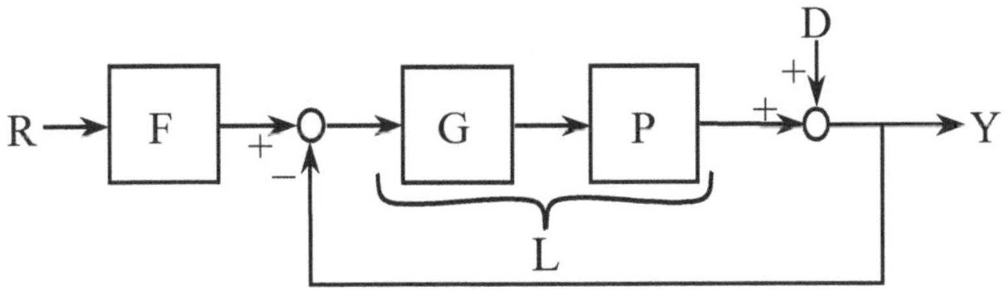

Fig. 3.28 MISO system.

where $m = 1$ for this example. (Note: In general, select $L_{o0}(s) = P(s)$. In addition, that the desired system type is achieved.) For the case where $T_D = P/(1 + L)$, it is desired that T_D have a zero at the origin so that $y(\infty) = 0$ for $d_1(t) = u_{-1}(t)$. Therefore $G(s)$ must have a pole at the origin. This requirement may place a severe restriction on trying to synthesize an L_o.

(b) Synthesis of $L_o(s)$. Using four cycle semi-log graph paper and the data in Table 3.6, construct or shape $L_o(s)$ so that it comes as close as possible to the *U*-contour. $\boldsymbol{L}_o(j\omega_i)$ must be as close as possible to $\boldsymbol{B}_o(j\omega_i)$ but never below it. For the approximate range $\omega_x \approx 15 \le \omega_i \le \omega_y$, the angle $\angle \boldsymbol{L}_o(j\omega_i) \ge -123^\circ$ *must* be satisfied. For this example the frequency ω_y is the value of $\boldsymbol{L}_o(j\omega_y)$ which results in $Lm\ \boldsymbol{L}_o(j\omega_i) \approx -36$ dB. These restrictions can be satisfied by assuming the format of the optimal transfer function as

$$\boldsymbol{L}_o(j\omega) = \boldsymbol{L}_{o0}(j\omega)\left[\frac{(j\omega - z_1)(j\omega - z_2)\cdots}{(j\omega - p_3)(j\omega - p_4)\cdots\left[\left(1 - \frac{\omega^2}{\omega_n^2}\right) + j2\varsigma\frac{\omega}{\omega_n}\right]}\right] \tag{3.67}$$

which has w real zeros, $n = w + \lambda - 2$ real poles, and a pair of complex-conjugate poles. For this example a table having the format shown in Table 3.6 is used to assist in obtaining the desired $L_o(s)$. The resulting loop transfer function is

$$L_o(s) = \frac{4.11\times10^{10}(s + 0.6)^2(s + 176)}{s^2(s + 1)(s + 200)(s + 14700 \pm j15000)} \tag{3.68}$$

where $\zeta = 0.7$, $\omega_n = 1000$ rad/sec, and $\omega_\phi = 200$ rad/sec. $\boldsymbol{L}_o(j\omega)$ is drawn on the Bode plot of Fig. 3.29. It is also plotted on the NC of Fig. 3.33. Since L_o is Type 2, then $T_D = 1/(1 + L)$ has two zeros at the origin.

TABLE 3.6 Data for loop shaping

ω	$\angle \boldsymbol{L}_{o1}$	**Angle contribution of poles (p) and zero(s) (z) in degrees**				$\angle \boldsymbol{L}_o$
		z_1	p_2	z_2	$p_{3,4}$	
1	−180°	32.0	−2.20	1.27	0	−149.0°
5	−180°	72.26	−10.89	6.34	−0.40	−112.7°
10	−180°	80.90	−21.04	12.53	−0.80	−108.4°
15	−180°	83.91	−29.98	18.43	−1.20	−108.8°
50	−180°	88.17	−62.53	48.01	−4.01	−110.4°
100	−180°	89.08	−75.43	65.77	−16.26	−108.6°
200	−180°	89.54	−82.59	77.32	−90.0	−111.99°
1000	−180°	89.90	−88.51	87.42		−181.2°
2000	−180°	89.95	−89.26	88.71	−136.98	227.6°

(c) Determination of $G(s)$. Use $L_o(s)$ and $P_o(s)$, where $P_o(s) = 1/s(s + 1)$ is the nominal plant, to obtain the transfer function $G(s) = L_o(s)/P_o(s)$.

Step 10. The input filter $F(s)$ is synthesized to yield the desired tracking of the input by the output $y(t)$ in the manner described in Sec. 3-15. The curves of $\boldsymbol{B}_U$, $\boldsymbol{B}_L$, $Lm\ T_{min}$, and $Lm\ T_{max}$ are plotted in Fig. 3.30. Since, for this example $|\boldsymbol{L}_o(j\omega)| >> 1$ over the desired BW, then $F(s) = T_R(s)$. Thus, the following $F(s)$ is synthesized to lie between $\boldsymbol{B}_U$ and $\boldsymbol{B}_L$ as shown in Fig. 3.30.

$$F(s) = \frac{18.85}{(s + 2.9)(s + 6.5)} \qquad \textbf{(3.69)}$$

The output responses for this example with a unit step input are all identical, as shown in Fig. 3.31 for the entire region of plant parameter uncertainty, where $M_p = 1.002$ and $t_s = 1.606\ s$.

Step 11. Time responses for a disturbance input: $d(t) = u_{-1}(t)$. For each point of Fig. 3.24, substitute the corresponding data into

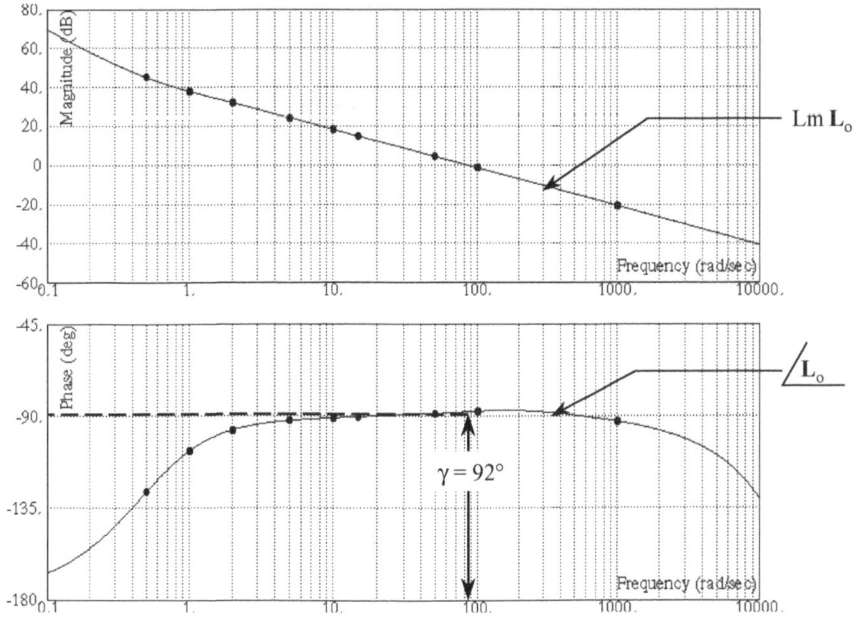

Fig. 3.29 Bode plot of $\boldsymbol{L}_o(j\omega)$ of Eq. (3.68).

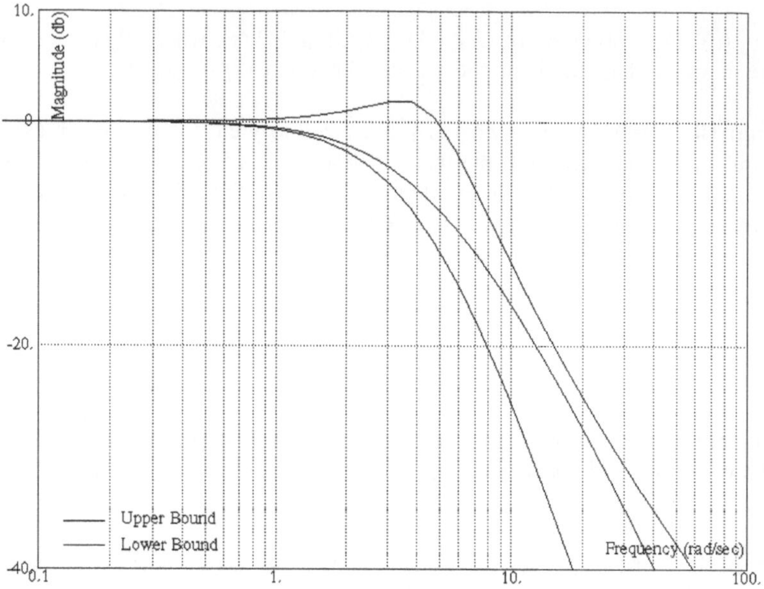

Fig. 3.30 Requirements on $F(s)$ for Design Example 1

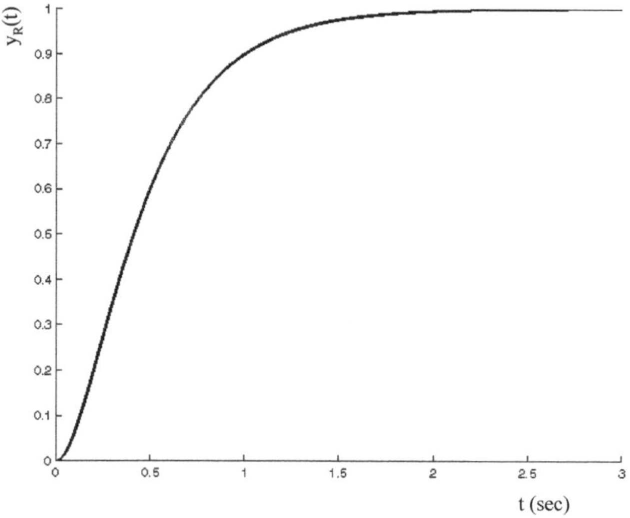

Fig. 3.31 Time responses to unit step input for all J plants.

$$Y_D(s) = \frac{1}{1 + P(s)G(s)} D(s) \qquad \textbf{(3.70)}$$

and determine $y_D(t)$ for each point. As indicated in Table 3.7, the disturbance time response characteristics for points A, B, C, and D satisfy the specification that $|y(t_x)| \leq 0.01$ for $t_x \geq 60\ ms$. The disturbance response plots, for the 8 cases, are shown in Fig. 3.32. As shown in the figure, although the initial peak overshoots exceed the magnitude of 0.01, the value of $|y_D(t)| \leq 0.01$ for $t_x \geq 60\ ms$ is achieved.

TABLE 3.7 Time response characteristics for Eq. (3.70)

Point	$y_D(t)$	Specified t_x	Actual t_x for $\lvert y_D(t)\rvert = 0.01$	$y(\infty)$
A	0.00998	0.06	0.058	0
B	0.0090	0.06	0.0074	0
C	0.00004	0.06	0.00226	0
D	0.000399	0.06	0.007	0

3-18 DESIGN EXAMPLE 2

The plant and tracking specifications of Design Example 1 are used for this example, with $r(t) = d_1(t) = u_{-1}(t)$, $d_2(t) = 0$, and $\alpha_p = 0.1$. Steps 1, 3-5, and 7 are the same as for example 1. For Step 2, $M_p \approx M_m = 0.1$; thus $Lm\ T_{D_1} = Lm\ M_{m_D} = -20$ dB. A QFT CAD (see Appendix C) is used to obtain Fig. 3.33 which shows the U-contour, $\boldsymbol{B}_o$ bounds, and $Lm\ \boldsymbol{L}_o(j\omega)$ given by Eq. (3.71). The synthesized $L_o(s)$ is

$$L_o(s) = \frac{9.93\times10^6 (s + 1.3)(s + 1.6)(s + 45)}{s^2 (s + 1)(s + 26)(s + 900 \pm j975)} \qquad \textbf{(3.71)}$$

Equations (3.57) and (3.59) and Fig. 3.33 are used to obtain the data for plotting $\boldsymbol{B}_U$, $\boldsymbol{B}_L$, Lm $\boldsymbol{T}_{min}$, and $Lm\ \boldsymbol{T}_{max}$ in Fig. 3.34. Since the tracking specifications and models are identical for both examples, the pre-filter for this example is the same as given by Eq. (3.69) and is plotted in Fig. 3.34. The simulation results are shown in Figs. 3.35 and 3.36. These figures reveal that all disturbance responses are below 0.03 and thus satisfy the specification $|y_D(t)_{max}| \leq 0.1$.

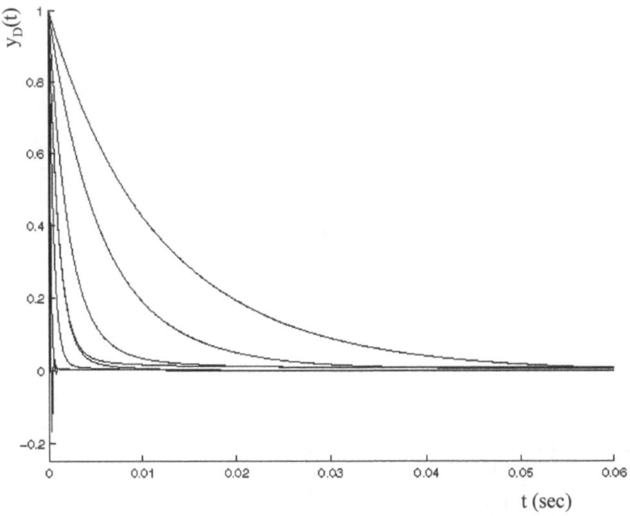

Fig. 3.32 Disturbance responses plots for all *J* plants.

Also, the tracking responses all lie in the range of $1 < M_p < 1.022$ and $t_s \leq 1.62\ s$. Thus, the desired performance specifications have all been met and a robust design has been achieved.

3-19 TEMPLATE GENERATION FOR UNSTABLE PLANTS

In the generation of templates for unstable plants, proper care must be exercised in analyzing the angular variation of $\boldsymbol{P}(j\omega)$ as the frequency is varied from zero to infinity. As an example, consider the plant

$$\boldsymbol{P}(j\omega) = \frac{K(j\omega - z_1)}{j\omega(j\omega - p_1)(j\omega - p_2)(j\omega - p_3)} \qquad \textbf{(3.72)}$$

where p_1 and p_2 are a complex-conjugate pair and

$$\angle \lim_{\omega \to \infty} \boldsymbol{P}(j\omega) = \angle K/(j\omega)^3 = -270^0 \qquad \textbf{(3.73)}$$

In Eq. (3.73) the convention used is that, in the limit as $\omega \to \infty$, each $j\omega$ term contributes an angle of $+90^o$. Two cases are analyzed for Eq. (3.72) as follows:

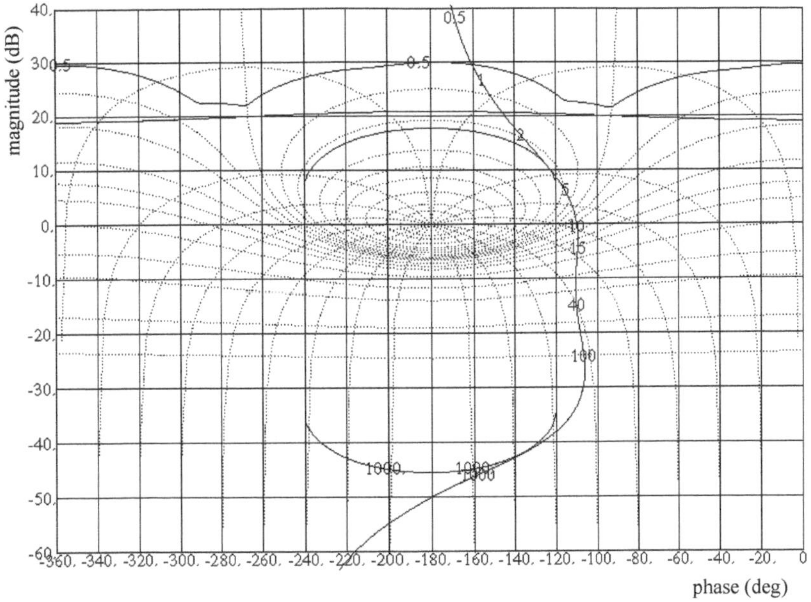

(a)

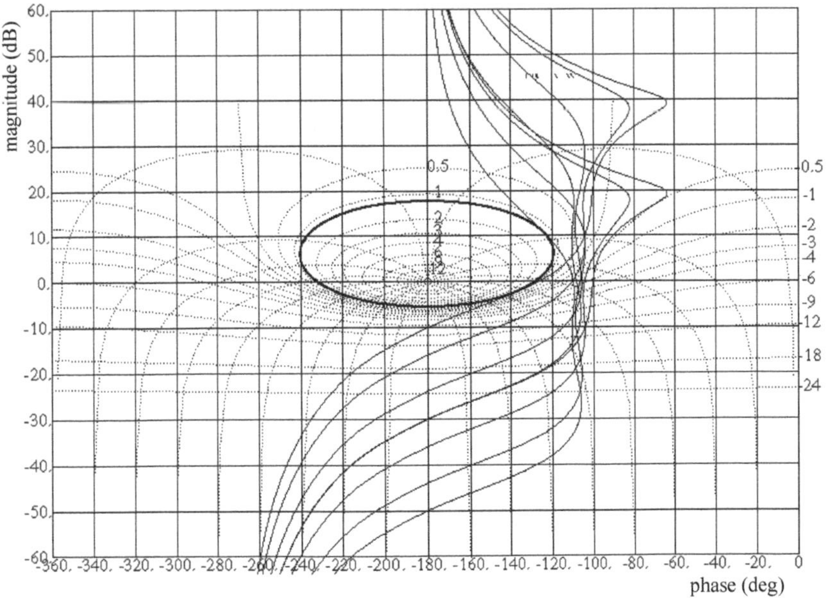

(b)

Fig. 3.33 (a) *U*-contour, $\boldsymbol{B}_o$ bounds, and $Lm\, \boldsymbol{L}_o(j\omega_i)$ for obtaining $\boldsymbol{L}_o(s)$; (b) All of the $L_i(j\omega)$ plotted on one Nichols plot along with the stability margin requirement.

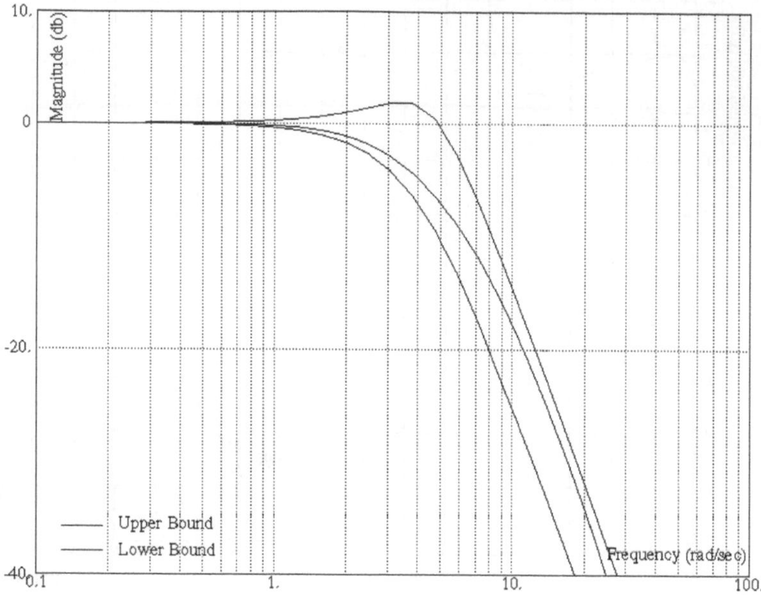

Fig. 3.34 Requirements and resulting pre-filter for design examples.

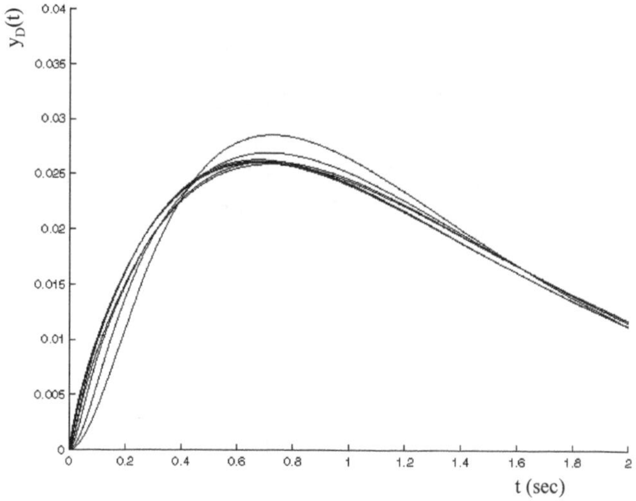

Fig. 3.35 Disturbance responses for Design Example 2.

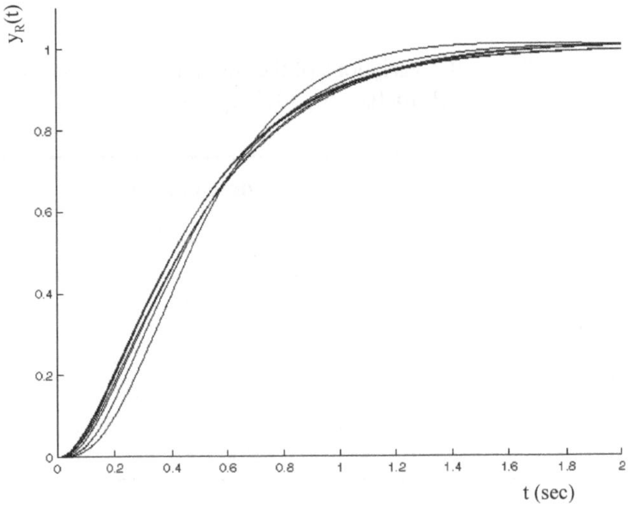

Fig. 3.36 Tracking responses for Design Example 2.

Case 1. The poles and zeros of Eq. (3.72) for a stable plant are plotted in Fig. 3.37*a*. In Eq. (3.72) the angle of $(j\omega_i - p_1)$ is negative clockwise (cw) for $\omega_i < \omega_{d_1}$ and is positive counterclockwise (ccw) for $\omega_i > \omega_{d_1}$. All other angles are positive (ccw) for all $0 \leq \omega \leq \infty$ as illustrated in Fig. 3.37a, and in Table 3.8.

Case 2. The poles and zeros of Eq. (3.72), for an unstable plant are plotted in Fig. 3.37b. For Eq. (3.72), all angles of first-order factors $(j\omega_i - p_i)$ are positive (ccw), as illustrated in Fig. 3.37b and in Table 3.8.

Note that the difference in the angle of $\boldsymbol{P}(j0)$ between the two cases is 360° and is necessary to account for the right-hand plane (RHP) poles. Thus, because all angles are measured ccw, the angle of $\boldsymbol{P}(j\omega)$ varies continuously in a given direction as the frequency is varied between zero and infinity. This feature is very important in obtaining a template where P contains both stable and unstable plants. The following guidelines should be used in the angular determination of $\angle \boldsymbol{P}(j\omega_i)$:

1. For stable plants, all angular directions are taken so that their values always lie within the range of -90° to $+90^\circ$.
2. For plants with RHP poles and zeros: (*a*) the angular directions as for the stable plant left-hand plane (LHP) poles and zeros of $P(s)$ are taken in the same manner; (*b*) the angular directions of all RHP poles and zeros are all taken ccw.

TABLE 3.8 Angular variations of the minimum-phase stable and unstable plants $P(j\omega)$ [see Eq. (3.70)] shown in Fig. 3.37

Case	ω_i	Angle in Degrees					
		$\angle p_0$	$\angle p_1$	$\angle p_2$	$\angle p_3$	$\angle z_1$	$\angle P(j\omega_i)$
1	0	+90	−45	+45	0	0	−90
	1	+90	0	+63.5	+63.5	+45	−172
	∞	+90	+90	+90	+90	+90	−270
2	0	+90	+225	+135	0	0	−450
	1	+90	+180	+116.5	+63.5	+45	−405
	∞	+90	+90	+90	+90	+90	−270

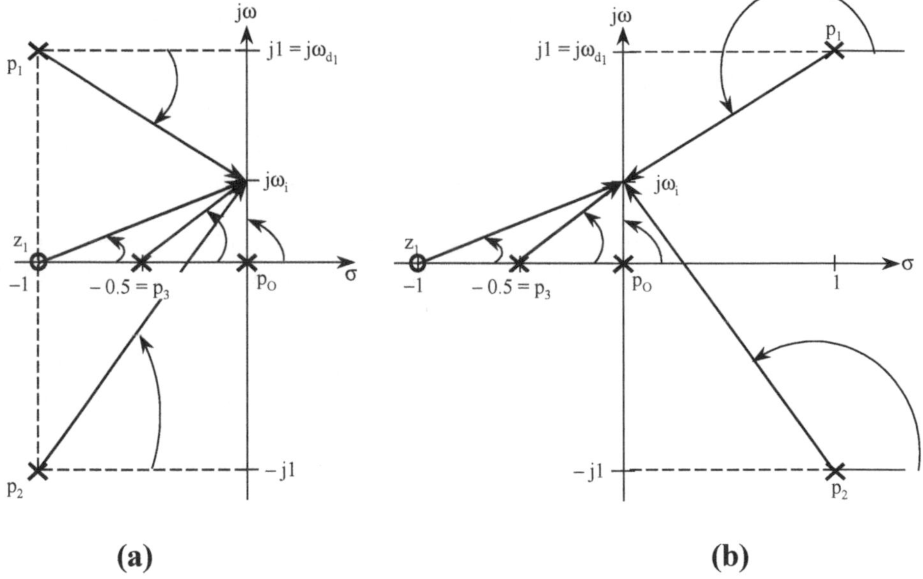

Fig. 3.37 Poles and zeros in the s-plane for (a) a stable plant; (b) an unstable plant.

3-20 SUMMARY

A general introduction to the MISO QFT technique is presented in this chapter. It is a frequency-response design method applied to the design of a MISO control system with an uncertain MISO plant P. As will be shown in Chapter 5 a MIMO system can be represented by an equivalent set of MISO systems. Two QFT MIMO system design approaches are available in which the equivalent MISO loops are designed according to the MISO design method presented in this chapter. This design is based upon

1. Specifying the tolerance in the ω-domain by means of the sets of plant transfer functions and closed-loop control ratios, $P = \{\mathbf{P}(j\omega)\}$ and $\mathfrak{I}(j\omega) = \{\mathbf{T}(j\omega)\}$, respectively.
2. Finding the resulting bounds on the loop transfer functions L_i and input filter transfer functions F of Fig. 3.27.

The robust design technique of this chapter permits the design of an analog control system that satisfies the desired performance specifications within the specified range of structured plant parameter variation. Since the QFT technique is based upon the design of the loop transmission function $L(s)$ of the MISO control system of Fig. 3.1, it is referred to as a MISO design technique. Although this chapter deals with m.p. plants, the QFT technique can be applied to n.m.p. plants and to digital control systems (see Chapter 4).[14,17,27,52]

4

DISCRETE QUANTITATIVE FEEDBACK TECHNIQUE[14]

4-1 INTRODUCTION[1]

This chapter focuses on the application of the QFT technique to MISO sampled-data control systems.[14] The QFT sampled-data (S-D) system design process is tuned to the bounds of uncertainty, the performance tolerances, and the sampling time T (or sampling frequency $\omega_s = 2\pi/T$). The QFT technique requires, as discussed in Sec. 4-5, the determination of the minimum sampling frequency $(\omega_s)_{min}$ bandwidth (BW) that is needed for a satisfactory design. The larger the plant uncertainty and the narrower the system performance tolerances, the larger must be the value of $(\omega_s)_{min}$. The use of the z- to the w' domain bilinear transformation[13] permits the analysis and design of sampled-data systems by the use of the *digitization (DIG) technique.*[13] That is, the w'-plane detailed QFT design procedure essentially parallels very closely that for continuous-time systems of Chapter 3, the difference being that the design must take into account the right-half-plane (RHP) zero(s) that result in the w' plant transfer function due to the bilinear transformation. Note, when a plant $P(s)$ is m.p. it becomes a n.m.p. plant when transformed into the w'-domain. Thus, proper care must be exercised in satisfying the stability bounds prescribed for the QFT design

The *pseudo-continuous*-time (PCT) approach, which is discussed in Sec. 4-8, is another DIG technique that allows the QFT design of the $D(z)$ controller to be done in the s-domain. Once the s-domain controller has been synthesized, in the manner described in Chapter 3, it is transformed into the z-domain by use of the *Tustin transformation*, a bilinear transformation, to obtain $D(z)$. The advantage of this approach, when the plant is m.p., is that it eliminates dealing with a n.m.p. plant and the problem associated in satisfying the stability bounds. Thus, the transformation either into the w'- or s-domain of the S-D MISO or MIMO control system enables the use of the MISO QFT analog design technique to be readily used, with minor exceptions, to perform the QFT design for the controller $D(w')$ or $D(s)$. If the w'- or s-domain simulations satisfy the desired performance specifications then by use of the bilinear transformation the z-domain controller $G(z)$ is obtained. With this z-domain controller a discrete-time domain simulation is obtained to verify the goodness of the design.

4-2 BILINEAR TRANSFORMATIONS[13]

The two DIG techniques, w'- or s-domain design, each require the use of a bilinear transformation. This section presents the z- to the w'-domain to the z-domain and the s- to the z-domain transformations.

4-2.1 *w*- AND *w'*- DOMAIN TRANSFORMATION

The QFT design of the sampled-data control system requires the use of the bilinear transformations

$$z = \frac{w+1}{-w+1} \tag{4.1}$$

$$w = \frac{z-1}{z+1} \tag{4.2}$$

Note that the w domain lacks the desirable property that as the sampling time T approaches zero, w should approach s; i.e.,

$$w\big|_{T\to 0} = \lim_{T\to 0}\left[\frac{z-1}{z+1} = \frac{e^{sT}-1}{e^{sT}+1} = \frac{sT+(sT)^2/2!+\cdots}{2+sT+(sT)^2/2!+\cdots}\right] = 0 \tag{4.3}$$

and where

$$z\,|_{T\to 0} = \lim_{T\to 0}[z = e^{sT}] = 1$$

This situation is overcome by defining

$$w' \equiv \frac{2}{T}w \equiv \frac{2}{T}\frac{sT+(sT)^2/2!+\cdots}{2+sT+(sT)^2/2!+\cdots} \tag{4.4}$$

Thus, in the w'plane the desirable property that $w' \to s$ as $T \to 0$ is achieved. This w'-plane property establishes the conceptual basis for defining a quantity in the w' domain, which is analogous to a quantity in the s domain. Substituting $w = Tw'/2$

into Eqs. (4.1) and (4.2) yields, respectively, the z to w' plane and the w'- to z-plane transformations as follows:

$$w' = \frac{2}{T}\frac{z-1}{z+1} \tag{4.5}$$

$$z = \frac{Tw' + 2}{-Tw' + 2} \tag{4.6}$$

Equation (4.6) represents an approximation of $z = e^{sT}$. The mapping of the z-plane via Eq. (4.5) into the w'-plane is shown in Fig. 4.1. From now on, *the w' transformation is used throughout the text and the prime designator is omitted* (*i.e.*, $w' = w$) except where noted in Table 4.1.

4-2.2 *s*-PLANE and *w*-PLANE RELATIONSHIP

The relationship between the s and w plane can be found by examination of Eq. (4.6) in terms of $z = e^{sT}$, i.e.,

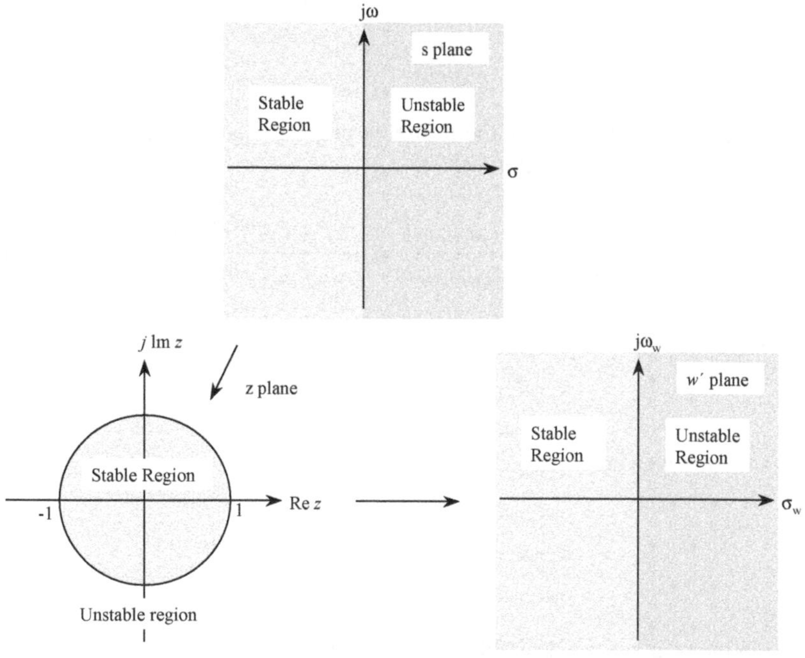

Fig. 4.1 Mapping of the s-plane into the z-plane by means of $z = e^{sT}$ of the z-plane into the w-plane by means of Eq. (4.5).

$$w = u + jv = \sigma_{wp} + j\omega_{wp}$$

$$= \frac{2}{T}\left[\frac{e^{sT}-1}{e^{sT}+1}\right]\frac{e^{-sT/2}}{e^{-sT/2}} = \frac{2}{T}\left[\frac{e^{sT/2}-e^{-sT/2}}{e^{sT/2}+e^{-sT/2}}\right] \tag{4.7}$$

which yields

$$w = \frac{2\tanh(sT/2)}{T} \tag{4.8}$$

For the s-plane imaginary axis, substitute $s = j\omega_{sp}$ into Eq. (4.8) to obtain

$$w = \sigma_{wp} + j\omega_{wp} = \frac{2\tanh(j\omega_{sp}T/2)}{T} = \frac{j2\tan(\omega_{sp}T/2)}{T} \tag{4.9}$$

or

$$v = \omega_{wp} = \frac{2\tan(\omega_{sp}T/2)}{T} \tag{4.10}$$

Thus, the imaginary axis in the primary strip of the s plane is mapped onto the entire imaginary axis of the w plane (see Fig. 4.1). If $\omega_{sp}T/2$ is small ($\omega_{sp}T/2 \le 0.297$), then from Eq. (4.10) obtain $v = \omega_{wp} \approx \omega_{sp}$.

For the real axis, substitute $s = \sigma_{sp}$ into Eq. (4.8) to obtain

$$w = \frac{2\tanh(\sigma_{sp}T/2)}{T} = \sigma_{wp} = u \tag{4.11}$$

By letting $\alpha = \sigma_{sp}T/2$, $\tanh \alpha$ can be expressed in the expanded form

$$\tanh\alpha = \frac{\alpha + \alpha^2/3! + \cdots}{1 + \alpha^2/2! + \cdots} \tag{4.12}$$

For $\alpha^2 << 2$, from Eqs. (4.11) and (4.12),

$$\sigma_{sp} \approx \sigma_{wp} \tag{4.13}$$

Thus, when the approximations are valid,

$$w = u + jv = \sigma_{wp} + j\omega_{wp} \approx \sigma_{sp} + j\omega_{sp} = s \tag{4.14}$$

If the approximations are not valid, then Eqs. (4.10) and (4.11) must be used to locate the *s*-plane poles and zeros properly in the *w*-plane. The mapping of the *s*-plane poles and zeros into the *w*-plane by use of these equations is referred to as *pre-warping of the s-plane poles and zeros*. The relationships of Eq. (4.14) are the basis of a QFT design method (DIG technique) in the *w*-plane. Table 4.1 illustrates the frequency range for which the degree of accuracy of the relationships of Eq. (4.14) is very good and how it is dependent upon the sampling time *T*.

A characteristic of a bilinear transformation is: in general, it transforms an unequal-order *z*-domain transfer function, one whose order of the numerator w_z and order of its denominator n_z are unequal, into one for which the order of the numerator (w_w) is equal to the order of its denominator (n_w). That is, $n_z \neq w_z$ in the *z* domain and $n_w = w_w$ in the *w* domain. Further, note that in transforming $G(z) = K_xG'(z)$ to $G(w) = K_wG'(w)$ by means of Eq. (4.5), the value of the gain constant K_w may be positive or negative.[13] The sign of K_w is determined by the coefficients of $G(z)$, which in turn are functions of *T*. This characteristic must be kept in mind when synthesizing $G(w)$ and $F(w)$.

TABLE 4.1 *w*- and *w′*-domain values of v_i for $\omega_s = 120$ rad/sec and $\omega_s = 240$ rad/sec

s-plane	**w-plane v_i values**		**w′-plane v_i values**		
ω_i values	**$\omega_s = 120$**	**$\omega_s = 240$**	**$\omega_s = 120$**	**$\omega_s = 240$**	**Valid approximation Range**
0.5	0.013	0.00655	0.5000	0.5000	↓
1.0	0.02618	0.01309	1.0002	1.0000	↓
2.0	0.0524	0.02619	2.0018	2.0004	↓
5.0	0.13165	0.06554	5.0287	5.0071	↓
10.0	0.26795	0.13165	10.235	10.057	↓
20.0	0.57735	0.26795	22.053	20.47	—
30.0	1.0	0.41420	38.2	31.63	—
40.0	1.7320	0.57735	66.15	44.11	—
50.0	3.73205	0.76733	142.55	58.619	—
60.0	—	1.0	—	76.394	—
70.0	—	1.3032	—	99.559	—
100.0	—	3.732	—	285.108	—
120.0	—	—	—	—	

4-2.3 *s*- TO *z*-PLANE TRANSFORMATION: TUSTIN TRANSFORMATION[13]

One of the most popular *s*- to *z*-plane transformations is the *Tustin algorithm.*[56,57] The function of z that is substituted for s^q in implementing the Tustin transformation is

$$s^q = \left(\frac{2}{T} \frac{1 - z^{-1}}{1 + z^{-1}} \right)^q \tag{4.15}$$

An advantage of the Tustin algorithm is that it is comparatively easy to implement. Also, the accuracy of the response of the Tustin *z*-domain transfer function is good compared with the response of the exact *z*-domain transfer function; i.e., the accuracy increases as the frequency increases.

The Tustin transformation for $q = 1$ is defined as

$$s \equiv \frac{2}{T} \frac{1 - z^{-1}}{1 + z^{-1}} = \frac{2}{T} \frac{z - 1}{z + 1} \tag{4.16}$$

which is a bilinear transformation and can be equated to the trapezoidal integration (s^{-1}) method. Note, for $q = 1$ Eq. (4.16) is identical to the *w*- to *z*- domain transformation of Eq. (4.16). Thus, both equations are from now on referred to as a *Tustin transformation.* Also, Eq. (4.16) can be derived by approximating $z = e^{sT}$ as a finite series. The operator equation (4.16) is as applicable to matrix equations as it is to scalar differential equations. The following discussion is useful therefore in understanding the mapping result for the vector model.

To represent functionally the *s*- to *z*-plane mapping, Eq. (4.16) is rearranged to yield

$$z = \frac{1 + sT/2}{1 - sT/2} \tag{4.17}$$

Utilizing the mathematical expression

$$\frac{1 + \mathrm{a}}{1 - \mathrm{a}} = \mathrm{e}^{\mathrm{j}2\tan^{-1}\mathrm{a}}$$

and letting $s = j\hat{\omega}_{sp}$ in Eq. (4.17) the following expression is obtained:[13]

$$z = \frac{1 + j\hat{\omega}_{sp}T/2}{1 - j\hat{\omega}_{sp}T/2} = e^{j2\tan^{-1}\hat{\omega}_{sp}T/2} \tag{4.18}$$

The exact $\mathcal{Z}$-transform yields $z = e^{j\omega_{sp}T}$, where ω_{sp} is an equivalent s-plane frequency. Thus, the following equation is obtained from Eq. (4.18):

$$e^{j\omega_{sp}T} = \exp\left(j2\tan^{-1}\frac{\hat{\omega}_{sp}T}{2} \right) \tag{4.19}$$

Equating the exponents yields

$$\frac{\omega_{sp}T}{2} = \tan^{-1}\frac{\hat{\omega}_{sp}T}{2} \tag{4.20}$$

or

$$\tan\frac{\omega_{sp}T}{2} = \frac{\hat{\omega}_{sp}T}{2} \tag{4.21}$$

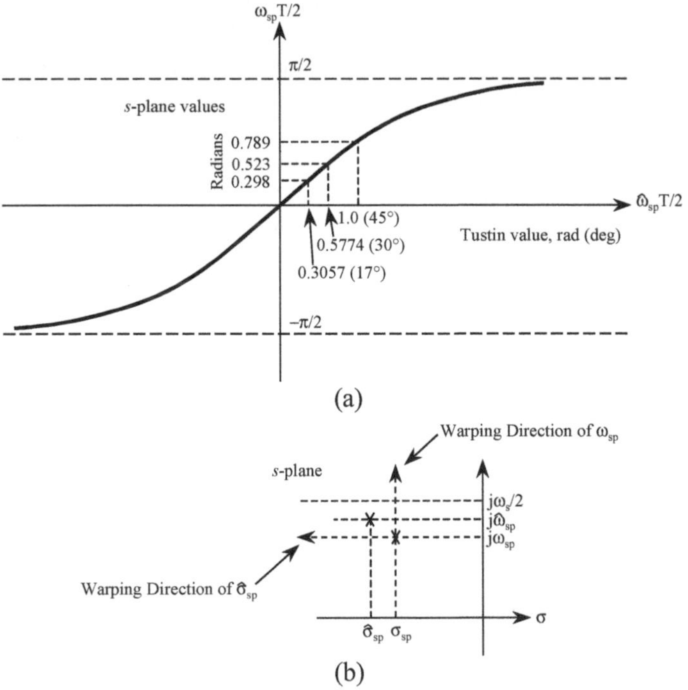

Fig. 4.2 Map of $\hat{\omega} = 2(\tan \omega_{sp}T/2)/T$. (a) Plot of Eq. (4.19) and (b) warping effect.

When $\omega_{sp}T/2 < 17^o$, or ≈ 0.30 rad, then

$$\omega_{sp} \approx \hat{\omega}_{sp} \tag{4.22}$$

Which means that in the frequency domain the Tustin approximation is good for small values of $\omega_{sp}T/2$.

Returning to Eq. (4.19), it is easy to realize that the s-plane imaginary axis is mapped into the *unit circle* (UC) in the z-plane as shown in Fig. 4.1. The left-half (LH) s-plane is mapped into the inside of the UC. The same stability regions exist for the exact $\mathcal{Z}$-transform and the Tustin approximation.

Also, in this approximation, the entire s-plane imaginary axis is mapped once and only once onto the UC. The Tustin approximation prevents pole and zero aliasing since the folding phenomenon does not occur with this method. However, there is again a warping penalty.[13] Compensation can be accomplished by using Eq. (4.21), which is depicted in Fig. 4.2. To compensate for the warping, prewarping of ω_{sp} by using Eq. (4.21) generates $\hat{\omega}_{sp}$. The continuous controller is mapped into the z-plane by means of Eq. (4.16) using the prewarped frequency $\hat{\omega}_{sp}$. The digital compensator (controller) must be tuned (i.e., its numerical coefficients adjusted) to finalize the design since approximations have been employed. As seen from Fig. 4.2, Eq. (4.22) is a good approximation when $\omega_{sp}T/2$ and $\hat{\omega}_{sp}T/2$ are both less than 0.3 rad.

The pre-warping approach for the Tustin approximation takes the s-plane imaginary axis and folds it back to $\pi/2$ to $-\pi/2$ as seen from Fig. 4.2. The spectrum of the input must also be taken into consideration when selecting an approximation procedure with or without pre-warping. It should be noted that in the previous discussion only the frequency has been pre-warped due to the interest in the controller frequency response. The real part of the s-plane pole influences such parameters as rise time, overshoot, and settling time. Thus, consideration of the warping of the real pole component is now analyzed as a fine-tuning approach. Proceeding in the same manner as used in deriving Eq. (4.22) substitute $z = e^{\sigma_{sp}T}$ and $s = \hat{\sigma}_{sp}$ into Eq. (4.17) to yield

$$e^{\sigma_{sp}T} = \frac{1 + \hat{\sigma}_{sp}T/2}{1 - \hat{\sigma}_{sp}T/2} \tag{4.23}$$

Replacing $e^{\sigma_{sp}T}$ by its exponential series and dividing the numerator by the denominator in Eq. (4.23) results in the expression

$$1 + \sigma_{sp}T + \frac{(\sigma_{sp}T)^2}{2} + \cdots = 1 + \frac{\hat{\sigma}_{sp}T}{1 - \hat{\sigma}_{sp}T/2} \tag{4.24}$$

If $|\sigma_{sp}T| >> (\sigma_{sp}T)^2/2$ (or $1 >> |\sigma_{sp}T/2|$) and $1 >> \left|\hat{\sigma}_{sp}T/2\right|$, then

$$\left|\hat{\sigma}_{sp}\right| \approx \left|\sigma_{sp}\right| << \frac{2}{T} \tag{4.25}$$

Thus, with Eqs. (4.22) and (4.25) satisfied, the Tustin approximation in the s domain is good for small magnitudes of the real and imaginary components of the variable s. The shaded area in Fig. 4.3 represents the allowable location of the poles and zeros in the s plane for a good Tustin approximation. Because of the mapping properties and its ease of use, the Tustin transformation is employed for the *DIG* technique in this text. Fig. 4.2*b* illustrates the warping effect of a pole (or zero) when the approximations are not satisfied.

A *matched* $\mathcal{Z}$*-transform* can be defined as a direct mapping of each s-domain root to a z-domain root; $s + a \rightarrow 1 - e^{-aT}z^{-1}$. The poles of $G(z)$ using this approach are identical to those resulting from the exact $\mathcal{Z}$ transformation of a given $G(s)$. However, the zeros and the d-c gain are usually different!

A characteristic of a bilinear transformation, as pointed out in Sec. 4-2.2, is that, in general, it transforms an unequal-order transfer function ($n_s \neq w_s$) in the s-domain into one for which the order of the numerator is equal to the order of its denominator ($n_z = w_z$) in the z-domain. This characteristic must be kept in mind when synthesizing $G(s)$ and $F(s)$.

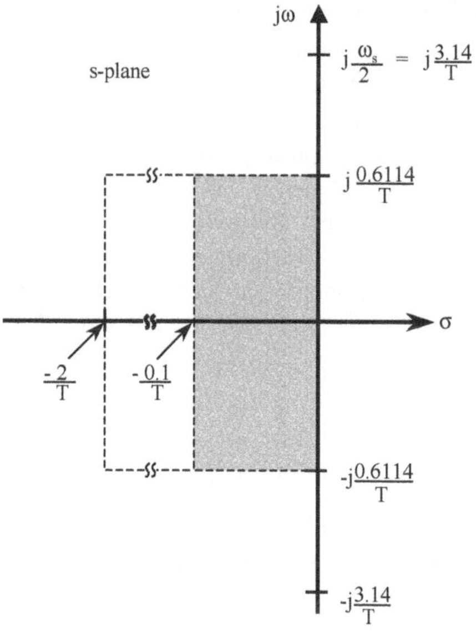

Fig. 4.3 Allowable location (shaded area) of dominant poles and zeros in s plane for a good Tustin approximation.

4-3 NON-MINIMUM PHASE ANALOG PLANT

The analog QFT design technique for m.p. plants can be modified when the plant is n.m.p. The case where the n.m.p. plant has only one RHP zero is considered in this section. For this case,

$$L(s) \equiv (1 - \tau s)L'_m(s) \tag{4.26}$$

where $L'_m(s)$ is a m.p. transfer function whose gain constant has a positive value. Eq. (4.26) is modified to

$$L(s) = \left[\frac{1-\tau s}{1+\tau s}\right][(1+\tau s)L'_m(s)] = A'(s)L_m(s) \tag{4.27}$$

where

$$A'(s) \equiv \frac{1-\tau s}{1+\tau s} \tag{4.28}$$

is n.m.p. and

$$L_m(s) = (1+\tau s)L'_m(s) \tag{4.29}$$

is m.p. The frequency domain characteristics of $A'(s)$ are determined by substituting $s = j\omega$ into Eq. (4.28). Thus,

$$A'(j\omega) = |A'|\angle A'(j\omega) \tag{4.30}$$

where

$$|A'| = \frac{\sqrt{1+(\tau\omega)^2}}{\sqrt{1+(\tau\omega)^2}} = 1 \tag{4.31}$$

$$\phi_{A'} = \angle A'(j\omega) = \tan^{-1}(-\omega\tau/1) - \tan^{-1}(\omega\tau/1) = \phi_N - \phi_D \tag{4.32}$$

and $\phi_N = -\phi_D < 0^\circ$ (see Fig. 4.4). Thus, Eqs. (4.31) and (4.32) reveal that $A'(s)$ is an all-pass-filter (a.p.f.) where its angular contribution

$$\phi_{A'} = \phi_N - \phi_D = 2\phi_N < 0 \tag{4.33}$$

contributes an "extra" phase lag to the m.p. function $L_m(s)$ as shown in Fig. 4.5.

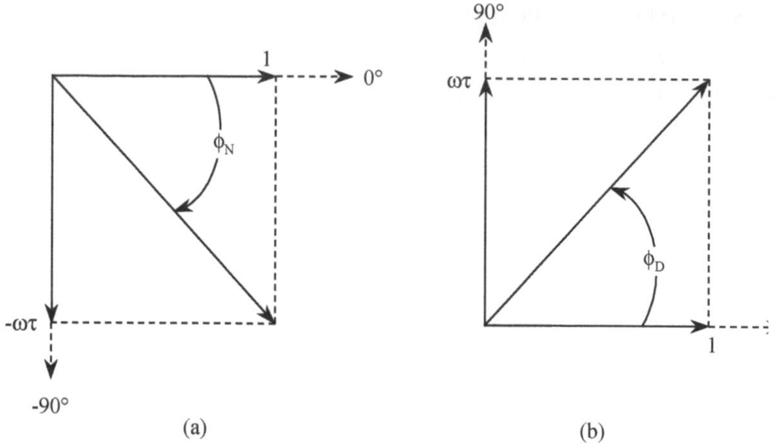

Fig. 4.4 Analysis of the angular contribution of Eq. 4.31: (a) numerator contribution $\phi_N < 0°$; (b) denominator contribution $\phi_D > 0°$.

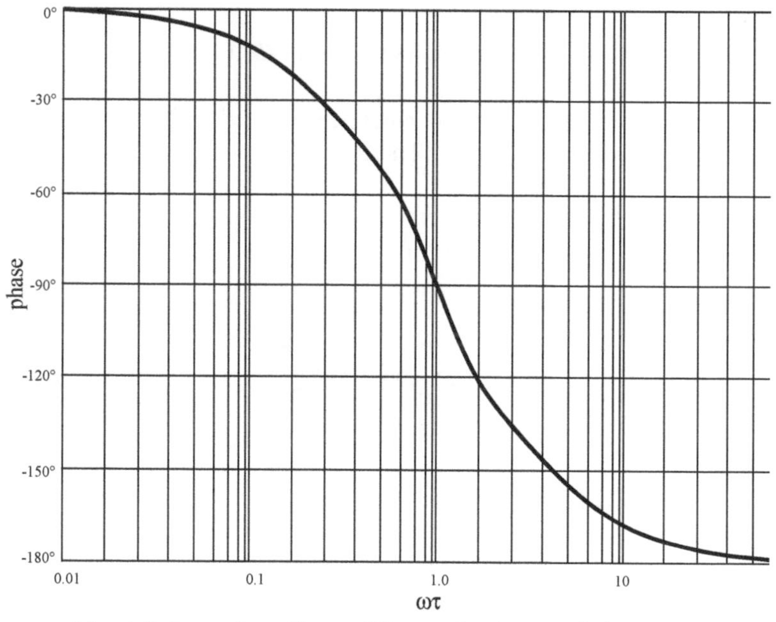

Fig. 4.5 An analog all-pass filter angle characteristic.

In the low frequency range, $0 \leq \omega_i \leq \omega_L$, where the magnitude of the angle of Eq. (4.33) is very small, the a.p.f. lag characteristic has essentially no effect on the determination of the optimal loop transmission function $L_o(s)$ or $L_{mo}(s)$. Whether the a.p.f.'s angular contribution is detrimental in achieving a satisfactory $L_{mo}(s)$ can be determined by the following approach: if the QFT design requires a phase margin angle $\gamma = 40^{\circ}$ at $\omega_i = \omega_\phi$, for which $\angle \mathbf{L}_{mo}(j\omega_\phi) = -90^{\circ}$, then

$$\angle \mathbf{L}_{mo}(j\omega_\phi) - \gamma = -130^{\circ}$$

This results in the restriction that $\phi_{A'} \geq -180^{\circ} - (-130^{\circ}) = -50^{\circ}$. If at $\omega = \omega_\phi$, $\phi_{A'} < -50^{\circ}$, then the actual phase margin angle will be less than 40°, i.e., $\gamma < 40^{\circ}$ at $\omega = \omega_\phi$ resulting in a larger value of M_m (or M_p). The limit on an achievable ω_ϕ for a n.m.p. plant having a RHP zero at z_i is $\omega_\phi = 0.5|z_i|$.[52]

4-3.1 ANALOG QFT DESIGN PROCEDURE FOR A n.m.p. PLANT

The analog design procedure of Chapter 3 is modified as follows for a n.m.p. analog plant:

1. For the nominal plant P_o, from Eq. (4.27) the nominal loop transmission function is denoted as

$$L_o(s) \equiv A'(s) L_{mo}(s) \tag{4.34}$$

2. Obtain the U-contour (B_h' – contour), and the B_D, B_R and the B_o' bounds corresponding to the template frequencies ω_i for $L_o(s)$ in the same manner as for the analog plant. Note that the B_h-contour and B_o bounds of Fig. 3.19 are now relabeled as B_h' – contours and B_o' bounds, respectively. Figure 4.6 shows the B_h' – contour and the $\mathbf{B}_o'(j\omega_1)$ bound. It is necessary to modify the angle characteristic of $\mathbf{L}_{mo}(j\omega)$ in order to compensate for $\phi_{A'} = \angle \mathbf{A}'(j\omega) < 0^{\circ}$. This is accomplished as follows: Eq. (4.34) is rearranged to

$$L_{mo}(s) = \frac{L_o(s)}{A'(s)} \tag{4.35}$$

which yields

$$\angle \mathbf{L}_{mo}(j\omega) = \angle \mathbf{L}_o(j\omega) - \phi_{A'}(j\omega) = \angle \mathbf{L}_o(j\omega) + [-\phi_{A'}(j\omega)] \tag{4.36}$$

Since $\phi_{A'}(j\omega) < 0^{\circ}$ then Eq. (4.36) implies that the B_o' bounds and the U-contour, for each value of frequency corresponding to the template frequencies, are shifted to the right by $|\phi_{A'}|$ to yield the corresponding

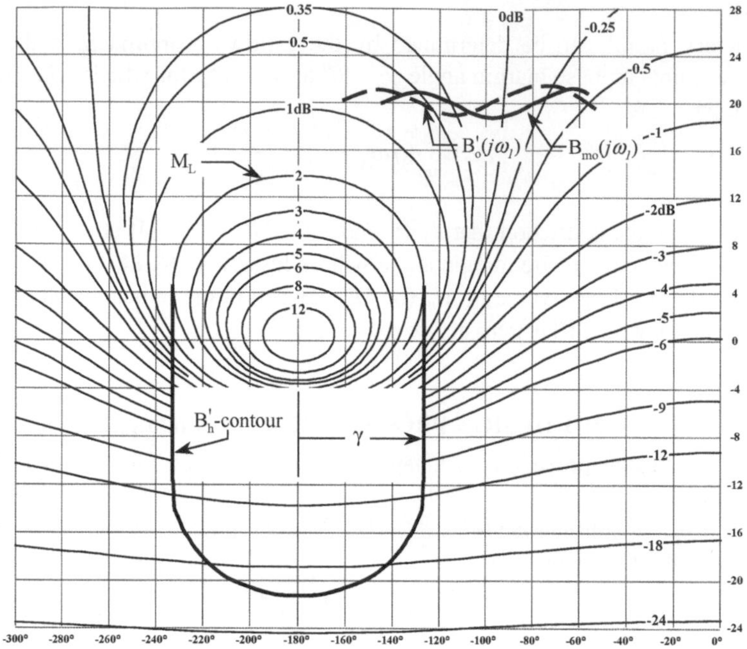

Fig. 4.6 Nichols chart characteristic for a non-minimum phase (n.m.p.) $L(s)$.

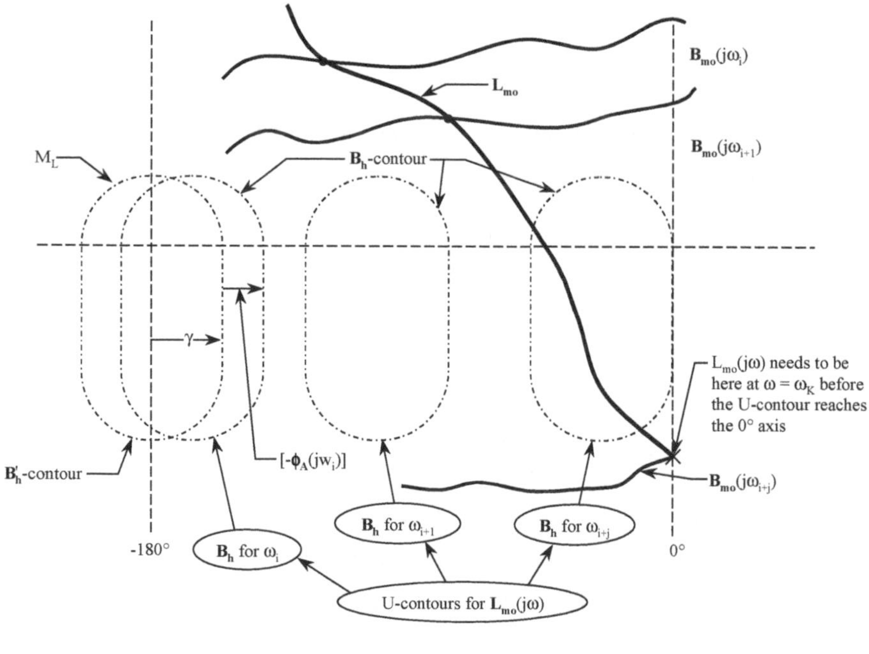

Fig. 4.7 The shifted U-contours for an n.m.p. $L_o(s)$.

B_{mo} bounds, as shown in Fig. 4.6, and the corresponding B_h-contours, as shown in Fig. 4.7.

3. Before proceeding with the design, the question that needs to be addressed is: can a realizable $L_o(s)$ be synthesized that satisfies all the bounds (B_{mo} bounds and B'_h-contours) and yet be able to go from a dB value of $Lm\ \mathbf{L}_{mo}(j\omega_i)$ to $Lm\ \mathbf{L}_{mo}(j\omega_K)$? For example, let

$$\Delta \text{dB} = [|\boldsymbol{B}(j\omega_i)| = 16\,\text{dB}] - [|\boldsymbol{B}(j\omega_{i+j})| = -50\,\text{dB}] = 66\,\text{dB}$$

where $\omega_i = 3$ and $\omega_{i+j} = 100$. Thus, in order to achieve this decrease, ΔdB $= 66$ dB, in one plus decade, the slope of $Lm\ \mathbf{L}_{mo}(j\omega)$ must be 60 dB/dec. This value for the slope results in a large "phase lag" characteristic which can force $Lm\ \mathbf{L}_{mo}(j\omega)$ to decrease so fast that it can not satisfy all the $\boldsymbol{B}_{mo}(j\omega_{i+1})$ bounds let alone yield a stable system. If this occurs, then a QFT design can be achieved to only meet the stability requirement, i.e., satisfy only the M_L specification. Accepting a larger value for M_L results in a "shrunken" U-contour and may allow a QFT design that satisfies all the specifications. That is, the synthesized $L_{mo}(s)$ lies on or above each $\boldsymbol{B}_{mo}(j\omega_i)$ bound and is to the "right" or to the "right bottom" of the shifted U-contour (B_h-contour).

4. In synthesizing $L_{mo}(s)$, for the case where only the stability M_L requirement is satisfied, one needs to exercise judgement as to where to try to locate the initial zero(s) that is (are) inserted into the synthesized function. In order to minimize the value of t_s locate the zero(s) as far left from the imaginary-axis as possible.

5. Form $L_o(s) = A'(s)L_{mo}(s)$ in order to obtain

$$G(s) = \frac{L_o(s)}{P_o(s)} \tag{4.37}$$

6. Design the filter and perform a simulation in the same manner as described in Chapter 3.

4-4 DISCRETE MISO MODEL WITH PLANT UNCERTAINTY

As the state-of-the-art of digital computers makes great strides, digital control systems are playing a more important and greater role today, such as digital flight control systems. Thus, it is important to extend the QFT continuous-time system design technique to an S-D control system such as represented by Fig. 4.8. In this figure the *ZOH* unit represents a *zero-order-hold* device whose transfer function is $G_{zo}(s) = (1 - e^{-sT})/s$. Also shown in this figure are switches representing an ideal sampler. The basic equations that describe the MISO system of Fig. 4.8 are

$$Y(s) = P(s)W(s) \qquad W(s) = M(s) + D(s) \qquad M(s) = G_{zo}(s)X^*(s)$$

$$X^*(s) = G_1^*(s)E^*(s) \qquad E^*(s) = V^*(s) - Y^*(s) \qquad V^*(s) = F^*(s)R^*(s)$$

Note, the starred functions represent impulse transfer functions; e.g., $Y^*(s)$ represents the ideal impulse transform of $Y(s)$. These equations are manipulated and transformed to yield the input/output z-domain relationships:

$$Y(z) = \frac{L(z)F(z)}{1 + L(z)}R(z) + \frac{PD(z)}{1 + L(z)} = Y_R(z) + Y_D(z) \tag{4.38}$$

where

$$L(z) = G_{zo}P(z)G_1(z) \tag{4.39}$$

$$P_z(z) = G_{ZO}P(z) = (1 - z^{-1})\mathcal{Z}\left[\frac{P(s)}{s}\right] = (1 - z^{-1})P_e(z) \tag{4.40}$$

$$P_e(z) \equiv \mathcal{Z}\left[\frac{P(s)}{s}\right] = \mathcal{Z}[P_e(s)] \tag{4.41}$$

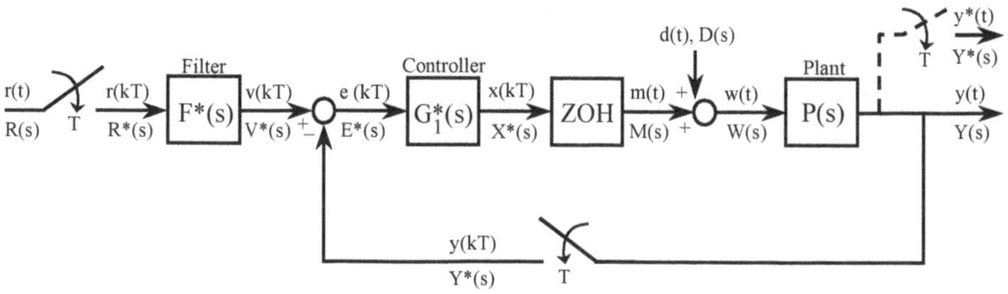

Fig. 4.8 A MISO sampled-data control system.

$$P_e(s) \equiv \frac{P(s)}{s} \tag{4.42}$$

$$PD(z) = \mathcal{Z}[P(s)D(s)] \tag{4.43}$$

$$T_R(z) = \frac{F(z)L(z)}{1 + L(z)} = F(z)T_R'(z) \tag{4.44}$$

$$T_R'(z) = \frac{L(z)}{1 + L(z)} \tag{4.45}$$

$$Y_D(z) = \frac{PD(z)}{1 + L(z)} \tag{4.46}$$

Substituting Eq. (4.40) into Eq. (4.39) yields

$$L(z) = G_1(z)\left(1 - z^{-1}\right)P_e(z) = G_1(z)P_z(z) \tag{4.47}$$

Note, for a unit step disturbance input function, $D(s) = 1/s$, that

$$P_e(s) = P(s)D(s) = P(s)/s \tag{4.48}$$

and thus

$$P_e(z) = PD(z) \tag{4.49}$$

The QFT design is based upon the uncertain plant being defined by Eq. (4.41). Once $L(z)$ has been synthesized then the controller $G_1(z)$, which is to be implemented, is readily determined. Analyzing and designing the S-D control system in the z-domain is referred to as the *direct* (*DIR*) design technique.[13]

4-5 QFT *w*-DOMAIN DIG DESIGN

As discussed in Sec. 4-1 the design techniques that have been highly developed for the s-domain can readily be applied in the w-domain. This feature enables the QFT design technique for analog systems to be utilized for the design of discrete systems in the w domain if certain conditions related to T, the sampling time, hold. The pertinent s-, z-, and w-plane relationships from Sec. 4-2 are repeated below:

$$s = \sigma_{sp} + j\omega_{sp} = \sigma + j\omega \tag{4.50}$$

$$z = \frac{Tw + 2}{-Tw + 2} \tag{4.51}$$

$$w = \sigma_{wp} + j\omega_{wp} = u + jv = \left(\frac{2}{T}\right)\left[\frac{z-1}{z+1}\right] \tag{4.52}$$

$$v = \left(\frac{2}{T}\right)\tan\left(\frac{\omega T}{2}\right) = \frac{\omega_s}{\pi}\tan\left(\frac{\omega\pi}{\omega_s}\right) \tag{4.53}$$

$$\omega_s = \frac{2\pi}{T} \tag{4.54}$$

$$z = e^{sT} = e^{\sigma T}\angle\omega T = |z|\angle\omega T \tag{4.55}$$

If

$$\alpha^2 = \left(\frac{\sigma_{sp}T}{2}\right)^2 << 2 \qquad and \qquad \frac{\omega_{sp}T}{2} \le 0.297 \tag{4.56}$$

are both satisfied then $s \approx w$. If for the range of parameter uncertainty and for a specified value of T, both conditions of Eq. (4.56) are satisfied in the low frequency range, then the approximation

$$\Im \boldsymbol{P}(s) \approx \Im \boldsymbol{P}(w) \tag{4.57}$$

is valid in the low frequency range. The w-domain QFT design procedure presented in this chapter is based upon a stable uncertain plant.

Example 4.1 For the plant

$$P(s) = \frac{K}{s(s+a)} \qquad where \ 1 \le K \le 10 \quad and \quad 1 \le a \le 4$$

let $T = 0.05$ *s*. Consider two cases: (*a*) $0 \le a \le 20$ and (*b*) $0 \le a \le 10$. The range of values for which Eq. (4.57) is valid is determined by utilizing Eq. (4.56) which yields:

$$(a) \quad \alpha^2 = \left(\frac{\sigma_{sp}T}{2}\right)^2 = \left(\frac{-aT}{2}\right)^2 = 0.25 << 2 \quad \rightarrow \quad 1 << 8$$

$$(b) \quad \alpha^2 = 0.0625 << 2 \quad \rightarrow \quad 1 << 32$$

For both ranges of a: $\omega_{sp}T/2 \le 0.297 \rightarrow \omega_{sp} \le 11.88$. Thus, for $0 \le a \le 10$ and $0 \le \omega_{sp} \le 11.88$ Eq. (4.57) is valid. As the value of a approaches 20, from the low side, then Eq. (4.57) is a "fair" approximation with some warping.

Example 4.2 Given the plant

$$P(s) = \left[\frac{K_x}{s(s+2)}\right]\left[\frac{K_y(s+a)}{s+b}\right] = P_xP_y \qquad \textbf{(4.58)}$$

obtained by using either $P(s)$ or $P_y(s)$ will be identical. Remember the $1/s$ in where P_x is the LTI portion and P_y is the plant uncertainty portion of $P(s)$. Note, the "s" in the denominator of Eq. (4.41) comes from the zero-order-hold (*ZOH*) unit and is not part of the plant $P(s)$.

Note, for simplification purposes, the double subscripts in Eq. (4.50) are dropped for the s-plane representation, and the $w = u + jv$ notation is used for the w-plane representation.

If ω_s is fixed a priori, then the QFT design can proceed in the w domain precisely in the same manner as for the continuous system.[1,2,14] The design can be successful only if ω_s is large enough for the specific plant uncertainty. Thus, for S-D control systems for which the minimum value of sampling time has not been specified, it is best to assume initially that ω_s is not known a priori and the derivation of the minimum ω_s needed becomes one of the principal design problems.[14] *For an achievable QFT design it is required that in the*

$$\lim_{v\rightarrow\infty}[P_{z_l}(jv)] \qquad \textbf{(4.59)}$$

the resulting values of gain for all J plants, where ι = 1,2,…,J, do not change sign.[29,58]

4-5.1 CLOSED-LOOP SYSTEM SPECIFICATIONS

Figure 3.4b represents the upper and lower bounds for the desired tracking responses and Fig. 3.5 represents the upper bound for the desired disturbance response. The $\delta_R(j\omega_i)$ specification suffices to control the time domain response for m.p. systems and those for which the RHP zeros are known.[19] The latter case applies for the w domain QFT design for a system whose plant $P(s)$ is m.p. As derived in Ref. 13, for ideal impulse sampling, that

$$H(j\omega) \approx TH^*(j\omega) \qquad \textbf{(4.60)}$$

where $\boldsymbol{H}(j\omega)$ is the analog transfer function, $\boldsymbol{H}^*(j\omega)$ is its corresponding impulsed sampled function, and $\omega << \omega_s$. Equation (4.60) is valid for $\omega \leq 0.1\omega_s$ where 0.1 is a good engineering rule-of-thumb value. Thus, where for the analog QFT design it is desired that

$$\left|\boldsymbol{B}_L(j\omega)\right| \leq Lm\boldsymbol{T}_R(j\omega) \leq \left|\boldsymbol{B}_U(j\omega)\right| \qquad \textbf{(4.61)}$$

and

$$\delta_R(j\omega) = \left|\boldsymbol{B}_U(j\omega)\right| - \left|\boldsymbol{B}_L(j\omega)\right| \geq \Delta\boldsymbol{T}_R(j\omega) \qquad \textbf{(4.62)}$$

for all P in $\mathcal{P}$ then for a discrete QFT designed system Eqs. (4.61) and (4.62) are modified to

$$\left|\boldsymbol{B}_L(j\omega)\right| < T[Lm\boldsymbol{T}_R^*(j\omega)] \leq \left|\boldsymbol{B}_U(j\omega)\right| \qquad \textbf{(4.63)}$$

$$\delta_R^*(j\omega) = \left|\boldsymbol{B}_U^*(j\omega)\right| - \left|\boldsymbol{B}_L^*(j\omega)\right| \geq \Delta\boldsymbol{T}_R^*(j\omega) \qquad \textbf{(4.64)}$$

$$\left|\boldsymbol{B}_L^*(j\omega)\right| = \left|\frac{\boldsymbol{B}_L(j\omega)}{T}\right| \leq Lm\boldsymbol{T}_R^*(j\omega) \leq \left|\frac{\boldsymbol{B}_U(j\omega)}{T}\right| = \left|\boldsymbol{B}_U^*(j\omega)\right| \qquad \textbf{(4.65)}$$

which is valid for a "low enough" part of the frequency spectrum.

The specified $\boldsymbol{B}_L$ and $\boldsymbol{B}_U$ bounds[1] for the design example to be used in this chapter is shown in Fig. 4.9. For the QFT technique it suffices if Eqs. (4.63) to (4.65) are valid up to where $\boldsymbol{B}_L$ is approximately $Lm\ \alpha_h \equiv -24$ dB ($\omega_h \equiv 10$ rad/sec in Fig. 4.9). The reason for this is that for $\omega > 10$, $\boldsymbol{B}_L$ decreases rapidly, so the allowable $|\boldsymbol{T}_R(j\omega)|$ uncertainty (i.e., $\boldsymbol{B}_U - \boldsymbol{B}_L$) soon exceeds the plant uncertainty.

The requirement for a reasonable stability margin independent of $\boldsymbol{B}_L$ and $\boldsymbol{B}_U$ is to maximize the value of ω at which the -24 dB value occurs on the $\boldsymbol{B}_L$ curve. For example: is the range 0 to 10 rad/sec "low enough?" In general, in feedback systems, the price paid for the benefits of feedback is in the BW of the loop transmission L of Eqs. (4.44) and (4.45).[40] This tends to force ω_s to be significantly larger than what is required by the sampling theorem. Thus, it is very reasonable to simply use Eqs. (4.63) to (4.65) for the sampled model which is based upon the continuous model.

A loop stability requirement similar to that done for the continuous system must be added to the disturbance response performance requirements, as indicated in Fig. 3.5.[1] This requirement (a damping constraint) is:

$$\left| \frac{\boldsymbol{L}^*}{1+\boldsymbol{L}^*} \right| < M_L \quad \textbf{(4.66)}$$

is a constant for all ω and for all P_e in $\mathcal{P}_e$. For the example in this chapter a value of $Lm\ M_L = 6$ dB is used. If conditionally stability is not allowed, then the following additional requirement is imposed:

$$\angle \boldsymbol{L}^* \geq -180^o + \gamma \quad \textbf{(4.67)}$$

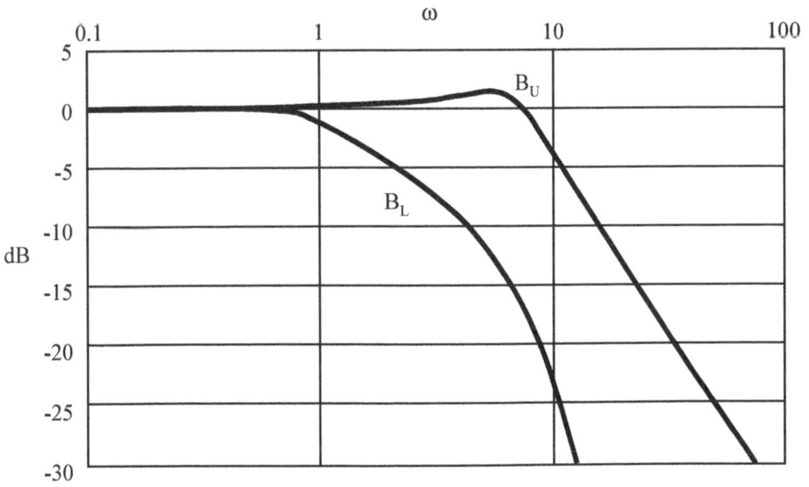

Note: Curves drawn approximately to scale.

Fig. 4.9 Bounds on $Lm\ \boldsymbol{T}_R(j\omega)$ used for $\boldsymbol{T}_R(j\omega)$ up to the -24 dB point.

where $\gamma > 0$ for $\omega < \omega_c$, ω_c is the crossover frequency $[|\mathbf{L}^*(j\omega_c)| = 1]$, and $|\mathbf{L}^*(j\omega)| < 1$ for $\omega > \omega_c$.

Thus, in order to proceed with a QFT design, it is necessary at the outset to determine the bounds and the values of M_L and γ. The change in T_R due to parameter variation, can be obtained (see Chapter 3) from Eqs.(4.44) and (4.45) in the logarithmic domain and is expressed as follows:

$$\Delta(Lm\mathbf{T}_R) = Lm\mathbf{T}_R - Lm\mathbf{F} = Lm\left[\frac{\mathbf{L}}{1+\mathbf{L}}\right] = Lm\mathbf{T}_R' \qquad \textbf{(4.68)}$$

As it is seen from Eq. (4.68), $\Delta[Lm\ T_R(z)]$ can be made arbitrarily small by choosing $L(z)$ sufficiently large. For the QFT technique, the designer finds the minimum $L(z)$ [or $\mathbf{L}(jv)$] to satisfy Eqs. (4.63) to (4.67).

4-5.2 PLANT TEMPLATES

As is required for continuous-time control systems, it is also necessary for S-D control systems to determine the plant templates $\Im\mathbf{P}_e(j\omega_i) = \{\mathbf{P}_e(j\omega_i)\}$ at a "sufficient" number of ω_i values. The number of plants J within the $\mathcal{P} = \{P_{e_\iota}\}$, where $\iota = 1,2,...,J$, is chosen to adequately describe the contour of the template which represents the region of plant uncertainty for each ω_i. For S-D systems, these "plant templates" are functions of the sampling frequency ω_s, but this becomes apparent only for $\omega > 0.25\omega_s \equiv \omega_{0.25}(v_i > v_{0.25})$ (an approximate engineering value, see Sec. 4-5.6). For the values of $\omega_i \le \omega_{0.25}$, the plant templates $\Im\mathbf{P}(j\omega_i) = \{\mathbf{P}(j\omega_i)\}$, for the continuous system, are essentially the same as the plant templates $\Im\mathbf{P}_e(jv_i) = \{\mathbf{P}_e(jv_i)\}$ for the S-D system. Since at this point of the design process the desired value of ω_s has yet to be determined, it is therefore reasonable to tentatively use the continuous $\Im\mathbf{P}(j\omega_i)$ templates in order to proceed with the design and eventually to be able to determine the required value of ω_s. A number of plant templates [for $v_i = (2/T)tan(\omega_i T/2)$] are shown in Fig. 4.10 for the plant of Example 4.1 calculated at $\omega_s = 60$ rad/sec. These templates are very similar to the templates for the continuous-time system and those for $\omega_s = 120$ and 240 rad/sec. To accentuate the value of doing a w-domain design [via Eq. (4.2)] vs. a w'-domain design [via Eq. (4.5)], a correlation is made between the values of

$$w = j\tan(\omega\pi/\omega_s) \quad \text{and} \quad w' = jv = j\frac{\omega_s}{\pi}\tan(\omega\pi/\omega_s)$$

for various values of ω_i in the frequency range $0 < \omega_i \le \omega_s/2$, in Table 4.1. Note, henceforth, once again the prime designator is omitted (i.e., $w' = w$).

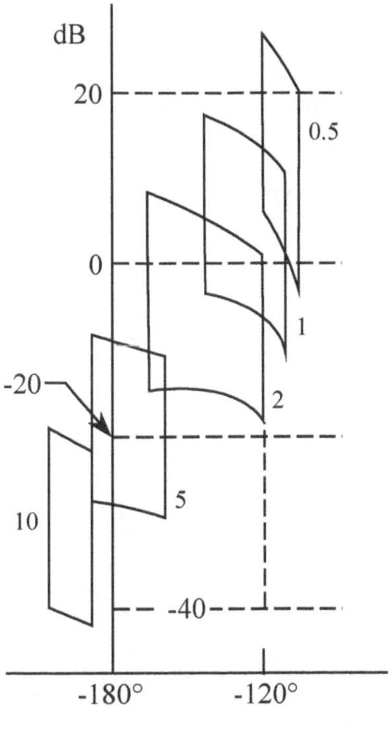

Fig. 4.10 Plant templates at $\omega_i = 0.5, 1, 2, 5$, and 10 rad/sec calculated for $\omega_s = 60$ rad/sec, but almost identical for $\omega_s = 120$ and 240 rad/sec and for the continuous-time system.[14]

4-5.3 BOUNDS $B(jv)$ ON $L_o(jv)$

Recall that the jv-axis of the w plane, where $0 \leq v \leq \infty$, corresponds to the upper half of the unit circle in the z plane, and to the primary strip $0 \leq j\omega_s/2$ in the s plane (see Sec. 4-2). Since QFT requires the shaping of L to satisfy certain bounds, then such shaping is difficult to do within the unit circle in the z-domain. Therefore, it is best to design F and G_1 in Fig. 4.8 in the $w = jv$ domain. As a consequence, the same design techniques that are used to perform loop shaping in the $s = j\omega$ domain can be used. The appropriate values of v_i (see Secs. 4-5.5 and 4-5.6) to be used for the templates are available from Eqs. (4.50) through (4.54), only after ω_s is chosen. Hence, the tentative value $\omega_s = 120$ is selected to initiate the design process. The choice is intentionally made low, in order to stress the relationship between $(\omega_s)_{min}$, the plant uncertainty, and the performance tolerances. After an individual achieves a sufficient understanding of the QFT technique, a good estimate can be made a priori for the value of $(\omega_s)_{min}$.

In order to determine the bounds, a nominal plant $P_{eo}(w)$ must be selected. Thus, $K = 1$ and $a = 1$ are selected as the nominal plant parameter values. The template $\Im \boldsymbol{P}_e(jv_i)$, with its corresponding $\delta_R(jv_i)$ value, is shifted on the NC to determine the bounds $\boldsymbol{B}(jv_i)$ for each value of v_i. Based upon Eqs. (4.40) and (4.47) the corresponding nominal $L_o(w)$ and $P_{zo}(w)$ are, respectively:

$$L_o(w) = G_1(w)P_{zo}(w) = G_1(w)\left[\frac{2w}{w+2/T}\right]P_{eo}(w) \tag{4.69}$$

$$P_{zo}(w) = \left[\frac{2w}{w+2/T}\right]P_{eo}(w) \tag{4.70}$$

where

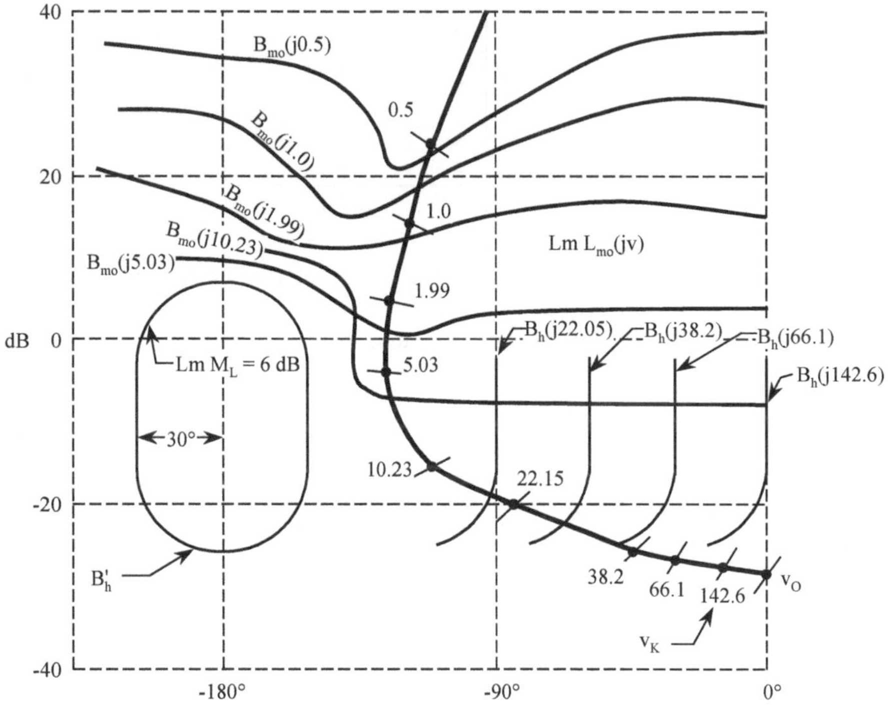

Note: Curves drawn approximately to scale

ω	0.5	1	5	10	20	30	40	50
p	0.013	0.026	0.131	0.27	0.58	1.00	1.73	3.73
v	0.5	1.00	5.03	10.23	22.05	38.2	66.1	142.6

Fig. 4.11 Unsuccessful design at $\omega_s = 120$ rad/sec.[14]

$$P_{eo}(w) = \frac{-2.99\times 10{-6}(w-38.2)(w+4377)(w+38.2)}{w2\,(w+0.9998)} \tag{4.71}$$

Thus Eq. (4.65) imposes the requirement that

$$\Delta T_R(jv) \le |\boldsymbol{B}_U(jv)| - |\boldsymbol{B}_L(jv)| \tag{4.72}$$

for all P_e in $\mathcal{P}_e$. The constraint imposed on *Lm* $\boldsymbol{L}_o(jv)$ by Eq. (4.72) is reflected on the NC by the bounds $\boldsymbol{B}_o(jv_i)$ on *Lm* $\boldsymbol{L}_o(jv_i)$ as discussed in Chapter 3. The NC is convenient because it contains loci of constant $|\boldsymbol{L}/(1+\boldsymbol{L})|$. For example, in Fig. 4.11, *Lm* $\boldsymbol{L}_o(j0.5)$, although not plotted, must be on or above $\boldsymbol{B}_{mo}(j0.5)$ ($\boldsymbol{B}_o$ shifted to the right as discussed in Sec. 4-5.4) in order to satisfy Eq. (4.72) for all P_e in $\mathcal{P}_e$.

A good rule in selecting the nominal plant P_{eo} from $\mathcal{P}_e$ is to select the P_{e_ι} in $\mathcal{P}_e$ that has the smallest magnitude and the largest phase lag over all other sP_e in $\mathcal{P}_e$, for all values of v_i, as P_{eo}. This selection criterion for P_{eo} (P_{zo}) minimizes the number of plant templates needed to determine the bounds.

4-5.4 NON-MINIMUM-PHASE $L_o(w)$

It is important to note that in the w domain any practical $L(w)$ is non-minimum-phase (n.m.p.) with a zero at $2/T$ [see Eqs. (4.69) and (4.70)]. [Note the zero in the "true"'w-domain is at 1 [Eq. (4.1)]. This result is due to the fact that any practical $L(z)$ has an excess of at least one pole over zeros which leads via Eq. (4.52) to the zero at $2/T$. Thus, the design technique, *for a stable uncertain plant* is modified[3] to incorporate the all-pass filter (a.p.f.)

$$A'(w) = \frac{2/T - w}{2/T + w} \tag{4.73}$$

as follows: Let the nominal loop transmission be defined as

$$L_o(w) \equiv -L_{mo}(w)A(w) = L_{mo}A'(w) \tag{4.74}$$

where

$$A(w) \equiv \frac{w - 2/T}{w + 2/T} = -A'(w) \tag{4.75}$$

and where $|\boldsymbol{A}'(jv)| = 1$ for $0 \le v \le \infty$ and L_{mo} is an m.p. function considered to have a positive value of gain. Since $|\boldsymbol{L}_o(jv) = |\boldsymbol{L}_{mo}(jv)|$ and

$$\angle L_o(jv) = \angle L_{mo}(jv) + \angle A'(jv) \tag{4.76}$$

or

$$\angle L_{mo}(jv) = \angle L_o(jv) - \angle A'(jv) \tag{4.77}$$

a bound $B'(jv)$ on $L_o(jv)$ becomes the bound $B(jv)$ on $L_{mo}(jv)$ by shifting $B'(jv)$ positively (to the right on the NC) by the angle

$$-\angle A'(jv) = 2\tan^{-1}\left(\frac{vT}{2}\right) = 2\tan^{-1} p > 0 \tag{4.78}$$

A non-dimensionalized $p = vT/2$ plot of Eq. (4.78) is shown in Fig. 4.12. The amount of shift is 90° at $p = 1$ ($v = 2/T$), 151.93° at $p = 4$ ($v = 8/T$), and it approaches 180° as $p(v)$ approaches infinity.

The inherent limitations on the feedback capabilities of an n.m.p. system are readily revealed as follows. At larger values of v_i (or ω_i), the templates tend to be vertical lines (due to the bilinear transformation characteristic) and Eqs. (4.63) through (4.65) tend to dominate, leading in m.p. systems to a single universal

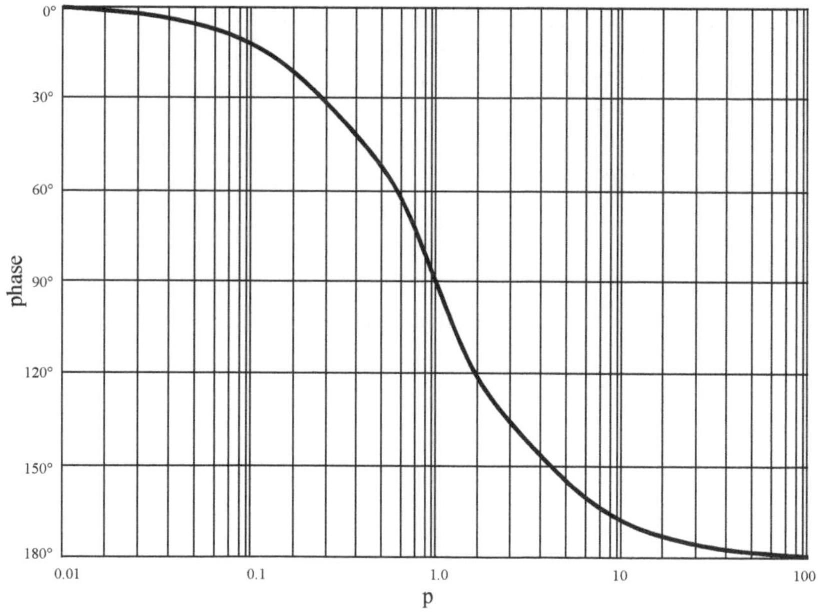

Fig. 4.12 Angle characteristic of Eq. (4.78).

high-frequency contour (bound) $\boldsymbol{B}'_h$ basically effective for all ω (or v) greater than some ω_h[3] (or v_h). For example, in Fig. 4.11 $\boldsymbol{B}'_h$ is shown with a width of $2(30^o)$ for $M_L = 6$ dB. However, in the n.m.p. system, the $\boldsymbol{B}'_h$- contour is shifted in the NC by $2\tan^{-1} vT/2$, to become $\boldsymbol{B}_h(jp)$ a function of p whose right boundary shifts to the vertical line $\angle \boldsymbol{L}_{mo}(jv) = 0^o$ at the value $p = p_K$. Thus, for Fig. 4.11, where $\gamma = 30^o$, then

$$2\tan^{-1} p_K = 180^{\circ} - 30^{\circ} = 150^{\circ} \tag{4.79}$$

results in $p_K = 3.73$ ($v_K = 142.55$ for $\omega_s = 120$) for this example. In general, although not done in this chapter, the $\boldsymbol{B}_h$ contours are denoted as a function of p in order for these contours to be independent of the sampling frequency. The Nyquist stability criterion dictates that the $\boldsymbol{L}_{mo}(jv)$ plot is on the "right side" or the "bottom right side" of the $\boldsymbol{B}_h(jp)$- or $\boldsymbol{B}_h(jv)$- contours for the frequency range of $0 \le v_i \le v_K$. It has been shown that:[3,19,40]

1. $\boldsymbol{L}_{mo}(jv)$ must reach the *right-hand bottom* of $\boldsymbol{B}_h(jv_K)$, i.e., approximately point K in Fig. 4.11, at a value of $v \le v_K$, and
2. $\boldsymbol{L}_{mo}(jv_K) < 0^o$ in order that there exist a practical L_{mo} which satisfies the bounds $\boldsymbol{B}(jv)$ and provides the required stability.

Requirement (2) is necessary in order to obtain an $\boldsymbol{L}_{mo}(jv)$ whose angle is always negative for all frequencies, a necessary requirement for a stable system. These requirements are very useful for determining the minimum ω_s needed for any specific uncertainty problem as discussed in Sec. 4-5.5.

For the situation where the uncertain plant parameter characteristics can yield an unstable plant(s), amongst the J plants being considered, if the nominal plant chosen is one of these unstable plants *then the a.p.f. to be used in the QFT design must include all RHP zeros of P_{zo}.* This situation is not discussed in this text.

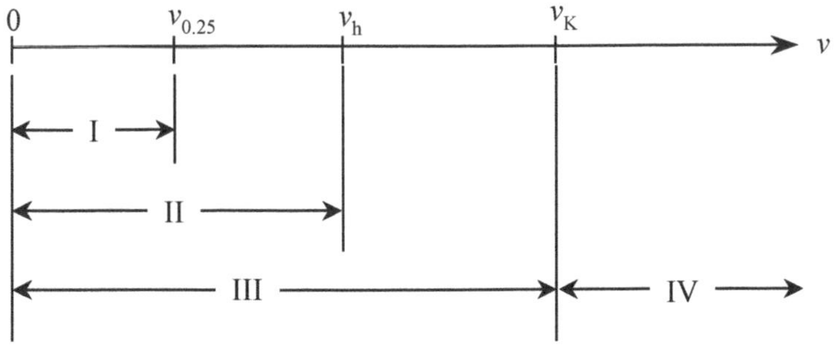

Fig. 4.13 Regions of the frequency spectrum for synthesizing $\boldsymbol{L}_{mo}(jv)$.

4-5.5 SYNTHESIZING $L_{mo}(w)$

Based upon the analysis made in the previous sections, it is possible to divide the frequency spectrum $0 < \nu_i < \infty$ into four general regions, as shown in Fig. 4.13, in order to formalize the design procedure. These regions are:

Region I. For the frequency range $0 < \nu_i < \nu_{0.25}$ where Eq. (4.56) is essentially satisfied use the analog templates, i.e., $\Im \mathbf{P}(j\omega_i) \approx \Im \mathbf{P}_e(j\nu_i)$. Note, depending on the value of T, ν_h may be less than $\nu_{0.25}$.

Region II. For the frequency range $\nu_{0.25} < \nu_i \le \nu_h$ use the w-domain templates $\Im \mathbf{P}_e(j\nu_i)$ to satisfy both the $\boldsymbol{B}_{mo}$ bounds and the $\boldsymbol{B}'_h$- and $\boldsymbol{B}_h$-contours, assuming T is small enough. If T is not small enough and can not be decreased use these templates to satisfy the stability requirement, i.e., satisfy only the $\boldsymbol{B}'_h$- and $\boldsymbol{B}_h$-contours. Depending on the value of T, the value of ν_h may be less than that of $\nu_{0.25}$.

Region III. For the frequency range $\nu_h < \nu_i \le \nu_K$ satisfy only the $\boldsymbol{B}_h$-contours (stability bounds).

Region IV. For the frequency range $\nu_i > \nu_K$ use the w-domain templates to satisfy only the M_L-contours and obtain the corresponding $\boldsymbol{B}_S$-contours.

A good engineering rule-of-thumb, based upon the definition of Region I, is to let $L_{mo1} \approx P_o|_{s=w}$ be the starting loop transmission function, with the appropriate gain adjustment, upon which the final synthesized $L_{mo}(w)$ is constructed. *Only the w-domain location of the s-domain poles of P_o should be used in setting $L_{mo1}(w) = P_o(s)|_{s=w}$.* Thus, from Eq. (4.69),

$$G_1(w) = \frac{L_o(w)}{P_{zo}(w)} \tag{4.80}$$

assuming perfect cancellation, results in a minimal order controller. It should be noted that for *a tracking control system $L_{mo}(w)$ must be a Type-1 or higher transfer function.* For example, from Eq. (4.68), the poles of $P_o(s) = 1/[s(s + 1)]$ are used for the starting L_{mo1}, i.e.,

$$L_{mo1} = \frac{K_{o1}}{w(w + 0.9998)} \tag{4.81}$$

Note: when starting with a Type m $P_o(s)$, in obtaining $L_{mo1}(w)$ by use of a CAD package, the CAD data may not result in m poles of $L_{mo1}(w)$ being *exactly at the origin.* If this occurs, the value of these m poles must be corrected in the CAD program to reflect their true value being *exactly at the origin.*

It should be noted that the discussion dealing with the discrete QFT design procedure is based upon the gain value of $P_{eo}(w)$ being a negative quantity. This results in the gain of $G_1(w)$ being a positive quantity. The sign of the gain value of $P_{eo}(w)$ is a function of the sampling time.[13] If the gain of $P_{eo}(w)$ is a positive quantity the same design procedure is followed, the only difference being that the gain of the resulting $G_I(w)$ is a negative quantity.

4-5.6 $\omega_s = 120$ IS TOO SMALL

The bounds on $\mathbf{L}_{mo}(jv)$ are shown in Fig. 4.11. Is there a practical $\mathbf{L}_{mo}(jv)$ that can satisfy these bounds? In other words, is there an $\mathbf{L}_{mo}(jv)$ that lies on or outside the $\mathbf{B}_{mo}(jv)$ bounds at small values of v but then decreases vs. v fast enough on the right side of the $\mathbf{B}_h(jv)$ contour in order to reach the "bottom" at $v < v_K$? The difficulty is that $d[Lm\ \mathbf{L}_{mo}(jv)]/d[Lm\ v]$ is a function of $\angle \mathbf{L}_{mo}(jv)$ resulting in an average of 20α dB/decade decrease for $\angle \mathbf{L}_{mo}(jv) = -90^o \alpha\ (\alpha = 1,2,\ldots)$. Thus, the more phase lag available for L_{mo}, the faster $|\mathbf{L}_{mo}(jv)|$ can decrease vs. v. However, the $\mathbf{B}_h(jv)$ contour steadily moves to the right vs. v, because of the n.m.p. $A(w)$, until at v_O there is no more phase lag for a decrease of $|\mathbf{L}_{mo}(jv)|$. Hence, it is necessary that $\mathbf{L}_{mo}(jv)$ reach the bottom of this moving barrier (contour), before the value of v reaches the value of v_O.

The following numerical procedure is convenient for determining the minimum ω_s that is required. Work backwards, that is, assign approximately -10° to $\angle \mathbf{L}_{mo}(jv_K)$, where $v_K < v_O$, and the corresponding right-side phase values of $\mathbf{B}_h(jv)$ to $\mathbf{L}_{mo}(jv)$ for $v < v_K$: see Fig. 4.11 for -30° at $v = 66.1$, -45° at $v = 38.2$, and -90° at 22.05. For $v < 22.05$, in Fig. 4.11, "reasonable" values of $-\angle A'(jv)$ [see Eq. (4.78)], based upon loop-shaping experience, are tried; e.g., the right-side phase angles of $\mathbf{B}_h(jv)$ of -120° occurs at $v = 10.3$, of -135.2^o occurs at $v = 4.97$, and similarly the corresponding angles for $v = 1.986, 0.993$ and 0.497 can be determined. As is well known,[1] $\mathbf{L}_{mo}(jv)$ determines $|\mathbf{L}_{mo}(jv)|$ within an arbitrary multiplier, and an algorithm due to Bode,[40] used in Gera and Horowitz,[59] is especially efficient in calculating numerically $|\mathbf{L}_{mo}(jv)|$ for values obtained from Eq. (4.77) for specified values of v_i.

The resulting $Lm\ \mathbf{L}_{mo}(jv)$, so obtained, is shifted vertically on the NC in Fig. 4.11 such that $Lm\ \mathbf{L}_{mo}(jv_K)$ is at point K and checked as to whether it satisfies the other $\mathbf{B}_h(jv_i)$ bounds. It is impossible to do so in this example, as seen in Fig. 4.11, since the bounds in the range of

$$0.497 < v < 22.15 \qquad (0.0065 < p < 0.58)$$

are violated. If $|\mathbf{L}_{mo}(jv)|$ is shifted upward so that it satisfies all the bounds, then it is above the critical point K at $v_K = 142.6\ (p = 3.73)$. Then for $v > v_K$, $\angle \mathbf{L}_{mo} > 0$ in order to satisfy the $\mathbf{B}_h(jv)$ contours for $v > v_K$, so $d|\mathbf{L}_{mo}|/dv > 0$ and $|\mathbf{L}_{mo}|$ must monotonically increase, which is impractical. That is, since the zeros of L_{mo} dominate, they yield a positive angle and a positive dB value for L_{mo}. Hence, $\omega_s = 120$ and, in turn, $\omega_s = 200$ are too small a sampling frequency for this uncertainty problem.

If the original plant is m.p., it is guaranteed that a satisfactory ω_s exists, because the sampled system approaches the continuous system as ω_s approaches infinity (T approaching zero). The challenge is to use the minimum ω_s value that is achievable; the value $\omega_s = 240$ is satisfactory for this example, as seen in Fig. 4.14, where the plot of $Lm\ \boldsymbol{L}_{mo1}(jv)$ satisfies all the bounds. The mechanism whereby a larger value of ω_s does this is seen in Figs. 4.11 and 4.14. The $\boldsymbol{B}_h(jv_i)$ high-frequency contours in these figures, for $v_i \geq 20$, are identical in shape and differ only in the frequency value associated for a given contour; i.e., the corresponding contours are related as follows:

$$\boldsymbol{B}_h(jv_i)_{\omega_s=120} = \boldsymbol{B}_h(j2v_i)_{\omega_s=240} \qquad (4.82)$$

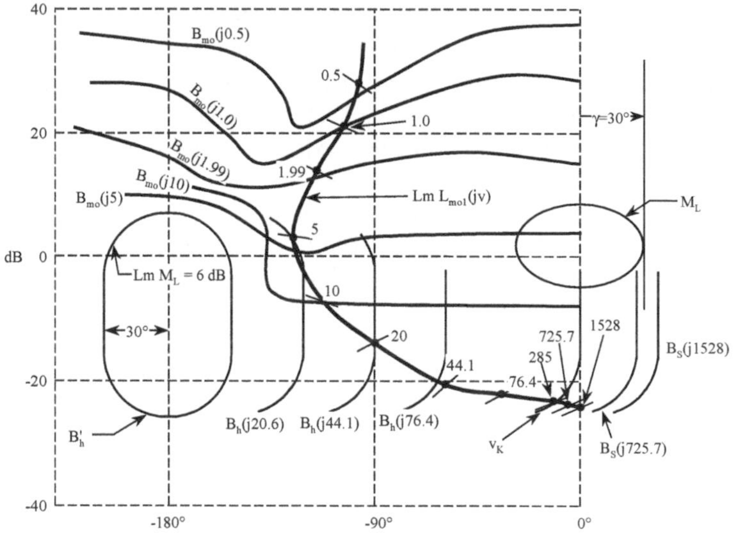

ω	0.5	1	5	10	20	30	40	50
p	0.065	0.0131	0.0655	0.132	0.2679	0.4142	0.5773	1.00
v	0.500	1.000	5.070	10.06	20.47	31.64	44.11	76.39

Fig. 4.14 A satisfactory design: L_{mo1} at $\omega_s = 240$.

The bounds $\boldsymbol{B}_{mo}(jv_i)$, for ω_s = *120* and *240*, are the same for $v_i \le 5.03$ because the low-frequency (ω) bounds are basically invariant with respect to ω. An analysis of these two figures reveals that the same decrease in log magnitude from about *20* dB to *−30* dB is achieved in the range $0.497 \le v_i \le 285$ in Fig. 4.14 instead of in the range $0.497 \le v_i \le 142.3$ in Fig. 4.11; i.e., an extra octave is available to achieve this decrease in log magnitude. In the event a preliminary analog QFT design is accomplished for the plant *P(s),* then a good rule-of-thumb in selecting a suitable sampling frequency is as follows: Select the value of ω_s to be two to three times the value of ω_c that results in $\angle \boldsymbol{L}_o(j\omega_c) = -180^o$. In order to use this rule it is assumed that the value of the analog gain-margin frequency ω_c is a practical or realistic value. For this example, the Bode algorithm[40] provides a satisfactory $Lm\ \boldsymbol{L}_{mo}(jv)$, labeled $Lm\ \boldsymbol{L}_{mo1}$ in Fig. 4.14.

In Fig. 4.14 the synthesized $L_{mo}(s)$ has to satisfy all the bounds $\boldsymbol{B}_{mo}(j\omega_i)$ through $\boldsymbol{B}_{mo}(j\omega_{i+j})$ and yet achieve the necessary dB decrease (Δ dB)in $Lm\ \boldsymbol{L}_{mo}(j\omega)$ in the frequency range ω_i to ω_K. For a S-D control system satisfying both requirements may not be obtainable for a given value of M_L and/or *T*. Increasing the value of M_L and/or decreasing the value of *T* widens the frequency bandwidth from v_i up to v_K. Examples of this situation are illustrated in Figs. 4.11 and 4.14. In order to determine if all requirements can be satisfied for a specified frequency bandwidth let

$$2^d = \frac{v_K}{v_i} \tag{4.83}$$

where d represents the number of octaves between v_i and v_K. One needs to estimate and select the dominant low-frequency bound $\boldsymbol{B}_{mo}(jv_i)$ and a starting point for $Lm\ \boldsymbol{L}_{mo}(jv_i)$ on this bound. Assuming an average phase change of $\theta_{av} = 75,^o$ then the average negative slope is given by

$$\left[\frac{\theta_{av}}{90^o}\right]\left[\frac{6\ dB}{Oct}\right] \tag{4.84}$$

Therefore the estimated decrease in dB (est. dB) is given by

$$\text{est. dB} = \left[\frac{\theta_{av}}{90^o}\right]\left[\frac{6\ \text{dB}}{Oct}\right]d \tag{4.85}$$

If the est. dB > Δ dB then the value of ω_s that has been chosen is satisfactory. Applying Eq. (4.83) to Fig. 4.11 results in

$$2^d = \left[\frac{v_K}{v_i}\right] = \left[\frac{142.6}{0.5}\right]$$

which yields $d = 8.153$ and an est. dB $= 40.76 < 43$ dB $= \Delta$ dB. Thus, the value of $\omega_s = 120$ is too small. If ω_s is increased to 240 it results in $d = 9.153$ and an est. dB $= 45.76$ (> 43 dB $= \Delta$ dB) then from Fig. 4.14:

$$2^d = \left[\frac{v_K}{v_i}\right] = \left[\frac{285.1}{0.5}\right]$$

which indicates that the value of $\omega_s = 240$ is satisfactory.

The numerator and denominator of the synthesized $\boldsymbol{L}_{mo1}(jv)$ w-plane transfer function must be of the same degree. The function numerically obtained in this manner, can be achieved as accurately as desired by a rational function $L_{mo}(w)$,[40] which can however, be of high degree. Hence, there is some trade-off between ω_s and the degree of $L_{mo}(w)$, which can be made at this numerical stage by allowing some over design. For the uncertain plant of Example 2.7, $K \in \{1,10\}$ and $a \in \{1,4\}$, the synthesized loop transmission is

$$\begin{aligned} L_{o1} &= \frac{-0.059559(w+12.987)(w+45.84)(w-76.39)}{w(w+3.056)(w+76.39)} \\ &= -L_{mo1}(w)A(w) = L_{mo1}(w)[A'(w)] \end{aligned} \qquad \textbf{(4.86)}$$

which is sketched in Fig. 4.14. Stability is assured when loop shaping for the m.p. $L_{mo}(w)$ is performed on the NC. When the n.m.p. function $L_o(w)$ is formed the proper sign on its gain term to assure a stable system *can also be ascertained from a polar plot analysis*. That is, the Nyquist stability criterion is applied to $L_o(w)$ in the polar plot domain.

If the disturbance response specification, for a unit-step disturbance input $d(t)$, is $y(\infty) = 0$, then $L_{mo}(w)$ and in-turn, $L_o(w)$ loop transmission functions must have at least two poles at the origin. *This ensures, for a nominal Type 1 P_{eo} plant, a Type 1 system characteristic for T_R and that $Y_D(w)$ will not have a pole at the origin.* Thus, the resulting *pseudo control ratio*

$$\left[\frac{Y_D(w)}{D(w)}\right]_P \qquad \textbf{(4.87)}$$

has a zero at the origin. Equation (4.87) is obtained by dividing the resulting $Y_D(w)$ by the bilinear transformation of the $\mathcal{Z}$-transform for a unit-step forcing function. For a Type m P_{eo} then $L_o(w)$ must have $m + 1$ poles *precisely* at the origin.

4-5.7 ERROR IN THE DESIGN

There is always the problem of choosing sufficient ω_i values, for the analog system, or v_i values, for the discrete system, especially at large values of ω or v. The poles and zeros of $P(s)$ or $P_e(w)$ are important, because the plant templates do not remain at almost vertical lines (giving the $\boldsymbol{B}_h$ type of bounds for all $\omega >$ some ω_h), until a frequency several times larger than the largest pole or zero of P_e. It is usually easy to detect the largest pole or zero in the continuous plant; as well as the largest pole in $P_e(w)$, because the poles of $P(s)$ map directly into $P_e(w)$. However, there may be large shifts in the zeros of $P_e(w)$, as was the case for this example, with a far-off zero at $w = -4377$ for the nominal plant $P_{eo}(w)$ of Eq. (4.70).

The result is almost vertical templates and hence $\boldsymbol{B}_h$-type of bounds from, approximately, $v = v_h = 20/0.9998 = 20/$[the pole -0.9998 of Eq. (4.71)] to $v = 4377/20 = 219 \approx v_K$ [the zero -4377 of Eq. (4.71)], (allowing a factor of 20 for the effect of a pole or zero on the phase). Note that 0.9998 is the smallest magnitude and 4377 is the largest magnitude of a pole or zero in Eq. (4.71). Since the templates $\Im \boldsymbol{P}_e(jv_i)$ broaden out again for $v > 219$, as shown in Fig. 4.15a, it is necessary to obtain the more stringent bounds B_S in Fig. 4.15b for $v = 1528, 15280, 1528000, \ldots$. These bounds, for $v_i > v_K$, are obtained by shifting the M_L contour to the right corresponding to the template frequencies, up to its maximum shift of 180°. That is, the maximum shift occurs when $-\angle \boldsymbol{A}'(jv) = 180^\circ$ as shown in Fig. 4.12. For $v > v_K$, the templates are translated to the right so that they are always tangent to their corresponding M_L contour in order to obtain sufficient points to determine the corresponding $\boldsymbol{B}_S$ bounds. Note, the templates are measured from the shifted M_L contours, not the shifted $\boldsymbol{B}_h$ contour. These high frequency M_L bounds are overlooked for the first trial design, with the "universal" type $\boldsymbol{B}_h$ assumed in Fig. 4.14 for all $v > 20.6$. The first design stops at point W in Fig. 4.15b, definitely violating the bounds for $v \geq 15{,}280$. The bound is moderately violated at $v = 15{,}280$ and violated by approximately ≈ 12 dB at $v = 152{,}800$, enough to cause instability. The L_{mo1} is augmented by a high frequency pole-zero pair to correct the situation. To determine the required pole-zero pair(s) the very-high-frequency (v.h.f.) templates of Fig. 4.15a are utilized to determine the $\boldsymbol{B}_S$ bounds shown in Fig. 4.15b. As illustrated in Fig. 4.16, these templates are shifted around the corresponding M_L contours to obtain the corresponding $\boldsymbol{B}_S$ contours. The pole-zero pair(s) that are selected must result in an $L_{mo2}(w)$ whose corresponding v.h.f. points lie on or below the corresponding $\boldsymbol{B}_S$ bounds. Thus

$$L_{mo2}(w) = \frac{1 + w/140566}{1 + w/8785.4} L_{mo1}(w) \tag{4.88}$$

which satisfies the $\boldsymbol{B}_S(jv)$ boundaries, as seen in Fig. 4.15b. Thus the final loop transmission function, in the w plane, is

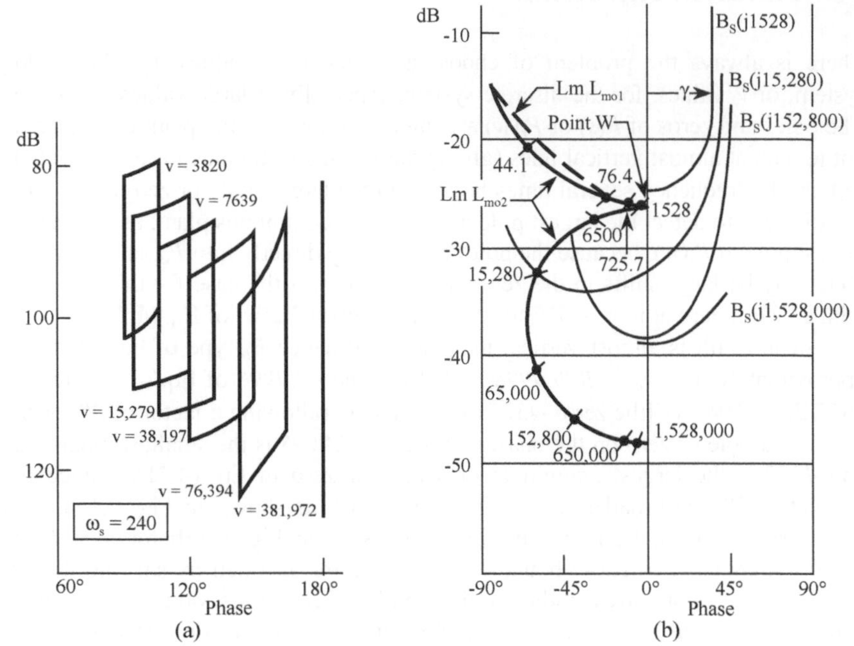

Fig. 4.15 (a) Neglected plant templates at higher v; (b) reason for unstable L_{mo1} design and successful L_{mo2}.

$$
\begin{aligned}
L_{o2}(w) &= -L_{mo2}(w)A(w) = L_{mo2}(w)A'(w) \\
&= G_1(w)P_{zo}(w)
\end{aligned}
\tag{4.89}
$$

Once $L_o(w)$ is determined then the w-plane controller transfer function is determined from Eq. (4.89), i.e.,

$$
G_1(w) = \frac{L_{o2}(w)}{P_{zo}(w)} \tag{4.90}
$$

where, for $\omega_s = 240$,

$$
P_{zo}(w) = \frac{-0.7476\times10^{-6}(w-76.39)(w+17{,}510)}{w(w+0.9999)} \tag{4.91}
$$

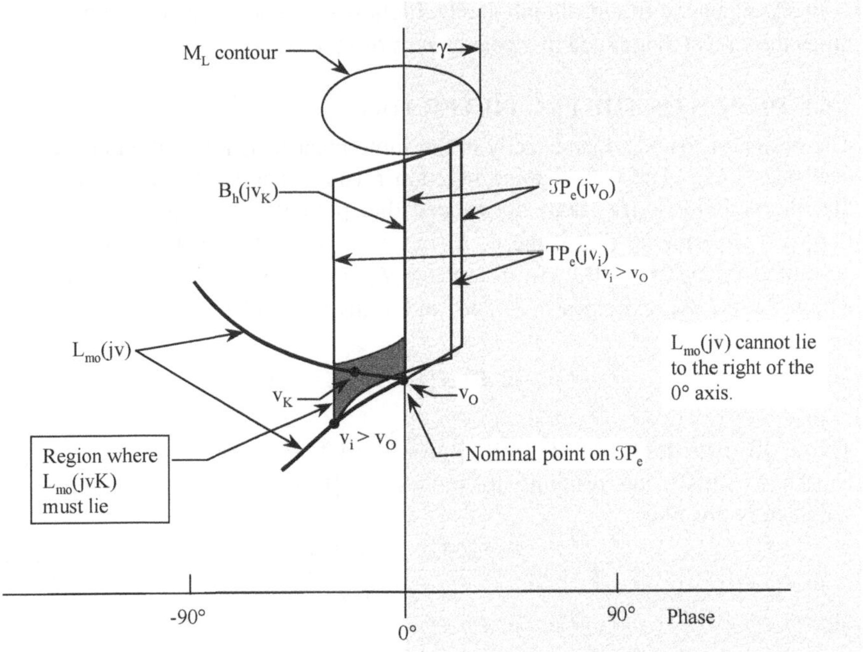

Fig. 4.16 Determination of the v.h.f. $\boldsymbol{B_S}$ bounds.

utilizing Eqs. (4.86), (4.88), (4.89), and (4.90) the controller transfer function is

$$G_1(w) = \frac{4979.19(w + 0.9999)(w + 12.987)(w + 45.84)(w + 140566)}{(w + 3.056)(w + 76.39)(w + 8785.4)(w + 17510)} \tag{4.92}$$

Substituting Eq. (4.51) into Eq. (4.88) yields

$$L_{mo2}(z) = \frac{0.10634(z - 0.7094)(z - 0.25)(z + 0.999891)}{(z - 1)(z - 0.9231)(z + 0.98276)} \tag{4.93}$$

The resulting loop shaping function is

$$L_o(z) = A(z)L_{mo2}(z)$$

$$= \frac{0.10634(z - 0.7094)(z - 0.25)(z + 0.99891)}{z(z - 1)(z - 0.9231)(z + 0.98276)} \tag{4.94}$$

where $A(z) = 1/z$.

As a general rule-of-thumb, let the highest frequency of concern for v_i be *10* times the largest magnitude of a pole or zero of $P_e(w) \in \mathcal{P}$.

4-5.8 DESIGN OF THE PRE-FILTER $F(w)$

The design of $F(w)$ is done exactly in the same manner as for continuous systems (see Sec. 3.15). The loop-shaping function $L_o(w) = A(w)L_{mo2}(w)$ only guarantees that the variation $\delta_R(jv_i)$ does not exceed that permitted by Eqs. (4.63) through (4.65). For example, if at some v_i, $\boldsymbol{B}_L(jv_i) = -6$ dB and $\boldsymbol{B}_U(jv_i) = 0$ dB then the permitted $\delta_R(jv_i) = 6$ dB. An acceptable $\boldsymbol{L}_o(jv)$ design may result in $|\boldsymbol{T'}_R(jv_i)|$, where $T_R = FT'_R$, extremes of *-4* dB and *1* dB with $\delta_R(jv_i) = 5$ dB. It is then necessary that

$$-2 \text{ dB} \leq |F(jv_i)| \leq -1 \text{ dB}$$

The *1* dB over design of $|\boldsymbol{T'}_R(jv_i)|$ allows *1* dB tolerances for $|\boldsymbol{F}(jv_i)|$. $F(w)$ is chosen to satisfy the resulting tolerances on $|\boldsymbol{F}(jv_i)|$. For this example, the satisfactory pre-filter

$$F(w) = \frac{0.033329\,(w + 229.2)}{w + 7.639} \tag{4.95}$$

is obtained. Note that the pre-filter $F(w) = 7.639/(w + 7.639)$ satisfies the w-domain but not the z-domain performance specifications. The reason for this is: in applying the bilinear transformation in order to transform this *simple* $F(w)$ to the z-domain results in an $F(z)$ whose numerator and denominator orders are the same. In other words, due to the transformation the zero of $F(z)$ will alter the desired QFT pre-filter frequency domain characteristics. In order to prevent this from occurring, it is necessary after the *simpler* $F(w)$ is synthesized, according to the QFT design procedure, that enough *non-dominant w-domain zeros are inserted into the simple* $F(w)$ *in order to create a pre-filter transfer function whose numerator and denominator are of the same order.* This was done in obtaining Eq. (4.95) whose zero at *-229.2* is non-dominant. By use of Eq. (4.51, the pre-filter in the z-domain is

$$F(z) = \frac{0.1212(z + 0.5)}{z - 0.8182} \tag{4.96}$$

Utilizing Eqs. (4.47) and (4.94), where

$$P_{eo} = P_{e1}(z) = \frac{3.397\times 10^{-4} z(z + 0.99131)}{(z - 1)^2 (z - 0.97416)} \tag{4.97}$$

yields

$$G_1(z) = \frac{313.05(z-0.7094)(z-0.25)(z+0.99891)(z-0.97416)}{z(z-0.9231)(z+0.98276)(z+0.99131)} \quad \textbf{(4.98)}$$

Since the zero at -0.99891 and the pole at -0.99131 are very close to one another, on the negative real-axis, then Eq. (4.98) can be approximated by

$$G_1(z) = \frac{313.05(z-0.7094)(z-0.25)(z-0.97416)}{z(z-0.9231)(z+0.98276)} \quad \textbf{(4.99)}$$

The poles of $G_l(z)$ at -0.98276 and -0.99131, since they are close to the unit (U.C.), can present a stability problem when $L_\iota(z) = G_l(z)P_{z_\iota}(z)$ is formed for all $\iota = 1,2,\ldots,J$ cases. The same situation can occur when the controller of Eq. (4.98) is utilized. The characteristic equation, $Q_\iota(z) = 1 + L_\iota(z)$ should be obtained in order to ascertain that a stable system exists for all J cases. This is illustrated in the next section. One must also ascertain that the Dahlin effect, as discussed in Ref. 13, is not a problem.

4-6 SIMULATION

As stressed in the previous sections, the essence of the QFT technique is the synthesis or loop shaping of the optimal loop transmission function. In applying the bilinear transformation to $G_l(w)$, to obtain $G_l(z)$, can result in significant warping of some of the poles and zeros of G_l. This "warping" is due to some pole and zeros of $G_l(w)$ lying outside the good Tustin (bilinear transformation) region of Fig. 4.3. This warping may result in a significant degradation of the desired loop shaping characteristics, i.e., the violation of the $\boldsymbol{B}_o$, $\boldsymbol{B}_h$, and $\boldsymbol{B}_S$ bounds in the z-domain. Whether warping is significant or not can be determined by obtaining the Bode plots of G_l(w) and $G_l(z)$. If the Bode plots, within the BW determined by where the -24 dB value occurs on $\boldsymbol{B}_L$ in Fig. 4.9 lie essentially on top of one another, then warping is insignificant with respect to satisfying the loop-shaping requirements in the z domain. The controller Bode plots of $G_l(w)$ and $G_l(z)$, for this example, within the BW of $0 \le \omega \le 10$ lie essentially on top of one another indicating insignificant warping. There is excellent correlation between these plots up to approximately ≈ 40 rad/sec.

A computer synthesis and simulation flow chart is presented in Appendix E. The pertinent w-domain transfer functions for the $\iota = 1,2,\ldots,J$ plants necessary to obtain the tracking control ratio $T_{R_\iota}(w)$, based upon Eqs. (4.39) through (4.49), and (4.87), are:

$$P_{e_\iota}(s) = P_\iota(s)D(s) \quad \textbf{(4.100)}$$

$$P_{e_\iota}(w) = P_\iota D(w) \tag{4.101}$$

$$P_{z_\iota}(w) = \frac{2wP_{e_\iota}(w)}{w + 2/T} \tag{4.102}$$

$$G(w) = \frac{2wG_1(w)}{w + 2/T} \tag{4.103}$$

$$L_\iota(w) = G(w)P_{e_\iota}(w) = P_{z_\iota}(w)G_1(w) \tag{4.104}$$

$$T_{R_\iota}(w) = \frac{F(w)L_\iota(w)}{1 + L_\iota(w)} \tag{4.105}$$

$$T_{R_\iota}(w) = \frac{F(w)L_\iota(w)}{1 + L_\iota(w)} \tag{4.106}$$

Substituting Eq. (4.104) into Eq. (4.105) yields

$$T_{R_\iota}(w) = F(w)\left[\frac{G_1(w)P_{z_\iota}(w)}{1 + G_1 P_{z_\iota}(w)}\right] \tag{4.107}$$

The pseudo disturbance control ratio $T_{D_\iota}(w)$, Eq. (4.87), is obtained based upon the *unit-step disturbance function* $d(t) = u_{-1}(t)$ and the following additional *w*-domain transfer functions:

$$Y_D(w) = \frac{P_\iota D(w)}{1 + L_\iota(w)} = \frac{P_\iota D(w)}{1 + P_{z_\iota}(w)G_1(w)} \tag{4.108}$$

$$D(w) = \frac{w + 2/T}{2w} \tag{4.109}$$

Note that Eq. (4.109) is the bilinear transformation of the $\mathcal{Z}$-transfer function for a unit-step function, $z/(z-1)$.

Thus, based upon these equations the pseudo disturbance control ratio is

$$T_{D_\iota}(w) = \frac{Y_{D_\iota}(w)}{[D(w)]_P} = \frac{1}{D(w)}\left[\frac{P_\iota D(w)}{1 + P_{z_\iota}(w)G_1(w)}\right] \tag{4.110}$$

Since, from Eqs. (4.102) and (4.109),

$$\left[\frac{P_\iota D(w)}{D(w)}\right] = \left[\frac{2w}{w+2/T}\right] P_\iota D(w) = P_{z_\iota}(w) \tag{4.111}$$

then Eq. (4.109) can be expressed as follows:

$$T_{D_\iota}(w) = \left[\frac{P_{z_\iota}(w)}{1+P_{z_\iota}(w)G_1(w)}\right] \tag{4.112}$$

From Eqs. (4.44) and (4.47), the z-domain tracking control ratio can be expressed as

$$T_{R_\iota}(z) = \left[\frac{F(z)L_\iota(z)}{1+L_\iota(z)}\right] = \left[\frac{F(z)G_1(z)P_{z_\iota}(z)}{1+P_{z_\iota}G_1(z)}\right] \tag{4.113}$$

The z-domain pseudo-disturbance control ratio, based upon Eqs. (4.40), (4.46), and (4.47), and $D(z) = z/(z-1)$, can be expressed as

$$\begin{aligned} T_{D_\iota}(z) &= \left[\frac{Y_{D_\iota}(z)}{D(z)}\right]_P = \left[\frac{z-1}{z}\right]\left[\frac{P_\iota D(z)}{1+P_{z_\iota}(z)G_1(z)}\right] \\ &= \left[\frac{[(z-1)/z]P_{e_\iota}(z)}{1+P_{z_\iota}(z)G_1(z)}\right] = \left[\frac{P_{z_\iota}(z)}{1+P_{z_\iota}(z)G_1(z)}\right] \end{aligned} \tag{4.114}$$

where

$$P_{z_\iota}(z) = \frac{z-1}{z}P_{e_\iota}(z)$$

The w-domain tracking step responses for five extreme cases are shown in Fig. 4.17. The w-domain figures of merit (FOM) shown in Tables 4.2 and 4.3 satisfy the specifications as represented by the boundaries in Fig. 4.14. The Bode plots for these cases all lie within the area bounded by the $\boldsymbol{B}_U$ and $\boldsymbol{B}_L$ bounds of Fig. 4.9. Since the specifications are met, the bilinear transformation of Eq. (4.51) is applied to Eq. (4.92) to obtain the z-domain controller transfer function of Eq. (4.98). The poles and zeros of $L_\iota(z) = P_{z_\iota}(z)G_l(z)$, for all five cases are shown in Table 4.4. A plot of the poles and zeros of $L_\iota(s)$ are shown in Fig. 4.18. Based upon the root-locus analysis of $L_\iota(z) = -1$, see Fig. 4.19, for all five cases, and the resulting characteristic equations $Q_\iota(z)$, it is determined for case *5* that one of the closed-loop poles is at *–1.008*. This unstable pole, see Fig. 4.19c, is a result of the $G_l(z)$ controller pole p_{g_y} at *–0.98276* and the plant zero $z_{z_5}(z)$ at *–0.9657*. By trial

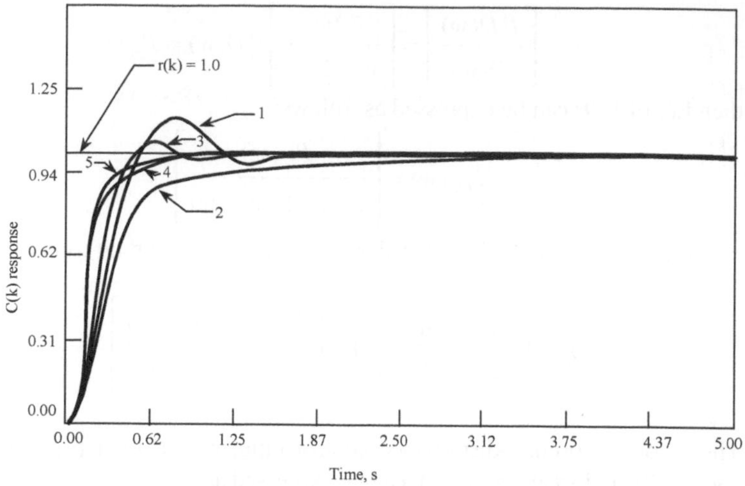

Fig. 4.17 Step responses for:

Case	1	2	3	4	5
a	1	4	2	2	4
k	1	1	3	10	10

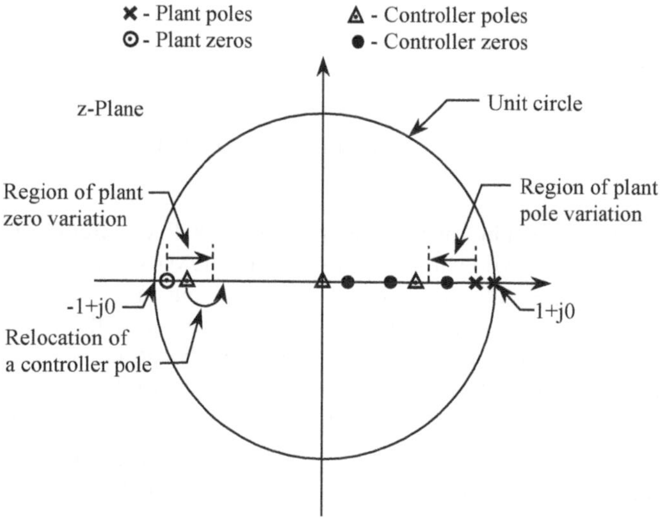

Fig. 4.18 A root-locus pole-zero location resulting in an unstable system pole.

and error, this controller pole is moved by an incremental amount toward the origin until all factors of $Q_t(z)$ are inside the Unit Circle (U.C.). For this example, this controller pole is replaced by a pole at -0.88 to yield the modified controller

$$G_1(z)_m = \frac{313.05(z-0.7094)(z-0.25)(z-0.97416)}{z(z-0.9231)(z+0.88)} \quad \textbf{(4.115)}$$

As Tables 4.2 and 4.3 indicate, the w-domain simulation results are essentially duplicated by the z-domain simulation utilizing the modified controller $G_1(z)_m$. Various CAD packages (Appendices C through E) include both analog and the discrete MISO system QFT routines which expedite the design process.

TABLE 4.2 Tracking FOM

Case (plants)	Domain	Controller	M_p	t_p s	t_s s
1	w	$G_1(w)$	1.139	0.664	1.38
	z	$G_1(z)$	1.14	0.6280	0.97
		$G_1(z)_m$	1.14	0.6283	1.335
2	w	$G_1(w)$	1.000	---	2.893
	z	$G_1(z)$	1.000	---	2.906
		$G_1(z)_m$	1.000	---	2.801
3	w	$G_1(w)$	1.016	0.451	0.353
	z	$G_1(z)$	1.017	0.445	0.340
		$G_1(z)_m$	1.013	0.44506	0.3403
4	w	$G_1(w)$	1.000	---	0.478
	z	$G_1(z)$	1.000	---	0.471
		$G_1(z)_m$	1.000	---	0.4711
5	w	$G_1(w)$	1.000	---	0.609
	z	$G_1(z)$	Unstable	---	---
		$G_1(z)_m$	1.000	---	0.6021

TABLE 4.3 Disturbance FOM#

Domain	*Controller*	**Case (Plant)**				
w	$G_1(w)$	**1**	**2**	**3**	**4**	**5**
		0.08619	0.08618	0.08665	0.08619	0.08618
z	$G_1(z)$	0.08648	0.08648	0.8648	0.08648	unstable
	$G_1(z)_m$	0.0820	0.0820	0.0820	0.0820	0.0820

#All responses are over-damped; only final values are shown.

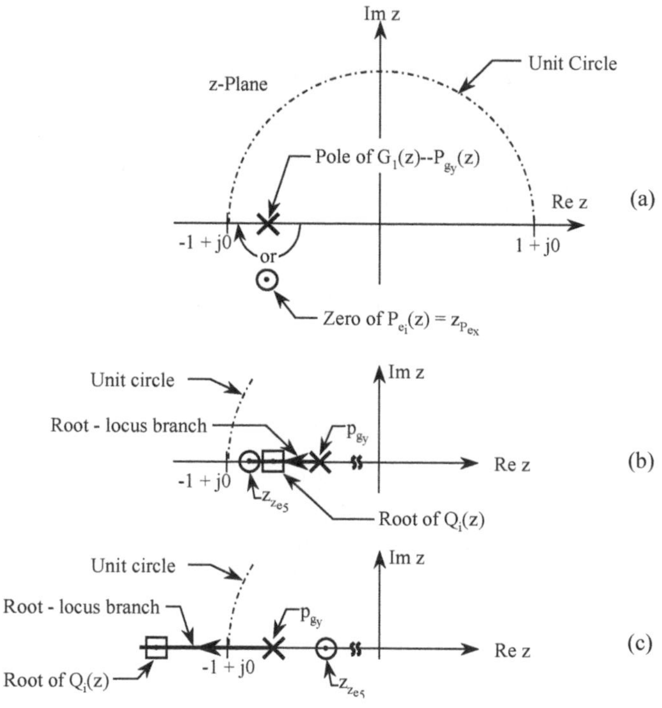

Fig. 4.19 Analysis of root-locus plot of $L(z) = G_I(z)P_{e_1}(z) = -1$: (a) pole-zero locations determining system stability; location yielding: (b) a stable system and (c) an unstable system.

TABLE 4.4 The poles and zeros (all real) of $L_\iota(z)$

	Case ι				
Item	**1**	**2**	**3**	**4**	**5**
Zeros	0.99131	0.9657	−0.9827	−0.99131	−0.9657
			0.7094		
			0.2500		
			0.97416		
Poles	0.97416	0.9006	0.9490	0.97416	0.9006
			0		
			1		
			0.9231		
			−0.98276		

4-7 BASIC DESIGN PROCEDURE FOR A MISO SAMPLED-DATA CONTROL SYSTEM

The design procedure of Chapter 3 is repeated in this section but amplified and modified, where necessary, to be applicable to S-D MISO systems. In order to minimize the warping effect, as discussed in the previous section, the smallest value of sampling time allowable should be selected. The following step numbers, in general, correspond to those of Chapter 3.

1. Synthesize the T_{R_U} and the T_{R_L} models in the s domain.

(a) The $T_{R_U}(s)$ model usually contains one zero and two poles thus it has an excess of poles over zeros of one. Since no ZOH device is involved then the $\mathcal{Z}$-transform of $T_{R_U}(s)$ can not be taken. If Eq. (4.14) is applicable, then the w-domain model is obtained as follows:

$$T_{R_U}(w) = T_{R_U}(s)\,|_{s=w} \tag{4.116}$$

$$T_{R_L}(w) = T_{R_L}(s)\,|_{s=w} \tag{4.117}$$

In general, for the value of T that is chosen and the model that is synthesized Eq. (4.14) will be satisfied. The reader must ascertain that $\alpha^2 << 2$ and $\omega_{sp}T/2 << 0.297$ in order to utilize Eqs. (4.116) and (4.117).

(b) If Eq. (4.16) is not applicable one can (1) select a lower value of T, if this is permissible, or (2) synthesize $T_{R_U}(s)$ that has an excess of poles over zeros of two in order to be able to obtain the $\mathcal{Z}$-transform of $T_{R_U}(s)$. In doing the latter, it is necessary to modify $T_{R_L}(s)$ to have an excess of four poles over zeros in order to maintain the requirement that as ω_i increases in value that $\delta_R(j\omega_i)$ also continually increases. Since the $T_{R_L}(s)$ model, by the QFT-imposed requirement, has at least an excess of two poles over zeros, its $\mathcal{Z}$-transform $T_{R_L}(z)$ can readily be obtained. Since the $\mathcal{Z}$-transform of the modified $T_{R_U}(s)$ and $T_{R_L}(s)$ can be obtained, $T_{R_U}(w)$ and $T_{R_L}(w)$ can be obtained by use of Eq. (4.5).

2. The $T_D(w)$ model can be obtained from $T_D(s)$ as follows:

(a) For the s-domain model that has only an excess of poles over zeros of one then the procedure outlined in Step 1(a) can be repeated to obtain $T_D(w)$.
(b) If Eq. (4.56) is not satisfied then $T_D(s)$ can be modified in the same manner as done for $T_{R_U}(s)$ in Step 1(b).
(c) For a discrete QFT design, it is best to select $T_D(s)$ = a constant value.

3. The frequency intervals for a QFT design may be divided (see Sec. 4-5) as follows:

Region I: Where Eq. (4.56) is satisfied for all J plants the analog QFT templates $\Im \boldsymbol{P}_e(j\omega)$ can be used to determine the appropriate bounds.
Region II: For the frequency range of 0 to v_h, assuming the value of T is small enough, all performance specifications ($\boldsymbol{B}_{mo}$, $\boldsymbol{B}_h$, $\boldsymbol{B}'_h$, and $\boldsymbol{B}_L$) can be satisfied.
Region III: For the frequency range of $v_h < v_i \leq v_K$, only the specifications associated with the $\boldsymbol{B}_h$ contours need to be satisfied.
Region IV: For the frequency range of $v_K < v_i$ only the specification associated with the B_S (M_L) stability contours need to be satisfied.

4. The number of templates that are necessary for the QFT design can be determined as follows:

(a) For the frequency range $0 < v_i < v_h$:

(1) If an analog QFT design has been accomplished for a given system, and if the *low frequency range* ($0 < v_i < v_{0.25}$), Eq. (4.56) is applicable then $\Im \boldsymbol{P}_e(j\omega_i) \approx \Im \boldsymbol{P}_e(jv_i)$.

(2) Whether or not Eq. (4.56) is applicable, the data for the templates. can be determined in the w domain in the same manner as is done for the analog design.

(b) For the frequency range $\nu_h < \nu_i < \nu_K$, the templates in the w domain have the same characteristic as that in the s domain, i.e., they approach a straight vertical line of V dB height.

(c) For the frequency range $\nu_K < \nu_i$, the templates [see Fig. 4.15a] broaden out again due to the “far out” pole or zero of $P_e(w)$ and then again approach a straight line. Sufficient $\boldsymbol{B}_S$ contours need to be obtained in order to ensure a stable system.

As a general rule, templates below ν_K should be approximately an octave apart. Above ν_K, the templates should be between two octaves to a decade apart. This can be modified depending on the evolving shaping characteristic of L_{mo}. See Sec. 4-4.6 for some additional guidelines.

5. Select the plant from the set of plants P_{e_ι} that has the smallest dB value and the largest (most negative) phase lag characteristics as the nominal plant P_{eo}.
6. Once the $\boldsymbol{B}'_h$ -contour has been constructed, based upon the M_L and ν_i values, construct the $\boldsymbol{B}_h(j\nu_i)$-contours corresponding to the ν_i values of the templates.
7. through item 9. The optimal bounds $\boldsymbol{B}_{mo}$ can be determined as follows:

   (a) For $0 < \nu_i < 20/$(smallest magnitude of a pole or zero; other than for the origin of P_{eo}) $= \nu_p$ determine the $\boldsymbol{B}_R$, $\mathbf{B}_D$, and $\boldsymbol{B}_{mo}$ bounds in the same manner of the analog QFT design.

   (b) For frequencies greater than ν_p, the $\boldsymbol{B}_h(j\nu_i)$-contours are determined by the appropriate angular shift to the right due to the a.p.f. characteristic for $\nu_p < \nu_i < \nu_K$.

10. For $\nu_i > \nu_K$, the stability bounds $\boldsymbol{B}_S$ are determined by using the corresponding templates and applying the appropriate shift to the right of the M_L contour due to the a.p.f. characteristic. The template is translated right or left while it is kept tangent to the corresponding M_L contour. For the selected tangent points, the nominal point on the template yields points for the corresponding $\boldsymbol{B}_S$ bound.
11. Synthesize $L_{mo}(w)$ in the same manner as done for the analog QFT design.

   (a) First trial design: Synthesize $L_{mo1} = P_o(s)|_{s=w}$ for the frequency range $0 < \nu_i < \nu_K$.

   (b) Second trial design: synthesize

$$L_{mo2} = [L_{mo1}]\left[\frac{(\ )\cdots(\ )}{(\ )\cdots(\ )}\right] \tag{4.118}$$

to satisfy the high-frequency ($v_i > v_K$) $\boldsymbol{B}_S$-contour bounds on their left or bottom sides [see Fig. 4.15a]. Once a satisfactory $L_{mo}(w)$ has been synthesized then the controller transfer function is obtained from Eq. (4.69), i.e.,

$$G_1(w) = (w + 2/T)\frac{L_o(w)}{2wP_{eo}(w)} \tag{4.119}$$

or

$$G(w) = \frac{L_o(w)}{P_{zo}(w)} = \frac{-L_{mo}(w)A(w)}{P_{eo}(w)} \tag{4.120}$$

where $A(w) = -A'(w)$.

12. Synthesize $F(w)$ in the same manner as for the analog QFT design employing the templates and the plot of $L_o(jv)$. *Remember $F(w)$ must be equal order over equal order.*
13. Simulation is first accomplished in the w domain in order to ensure that all system performance specifications have been satisfied. If the w-domain design is not satisfactory then do another w-domain design to try to achieve the desired performance. Once a satisfactory w-domain design has been achieved, transform $G(w)$ into the z-domain by using the bilinear transformation of Eq. (4.6) to obtain $G(z)$. In order to validate that the desired loop shaping that has been achieved in the w domain has been maintained in the z domain, it is necessary to validate that the Bode plots $Lm\ \mathbf{G}(jv_i)$ and $\angle \mathbf{G}(jv_i)$ *vs.* v_i are essentially the same as the Bode plots of $Lm\ \mathbf{G}(z_i)$ and $\angle \mathbf{G}(z_i)$ *vs.* z_i in the frequency range of $0 < \omega_i < \omega_s/2$ where $z = e^{j\omega_i T}$. If these plots are not "reasonably" close to one another, then warping has been sufficient enough to degrade the desired z-domain loop shaping characteristics. If this occurs, it will be necessary to modify $G(z)$ until the z-domain Bode plots of $G(z)$ are essentially the same as those for $G(w)$. Once a satisfactory $G(z)$ has been achieved, a discrete-time domain simulation is performed in order to validate that the desired S-D control system performance specifications have been achieved. Before proceeding with the z-domain simulation, the factors of the characteristic equation $Q_i(z)$

should be determined to ascertain that stable responses are achievable for all of the J cases as discussed in Sec. 4-6.

4-8 QFT TECHNIQUE APPLIED TO THE PSEUDO-CONTINUOUS-TIME (PCT) SYSTEM

As noted in the preceding sections, the n.m.p. characteristic of a w-domain plant transfer function requires the use of an a.p.f. in order to apply the QFT technique. The resulting modification of the m.p. analog QFT design procedure of Chapter 3, in order to take into account the use of this filter, results in a more involved w-domain QFT design procedure. When the requirements in Sec. 4-2.3 for a *pseudo-continuous-time* (PCT) representation of a S-D system are satisfied and if $P(s)$ is m.p., then the simpler analog QFT design procedure of Chapter 3 can be applied to the m.p. PCT system. This PCT design approach is also referred to as a DIG design method. Using the minimum practical sampling time allowable enhances the possibility for a given m.p. plant to satisfy these requirements.

4-8.1 INTRODUCTION TO PSEUDO-CONTINUOUS-TIME SYSTEM DIG TECHNIQUE[13]

The DIG method of designing a S-D system, in the complex-frequency s plane, requires a satisfactory pseudo-continuous-time (PCT) model of the S-D system. In other words, for the S-D system of Fig. 4.20, the sampler and the ZOH units must be approximated by a linear continuous-time unit $G_A(s)$, as shown in Fig. 4.21c. The DIG method requires that the dominant poles and zeros of the PCT model should lie in the shaded area of Fig. 4.3. for a high level of correlation with the S-D system. To determine $G_A(s)$, first consider the frequency component $\boldsymbol{E}^*(j\omega)$ representing the continuous-time signal $\boldsymbol{E}(j\omega)$, where all its sidebands are multiplied by $1/T$ [see Eq. (6.9) of Ref. 13]. Because of the low-pass filtering characteristics of a S-D system, only the primary component needs to be considered in the analysis of the system. Therefore, the PCT approximation of the sampler of Fig. 4.20 is shown in Fig. 4.21b.

Using the first-order Pade′ approximation (see Ref.13), the transfer function of the ZOH device, when the value of T is small enough, is approximated as follows:

$$G_{zo}(s) = \frac{1 - e^{-Ts}}{s} \approx \frac{2T}{Ts + 2} = G_{pa}(s) \qquad \textbf{(4.121)}$$

Thus, the Pade′ approximation $G_{pa}(s)$ is used to replace $G_{zo}(s)$ as shown in Fig. 4.21*a*. This approximation is good for $\omega_c \le \omega_s/10$, whereas the second-order approximation is good for $\omega_c \le \omega_s/3$ (Ref. 60). Therefore, the sampler and ZOH units of a S-D system are approximated in the PCT system of Fig. 4.21*c* by the transfer function

$$G_A(s) = \frac{1}{T} G_{pa}(s) = \frac{2}{Ts + 2} \tag{4.122}$$

Since Eq. (4.122) satisfies the condition $\lim_{T \to 0} G_A(s) = 1$ it is an accurate PCT representation of the sampler and ZOH units, because it satisfies the requirement that as $T \to 0$ the output of $G_A(s)$ must equal its input. Further, note that in the frequency domain as $\omega_s \to \infty$ ($T \to 0$), then the primary strip becomes the entire frequency-spectrum domain, which is the representation for the continuous-time system.[13]

Note that in obtaining PCT systems for a S-D system the factor 1/T replaces only the sampler that is sampling the continuous-time signal. This multiplier of *1/T* attenuates the fundamental frequency of the sampled signal and all its harmonics are attenuated. To illustrate the effect of the value of *T* on the validity of the results obtained by the DIG method, consider the S-D closed-loop control system of Fig. 4.20 where

$$G_x(s) = \frac{K_x}{s(s+1)(s+5)} \tag{4.123}$$

The closed-loop system performance for three values of *T* are determined in both the *s*- and *z*-domains, i.e., the DIG technique and the *direct* (DIR) *technique* (the *z*-analysis), respectively. Table 4.5 presents the required value of K_x and time-response characteristics for each value of *T*. Note that for $T \leq 0.1$ s there is a high

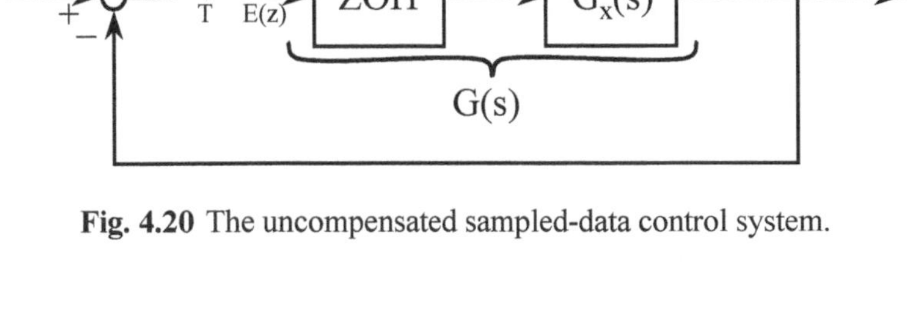

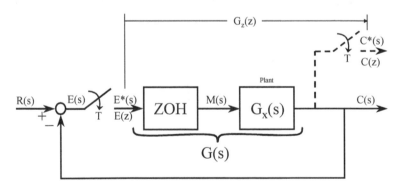

Fig. 4.20 The uncompensated sampled-data control system.

level of correlation between the DIG and DIR models. For $T \leq 1\ s$ there is still a relatively good correlation. (The designer needs to specify, for a given application, what is considered to be "good correlation.")

TABLE 4.5 Analysis of a PCT system representing a sampled-data control system for $\zeta = 0.45$

Method	T, s	Domain	K_x	M_p	T_p, s	T_s, s
DIR	0.01	z	4.147	1.202	4.16	9.53
DIG		s	4.215	1.206	4.11	9.478
DIR	0.1	z	3.892	1.202	4.2-4.3	9.8+
DIG		s	3.906	1.203	4.33^-	9.90+
DIR	1	z	2.4393	1.199	6	13-14
DIG		s	2.496	1.200	6.18	13.76

4-8.2 SIMPLE PCT EXAMPLE

Figure 4.20 represents a basic or uncompensated S-D control system where

Case 1

$$G_x(s) = \frac{K_x}{s(s+1)} \tag{4.124}$$

is used to illustrate the approaches for improving the performance of a basic system. The root-locus plot $G_x(s) = -1$ shown in Fig. 4.22*a* yields, for $\zeta = 0.7071$, $K_x = 0.4767$.

One approach for designing a S-D unity-feedback control system is first to obtain a suitable closed-loop model $[C(s)/R(s)]_T$ for the PCT unity-feedback control system of Fig. 4.21, utilizing the plant of the S-D control system. This model is then used as a guide for selecting an acceptable $C(z)/R(z)$. Thus, for the plant of Eq. (4.124)

$$G_{PC}(s) = G_A(s)G_x(s) = \frac{2K_x / T}{s(s+1)(s+2/T)} \tag{4.125}$$

and for $T = 0.1\ s$,

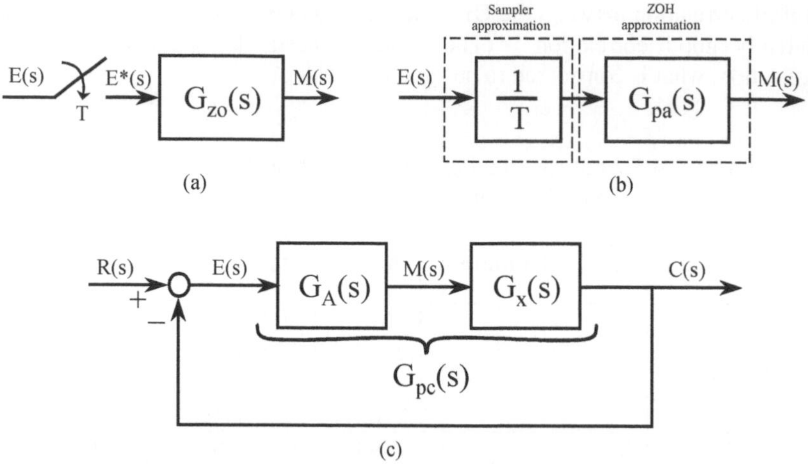

Fig. 4.21 (a) Sampler and ZOH; (b) approximations of the sampler and ZOH; (c) the approximate continuous-time control system equivalent of Fig. 4.20.

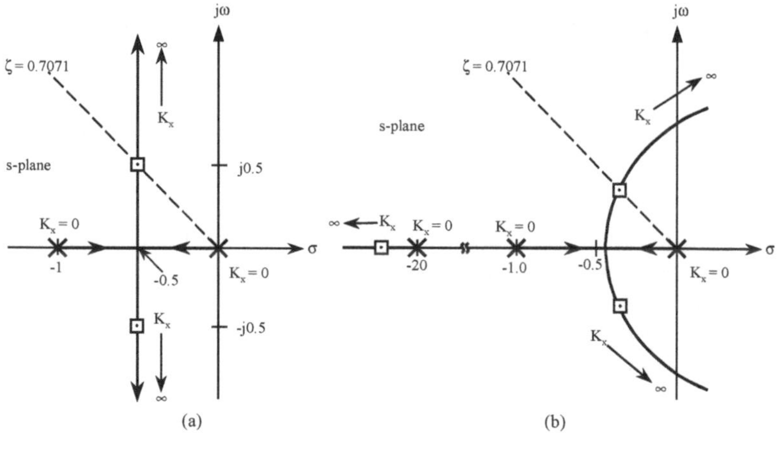

Fig. 4.22 Root locus for (a) Case 1, Eq. (4.124); (b) Case 2, Eq. (4.125).

$$\left[\frac{C(s)}{R(s)}\right]_T = \frac{G_{PC}(s)}{1+G_{PC}(s)} = \frac{20K_x}{s^3 + 21s^2 + 20s + 20K_x} \tag{4.126}$$

The root locus for $G_{PC}(s) = -1$ is shown in Fig. 4.22b. For comparison purposes the root-locus plot for $G_x(s) = -1$ is shown in Fig. 4.22a These figures illustrate the effect of inserting a lag network in cascade in a feedback control system; i.e., the lag characteristic of $G_{zo}(s)$ reduces the degree of system stability, as illustrated by Fig. 4.22, which transformed a completely stable system into a conditionally stable system. Thus, for a given value of ζ, the values of t_p and t_s (and T_s) are increased. Therefore, as stated previously, the ZOH unit degrades the degree of system stability.

For the model it is assumed that the desired value of the damping ratio ζ for the dominant roots is *0.7071*. Thus, for a unit-step input,

Case 2

$$[C(s)]_T = \frac{9.534}{s(s^3 + 21s^2 + 20s + 9.534)}$$
$$= \frac{9.534}{s(s + 0.4875 \pm j0.4883)(s + 20.03)} \qquad \textbf{(4.127)}$$

where $K_x = 0.4767$. The real and imaginary parts of the desired roots of Eq. (4.127), for $T = 0.1\ s$, lie in the acceptable region of Fig. 4.3 for a good Tustin approximation. Note that the pole at *–20.03* is due to the Pade′ approximation for $G_{zo}(s)$ [see Eq. (4.121)].

4-8.3 THE SAMPLED-DATA CONTROL SYSTEM EXAMPLE

The determination of the time-domain performance of the S-D control system of Fig. 4.20 may be achieved by either obtaining the exact expression for $C(z)$ or applying the Tustin transformation to Eq. (4.126) to obtain the approximate expression for $C(z)$. Proceeding with the exact approach first requires the $\mathcal{Z}$-transfer function of the forward loop of Fig. 4.20. For the plant transfer function of Eq. (4.124),

$$G_z(z) = \mathcal{Z}\left[\frac{K_x(1 - e^{-sT})}{s^2(s+1)}\right] = (1 - z^{-1})\mathcal{Z}\left[\frac{K_x}{s^2(s+1)}\right]$$
$$= \frac{K_x[(T - 1 + e^{-T})z + (1 - Te^{-T} - e^{-T})]}{z^2 - (1 + e^{-T} + K_x - TK_x - K_x e^{-T})z + e^{-T} + K_x - K_x(T+1)e^{-T}} \qquad \textbf{(4.128)}$$

Thus, for $T = 0.1$ s and $K_x = 0.4767$,

Case 3

$$G_Z(z) = \frac{0.002306(z + 0.9672)}{(z - 1)(z - 0.9048)} \tag{4.129}$$

or

$$\frac{C(z)}{R(z)} = \frac{0.002306(z + 0.9672)}{(z - 0.9513 \pm j0.04649)} \tag{4.130}$$

The DIG technique requires that the s-domain model control ratio be transformed into a z-domain model. Applying the Tustin transformation to

$$\left[\frac{C(s)}{R(s)}\right]_T = \frac{G_{PC}(s)}{1 + G_{PC}(s)} = F_T(s) \tag{4.131}$$

yields

$$\frac{[C(z)]_{TU}}{[R(z)]_{TU}} = [F(z)]_{TU} \tag{4.132}$$

This equation is rearranged to

$$[C(z)]_{TU} = [F(z)]_{TU}[R(z)]_{TU} \tag{4.133}$$

As stated in Sec. 4-8.1

$$R(z) = \mathcal{Z}[r^*(t)] = \frac{1}{T}[R(z)]_{TU} \tag{4.134}$$

$$C(z) = \mathcal{Z}[c^*(t)] = \frac{1}{T}[C(z)]_{TU} \qquad 4.135)$$

Substituting from Eqs. (4.134) and (4.135) into Eq. (4.132) yields

$$\frac{C(z)}{R(z)} = [F(z)]_{TU} \tag{4.136}$$

Substituting from Eq. (4.135) into Eq. (4.133) and rearranging yields

$$C(z) = \frac{1}{T}[F(z)]_{TU}[R(z)]_{TU} = \frac{1}{T}[\text{Tustin of } F_T(s)R(s)] \quad \textbf{(4.137)}$$

Thus, based upon Eq. (4.136), the Tustin transformation of Eq. (4.126), with $K_x = 0.4767$, results in a Tustin model of the control ratio as follows:

Case 4

$$\frac{C(z)}{R(z)} = \left[\frac{C(z)}{R(z)}\right]_{TU} = \frac{5.672\times 10^{-4}(z+1)^3}{(z-0.9513 \pm j0.04651)(z+6.252\times 10^{-4})} \quad \textbf{(4.138)}$$

Note that the dominant poles of Eq. (4.138) are essentially the same as those of Eq. (4.130) due to the value of T used which resulted in the dominant roots lying in the good Tustin region of Fig. 4.3. The non-dominant pole is due to the *Padé* approximation of $G_{zo}(s)$. In using the exact $\mathcal{Z}$ transformation the order of the numerator polynomial of $C(z)/R(z)$, Eq. (4.130) is one less than the order of its denominator polynomial. When using the Tustin transformation, the order of the numerator polynomial of the resulting $[C(z)/R(z)]_{TU}$ is in general equal to the order of its corresponding denominator polynomial [see Eq. (4.138)]. Thus, $[C(z)]_{TU}$ results in a value of $c^*(t) \neq 0$ at $t = 0$, which is in error based upon zero initial conditions. Table 4.6 illustrates the effect of this characteristic of the Tustin transformation on the time response due to a unit-step forcing function. The degradation of the time response by use of the Tustin transformation is minimal; i.e., the resulting values of the FOM are in close agreement to those obtained by exact $\mathcal{Z}$ transformation. Therefore, the Tustin transformation is a valid design tool when the dominant zeros and poles of $[C(s)/R(s)]_{TU}$ lie in the acceptable Tustin region of Fig. 4.3.

Table 4.7 summarizes the time-response FOM for a unit-step forcing function of (1) the continuous-time system of Fig. 4.21 for the two cases of $G(s) = G_x(s)$ [with $G_A(s)$ removed] and $G(s) = G_A(s)G_x(s)$ and (2) the S-D system of Fig. 4.20 based upon the exact and Tustin expressions for $C(z)$. Note that the value of M_p occurs between 6.4 and 6.5 s and the value of t_s occurs between 8.6 and 8.7 s. The table reveals that:

1. In converting a continuous-time system into a sampled-data system the time-response characteristics are degraded.

TABLE 4.6 Comparison of time responses between $C(z)$ and $[C(z)]_{TU}$ for a unit-step input and $T = 0.1\ s$

k	c(kT)	
	Case 3 (exact), $C(z)$	**Case 4 (Tustin),** $[C(z)]_{TU}$
0	0.	0.5672E-03
2	0.8924E-02	0.9823E-02
4	0.3340E-01	0.3403E-01
6	0.7024E-01	0.7064E-01
8	0.1166	0.1168
10	0.1701	0.1701
12	0.2284	0.2283
14	0.2897	0.2894
18	0.4153	0.4148
22	0.5370	0.5364
26	0.6485	0.6478
30	0.7461	0.7453
34	0.8281	0.8273
38	0.8944	0.8936
42	0.9460	0.9452
46	0.9844	0.9836
50	1.011	1.011
54	1.029	1.028
58	1.039	1.038
60	1.042	1.041
61	1.042	1.042
62	1.043	1.043
63	1.043	1.043
64	1.043	1.043
65	1.043	1.043
66	1.043	1.043
67	1.043	1.043
68	1.042	1.042
85	1.022	1.022
86	1.021	1.021
87	1.019	1.019

TABLE 4.7 Time-response characteristics of the uncompensated system

	M_p	t_p, s	t_s, s	K_1, s^{-1}	Case
			Continuous-time system		
$G(s) = G_x(s)$	1.04821	6.3	8.40 to 8.45	0.4767	1
$G_M(s) = G_A(s)G_x(s)$	1.04342	6.48	8.69		2
			Sampled-data system		
$C(z)$	1.043	6.45	8.75	0.4765	3
$[C(z)]_{TU}$	1.043	6.45	8.75		4

2. The time-response characteristics of the S-D system, using the values of gain obtained from the continuous-time model, agrees favorably with those of the continuous-time model. As may be expected, there is some variation in the values obtained when utilizing the exact $C(z)$ and $[C(z)]_{TU}$.

4-8.4 THE PCT SYSTEM OF FIG. 4.8

The PCT system representing the MISO digital control system of Fig. 4.8 is shown in Fig. 4.23a. Note that *the sampler sampling the forcing function and the sampler in the system's output $y(t)$ are replaced by a factor of $1/T$.* This diagram is simplified to the one shown in Fig. 4.23b and is the structure that is used for the QFT design.

4-8.5 PCT DESIGN SUMMARY

Once a satisfactory controller $D_c(s)$ (see Fig. 4.23b), whose order of the numerator and denominator are, respectively, w_s and n_s, has been achieved, then (a) if n_s is greater than $w_s + 2$ use the exact $\mathcal{Z}$-transform to obtain the discrete controller $D_c(z)$, or (b) if $w_s < n_s < w_s + 2$ use the Tustin transformation to obtain $[D_c(z)]_{TU}$.[13]

A final check should be made before simulating the discrete design; i.e., the Bode plots of $D_c(s)|_{s=j\omega}$ and $D_c(z)|_{z=e^{(j\omega T)}}$ should be essentially the same within the desired BW in order to ascertain that the discrete-time system response characteristics will be essentially the same as those obtained for the analog PCT system response. If the plots differ appreciably it implies that warping has occurred and the desired discrete-time response characteristics may not be achieved (depending on the degree of the warping).

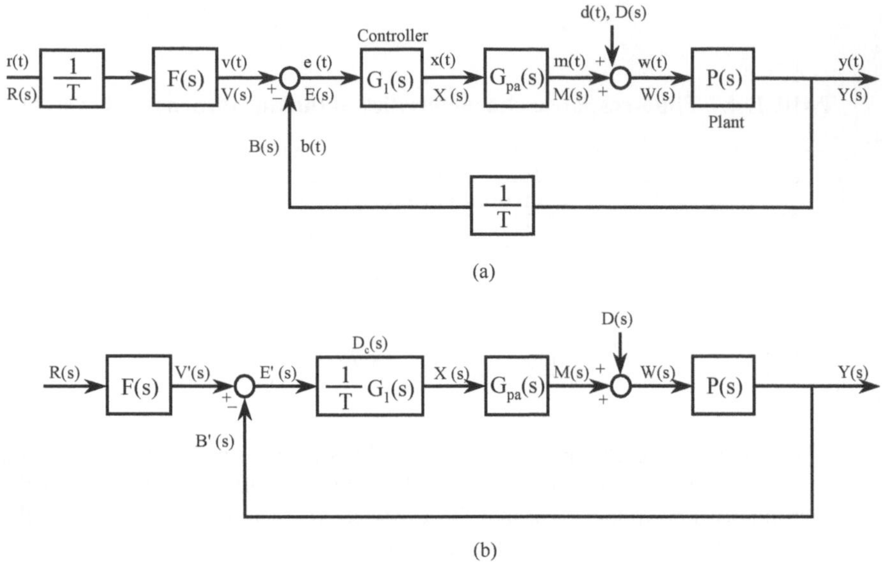

Fig. 4.23 The PCT equivalents of Fig. 4.8.

4-9 APPLICABILITY OF DESIGN TECHNIQUE TO OTHER PLANTS

As stated in Chapter 1, the QFT design techniques can be applied to an *mxm* MIMO control system by transforming this system into an m^2 equivalent MISO systems (see Chapter 5). Each MISO system has two inputs, one desired (tracking) and one unwanted (cross-coupling effects) inputs, and one output. The solution of these MISO systems are guaranteed to satisfy the original MIMO problem. Thus, the MISO loop QFT design techniques of Chapters 3 and 4 can be applied to the QFT design of the m^2 equivalent MISO systems as discussed in the remaining chapters of the text.

4-10 DESIGNING *L*(*w*) DIRECTLY

It is possible to perform the loop shaping of the n.m.p. $L(w)$ by the method of Chapter 3 with the aid of the MIMO QFT CAD package of Appendix C. Since the QFT design is done on the NC, it is necessary to verify all the J characteristic equations in order to determine if the sign of the controller gain is correct for achieving a stable system. This approach is simpler than using the approach of Sec. 4-5.

4-11 SUMMARY

4-11.1 MINIMUM-PHASE, NON-MINIMUM PHASE, AND UNSTABLE $P(s)$

As pointed out earlier, the minimum required sampling frequency ω_s *is very strongly a function of the uncertainty problem, that is, to the extent of the plant uncertainty and of the system's performance tolerances. If the continuous plant* $P(s)$ *is m.p., then the sampled loop transmission function* $L(w)$ *need have only one RHP zero at* $w = 2/T$. *This can be asserted even though the zeros of* $P(s)$, *unlike the poles, do not map directly into the zeros of* $P_e(w)$. *The reason is that the range of the s plane over which the mapping is almost directly so, can be made as large as desired by increasing* ω_s. *The larger* ω_s, *the smaller the importance of the surplus phase lag of the a.p.f.* $A(w) = (-w + 2/T)/(w + 2/T)$, *see Fig. 4.12, because the bandwidth of importance recedes into lower v values (compare the low v bounds of Figs. 4.11 and 4.12.*

However, if $P(s)$ is n.m.p., then it may be impossible to achieve the desired performance for all P in $\mathcal{P}$, precisely as for the continuous systems.[3] One should then first check to verify that the problem is solvable as a QFT continuous system design. If so, then a QFT discrete system design is achievable by use of a large enough value for ω_s. If it is barely solvable in the former, ω_s will have to be very large. If it is not solvable for a continuous system design, it is certainly not solvable as a discrete system design. Unstable but m.p. plants pose no problem, even though the sampled $L(w)$ has RHP poles and the RHP zero at $2/T$. The reason is that with large enough ω_s, the RHP poles are arbitrarily close to the origin in the w-plane.[14] Unstable n.m.p. $P(s)$ do pose a great problem,[29] but it has been shown how time-varying compensation may be used for such plants to achieve stability (but not small system sensitivity) over large plant parameter uncertainity.[14]

4-11.2 DIGITAL CONTROLLER IMPLEMENTATION

This chapter can not be concluded without mentioning the importance of the implementation issue as stressed in Fig. 1.1. In designing the digital controller to achieve the desired control system performance requirements, the engineer must be aware of the factors that play an important role in the implementation of the controller. For example, how is maximum computational accuracy achieved? This particular aspect is discussed in Chapter 9. The reader is referred to the technical literature that discuss the other factors involved in controller implementation; e.g., Ref. 13.

4-11.3 CONCLUSIONS

The basic feedback concepts and quantitative design procedures are the same in the w domain for the sampled-data systems as in the s domain for the continuous systems, if the design is executed in the v domain ($w = jv$) for the former. The price paid is in the loop bandwidth for both, i.e., the desired bandwidth may be achievable in the former by making ω_s large enough. Any practical sampled-data system is n.m.p. in the w domain, but the harmful effect of the RHP zero at $2/T$ can be made arbitrarily small, by making ω_s large enough. The main difference in quantitative design is the need for some experimentation to find $(\omega_s)_{min}$. When applicable, as determined by the plant characteristics and the minimum practical sampling that can be used, the QFT technique can be applied to the PCT system representation of the sampled-data control system. For an m.p. plant, the resulting PCT system is also m.p. eliminating the need for the use of an a.p.f. The QFT design procedure of Chapter 3 can be applied to the PCT system, and the resulting $G_1(s) = D_c(s)$, by means of the Tustin transformation, is transformed into the z- domain to yield $D_c(z)$. Some of the characteristics involved in the use of a bilinear transformation, which must be kept in mind by a control system designer, are stressed in Fig. 4.24. As pointed out in Sec. 4-10, with the aid of the MIMO QFT CAD package of Appendix C it is possible to apply the method of Chapter 3 directly to obtain the n.m.p. $L(w)$ with the requirement to verify all J characteristic equations to assure that the sign of the controller gain is correct for a stable robust system design.

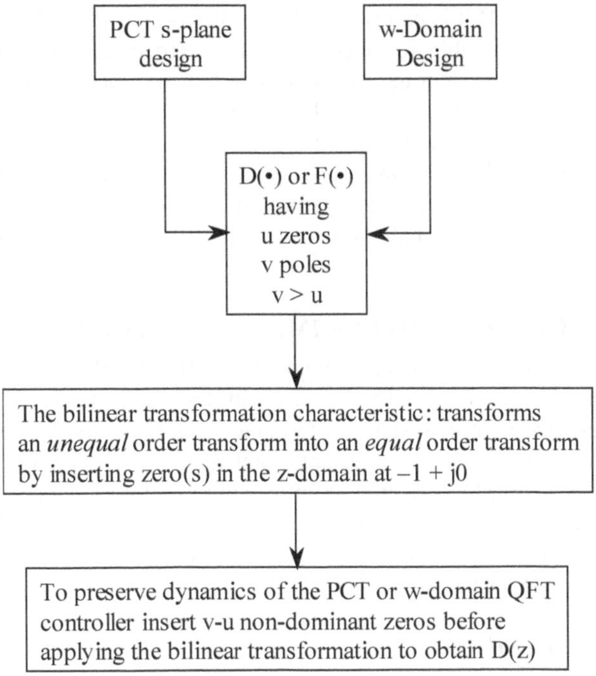

Fig. 4.24 Bilinear transformation characteristics.

5

MULTIPLE-INPUT MULTIPLE-OUTPUT (MIMO) PLANTS: STRUCTURED PLANT PARAMETER UNCERTAINTY

5-1 INTRODUCTION

The previous chapters present the development and design procedures of the quantitative feedback theory (QFT) synthesis technique for MISO control systems. This QFT MISO design technique is extended in this and following chapters to enable the design of MIMO control systems. This extension is based upon the mathematical development in Sec. 5-2 which results in the representation of a MIMO control system by m^2 *MISO equivalent control systems.*

The highly structured uncertain linear-time-invariant (LTI) MIMO plant has the following features:[4]

1. It is quantitative in nature. The extent of parameter uncertainty is defined a priori and so are the tolerances on the system responses to the cross-coupling interaction c_{ij} and to external disturbance functions $(d_e)_{ij}$.
2. The synthesis problem is converted into a number of single-loop problems, in which structured parameter uncertainty, external disturbance, and performance tolerances are derived from the original MIMO problem. The solutions to these single-loop problems are guaranteed to work for the MIMO plant. It is not necessary to consider the system characteristic equation. Any technique may be used for the single-loop design problems – state space, frequency response, or even cut and try.

3. The design is tuned to the extent of the uncertainty and the performance tolerances. The design for a MIMO system, as stated previously, involves the design of an equivalent set of MIS0 system feedback loops. The design process for these individual loops is the same as the design of a MISO system described in Chapters 3 and 4 for analog and digital control systems, respectively. In general, an $m x \ell$ open-loop MIMO plant can be represented in a matrix notation as

$$y(t) = \boldsymbol{P}\boldsymbol{u}(t) \tag{5.1}$$

where $y(t)$ = m-dimensional plant output vector
$\boldsymbol{u}(t)$ = ℓ-dimensional plant input vector
$\boldsymbol{P}$ = $m x \ell$ plant transfer function matrix relating $\boldsymbol{u}(t)$ to $y(t)$

Consider the MIMO plant of Fig 5.1 for which $\boldsymbol{P}$ is a member of the plant set $\mathcal{P}$ ($\boldsymbol{P} \in \mathcal{P}$), the set of all plant transfer functions relating each input to each output.[7,52] When a system has variable parameters, which are known or unknown, the transfer function matrix of the plant may be represented by the associated set $\{p_1, p_2, \ldots\}$. This set of transfer function matrices is contained in $\mathcal{P}$. The design method requires that the uncertainty in $\boldsymbol{P}$ be known or is at least bounded. In any MIMO system with m inputs there are at most m outputs which can be independently controlled.[19] Therefore, the same dimensions, $\ell = m$, are used for both the input and output vectors in the design procedure presented here. If the model defines an unequal number of inputs and outputs, the first step is to modify the model so that the dimensions of the input and output are the same. The system is then defined as being of order $m \times m$.

By using fixed point theory[7,21] Horowitz has shown, (see Sec. 5-6) that the MIMO problem for an $m \times m$ system can be separated into m equivalent single-loop MISO systems and m^2 pre-filter/cross-coupling problems, which are each designed as outlined in Chapter 3.[4,18] The cross-coupling is akin to the disturbance D_1 of Fig. 3.1.

5-2 THE MIMO PLANT

The $\boldsymbol{P}$ matrix may be formed from either the system state space matrix representation or from the system linear differential equations. The state space representation for a LTI MIMO system is:

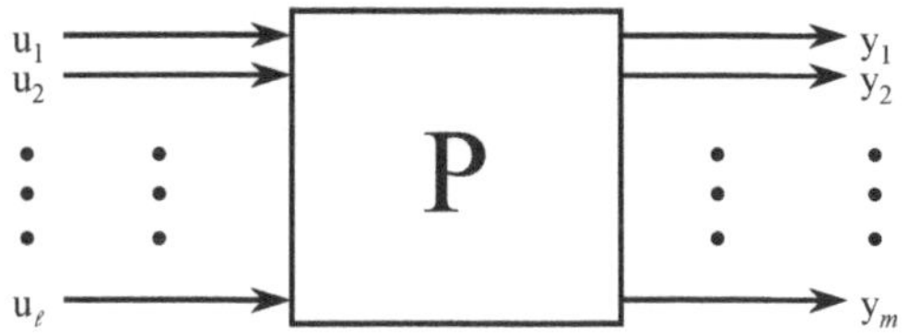

Fig. 5.1 A MIMO plant.

$$\dot{x}(t) = \boldsymbol{A}x(t) + \boldsymbol{B}u(t)$$
$$y(\mathrm{t}) = \boldsymbol{C}x(t) \tag{5.2}$$

where the $\boldsymbol{A}$, $\boldsymbol{B}$, and $\boldsymbol{C}$ are constant matrices and x is an m-dimensional vector. The plant transfer function matrix $\boldsymbol{P}$ is evaluated as

$$\boldsymbol{P}(\mathrm{s}) = \boldsymbol{C}\left[\mathrm{s}\boldsymbol{I} - \boldsymbol{A}\right]^{-1}\boldsymbol{B} \tag{5.3}$$

If the plant model consists of m coupled linear-time-invariant differential equations, the general plant model for a MIMO system with two inputs and two outputs has the form:

$$a(s)\, y_1(s) + b(s)\, y_2(s) = f(s)u_1(s) + g(s)u_2(s)$$
$$c(s)\, y_1(s) + d(s)\, y_2(s) = h(s)u_1(s) + k(s)(s) \tag{5.4}$$

where $a(s)$ through $d(s)$ are polynomials in s, f through k are constant coefficients, $y_1(s)$ and $y_2(s)$ are the outputs, and $u_1(s)$ and $u_2(s)$ are the inputs. In matrix notation the system is represented by:

$$\begin{bmatrix} a(s) & b(s) \\ c(s) & d(s) \end{bmatrix} \mathbf{Y}(s) = \begin{bmatrix} f(s) & g(s) \\ h(s) & k(s) \end{bmatrix} \mathbf{U}(s) \tag{5.5}$$

This is defined as a $2{\times}2$ system. In the general case with m inputs and m outputs, the system is defined as $m{\times}m$. Let the matrix pre-multiplying the output vector $\boldsymbol{Y}(s)$ be $\boldsymbol{D}(s)$ and the matrix pre-multiplying the input vector $\boldsymbol{U}(s)$ be $\boldsymbol{N(s)}$. Thus, Eq. (5.5) may then be written as:

$$\boldsymbol{D}(\mathrm{s})\boldsymbol{Y}(\mathrm{s}) = \boldsymbol{N(s)}\boldsymbol{U}(\mathrm{s}) \tag{5.6}$$

The solution of Eq. (5.6) for the output $\boldsymbol{Y}(s)$, where $\boldsymbol{D}$ must be nonsingular, yields:

$$\mathbf{Y}(s) = \boldsymbol{D}^{-1}(s)\mathbf{N(s)}\mathbf{U}(s) = \mathbf{P}(s)\mathbf{U}(s) \tag{5.7}$$

Thus, the $m{\times}m$ plant transfer function matrix, $\boldsymbol{P}(s)$, is:

$$\mathbf{P}(s) = \boldsymbol{D}^{-1}(s)\mathbf{N}(s) \tag{5.8}$$

This plant matrix $\boldsymbol{P}(s) = [p_{ij}(s)]$ is a member of the set $\mathcal{P} = \{\boldsymbol{P}(s)\}$ of possible plant matrices which are functions of the structured uncertainty in the plant parameters. In practice, only finite set of $\boldsymbol{P}$ matrices are formed, representing the extreme boundaries of the plant uncertainty under varying conditions. Only LTI systems are considered in this text. By obtaining these LTI plants that describe the extremities of the uncertainty (the non-linearity of the system) the QFT technique can achieve,

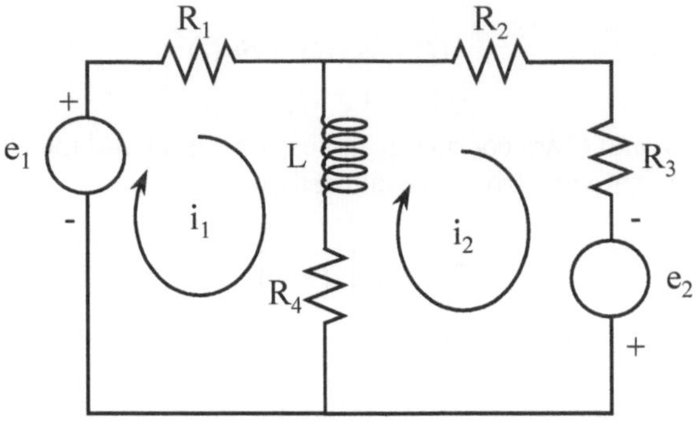

Fig. 5.2 An electrical network.

for many real world nonlinear problems, a satisfactory control system design based on linear techniques.

Example 5.1 The Kirchoff's Voltage law (loop method) is applied to the electrical network of Fig. 5.2 to yield the following differential equations (the "D" operator notation for "d/dt" is used):

$$e_1(t) = (R_1 + R_4 + LD)\, i_1(t) - (R_4 + LD)\, i_2(t) \tag{5.9}$$

$$e_2(t) = -(R_4 + LD)\, i_1(t) + (R_2 + R_3 + R_4 + LD)\, i_2(t) \tag{5.10}$$

The Laplace transform of these equations are

$$(R_1 + R_4 + Ls)\, I_1(s) - (R_4 + Ls)\, I_2(s) = E_1(s) \tag{5.11}$$

$$-(R_4 + Ls)\, I_1(s) + (R_2 + R_3 + R_4 + Ls)\, I_2(s) = E_2(s) \tag{5.12}$$

assuming zero initial conditions. Let

$$\mathbf{U} = \begin{bmatrix} u_1 \\ u_2 \end{bmatrix} = \begin{bmatrix} e_1 \\ e_2 \end{bmatrix}$$

m Inputs

and

$$\mathbf{Y} = \begin{bmatrix} y_1 \\ y_2 \end{bmatrix} = \begin{bmatrix} i_1 \\ i_2 \end{bmatrix}$$

m Outputs

Thus, Eqs. (5.11) and (5.12) are of the form:

$$d_{11}(s)Y_1(s) + d_{12}(s)Y_2(s) = n_{11}(s)U_1(s) + n_{12}(s)U_2(s) \tag{5.13}$$

$$d_{21}(s)Y_1(s) + d_{22}(s)Y_2(s) = n_{21}(s)U_1(s) + n_{22}(s)U_2(s) \tag{5.14}$$

where, for this example, $n_{12} = n_{21} = 0$. These equations are of the general form:

$$d_{i1}(s)Y_1(s) + \ldots + d_{im}(s)Y_m(s) = n_{i1}(s)U_1(s) + \ldots + n_{im}(s)U_m(s) \tag{5.15}$$

where $i = 1, 2, \ldots, m$

$$\mathbf{D} = \begin{bmatrix} d_{11} & d_{12} & \cdots & d_{1m} \\ d_{21} & d_{22} & \cdots & d_{2m} \\ \vdots & \vdots & & \vdots \\ d_{m1} & d_{m2} & \cdots & d_{mm} \end{bmatrix} \tag{5.16}$$

$$\mathbf{N} = \begin{bmatrix} n_{11} & n_{12} & \cdots & n_{1m} \\ n_{21} & n_{22} & \cdots & n_{2m} \\ \vdots & \vdots & & \vdots \\ n_{m1} & n_{m2} & \cdots & n_{mm} \end{bmatrix} \tag{5.17}$$

Thus, Eqs. (5.13) and (5.14) can be expressed as follows:

$$\underset{m\times m}{\mathbf{D}} \underset{m\times 1}{\begin{bmatrix} Y_1 \\ Y_2 \\ \vdots \\ Y_m \end{bmatrix}} = \underset{m\times m}{\mathbf{N}} \underset{m\times 1}{\begin{bmatrix} U_1 \\ U_2 \\ \vdots \\ U_m \end{bmatrix}} \tag{5.18}$$

which is of the form of Eq. (5.6). For this example, based on Eq. (5.7) and where $m = 2$, the expression for $\boldsymbol{Y}$ is:

$$\boldsymbol{Y}(s) = \begin{bmatrix} p_{11}(s) & p_{12}(s) \\ p_{21}(s) & p_{22}(s) \end{bmatrix} \begin{bmatrix} U_1(s) \\ U_2(s) \end{bmatrix} = \begin{bmatrix} p_{11}(s)U_1(s) & p_{12}(s)U_2(s) \\ p_{21}(s)U_1(s) & p_{22}(s)U_2(s) \end{bmatrix} \tag{5.19}$$

Suppose that $p_{11}p_{22} = p_{12}p_{21}$, resulting in

$$|\boldsymbol{P}| = \begin{vmatrix} p_{11} & p_{12} \\ p_{21} & p_{22} \end{vmatrix} = p_{11}p_{22} - p_{12}p_{21} = 0 \tag{5.20}$$

Thus, for this example, $|\boldsymbol{P}|$ is singular. Equation (5.18) yields:

$$\begin{aligned} Y_2 &= p_{21}U_1 + p_{22}U_2 = p_{21}U_1 + \left(\frac{p_{12}p_{21}}{p_{11}}\right)U_2 \\ &= \frac{p_{21}(p_{11}U_1 + p_{12}U_2)}{p_{11}} = \left(\frac{p_{21}}{p_{11}}\right)Y_1 \end{aligned} \tag{5.21}$$

Equation (5.21) reveals that Y_1 and Y_2 are not independent of each other. Thus, they can not be controlled independently, i.e., an uncontrollable system. Therefore, *when **P** is singular the system is uncontrollable.*

From Eq. (5.18) the signal flow graph (SFG) of Fig. 5.3a is obtained which represents a plant with structured plant parameter uncertainty with no cross-coupling effects. Figure 5.3b is the SFG of the compensated MIMO closed-loop control system where the compensator and pre-filter matrices are, respectively:

$$\boldsymbol{G}(s) = \begin{bmatrix} g_{11} & g_{12} \\ g_{21} & g_{22} \end{bmatrix} \qquad F(s) = \begin{bmatrix} f_{11} & f_{12} \\ f_{21} & f_{22} \end{bmatrix}$$

The control ratio matrix $\boldsymbol{T}$ is:

$$\boldsymbol{T} = \begin{bmatrix} t_{11} & t_{12} \\ t_{21} & t_{22} \end{bmatrix} \tag{5.22}$$

where the closed-loop system control ratio $t_{ij} = y_i/r_j$ relates the ith *output to the* j^{th} *input* and the *tolerance matrix* is given by

$$\Im(s) = \begin{bmatrix} \tau_{11}(s) & \cdots & \tau_{1m}(s) \\ \vdots & \ddots & \vdots \\ \tau_{m1}(s) & \cdots & \tau_{mm}(s) \end{bmatrix}$$

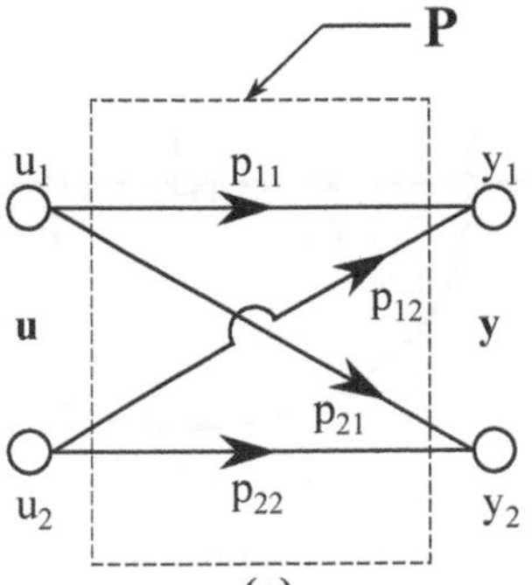

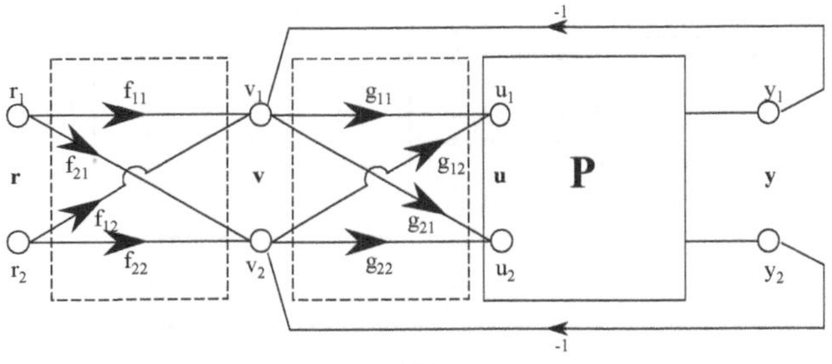

Fig. 5.3 The signal flow graph of: (a) Eq. (5.14), and (b) the compensated MIMO control system.

and where its elements of the $\mathcal{T}(s)$ matrix are given by $\tau_{ij}(s) \rightarrow b_{ij} \geq t_{ij} \geq a_{ij}$. Figure 5.3b may be represented by the simplified SFG of Fig. 5.4.

5-3 INTRODUCTION TO MIMO COMPENSATION

Figure 5.3*b* has the *mxm* closed-loop MIMO feedback control structure of Fig. 5.5 in which ***F***, ***G***, ***P***, ***T*** are each $m \times m$ matrices, and $\mathcal{P} = \{\boldsymbol{P}\}$ is a set of matrices due to plant uncertainty. There are m^2 closed-loop system transfer functions (transmissions) $t_{ij}(s)$ relating the outputs $y_i(s)$ to the inputs $r_j(s)$, i.e., $y_i(s) = t_{ij}(s)r_j(s)$. In a quantitative problem statement, there are *tolerance bounds* on each $t_{ij}(s)$, giving m^2 sets of acceptable regions $\tau_{ij}(s)$ which are to be specified in the design, thus $t_{ij}(s) \in \tau_{ij}(s)$ and $\mathfrak{T}(s) = \{\tau_{ij}(s)\}$. The application of QFT to 2×2 and 3×3 systems has been highly developed and is illustrated in later sections of this chapter.

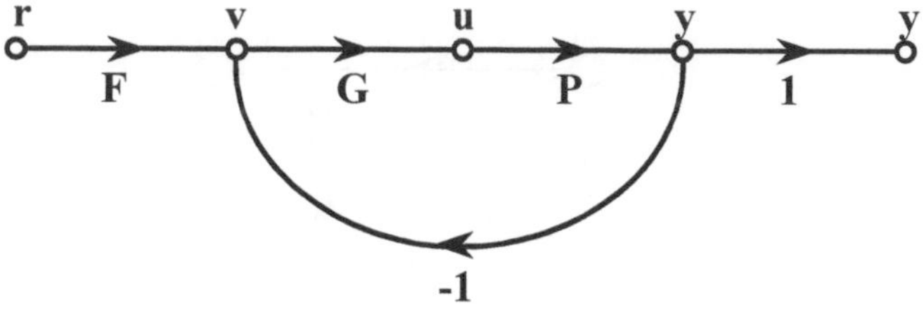

Fig. 5.4 The simplified SFG of Fig. 5.3b.

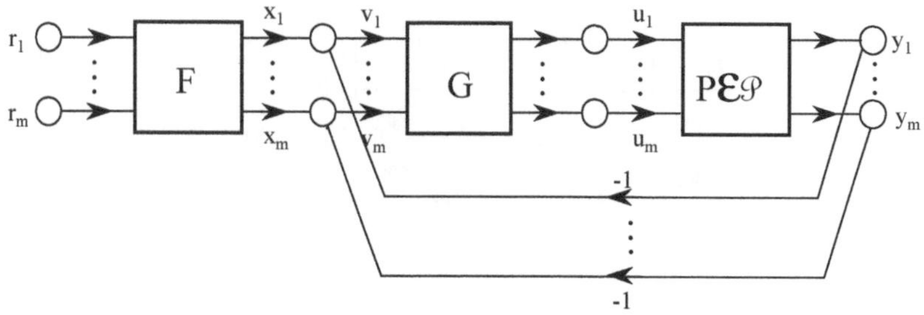

Fig. 5.5 MIMO feedback structure.

In Fig.5.5, the compensator ***G*** may be characterized by either as a diagonal or as a non-diagonal matrix. The use of a fully populated (non-diagonal) matrix allows the control system designer much more design flexibility to achieve the maximum number of the desired performance specification 9see the last paragraph of Sec. 7-9). In other words, the use of non-diagonal elements of ***G*** may ease the diagonal compensator design problem. Unless otherwise stated a diagonal ***G*** is used in Chapters 5 - 9. Chapter 10 discusses the use of a non-diagonal compensator for a MIMO QFT design.

From Fig. 5.5 the following equations can be written:

$$\mathbf{y} = \mathbf{Pu} \qquad \mathbf{u} = \mathbf{Gv} \qquad \mathbf{v} = \mathbf{x} - \mathbf{y} \qquad \mathbf{x} = \mathbf{Fr}$$

In these equations $\boldsymbol{P}(s) = [p_{ij}(s)]$ is the matrix of plant transfer functions, $\boldsymbol{G}(s)$ is the matrix of compensator transfer functions and is often simplified so that it is diagonal, that is, $\boldsymbol{G}(s) = \text{diag}\ \{g_i(s)\}$, and $\boldsymbol{F}(s) = \{f_{ij}(s)\}$ is the matrix of pre-filter transfer functions which may also be a diagonal matrix. The first two expressions yield:

$$y = PGv$$

which is utilized with the remaining two expressions to obtain

$$y = PG[x - y] = PG[Fr - y]$$

This equation is rearranged to yield:

$$y = [I + PG]^{-1}PGFr \tag{5.23}$$

where the system control ratio relating **r** to **y** is:

$$\mathbf{T} = [I + PG]^{-1}PGF \tag{5.24}$$

To appreciate the difficulty of the design problem, note the very complex expression for t_{11} given by Eq. (5.25), for the case $m = 3$ with a diagonal $\boldsymbol{G}$ matrix. However, the QFT design procedure systematizes and simplifies the manner of achieving a satisfactory system design.

$$\begin{aligned} t_{11} = (&[p_{11}f_{11}g_1 + p_{12}f_{21}g_2 + p_{13}f_{31}g_3][(1 + p_{22}g_2)(1 + \\ &\qquad p_{33}g_3 - p_{23}p_{32}g_2g_3] \\ &- [p_{21}f_{11}g_1 + p_{22}f_{21}g_2 + p_{23}f_{31}g_3][p_{12}g_2(1 + p_{33}g_3) - p_{32}p_{13}g_2g_3] \\ &+ [p_{31}f_{11}g_1 + p_{32}f_{21}g_2 + p_{33}f_{31}g_3][p_{23}p_{12}g_2g_3 - (1 + p_{22}g_2)p_{13}g_3])/ \\ &((1 + p_{11}g_1)[(1 + p_{22}g_2)(1 + p_{33}g_3) - p_{23}p_{32}g_2g_3] \\ &- p_{21}g_1[p_{12}g_2(1 + p_{33}g_3) - p_{32}p_{13}g_2g_3] + p_{31}g_1[p_{12}p_{23}g_2g_3 \\ &- p_{13}g_3(1 + p_{22}g_2)]) \end{aligned} \tag{5.25}$$

There are $m^2 = 9$ such that the $t_{ij}(s)$ expressions all have the same denominator, and there may be considerable uncertainty in the nine plant transfer functions $p_{ij}(s)$. The design objective is a system which behaves as desired for the entire range of uncertainty. This requires finding nine $f_{ij}(s)$ and three $g_i(s)$ such that each $t_{ij}(s)$ stays within its acceptable region $\tau_{ij}(s)$ no matter how the $p_{ij}(s)$ may vary. Clearly, this is a very difficult problem. Even the stability problem alone, ensuring that the characteristic polynomial [the denominator of Eq. (5.24)] has no factors in the RHP for all possible $p_{ij}(s)$, is extremely difficult. Most design approaches treat stability for fixed parameter set, neglecting uncertainty, and attempting to cope with the plant uncertainty by trying to design the system to have conservative stability margins. Two highly developed QFT design techniques, *Method 1* and *Method 2*, exist for the design of such systems and are presented in this chapter. In both approaches the MIMO system is converted into an equivalent set of single-loop systems. Methods 1 and 2 utilize the MISO design method of Chapter 3. Method 2, "the improved method," is an outgrowth of Method 1 in which the designed components of the previously designed loop are used in the design of the succeeding loops.

5-4 MIMO COMPENSATION

The basic MIMO compensation structure for a two-by-two MIMO system is shown in Fig. 5.6. The structure for a three-by-three MIMO system is shown in Fig. 5.7. They consist of the uncertain plant matrix $\boldsymbol{P}$, the diagonal compensation matrix $\boldsymbol{G}$, and the pre-filter matrix $\boldsymbol{F}$. This chapter considers only a diagonal $\boldsymbol{G}$ matrix, though a non-diagonal $\boldsymbol{G}$ matrix (see Chapter 10) allows the designer much more design flexibility.[24,81,82] These matrices are defined as follows:

$$\boldsymbol{G}=\begin{bmatrix} g_1 & 0 & \cdots & 0 \\ 0 & g_2 & \cdots & 0 \\ \cdot & \cdot & & \cdot \\ \cdot & \cdot & \ddots & \cdot \\ \cdot & \cdot & & \cdot \\ 0 & 0 & \cdots & g_m \end{bmatrix} \quad \boldsymbol{F}=\begin{bmatrix} f_{11} & f_{12} & \cdots & f_{1m} \\ f_{21} & f_{22} & \cdots & f_{2m} \\ \cdot & \cdot & & \cdot \\ \cdot & \cdot & \ddots & \cdot \\ \cdot & \cdot & & \cdot \\ f_{m1} & f_{m2} & \cdots & f_{mm} \end{bmatrix} \quad \boldsymbol{P}=\begin{bmatrix} p_{11} & p_{12} & \cdots & p_{1m} \\ p_{21} & p_{22} & \cdots & p_{2m} \\ \cdot & \cdot & & \cdot \\ \cdot & \cdot & \ddots & \cdot \\ \cdot & \cdot & & \cdot \\ p_{m1} & p_{m2} & \cdots & p_{mm} \end{bmatrix} \tag{5.26}$$

The dashes in Eq. (5.26) denote the $\boldsymbol{G}$, $\boldsymbol{F}$, and $\boldsymbol{P}$ matrices for a $2x2$ system. Substituting these matrices into Eq. (5.23) yields the $t_{ij}(s)$ control ratios relating the ith output to the jth input. From these $t_{ij}(s)$ expressions the SFG of Fig. 5.8 is obtained. The SFG of Fig. 5.9 for a 3x3 system is obtained in a similar manner.

5-5 INTRODUCTION TO MISO EQUIVALENTS

From Eq. (5.23) obtain[1]

$$\boldsymbol{T}=\frac{(\text{adj}[\boldsymbol{I}+\boldsymbol{PG}])\boldsymbol{PGF}}{\det|\boldsymbol{I}+\boldsymbol{PG}|}$$

For a 3x3 plant and a diagonal $\boldsymbol{G}$ matrix, the denominator of this equation becomes

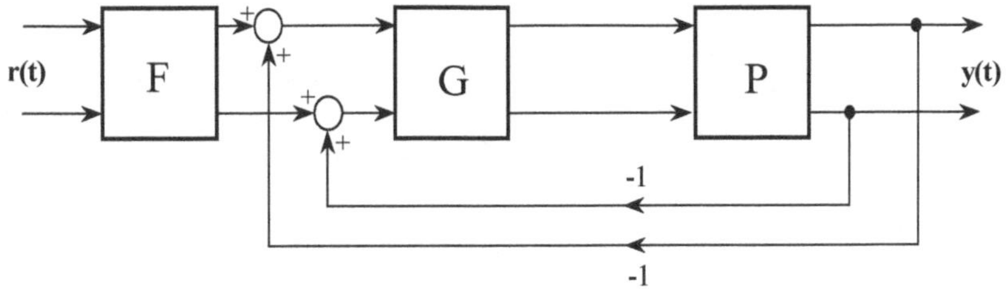

Fig. 5.6 MIMO control structure two-by-two system (*2x2*).

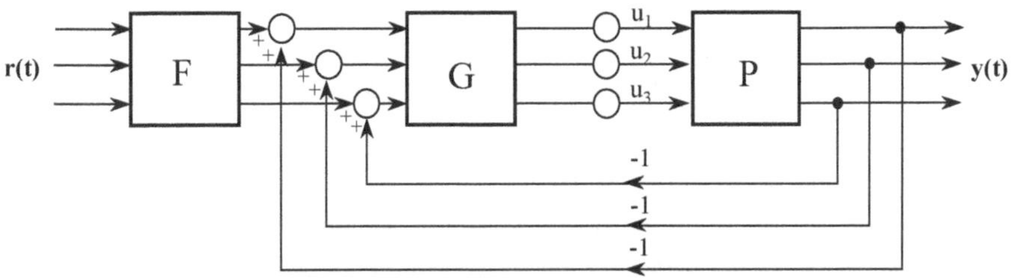

Fig. 5.7 MIMO control structure three-by-three system (*3x3*)

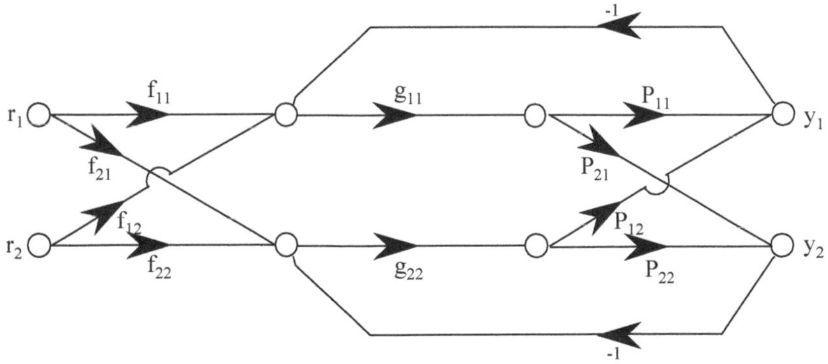

Fig. 5.8 Two-by-two (*2x2*) MIMO signal flow graph.

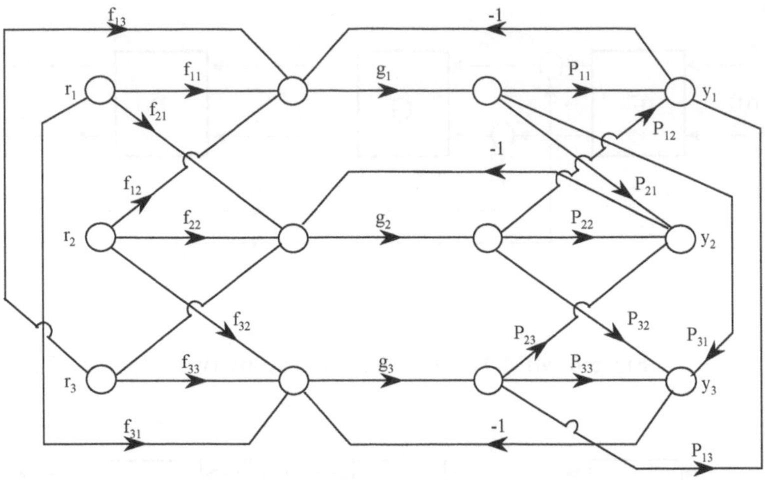

Fig. 5.9 Three-by-three (*3x3*) MIMO signal flow graph.

$$\det|\mathbf{I}+\mathbf{PG}| = \begin{vmatrix} 1+p_{11}g_1 & p_{12}g_2 & p_{13}g_3 \\ p_{21}g_1 & 1+p_{22}g_2 & p_{23}g_3 \\ p_{31}g_1 & p_{32}g_2 & 1+p_{33}g_3 \end{vmatrix}$$

which is a very "messy" equation for the purpose of analysis and synthesis in achieving a satisfactory design of the control system. In analyzing Eq. (5.23), it is noted that if $|\boldsymbol{P}(j\omega)\boldsymbol{G}(j\omega)| >> \boldsymbol{I}$ then $\boldsymbol{T} \approx \boldsymbol{F}$ and the system becomes insensitive to the parameter variations in $\boldsymbol{P}$.

5-5.1 EFFECTIVE MISO EQUIVALENTS

The objective of this section is to find a suitable mapping that permits the analysis and synthesis of a MIMO control system by a set of equivalent MISO control systems. This mapping results in m^2 equivalent systems, each with two inputs and one output. One input is designated as a "desired" input and the other as an "unwanted" input (cross-coupling effects and/or external system disturbances). First, Eq. (5.23) is pre-multiplied by $[\boldsymbol{I} + \boldsymbol{PG}]$ to obtain

$$[\boldsymbol{I} + \boldsymbol{PG}]\boldsymbol{T} = \boldsymbol{PGF} \tag{5.27}$$

When $\boldsymbol{P}$ is nonsingular, then pre-multiplying both sides of this equation by $\boldsymbol{P}^{-1}$ yields

$$[\boldsymbol{P}^{-1} + \boldsymbol{G}]\boldsymbol{T} = \boldsymbol{G}\boldsymbol{F} \tag{5.28}$$

which *puts the constrained or the structured parametric uncertainty in one place in the equation.* Let

$$\boldsymbol{P}^{-1} = \begin{bmatrix} p^*_{11} & p^*_{12} & \dots p^*_{1m} \\ p^*_{21} & p^*_{22} & \dots p^*_{2m} \\ \cdot & \cdot & \cdot \\ \cdot & \cdot & \ddots \quad \cdot \\ \cdot & \cdot & \cdot \\ p^*_{m1} & p^*_{m2} & \dots p^*_{mm} \end{bmatrix} \tag{5.29}$$

The m^2 effective plant transfer functions are based upon defining:

$$q_{ij} \equiv 1/p^*_{ij} = \frac{\det[\boldsymbol{P}]}{Adj_{ij}\boldsymbol{P}} \tag{5.30}$$

The $\boldsymbol{Q}$ matrix is then formed as

$$\boldsymbol{Q} = \begin{bmatrix} q_{11} & q_{12} & \cdots q_{1m} \\ q_{21} & q_{22} & \cdots q_{2m} \\ \cdot & \cdot & \cdot \\ \cdot & \cdot & \ddots \quad \cdot \\ \cdot & \cdot & \cdot \\ q_{m1} & q_{m2} & \cdots q_{mm} \end{bmatrix} = \begin{bmatrix} 1/p^*_{11} & 1/p^*_{12} & \dots 1/p^*_{1m} \\ 1/p^*_{21} & 1/p^*_{22} & \dots 1/p^*_{2m} \\ \cdot & \cdot & \cdot \\ \cdot & \cdot & \ddots \quad \cdot \\ \cdot & \cdot & \cdot \\ 1/p^*_{m1} & 1/p^*_{m2} & \dots 1/p^*_{mm} \end{bmatrix} \tag{5.31}$$

where

$$\boldsymbol{P} = [p_{ij}],\ \boldsymbol{P}^{-1} = [p^*_{ij}] = [1/q_{ij}],\ \boldsymbol{Q} = [q_{ij}] = [1/p^*_{ij}]$$

The matrix $\boldsymbol{P}^{-1}$ is partitioned to the form

$$P^{-1} = [p^*_{ij}] = [1/q_{ij}] = \Lambda + B \qquad \textbf{(5.32)}$$

where Λ is the diagonal part and B is the balance of P^{-1}, thus $\lambda_{ii} = 1/q_{ii} = p^*_{ii}$, $b_{ii} = 0$, and $b_{ij} = 1/q_{ij} = p^*_{ij}$ for $i \neq j$. Next, rewrite Eq. (5.28) using Eq. (5.32) and with G diagonal. This yields $[\Lambda + G]T = GF - BT$ which produces

$$T = [\Lambda + G]^{-1}[GF - BT] \qquad \textbf{(5.33)}$$

This expression is used to define the desired fixed point mapping, based upon *unit impulse functions*, where each of the m^2 matrix elements on the right side of Eq. (5.33) can be interpreted as a MISO problem. Proof of the fact that design of the individual MISO feedback loops will yield a satisfactory MIMO design is based on the Schauder fixed point theorem.[7] This theorem is described by defining a mapping on $\mathfrak{I}$ as follows:

$$Y(T_i) \equiv [\Lambda + G]^{-1}[GF - BT_i] \equiv T_j \qquad \textbf{(5.34)}$$

where each T_i and T_j is from the acceptable set $\mathfrak{I}$. If this mapping has a fixed point, i.e., $T_i, T_j \in \mathfrak{I}$ such that $Y(T_i) = T_j$ (see Fig. 5.10), then a solution to the robust control problem has been achieved yielding a solution in the acceptable set $\mathfrak{I}$. Recalling that Λ and G in Eq. (5.33) are both diagonal, the *1,1* element on the right side of Eq. (5.34) for the *3×3* case, for a *unit impulse input*, yields the output

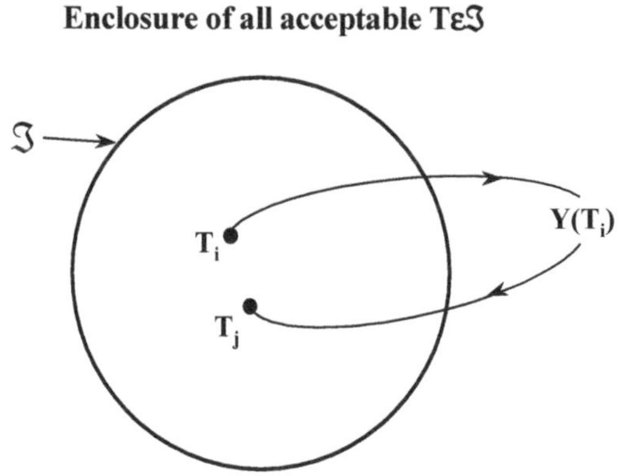

Fig. 5.10 Schauder fixed point mapping.

$$y_{11} = \frac{q_{11}}{1+g_1 q_{11}}\left[g_1 f_{11} - \left(\frac{t_{21}}{q_{12}} + \frac{t_{31}}{q_{13}}\right)\right] \tag{5.35}$$

This corresponds precisely to the first structure in Fig. 5.11 and is the *control ratio* that relates the ith output to the jth input, where $i = j = 1$ in Eq. (5.35). Similarly each of the nine structures in Fig. 5.11 corresponds to one of the elements of $\boldsymbol{Y}(\boldsymbol{T})$ of Eq. (5.33). The general transformation result of m^2 MISO system loops is shown in Fig. 5.12. Figure 5.11 shows the four effective MISO loops (in the boxed area) resulting from a 2×2 system and the nine effective MISO loops resulting from a 3×3 system.[4] The control ratios, for unit impulse inputs, for the mxm system of Fig. 5.12 obtained from Eq. (5.34) have the form

$$y_{ij} = w_{ii}(v_{ij} + c_{ij}) \tag{5.36}$$

where

$$w_{ii} = q_{ii}/(1 + g_i q_{ii}) \tag{5.37}$$

$$v_{ij} = g_i f_{ij} \tag{5.38}$$

and

$$c_{ij} = -\sum_{k \neq i}\left[\frac{t_{kj}}{q_{ik}}\right], \quad k = 1,2,\ldots,m \tag{5.39}$$

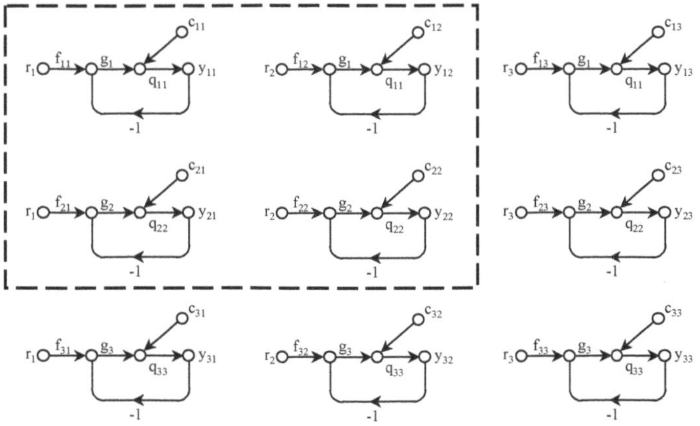

Fig. 5.11 Effective MISO loops two-by-two (boxed in loops) and three-by-three (all 9 loops).

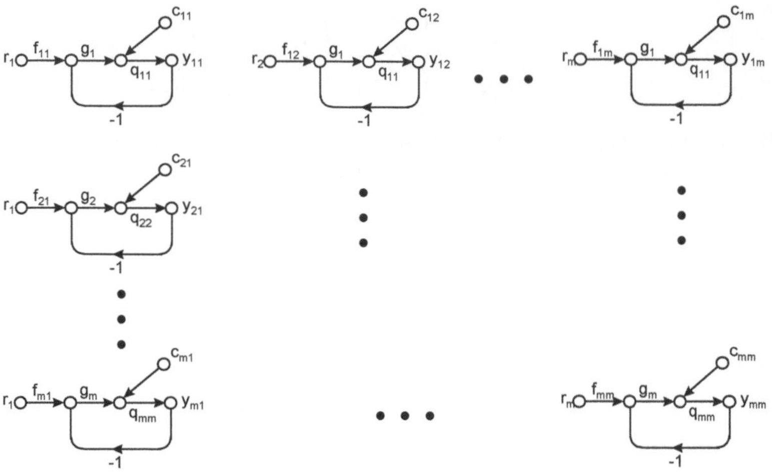

Fig. 5.12 Effective MISO loops (in general).

Equation (5.39) represents the interaction (cross-coupling) between the loops.

Thus, Eq. (5.36) represents the control ratio of the ith MISO loop where the transfer function $w_{ii}v_{ij}$ relates the ith output to i^{th} "desired" input r_i and the transfer function $w_{ii}c_{ij}$ relates the ith output to the jth "cross-coupling effect" input c_{ij}. The transfer function of Eq. (5.36) can thus be expressed, with $r_i(s) = 1$ (*a unit impulse function*), as

$$y_{ij} = (y_{ij})_{r_j} + (y_{ij})_{c_{ij}} = y_{r_j} + y_{c_{ij}} \tag{5.40}$$

or

$$t_{ij} = t_{r_{ij}} + t_{c_{ij}} \tag{5.41}$$

where

$$t_{r_{ij}} = y_{r_j} = w_{ii}\, v_{ij} \tag{5.42}$$

$$t_{c_{ij}} = y_{c_{ij}} = w_{ii}\, c_{ij} \tag{5.43}$$

and where now the upper bound, in the low frequency range is expressed as

$$b_{ij} = b'_{ij} + \tau_{c_{ij}} \tag{5.44}$$

Thus

$$\tau_{c_{ij}} = b_{ij} - b'_{ij} \tag{5.45}$$

represents the maximum portion of b_{ij} allocated towards cross-coupling effect rejection and b'_{ij} represents the upper bound for the tracking portion of t_{ij}. For any particular loop there is a cross-coupling effect input which is a function of all the other loop outputs. The object of the design is to have each loop track its desired input while minimizing the outputs due to the cross-coupling effects.

In each of the nine structures of Fig. 5.11 it is necessary that the control ratio $y_{ij}(s)$, with $r(s) = 1$, must be a member of the acceptable set $t_{ij} \in \mathfrak{I}_{ij}(s)$ (see Fig. 5.10). All the $g_i(s)$, $f_{ij}(s)$ must be chosen to ensure this condition is satisfied, thus constituting nine MISO design problems. If all of these MISO problems are solved, there exists a fixed point, and then $y_{ij}(s)$ on the left side of Eq. (5.34) may be replaced by a t_{ij} and all the elements of $\boldsymbol{T}$ on the right side of the same equation by t_{kj}. This means that there exist nine t_{ik} and t_{kj}, each in its acceptable set, which is a solution to Fig. 5.5. If each element is *1:1*, then this solution must be unique. A more formal and detailed treatment is given in Ref. 7. Note that if the plant has transmission zeros in the right-half-plane (RHP) it only indicates that q_{ij} *may* be n.m.p. or the *det* $\boldsymbol{P}$ may have zeros in the RHP (see Sec. 6-2). For a controllable and observable plant, the transmission zeros can be computed from the determinant of the *system matrix* which is defined [1,75] by

$$\begin{vmatrix} s\mathbf{I} - \mathbf{A} & \mathbf{B} \\ -\mathbf{C} & \mathbf{0} \end{vmatrix} = 0$$

where $\boldsymbol{A}$, $\boldsymbol{B}$, $\boldsymbol{C}$, and $\boldsymbol{I}$ are *mxm* matrices. The number of transmission zeros[26] for a system having the same number of inputs as outputs is equal to d where d is the rank deficiency of the matrix product of $\boldsymbol{CB}$.

5-6 EFFECTIVE MISO LOOPS OF THE MIMO SYSTEM

There are two design methods for designing MIMO systems. In the first method each MISO loop in Figs. 5.11 and 5.12 is treated as an individual MISO design problem, which is solved using the procedures explained in Chapters 3 and 4. The $f_{ij}(s)$ and $g_i(s)$ are the compensator elements of the $\boldsymbol{F}(s)$ and $\boldsymbol{G}(s)$ matrices described previously.

The cross-coupling effect $c_{ij}(s)$ expressed by Eq. (5.39) represents the interaction between the loops, i.e.,

$$c_{ij} = -\sum_{k \neq i} \left[\frac{b_{kj}}{q_{ik}} \right], \quad k = 1, 2, \dots, m \tag{5.46}$$

where the numerator b_{kj} is the upper response bound, T_{R_U} or T_D in Fig. 3.4b or 3.5, for the respective output/input relationship. These are obtained from the design specifications.[4] The first subscript k refers to the output variable, and the second subscript j refers to the input variable. Therefore, b_{kj} is a function of the response requirements on the output y_k due to the input r_j. The lower bound a_{kj} needs defining only when there is a command input. It should be noted that *if the phase margin frequencies ω_ϕ of the loops are not widely separated there is considerable interaction between the loops.*[52]

5-6.1 EXAMPLE: THE 2x2 PLANT

For this example a diagonal $\boldsymbol{G}$ matrix is utilized. The use of a diagonal matrix results in restricting the design freedom available to achieve the desired performance specifications. This is offset by the resulting simplified design process. The elements of a diagonal $\boldsymbol{G}$ are denoted with a single subscript, i.e., g_i. The $\boldsymbol{P}$ and $\boldsymbol{P}^{-1}$ matrices are, respectively:

$$\boldsymbol{P} = \begin{bmatrix} p_{11} & p_{12} \\ p_{21} & p_{22} \end{bmatrix} \quad \textbf{(5.47)}$$

$$\boldsymbol{P}^{-1} = \frac{1}{\Delta}\begin{bmatrix} p_{22} & -p_{12} \\ -p_{21} & p_{11} \end{bmatrix} \quad \textbf{(5.48)}$$

where $\Delta = p_{11}p_{22} - p_{12}p_{21}$. From Eq. (5.30):

$$\boldsymbol{P}^{-1} = \begin{bmatrix} \dfrac{1}{q_{11}} & \dfrac{1}{q_{12}} \\[2ex] \dfrac{1}{q_{21}} & \dfrac{1}{q_{22}} \end{bmatrix} \quad \textbf{(5.49)}$$

where

$$q_{11} = \frac{\Delta}{p_{22}}, \quad q_{12} = \frac{-\Delta}{p_{12}}, \qquad q_{21} = \frac{-\Delta}{p_{21}}, \quad q_{22} = \frac{\Delta}{p_{11}} \quad \textbf{(5.50)}$$

Substituting Eq. (5.49) into Eq. (5.28) yields:

$$\begin{bmatrix} \dfrac{1}{q_{11}} + g_1 & \dfrac{1}{q_{12}} \\ \dfrac{1}{q_{21}} & \dfrac{1}{q_{22}} + g_2 \end{bmatrix} \begin{bmatrix} t_{11} & t_{12} \\ t_{21} & t_{22} \end{bmatrix} = \begin{bmatrix} g_1 f_{11} & g_1 f_{12} \\ g_2 f_{21} & g_2 f_{22} \end{bmatrix} \tag{5.51}$$

The responses due to *input 1*, obtained from Eq. (5.51), are:

$$\begin{aligned} \left(\frac{1}{q_{11}} + g_1\right)t_{11} + \frac{t_{21}}{q_{12}} &= g_1 f_{11} \\ \frac{t_{11}}{q_{21}} + \left(\frac{1}{q_{22}} + g_2\right)t_{21} &= g_2 f_{21} \end{aligned} \tag{5.52}$$

The responses caused by *input 2*, obtained from Eq. (5.51), are:

$$\begin{aligned} \left(\frac{1}{q_{11}} + g_1\right)t_{12} + \frac{t_{22}}{q_{12}} &= g_1 f_{12} \\ \frac{t_{12}}{q_{21}} + \left(\frac{1}{q_{22}} + g_2\right)t_{22} &= g_2 f_{22} \end{aligned} \tag{5.53}$$

These equations are rearranged into a format that readily permits the synthesis of the g_i's and the f_i's that will result in the MIMO control system achieving the desired system performance specifications. Equations (5.52) and (5.53) are manipulated to achieve the following format:

$$\underline{\text{For input } r_1}: t_{11} = \frac{g_1 f_{11} - \dfrac{t_{21}}{q_{12}}}{\dfrac{1}{q_{11}} + g_1} \qquad t_{21} = \frac{g_2 f_{21} - \dfrac{t_{11}}{q_{21}}}{\dfrac{1}{q_{22}} + g_2}$$

$$\underline{\text{For input } r_2}: t_{12} = \frac{g_1 f_{12} - \dfrac{t_{22}}{q_{12}}}{\dfrac{1}{q_{11}} + g_1} \qquad t_{22} = \frac{g_2 f_{22} - \dfrac{t_{12}}{q_{21}}}{\dfrac{1}{q_{22}} + g_2} \tag{5.54}$$

Multiplying the t_{11} and t_{12} equations by q_{11} and the t_{21} and t_{22} equations by q_{22} in Eq. (5.54), respectively, yields the equations shown in Fig. 5.13. Associated with each equation in this figure is its corresponding SFG.

$$t_{11} = \frac{\underbrace{q_{11} g_1}_{=L_1} f_{11} + \underbrace{\left(\frac{-t_{21}}{q_{12}}\right) q_{11}}_{=c_{11}}}{1 + \underbrace{g_1 q_{11}}_{L_1}} \tag{5.55}$$

$$t_{21} = \frac{L_2 f_{12} + \underbrace{\left(\frac{-t_{11}}{q_{21}}\right) q_{22}}_{=c_{21}}}{1 + \underbrace{L_2}_{=q_{22} g_2}} \tag{5.56}$$

$$t_{12} = \frac{L_1 f_{21} + \left(\frac{-t_{22}}{q_{12}}\right) q_{11}}{1 + L_1} \tag{5.57}$$

$$t_{22} = \frac{L_2 f_{22} + \underbrace{\left(\frac{-t_{12}}{q_{21}}\right) q_{22}}_{=c_{22}}}{1 + L_2} \tag{5.58}$$

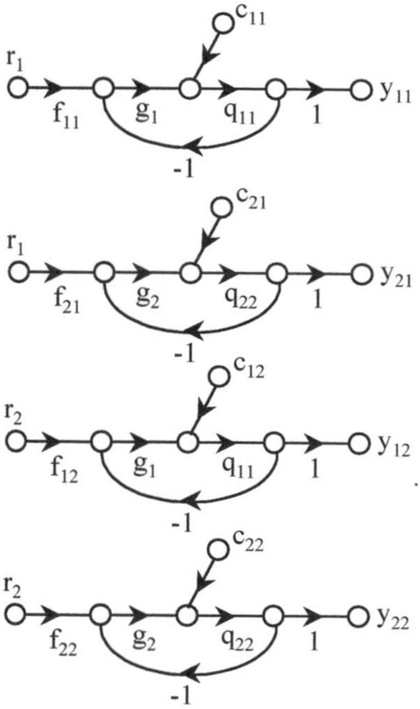

Fig. 5.13 2x2 MISO structures and their respective t_{ij} equations.

Equations (5.55) through (5.58) are of the format of Eqs. (5.36) through (5.45). Note:

1. t_{22} and t_{21} can automatically be obtained from the expressions for t_{11} and t_{12} by interchanging $1 \rightarrow 2$ and $2 \rightarrow 1$ in the equations for t_{11} and t_{12}.
2. The c_{ij} terms in these equations represent the cross-coupling effect from the other loops. These *terms are functions of* the other t_{kj}*'s* and the structured parameter uncertainty of the plant.
3. Theoretically, by making $|\boldsymbol{L}_i(j\omega)|$ "large enough," so that $c_{ij} \approx 0$ a "decoupled system" is achieved.

In a similar fashion, the t_{ij} expressions and their corresponding SFG may be obtained for any *mxm* control system.

5-6.2 PERFORMANCE BOUNDS

Based upon *unit impulse inputs*, from Eq. (5.41) obtain:

$$t_{ij} = t_{r_{ij}} + t_{c_{ij}} \tag{5.59}$$

Let

$$\phi_{ij} \textit{ be the actual value of } t_{ij}$$

$$\tau_{r_{ij}} \textit{ be the actual value of } t_{r_{ij}}$$

$$\tau_{c_{ij}} \textit{ be the actual value of } t_{c_{ij}}$$

A *2x2* control system is used to illustrate the concept of performance bounds. The "actual value" expression corresponding to the t_{11} expression of Eq. (5.55) is

$$\phi_{11} = \tau_{r_{11}} + \tau_{c11}$$

where the τ_r term represents the transmission due to the command input r_1 and the τ_c term represents the transmission due to the cross-coupling effects. For LTI system, the linear superposition theorem is utilized in the development of the performance bounds.

For all $\boldsymbol{P} \in \mathcal{P}$ and $t_{21} \in \tau_{21}$, and $c_{11} = -t_{21}/q_{12}$, the output ϕ_{11} must satisfy the performance specifications on t_{11}. Thus, it is necessary to specify a priori the closed-loop transfer functions t_{ij}. Consider the specifications b_{11} and a_{11} on t_{11}, as illustrated in Fig. 5.14 [see Eqs. (5.44) and (5.45)] *for an m.p. system*. Only magnitudes need be considered for the m.p. system since the magnitude determines the phase.

Thus, in terms of the actual values:

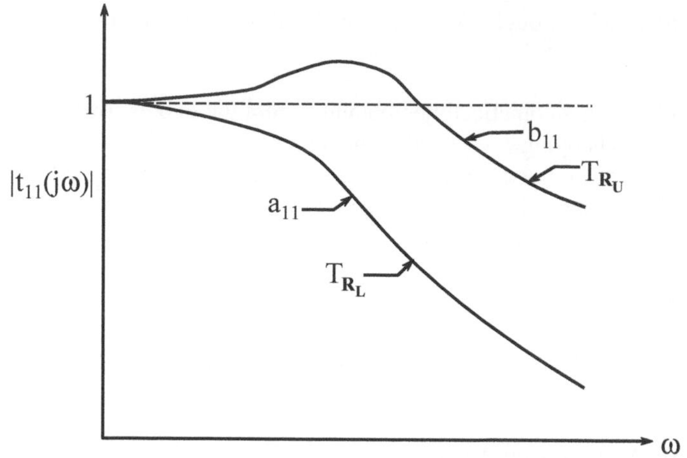

Fig. 5.14 Upper and lower tracking bounds.

$$a_{11}(j\omega) \le |\phi_{11}(j\omega)| \le b_{11}(j\omega)$$
$$\le |\tau_{r_{11}} + \tau_{c_{11}}| \le \qquad \textbf{(5.60)}$$

⇑ *lower bound* ⇑ *upper bound*

Since the relative phases of the τ's are not known and not required for m.p. systems, then to ensure the achievement of the desired performance specifications Eq. (5.60) is expressed as follows:

$$a_{11}(j\omega) \le ||\tau_{r_{11}}| - |\tau_{c_{11}}|| \text{ -- } \textit{the smallest bound} \qquad \textbf{(5.61)}$$

and

$$|\tau_{r_{11}}| + |\tau_{c_{11}}| \le b_{11}(j\omega) \text{ -- } \textit{the largest bound} \qquad \textbf{(5.62)}$$

These represent an over design since they result in a more restrictive performance. A pictorial representation of Eqs. (5.61) and (5.62), for $\omega = \omega_i$, is shown in Fig. 5.15. Note that the τ's in this figure and in the remaining discussion in this section represent only magnitudes. From this figure the following expression is obtained:

$$\Delta\tau = \Delta\tau_{r_{11}} + 2\tau_{c_{11}} = b_{11} - a_{11} \tag{5.63}$$

In the "low frequency range" the bandwidth (BW) of concern, $0 < \omega < \omega_h$, $\Delta\tau$ is split up based upon the desired performance specifications. As is discussed later, for the high frequency range only an upper bound is of concern; thus, there is no need to be concerned with a "split."

Example 5.2 Consider the bound determinations b_{ii}, b'_{ii}, a_{ii}, a'_{ii}, and $b_{c_{ii}}$ *on* $L_1 = g_1 q_{11}$, for a *2x2* system. Thus, $b_{r_{11}}$, based upon

$$a'_{ii} = (\tau_{r_{11}})_L \le |t_{r_{11}}| \le (\tau_{r_{11}})_U = b'_{ii} \tag{5.64}$$

and Fig. 5.15, for t_{11} can be determined in the same manner as for a MISO system (see Chapter 3). In Fig. 5.15, for illustrative purposes only, $\tau_{c_{11}} = 0.05$ and $\Delta\tau_{r_{11}}$ = 0.2 at $\omega = \omega_i$. Referring to Fig. 5.13 and to Eq. (5.55), the bound on the cross-coupling effect $b_{c_{ij}}$, is determined as follows:

$$|\mathbf{t}_{c_{11}}| = \left|\frac{\mathbf{c}_{11}\mathbf{q}_{11}}{1+\mathbf{L}_1}\right| \le \tau_{c_{ij}} = \tau_{c_{11}} = 0.05 \tag{5.65}$$

where for the cross-coupling effect the *upper bound* for $t_{c_{11}}$ is given by

$$|t_{c_{ij}}| = |t_{c_{11}}| \le b_{c_{ij}} = b_{c_{11}} \quad where\ i = j = 1$$

Substituting the expression for c_{11} [see Eq. (5.46)] into Eq. (5.65) and then manipulating this equation the following constraint on $\boldsymbol{L}_1(j\omega_i)$, where b_{21} is the upper bound on t_{21}, is obtained:

$$|1+\mathbf{L}_1| \ge \frac{|\mathbf{c}_{11}\mathbf{q}_{11}|}{\tau_{c_{11}} = 0.05} = 20\left|\frac{\mathbf{t}_{21}\mathbf{q}_{11}}{\mathbf{q}_{12}}\right| = 20\left|\frac{\mathbf{b}_{21}\mathbf{q}_{11}}{\mathbf{q}_{12}}\right| \tag{5.66}$$

where the upper bound b_{21} is inserted for t_{21}. It is necessary to manipulate this equation in a manner that permits the utilization of the Nichols Chart (NC). This is accomplished by making the substitution of $L = 1/\ell$ into this equation. Thus,

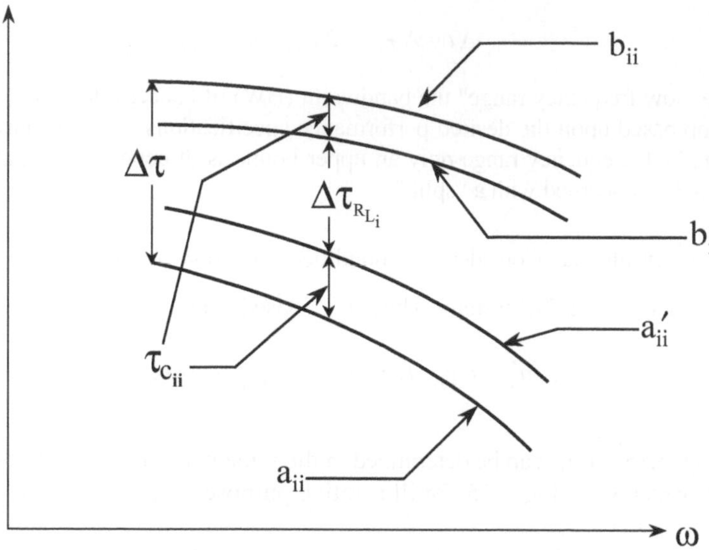

Fig. 5.15 Allocation for tracking and cross-coupling specifications for t_{ii} responses.

$$\left|1+\frac{1}{\ell_1}\right| = \left|\frac{1+\ell_1}{\ell_1}\right| \geq 20\left|\frac{b_{21}q_{11}}{q_{12}}\right| \tag{5.67}$$

Inverting this equation yields:

$$\left|\frac{\ell_1}{1+\ell_1}\right| \leq \frac{1}{20}\left|\frac{q_{12}}{b_{21}q_{11}}\right| \tag{5.68}$$

which is of the mathematical format that allows the use of the NC for the graphical determination of the cross-coupling bound $b_{c_{11}}$ (see Sec. 3-10). Equations (5.67) and Eq. (5.68) are, respectively, of the following mathematical format:

$$|\boldsymbol{A}| \geq |\boldsymbol{B}| \quad (a)$$
$$|\boldsymbol{C}| \leq |\boldsymbol{D}| \quad (b) \tag{5.69}$$

In determining these bounds, it is necessary to insert the actual plant parameters into these equations. That is, for each of the J plant models, $\boldsymbol{P}_\iota$ ($\iota = 1,2,\ldots,J$), insert the corresponding plant parameters into these equations and determine the magnitudes A_ι, B_ι, C_ι, and D_ι. The following magnitudes, for each value of ω_i, are used from these J sets of values to determine the cross-coupling bound:

$$\begin{aligned} |\boldsymbol{A}_\iota|_{min} &\geq |\boldsymbol{B}_\iota|_{max} \\ |\boldsymbol{C}_\iota|_{max} &\leq |\boldsymbol{D}_\iota|_{min} \end{aligned} \qquad \textbf{(5.70)}$$

"High bounds" on the NC require h.g., thus, in order to *minimize* the required compensator gain, the *optimum* choice of the

$$\Delta\tau_{r_{11}}, \tau_{c_{11}} \qquad \textbf{(5.71)}$$

specifications are those that result in achieving essentially the same tracking and cross-coupling bounds, i.e.,

$$\boldsymbol{b}_{c11} \approx \boldsymbol{b}_{r11} \qquad \textbf{(5.72)}$$

A recommended method for determining an appropriate set of constraints (specifications) on Eq. (5.71) is to do a design initially based on the following assumption:

$$\Delta\tau_{r_{11}} = b'_{11} - a'_{11}$$

With this design, determine how big $\tau_{c_{11}}$ is and then, by trial and error if a CAD package is not available, adjust $\tau_{c_{11}}$ until the condition

$$\Delta\tau_{r_{11}} \approx 2\tau_{c_{11}}$$

is satisfied. The MIMO/QFT CAD discussed in Appendix C automates this procedure. This procedure is illustrated in Fig. 5.16. Depending on the starting quantities, by decreasing (increasing) one quantity and increasing (decreasing) the other quantity expedites achieving the condition of Eq. (5.72). By the procedure just described, the optimal bound $\boldsymbol{b}_{o_1}$ is given by Eq. (5.72).

5-6.3 QFT DESIGN METHOD 1

In the manner described in Sec. 5-6.2, for determining the bounds of the first loop to be designed, determine the other bounds required for the other L_i's. This may require trial and error; i.e., in loop shaping L_1, based on Eqs. (5.42) through (5.45) and (5.57), it is necessary to obtain the bounds for L_{o_1} which satisfy the specifications for both t_{11} and t_{12}. Once these specifications are satisfied then proceed to do the looping shaping to determine L_{o_ι}.[4]

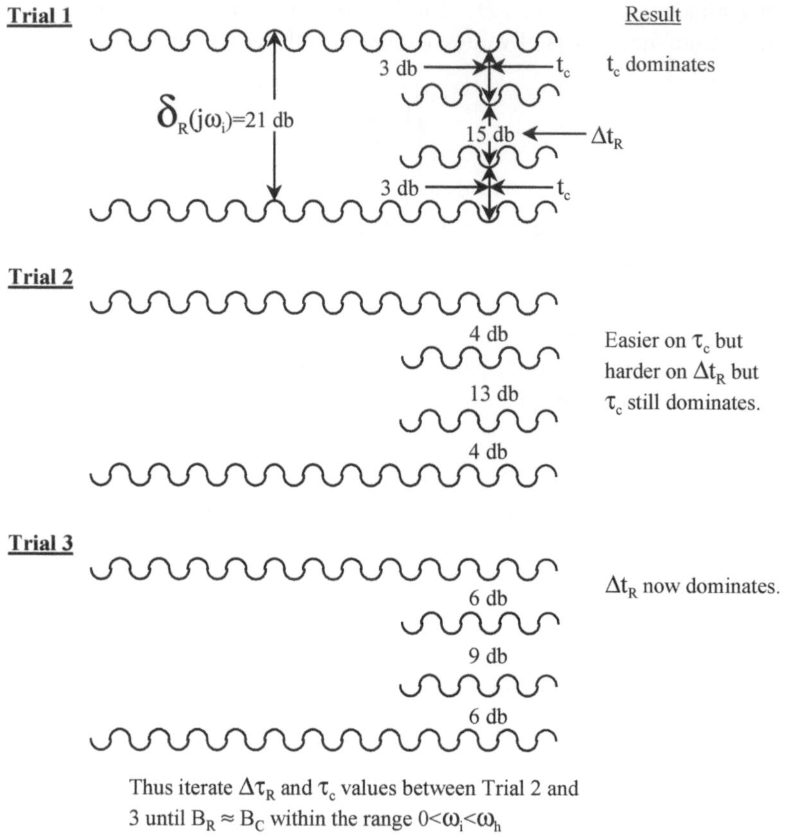

Fig. 5.16 Procedure pictorial representation to achieve $B_r \approx B_c$.

5-6.4 QFT DESIGN METHOD 2

The QFT design Method 2, for many problems, may yield a better control system design. This method is an improvement over Method 1 in that it utilizes the resulting designed g_i's and f_{ij}'s of the first MISO equivalent loop that is designed in the design of the succeeding loops, etc. This feature of Method 2 reduces the over-design in the early part of the design process. Method 2 may involve a trade-off in the design parameters. The final MISO equivalent loop to be designed uses the exact transfer functions of the previously designed g_i's and f_{ij}'s thus this loop has the least amount of over-design.[4]

The order in which the MISO loops are designed is important. Any order may be chosen but some orders may produce less over design (lower bandwidth) than others. The general rule in the choice of the design order of the loops is *that the most constrained loop is chosen as the starting loop.* That is, the P.S. to be satisfied has placed a *constraint* on which loop is to be chosen as the starting loop. For example, the BW specifications for each loop, $\omega_{h_{ii}}$, to be satisfied places the constraint on which loop is to be chosen as the starting loop. Thus, choose the "starting" loop i for which it is most important to minimize the BW requirements. Some of the factors involved in determining the BW requirements are:

1. Sensor noise
2. Loop i has *severe* bending mode problems that other loops do not have.
3. The "high frequency gain" (h.f.g.) uncertainty of q_{ii} may be very much greater than for the other loop(s). The h.f.g. uncertainty may effect the size of the templates.

Professor Horowitz provides the following insight:

(1) Analyze the various q_{ii} templates over a reasonable range of frequencies: almost vertical at low and high (for the analog case) frequencies. For the discrete case, up to the frequency range in which the templates first narrow, before widening again for $\omega_i > \omega_s/2$ (ω_s sampling frequency), rather than the final frequency range ($\omega_i > \omega_s/2$) in which the templates approach a vertical line (see Chapter 4).

(2) If the feedback requirements per loop are roughly the same, which indicates the t_{ii} specifications are about the same, and the t_{ij}, $i \neq j$, are also roughly similar for the different loops, then the loop with the smallest q_{ii} templates should be the "starting loop." That is, the loop with the smallest amount of feedback should be chosen as the starting loop. The reason for this choice is that there is a tendency towards BW propagation as the design proceeds from the first loop to the final loop that is designed. Therefore, for this case the loop with the smallest BW requirement becomes the most constrained loop due to the BW propagation effect. Thus, it is the loop to be designed first and the second loop to be designed should be the one having the second smallest feedback requirement, etc. Also, see the discussion in Chapter 9.

All of these factors emerge from the *transparency* of QFT which helps to reduce the trial and error that is involved in achieving a satisfactory design. If, in choosing the loop with the most severe BW limitation causes a design problem in the succeeding designed loops, then try a different starting loop selection. This is based on the knowledge, that in general, the BW of the succeeding designed loops are higher than the BW of the previously designed loops. Further insight into satisfying the BW requirement is given in Chapter 9.

If Method 2 can not satisfy the BW requirements for all loops then Method 1 must be used, assuming the *diagonal dominance* condition (see Sec. 5-7) is satisfied. If Method 1 can not be used then it is necessary to reevaluate the performance specifications, etc.

5-6.5 SUMMARY

For the cross-coupling effect rejections problem, the responses must be less than some bound, that is:

$$(y_{ij})_{c_{ij}} \le \tau_{c_{ij}} = b_{c_{ij}}$$

Thus the loop equations become

$$\tau_{ij} \ge (y_{ij})_{c_{ij}} = \frac{c_{ij}\,|\mathbf{q}_{ii}|}{|1 + \mathbf{g}_i\mathbf{q}_{ii}|} \tag{5.73}$$

where $\boldsymbol{g_i q_{ii}} = \boldsymbol{L_i}$. Equation (5.73) is manipulated to yield:

$$|1 + \boldsymbol{L}_i| \ge \frac{c_{ij}\,|\boldsymbol{q}_{ii}|}{\tau_{ij}} \tag{5.74}$$

Substituting Eq. (5.46) into Eq. (5.74) yields:

$$|1 + \mathbf{L}_i| \ge \left| -\sum_{k \ne i} \frac{b_{kj}}{q_{ik}} \right| \frac{|q_{ii}|}{\tau_{ij}} \tag{5.75}$$

For example, in a $3x3$ system for the first loop, L_1, where $i = 1, j = 2$, and in the first term $k = 2$ and in the second term $k = 3$, Eq. (5.75) becomes:

$$|1 + \mathbf{L}_1| \ge \frac{\left| \dfrac{b_{22}}{|\mathbf{q}_{12}|} + \dfrac{b_{32}}{|\mathbf{q}_{13}|} \right| |\mathbf{q}_{11}|}{\tau_{c12}} \tag{5.76}$$

Remember to use only the magnitude in the cross-coupling calculations. This assumes the worst case.

5-7 CONSTRAINTS ON THE PLANT MATRIX[21]

In order to use the QFT technique the following critical condition must be satisfied.

> **Condition 1**: ***P*** must be nonsingular for any combination of possible plant parameters, to ensure that $\boldsymbol{P}^{-1}$ exists.

In the high frequency range $y_{r_i}(j\omega)$ approaches zero as $\omega \rightarrow \infty$. Therefore, Eqs. (5.40) and (5.41) can be approximated by

$$t_{ij} \approx y_{c_{ij}} \tag{5.77}$$

Thus from Eqs. (5.36), (5.37) and (5.46), with impulse input functions, Eq. (5.77) can be expressed as

$$y_{c_{ij}} = \frac{c_{ij}\, q_{ii}}{1 + g_i q_{ii}} \tag{5.78}$$

Consider first the $2x2$ plant, i.e., $m = 2$. Specifying that $|y_{c_{11}}| \leq b_{ii}$ (the given upper bound in the high frequency range), Eqs. (5.46) and (5.78) yield, for $i = j = 1$:

$$b_{11} \geq \left|\frac{-b_{21}}{q_{12}}\right| \left|\frac{q_{11}}{1 + L_1}\right| \tag{5.79}$$

For $i = 2, j = 1$, where it is specified that $(y_c)_{21} \leq b_{21}$ (the given cross-coupling upper bound). Henceforth, the cross-coupling bound notation $\boldsymbol{b}_{c_{ij}}$ is simplified to $\boldsymbol{b}_{ij}$ for $i \neq j$, thus Eqs. (5.46) and (5.79) yield:

$$b_{21} \geq \left|\frac{-b_{11}}{q_{21}}\right| \left|\frac{q_{22}}{1 + L_2}\right| \tag{5.80}$$

Equations (5.79) and (5.80) are rearranged to

$$|1 + L_1| \geq \frac{b_{21}}{b_{11}} \left|\frac{q_{11}}{q_{12}}\right| \tag{5.81}$$

$$|1+L_2| \geq \frac{b_{11}}{b_{21}}\left|\frac{q_{22}}{q_{21}}\right| \tag{5.82}$$

Multiplying Eqs. (5.81) by Eq. (5.82), where $L_1 = L_2 \approx 0$ in the high frequency range, results in

$$1 \geq \left|\frac{q_{11}q_{22}}{q_{12}q_{21}}\right| \tag{5.83}$$

Since

$$P = \begin{bmatrix} p_{11} & p_{12} \\ p_{21} & p_{22} \end{bmatrix}, \quad P^{-1} = \left[p_{ij}^*\right] = \frac{1}{\Delta}\begin{bmatrix} p_{11} & -p_{12} \\ -p_{21} & p_{11} \end{bmatrix}, \quad Q = [q_{ij}] = \begin{bmatrix} q_{11} & q_{12} \\ q_{21} & q_{22} \end{bmatrix}$$

and

$$q_{ij} = \frac{1}{p_{ij}^*} = \frac{\Delta}{p_{ij}}$$

then

$$q_{11} = \frac{1}{p_{11}^*} = \frac{\Delta}{p_{11}^*} \qquad q_{12} = \frac{1}{p_{12}^*} = \frac{-\Delta}{p_{12}^*}$$

$$q_{21} = \frac{1}{p_{21}^*} = \frac{-\Delta}{p_{21}^*} \qquad q_{22} = \frac{1}{p_{22}^*} = \frac{\Delta}{p_{22}^*} \tag{5.84}$$

Substituting Eqs. (5.84) into Eq. (5.83) yields Condition 2.

Condition 2 (*2x2* plant): As $\omega \to \infty$

$$|p_{11}p_{22}| > |p_{12}p_{21}| \quad \textbf{(a)}$$
$$or$$
$$|p_{11}p_{22}| - |p_{12}p_{21}| > 0 \quad \textbf{(b)} \tag{5.85}$$

Since p_{11} and p_{22} are elements of the principal diagonal of $\boldsymbol{P}$, then Eq (5.85) is the *diagonal dominance condition* for the *2x2* plant. Equation (5.85(a)) must also be satisfied for all frequencies. This condition is obtained considering only the left column of the MISO loops for the *2x2* plant of Fig 5.11. This condition may also be obtained by using the right column of the MISO loops in Fig 5.11 since the loop transmissions L_1 and L_2 are again involved in the derivation. If Eq. (5.85) is not satisfied then refer to Item 3 in the summary of Example 6.6.

Next consider the $3x3$ plant, i.e., $m = 3$. For $i = j = 1$ (the left column of Fig. 5.11), where $\mathbf{L}_1 = \mathbf{L}_2 = \mathbf{L}_3 \approx 0$ as $\omega \rightarrow \infty$ and where it is specified that

$$|y_{c11}| \leq b_{11}, \quad |y_{c21}| \leq b_{21}, \quad |y_{c31}| \leq b_{31}$$

Eqs. (5.46) and (5.78) yield, for $i = 1,2,3$ respectively,

$$1 \geq \left|\frac{q_{11}}{b_{11}}\right| \left[\left|\frac{b_{21}}{q_{12}}\right| + \left|\frac{b_{31}}{q_{13}}\right| \right] \tag{5.86}$$

$$1 \geq \left|\frac{q_{22}}{b_{21}}\right| \left[\left|\frac{b_{11}}{q_{21}}\right| + \left|\frac{b_{31}}{q_{23}}\right| \right] \tag{5.87}$$

$$1 \geq \left|\frac{q_{33}}{b_{31}}\right| \left[\left|\frac{b_{11}}{q_{31}}\right| + \left|\frac{b_{21}}{q_{32}}\right| \right] \tag{5.88}$$

Letting

$$\lambda_1 = \frac{b_{21}}{b_{11}} \qquad \lambda_2 = \frac{b_{31}}{b_{11}} \tag{5.89}$$

Equations (5.86) through (5.88) become, respectively,

$$1 \geq |q_{11}| \left[\left|\frac{\lambda_1}{q_{12}}\right| + \left|\frac{\lambda_2}{q_{13}}\right| \right] \tag{5.90}$$

$$\lambda_1 \geq |q_{22}| \left[\left|\frac{1}{q_{21}}\right| + \left|\frac{\lambda_2}{q_{23}}\right| \right] \tag{5.91}$$

$$\lambda_2 \geq |q_{33}| \left[\left|\frac{1}{q_{31}}\right| + \left|\frac{\lambda_1}{q_{32}}\right| \right] \tag{5.92}$$

From Eqs. (5.91) and (5.92) and using

$$\gamma_{ij} = \left[\frac{q_{ii}\, q_{jj}}{q_{ij}\, q_{ji}}\right]_{\substack{i=2\\ j=3}} = \gamma_{23} = \frac{q_{22}\, q_{33}}{q_{23}\, q_{32}} \tag{5.93}$$

the following expressions for λ_1 and λ_2 are obtained

$$\lambda_1 \geq \frac{\left|\dfrac{q_{22}}{q_{21}}\right| + \left|\dfrac{q_{22} q_{33}}{q_{23} q_{31}}\right|}{|1 - \gamma_{23}|} \tag{5.94}$$

$$\lambda_2 \geq \frac{\left|\dfrac{q_{33}}{q_{31}}\right| + \left|\dfrac{q_{22} q_{33}}{q_{21} q_{32}}\right|}{|1 - \gamma_{23}|} \tag{5.95}$$

Substitute Eqs. (5.94) and (5.95) into Eq. (5.90) to obtain

$$\left|\frac{q_{11}}{q_{12}}\right| \left[\, \left|\frac{q_{22}}{q_{21}}\right| + \left|\frac{q_{22}\, q_{33}}{q_{23}\, q_{31}}\right| \,\right] + \left|\frac{q_{11}}{q_{13}}\right| \left[\, \left|\frac{q_{33}}{q_{31}}\right| + \left|\frac{q_{33}\, q_{22}}{q_{21}\, q_{32}}\right| \,\right] \leq |1 - \gamma_{23}| \tag{5.96}$$

Substituting $q_{ij} = 1/\, p^{*}_{ij}$ into Eq. (5.96) yields:

$$\left|\frac{p_{12}^*}{p_{11}^*}\right|\left[\left|\frac{p_{21}^*}{p_{22}^*}\right|+\left|\frac{p_{23}^* p_{31}^*}{p_{22}^* p_{33}^*}\right|\right]+ \tag{5.97}$$

$$\left|\frac{p_{13}^*}{p_{11}^*}\right|\left[\left|\frac{p_{31}^*}{p_{33}^*}\right|+\left|\frac{p_{21}^* p_{32}^*}{p_{22}^* p_{33}^*}\right|\right]\leq\left|1-\frac{p_{23}^* p_{32}^*}{p_{22}^* p_{33}^*}\right|$$

Multiplying Eq. (5.97) by

$$p_{11}^* p_{22}^* p_{33}^* = p_{123}^*$$

yields Condition 3.

Condition 3 (*3x3* plant; applies only for QFT design Method 1)

As $\omega \rightarrow \infty$

$$|p_{123}^*| \geq |p_{12}^*| \left[|p_{21}^* p_{33}^*| + |p_{23}^* p_{31}^*|\right] + \tag{5.98}$$

$$|p_{13}^*| \left[|p_{22}^* p_{31}^*| + |p_{21}^* p_{32}^*|\right] + |p_{11}^* p_{23}^* p_{32}^*|$$

See Ref. 7 for higher order plants. This condition is necessary only if Eq. (5.8) is used to generate the plant.

Condition 1 ensures controllability of the plant since the inverse of ***P*** produces the effective transfer functions used in the design. If the **P** matrix, resulting from the original ordering of the elements of the input and output vectors does not satisfy Condition 2, then a reordering of the input and output vectors may result in satisfying these conditions.

5-8 BASICALLY NONINTERACTING (BNIC) LOOPS

A basically non-interacting (BNIC) loop[4] is one in which the output $y_k(s)$ due to the input $r_j(s)$ is ideally zero. Plant uncertainty and loop interaction (cross-coupling) makes the ideal response unachievable. Thus, the system performance specifications describe a range of acceptable responses for the commanded output and a maximum tolerable response for the un-commanded outputs. The un-commanded outputs are treated as cross-coupling effects (akin to disturbances).

For an LTI plant, having no parameter uncertainty, it is possible to essentially achieve zero cross-coupling effects, i.e., the output $y_k \approx 0$ due to c_{ij}. This desired result can be achieved by post multiplying $\boldsymbol{P}$ by a matrix $\boldsymbol{W}$ to yield:

$$\mathbf{P}_n = \mathbf{PW} = [p_{ii_n}] \text{ where } p_{ij_n} = 0 \text{ for } i \neq j$$

resulting in a diagonal $\boldsymbol{P}_n$ matrix for $\boldsymbol{P}$ representing the nominal plant case in the set $\mathcal{P}$. With plant uncertainty the off-diagonal terms of $\boldsymbol{P}_n$ will not be zero but "very small" in comparison to $\boldsymbol{P}$, for the non-nominal plant cases in $\mathcal{P}$. In some design problems it may be necessary or desired to determine a $\boldsymbol{P}_n$ upon which the QFT design is accomplished. Doing this minimizes the effort required to achieve the desired BW and minimizes the cross-coupling effects. Since $|t_{ij}(j\omega)| \leq b_{ij}(j\omega)$, $i \neq j$, for all ω, it is clearly best to let $f_{ij} = 0$, for $i \neq j$. Thus,

$$t_{ij} = t_{c_{ij}} = \frac{c_{ij}\, q_{ii}}{1 + L_i} \quad \text{for all } i \neq j$$

This was done on an AFTI-16 design by Horowitz[30] as shown in Fig.5.17. In general, an upper bound b_{ij}, $i \neq j$, is specified in order to achieve the performance specification:

$$|\tau_{c_{ij}}| \leq b_{ij} \quad \text{for all } P \,\varepsilon\, \mathcal{P}$$

A system designed to this specification is called BNIC.

5-9 SUMMARY

This chapter describes the multiple-input multiple-output closed-loop system and the plant matrix. Guidelines for finding the $\boldsymbol{P}$ matrix, which relates the input vector to the output vector, are given.

The method of representing a MIMO system by an equivalent set of MISO systems is presented using $\boldsymbol{P}^{-1}$. Two design approaches, which are discussed in detail in Chapters 6 and 7, respectively, are available in which the equivalent MISO loops are designed according to the MISO design method outlined in Chapters 3 and 4. Since the g_i's are the same for all m MISO systems in each row of Fig. 5.12, the compensator must be "good enough" (large enough gain over the desired BW) to handle the "feedback needs" of the *worst* of the m systems for each value of ω_i within the BW. If it is possible to find the f_{ij}'s and the g_i's which satisfy the performance specifications for the m^2 systems then it is *guaranteed* that these pre-filters and compensators, when used as elements in the

MIMO systems of Fig. 5.8 or 5.9, satisfy the design specifications of the original MIMO system. *Thus, the mxm MIMO has been converted into m design problems.*

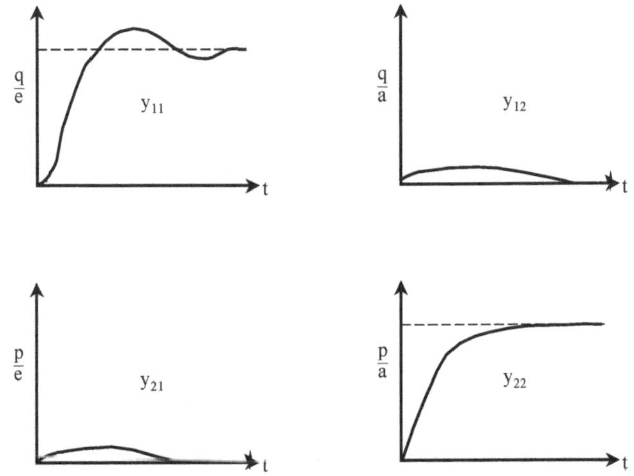

Fig. 5.17 Output time response sketch for *2x2* plant: a - aileron deflection, *q* - pitch rate, *e* - elevator deflection, and *p* - roll rate.

6

DESIGN METHOD 1 –THE SINGLE-LOOP (MISO) EQUIVALENTS[18]

6-1 INTRODUCTION

The QFT technique developed in Chapters 2 through 4 is based upon a MISO system structure having two or three external inputs and one output. In Chapter 5 it is shown that an *mxm* MIMO system can be represented by m^2 MISO equivalent systems having a desired tracking input command, one cross-coupling effect input signal, and one output. Thus, the QFT design technique developed in Chapters 3 and 4 can now be applied, with the appropriate modifications indicated in Chapter 5, to the design of the m^2 MISO equivalent systems. This chapter presents the details of the QFT Design Method 1, sensitivity analysis, and design performance trade-offs. The following chapter presents the details of the QFT Design Method 2.

As indicated in Chapter 5, let $\tau_{ij}(s)$ be the set of acceptable transfer functions $t_{ij}(s) = y_i(s)/r_j(s)$ [output $y_i(s)$ in response to the command input $r_j(s)$ in Fig. 5.5]. Let the plant inverse *matrix* $\boldsymbol{P}^{-1} = [l/q_{ij}]$, so the plant set $\boldsymbol{P}^{-1}$ generates sets of $\mathcal{q}_{ij} = \{q_{ij}\}$. Replace the *mxm* MIMO problem by *m* single loops and m^2 pre-filters (see Fig. 5.11 for $m = 3$). In Fig. 5.11, $\boldsymbol{P}^{-1} = [l/q_{ij}]$, and the uncertainty in $\boldsymbol{P}$ generates sets $Q = \{q_{ij}\}$. The cross-coupling effect

$$c_{11} = -\left(\frac{t_{21}}{q_{12}} + \frac{t_{31}}{q_{13}}\right) \qquad t_{k1} \in \tau_{k1} \tag{6.1}$$

in Fig. 5.11 is any member of a set c_{11} generated by the t_{k1} in τ_{k1} and the q_{ik} in $\mathcal{q}_{ik}$, $k = 2,3$.

In general in Fig. 5.11,

$$\mathscr{C}_{ij} = \{c_{ij}\}, \quad c_{ij} = -\sum_{k \neq i} \left(\frac{t_{kj}}{q_{ik}} \right), \quad t_{kj} \in \tau_{kj} \tag{6.2}$$

for the set of acceptable t_{kj} transfer functions. For the top row of MISO loops in Fig. 5.11, the MISO design problem is to find $L_1 = g_1 q_{11}$ and f_{11} such that the output is a member of the set τ_{11} for all q_{11} in $\mathscr{Q}_{11}$ and for all c_{11} in $\mathscr{C}$. Similarly, for the middle MISO loop in the top row of Fig. 5.11 find $L_1 = g_1 q_{11}$ and f_{12} so that its output is in τ_{12} for all q_{11} in $\mathscr{Q}_{11}$ and all c_{12} in $\mathscr{C}_{11}$, etc. Note that L_1 is the same for all the MISO structures in the first row of Fig. 5.11, etc. In each of these three structures, the uncertainty problem (due to the $\mathscr{Q}_{11}$, $\mathscr{C}_{11}$) gives bounds which in-turn result in the optimal bound b_{o11} the level of feedback L_1 needs. Thus, in the manner described in Sec. 3-12, the optimal bound b_{o11} is derived and for which the toughest of these bounds must be satisfied by L_1. That is, for each row there will exist, assuming a diagonal pre-filter matrix $\boldsymbol{F}$, three cross-coupling bounds and one tracking bound; thus, the toughest portions of these four bounds are combined to form the optimal bound to be used in loop shaping L_i.

If the designer designs these MISO systems to satisfy their above stated specifications, then it is guaranteed that these same f_{ij}, g_i satisfy the MIMO uncertainty problem. It is not necessary to consider the highly complex system characteristic equation [denominator of Eq. (5.25)] with its uncertainty p_{ij} plant parameters. System stability (and much more than that) for all $\boldsymbol{P}$ in $\mathscr{P}$ is automatically guaranteed. It is easier to present the important ideas by means of a design example.

6-2 DESIGN EXAMPLE

The power of the technique is illustrated by presenting the results of Ref. 4, which, in the words of Professor Horowitz, "was done as a Master's thesis by a typical graduate student who like nearly all control graduates, has had no courses dealing with uncertainty." The plant and uncertainties are

$$p_{ij} = \frac{A_{ij}s + B_{ij}}{s^2 + Es + F}$$

$A_{11} \in [2, 4]$	$A_{12} \in [-0.5, 1.1]$	$A_{22} \in [5, 101]$
$A_{31} \in [-0.8, -1.8]$	all other $A_{ij} = 0$	
$B_{11} \in [-0.15, 1]$	$B_{12} \in [-1, 2]$	$B_{13} \in [1,4]$
$B_{21} \in [1, 2]$	$B_{22} \in [5, 10]$	$B_{23} \in [-1, 4]$
$B_{31} \in [-1, 2]$	$B_{32} \in [15, 25]$	$B_{33} \in [10, 20]$
$E \in [-0.2, 2]$	$F \in [0.5, 2]$	**(6.3)**

Note that p_{12} is always non-minimum phase (n.m.p.) and p_{11} is also n.m.p. for part of the structured parameter uncertainty range. Also, the plant is unstable for part of the parameter range. At small ω there is no diagonal dominance in any row, nor in columns *1,2*. Hence, Rosenbrock's technique[61] can not be used even if there was no parameter uncertainty.

Since the MISO plants in Fig. 5.11 are $q_{11} = [det\ \boldsymbol{P}]/[Adj_{11}\ \boldsymbol{P}]$, it follows that if *det* $\boldsymbol{P}(s)$ has RHP zeros, the equivalent single-loop transmissions are n.m.p. (see Sec. 7-10). The technique can still be used, but success can be guaranteed only if the performance tolerances are compatible with the n.m.p. character, just as in the single-loop system (see Chapter 4 and Appendix F-2).[37] For this design example, *det* $\boldsymbol{P}(s)$ has all its zeros in the interior of the LHP, so this problem does not arise.

6-2.1 PERFORMANCE TOLERANCES

Performance tolerances are on the magnitudes of the elements of the *3x3* closed loop system transmission matrix $\boldsymbol{T} = [t_{ij}(s)]$, thus, in the frequency domain $|t_{ij}(j\omega)|$ suffices when they are m.p. In fact, as previously noted,[62] time-domain specifications on the system output and on as many of its derivatives as desired, may then be achieved by means of bounds on $|t_{ij}(j\omega)|$. They are shown in Fig. 6.la

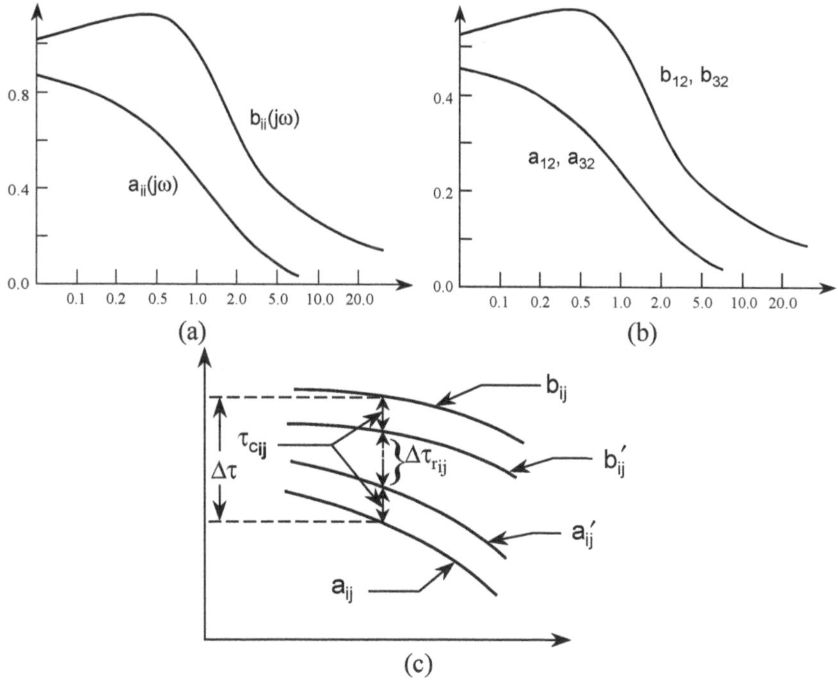

Fig. 6.1 Tolerances (a) on $|t_{ii}(j\omega)|$, (b) on $|t_{12}(j\omega)|$, $|t_{32}(j\omega)|$, and (c) allocation.

for $|t_{ii}(j\omega)|$, $i = 1,2,3,$ and in Fig. 6.lb for $|t_{12}|$, $|t_{32}|$. The four remaining specifications are $|t_{ij}(j\omega)| \le 0.1$ for all ω.

The first five are called *interacting* and the latter four *basically non-interacting* (BNIC), because ideally zero t_{13}, t_{21}, t_{23}, t_{31} are desired for all ω. But this is impossible because of parameter uncertainty, so upper bounds must be assigned. The matter of m.p. or n.m.p. character of the BNIC elements is of no concern; t_{12}, t_{32} are deliberately chosen interacting in order to obtain a design problem with considerable variety. The functions which satisfy the assigned tolerances on t_{ij} constitute a set τ_{ij}.

6-2.2 SENSITIVITY ANALYSIS[40]

An important aspect in the design of a control system is the insensitivity of the system outputs to items such as: sensor noise, parameter uncertainty, cross-coupling effects, and external system disturbances (see Chapter 8). The effect of these items on system performance can be expressed in terms of the *sensitivity function*[1]

$$S_{\delta}^{T} = \frac{\delta}{T}\left[\frac{\partial T}{\partial \delta}\right] \tag{6.4}$$

where δ represents the variable parameter in T. Figure 6.2 is used for the purpose of analyzing the sensitivity of a system to three of these items.

Using the linear superposition theorem, where

$$Y = Y_R + Y_C + Y_N$$

and

$$T_R = \frac{Y_R}{R} = \frac{FL}{1+L} \quad \text{(a)}$$

$$T_N = \frac{Y_N}{N} = \frac{-L}{1+L} \quad \text{(b)} \tag{6.5}$$

$$T_C = \frac{Y_C}{C} = \frac{P}{1+L} \quad \text{(c)}$$

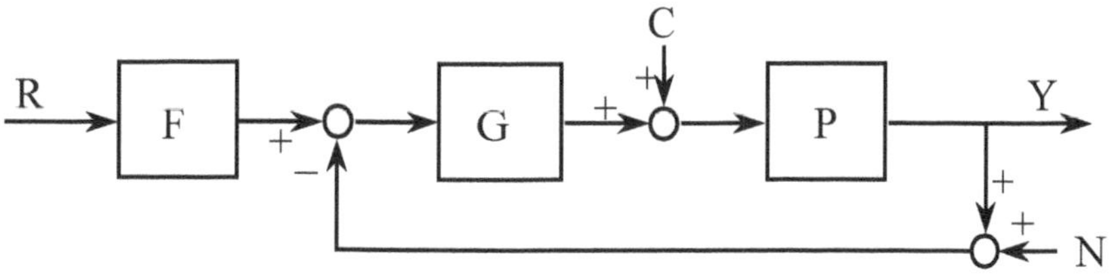

Fig. 6.2 An example of system sensitivity analysis.

and $L = GP$, the following transfer functions and sensitivity functions (where $\delta = P$) are obtained, respectively:

$$S_P^{T_R} = \frac{FG}{1+L} \quad \text{(a)}$$

$$S_P^{T_N} = \frac{-G}{1+L} \quad \text{(b)} \qquad \textbf{(6.6)}$$

$$S_P^{T_C} = \frac{1}{1+L} \quad \text{(c)}$$

Since sensitivity is a function of frequency, it is necessary to achieve a slope for $Lm\,\mathbf{L}_o(j\omega)$ that minimizes the effect on the system due to sensor noise. This is the most important case, since the "minimum BW" of Eq. (6.5b) tends to be greater than the BW of Eq. (6.5a), as illustrated in Fig. 6.3. Based on the magnitude characteristic of $\mathbf{L}_o$ for low and high frequency ranges, then:

For the low frequency range, where $|\mathbf{L}(j\omega)| >> 1$, from Eq. (6.6b)

$$\left|\mathbf{S}_{\mathrm{P}}^{\mathrm{T_N}}\right| \approx \left|\frac{-1}{\mathbf{P}(j\omega)}\right| \qquad \textbf{(6.7)}$$

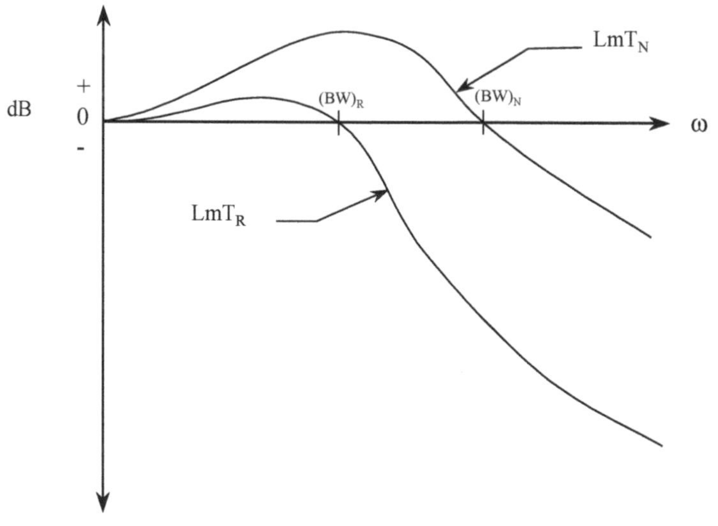

Fig. 6.3 Frequency response characteristics for the system of Fig. 6.2.

For the high frequency range, where $|L(j\omega)| << 1$, from Eq. (6.6b)

$$\left|\mathbf{S}_{\mathbf{P}}^{\mathrm{T_N}}\right| \approx |\,\mathbf{-G(j\omega)}\,| = \left|\frac{\mathbf{-L(j\omega)}}{\mathbf{P(j\omega)}}\right| \tag{6.8}$$

The BW characteristics of the open-loop function $L(j\omega)$, with respect to sensitivity, are illustrated in Fig. 6.4. As seen from this figure, the low frequency sensitivity given by Eq. (6.7) is satisfactory but the high frequency sensitivity given by Eq. (6.8) is unsatisfactory since it can present a serious noise rejection problem.

Based upon the analysis of Fig. 6.4, it is necessary to try to make the phase margin frequency ω_ϕ (the loop transmission BW), small enough in order to minimize the sensor noise effect on the system's output. For most practical systems $n \geq w + 2$ and

$$\int_0^\infty \log\,[S_P^T]d\omega = 0 \tag{6.9}$$

$$\begin{aligned} |S_P^T N| < 1 \;\rightarrow\; \log[S_p^T N] < 0 \quad (a) \\ |S_P^T N| > 1 \;\rightarrow\; \log[S_P^T N] > 0 \quad (b) \end{aligned} \tag{6.10}$$

Thus, the designer must try to locate the condition of Eq. (6.10b) in the "high

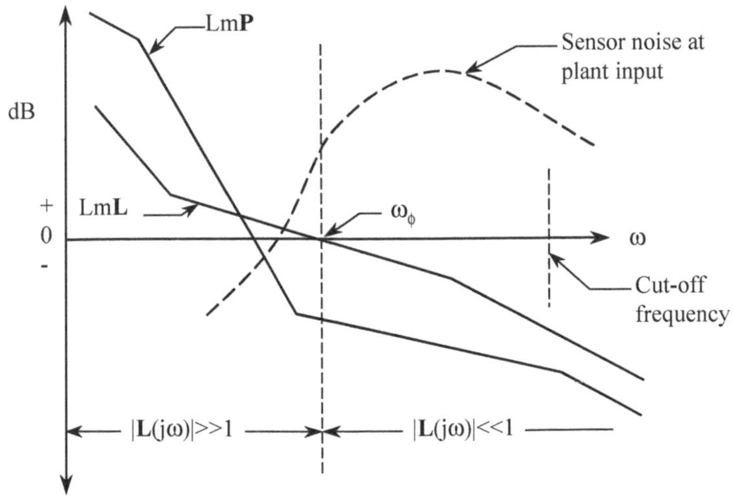

Fig. 6.4 Bandwidth characteristics of Fig. 6.2.

frequency" (h.f.) range where the system performance is not of concern, i.e., the noise effect on the output is negligible.

The analysis for external disturbance effect (see Chapter 8) on the system output is identical to that for cross-coupling effects. For either case, low sensitivity is conducive to their rejection.

6-2.3 SIMPLIFICATION OF THE SINGLE-LOOP STRUCTURES

For the second row of Fig. 5.11 $y_{ij}(s)$ has the two components due to $r_j(s) = 1$ and to $c_{ij}(s)$. Their respective outputs are:

$$y_{r_i} = \frac{f_{ij} L_i}{1 + L_i} \quad (a) \qquad y_{c_{ij}} = \frac{c_{ij} q_{ii}}{1 + L_i} \quad (b) \qquad where\ L_i = g_i q_{ii} \quad (c) \tag{6.11}$$

At a fixed $\boldsymbol{P} \in \mathcal{P}$ (and hence $m^2\ q_{ij}$) $|\ \boldsymbol{y}_{cij}|_{max}$ occurs at

$$|\ \boldsymbol{c}_{ij}|_{max} = \left| \sum_{k \neq i} \frac{\boldsymbol{t}_{kj}}{\boldsymbol{q}_{ik}} \right|_{max}$$

The relative phases of the $\boldsymbol{t}_{kj}/\boldsymbol{q}_{ik}$ are not known so the extreme magnitude of

$$|\ c_{ij}|_{max} = \left| \sum_{k \neq i} \frac{\mathbf{t}_{kj}}{\mathbf{q}_{ik}} \right|_{max} = \left| \sum_{k \neq i} \frac{b_{kj}}{\mathbf{q}_{ik}} \right|_{max} \tag{6.12}$$

resulting from among the J plants must be used in the design. That is, where b_{ii} and b_{ki} represent the upper bounds of the control ratio t_{ii} and t_{ki}, respectively, in Fig. 6.1a and b. Also shown in this figure are the corresponding lower bounds a_{ii} and a_{ki}, respectively.

There are two kinds of performance tolerances, i.e., tolerances on t_{ii} and on t_{ij} for $i \neq j$ (BNIC). For the BNIC ($k,i = 1,3;\ 2,1;\ 2,3$; etc.) it is necessary that

$$|\ \mathbf{y}_{r_i} + \mathbf{y}_{c_{ij}}| = |\ \tau_{ij}(j\omega) + \tau_{c_{ij}}|_{max} \leq |\ \tau_{ij}|_{max} + \tag{6.13}$$

$$|\ \tau_{c_{ij}}|_{max} = |\ \tau_{c_{ij}}|_{max} \leq b_{ij}(j\omega)$$

because the relative phase of the two terms is not known, so it is clearly best to force $\tau_{ij} = 0$ by setting $f_{ij} = 0\ (i \neq j)$ in Eq. (6.11) for ($1,3;\ 2,1;\ 2,3;\ 3,1$). Let some fixed $\boldsymbol{P}$ be chosen as the nominal plant matrix $\boldsymbol{P}_o = [p_{ij}]$, generating a nominal $\boldsymbol{P}_o^{-1} = [1/q_{ijo}]$ and the nominal loop transmissions $L_{i_o} = g_i q_{ii_o}$, so $L_i = L_{i_o} q_{ii}/q_{ii_o}$. In this example, the nominal plant parameters of Eq. (6.3) are assumed as:

$$\begin{array}{lll} A_{11} = 2 & B_{11} = 0.1 & A_{12} = 1.1 \\ B_{12} = -1 & B_{13} = 4 & B_{21} = 2 \\ A_{22} = 5 & B_{22} = 5 & B_{23} = 4 \\ A_{31} = 1.8 & B_{31} = 1 & B_{32} = 20 \\ B33 = 10 & E = 2 & F = 2 \end{array} \qquad \textbf{(6.14)}$$

Using Eqs. (6.11b) and (6.12) the condition given by Eq. (6.13) can now be written (for the BNIC t_{ij}) to yield the following design specification:

$$\left| \frac{1}{1 + L_{i_o} \dfrac{q_{ii}}{q_{ii_o}}} \right| \le \frac{b_{ij}}{c_{ij}\, q_{ii}} = \frac{b_{ij}}{\sum\limits_{k \ne i} \left| \dfrac{q_{ii}}{q_{ik}} \right| b_{ki}} \qquad \textbf{(6.15)}$$

Specification type described by Eq. (6.15) is denoted as a *D-type*, specifically D_{ij}.

The specifications on the interacting t_{ij} terms (all t_{ii} and t_{12}, t_{32}) are, using Eq. (6.11),

$$a_{ij}(\omega) \le |\tau_{ij}(j\omega) + \tau_{Cik}(j\omega)| \le b_{ij}(\omega)$$

with a_{ij}, b_{ij} being the designated bounds shown in Fig. 6.1. Suppose that due to the uncertainty in q_{ii}, $|\tau_{ij}|$ of Eq. (6.11) is in the interval $[a'_{ij}, b'_{ij}]$ [see Fig. 6.1c]. Since the relative phases of of $\tau_{ij}, \tau_{c_{ij}}$ are not known a priori, it is necessary that:

$$|b'_{ij} + \tau_{cij}| \le b_{ij}, \qquad |a'_{ij} - \tau_{cij}| \ge a_{ij} \qquad \textbf{(6.16)}$$

From Eq. (6.11) the above may be summarized by Eq. (6.16) and

$$\Delta \left| \frac{L_i}{1 + L_i} \right| \le \frac{b'_{ij}}{a'_{ij}} \quad \text{(a)} \qquad \left| \frac{1}{1 + L_i} \right| \le \left| \frac{\tau_{cij}}{c_{ij} q_{ii}} \right|_{max} = \frac{|\tau_{cij}|}{\sum\limits_{k \ne i} \left| \dfrac{q_{ii}}{q_{ik}} \right|_{max} b_{ki}} \quad \text{(b)} \qquad \textbf{(6.17)}$$

A_{11} , A_{12}	B_{11} , B_{12} , D_{13}
$\Delta\left\lvert\frac{\underline{L}_1}{1+\underline{L}_1}\right\rvert \le \frac{b'_{11}}{a'_{11}}, \frac{b'_{12}}{a'_{12}}$	$\frac{1}{\lvert 1+\underline{L}_1\rvert} \le \frac{\tau_{c_{11}}}{b_{21}\left\lvert\frac{\underline{q}_{11}}{\underline{q}_{12}}\right\rvert + b_{31}\left\lvert\frac{\underline{q}_{11}}{\underline{q}_{13}}\right\rvert}, \frac{\tau_{c_{12}}}{b_{22}\left\lvert\frac{\underline{q}_{11}}{\underline{q}_{12}}\right\rvert + b_{32}\left\lvert\frac{\underline{q}_{11}}{\underline{q}_{13}}\right\rvert}, \frac{b_{13}}{b_{23}\left\lvert\frac{\underline{q}_{11}}{\underline{q}_{12}}\right\rvert + b_{33}\left\lvert\frac{\underline{q}_{11}}{\underline{q}_{13}}\right\rvert}$

D_{21}	A_{22}	B_{22} , D_{23}
$\frac{1}{\lvert 1+\underline{L}_2\rvert} \le \frac{b_{21}}{b_{11}\left\lvert\frac{\underline{q}_{22}}{\underline{q}_{21}}\right\rvert + b_{31}\left\lvert\frac{\underline{q}_{22}}{\underline{q}_{23}}\right\rvert}$	$\Delta\left\lvert\frac{\underline{L}_2}{1+\underline{L}_2}\right\rvert \le \frac{b'_{22}}{a'_{22}}$	$\frac{1}{\lvert 1+\underline{L}_2\rvert} \le \frac{\tau_{c_{22}}}{b_{12}\left\lvert\frac{\underline{q}_{22}}{\underline{q}_{21}}\right\rvert + b_{32}\left\lvert\frac{\underline{q}_{22}}{\underline{q}_{23}}\right\rvert}, \frac{b_{23}}{b_{13}\left\lvert\frac{\underline{q}_{22}}{\underline{q}_{21}}\right\rvert + b_{33}\left\lvert\frac{\underline{q}_{22}}{\underline{q}_{23}}\right\rvert}$

D_{31} , B_{32}	A_{32} , A_{33}	B_{33}
$\frac{1}{\lvert 1+\underline{L}_3\rvert} \le \frac{b_{31}}{b_{11}\left\lvert\frac{\underline{q}_{33}}{\underline{q}_{31}}\right\rvert + b_{21}\left\lvert\frac{\underline{q}_{33}}{\underline{q}_{32}}\right\rvert}, \frac{\tau_{c_{32}}}{b_{12}\left\lvert\frac{\underline{q}_{33}}{\underline{q}_{31}}\right\rvert + b_{22}\left\lvert\frac{\underline{q}_{33}}{\underline{q}_{32}}\right\rvert}$	$\Delta\left\lvert\frac{\underline{L}_3}{1+\underline{L}_3}\right\rvert \le \frac{b'_{32}}{a'_{32}}, \frac{b'_{33}}{a'_{33}}$	$\frac{1}{\lvert 1+\underline{L}_3\rvert} \le \frac{\tau_{c_{33}}}{b_{13}\left\lvert\frac{\underline{q}_{33}}{\underline{q}_{31}}\right\rvert + b_{23}\left\lvert\frac{\underline{q}_{33}}{\underline{q}_{32}}\right\rvert}$

Table 6.1 Constraints on the equivalent single-loop structures

Equation (6.17a) is denoted as A_{ij}, Eq. (6.17b) as B_{ij}. Note that Eq. (6.17b) is almost the same as D_{ij} of Eq. (6.15) except that $\iota_{c_{ij}}$ is not known and must be suitably chosen together with a'_{ij}, b'_{ij} subject to Eq. (6.16). The results of this section are summarized in Table 6.1.

6-3 HIGH FREQUENCY RANGE ANALYSIS

The analysis in the h.f. range, for the case of $i = j$, can be based upon the fact that the filter f_{ii} when designed to yield the desired response has the value $|f_{ii}(j\omega)| \approx 0$. Thus, $t_{r_{ij}}$ in Eqs. (5.41) and (5.42) can be neglected. For a *2x2* system, consider the bounds shown in Fig. 6.5, for $i = j = 1$. Based upon the knowledge of the control system to be designed, the designer selects the value of $\omega_{h.f.}$ above which the cross-coupling effects have no effect on system performance. Thus, as shown in Fig. 6.5, the lower bound a_{11} is "dropped" to zero and only the upper bound b_{11} is used – eliminating the need to determine b'_{11}. Thus, from Eq. (5.81):

$$|1 + \boldsymbol{L}_1| \geq \frac{b_{21}}{b_{11}}\left|\frac{q_{11}}{q_{12}}\right| = \lambda\left|\frac{q_{11}}{q_{12}}\right| \quad \text{where } \lambda = \frac{b_{21}}{b_{11}} \tag{6.18}$$

In a similar manner, for L_2, from Eq. (5.82), obtain:

$$|1 + \boldsymbol{L}_2| \geq \frac{b_{11}}{b_{21}}\left|\frac{q_{22}}{q_{21}}\right| = \frac{1}{\lambda}\left|\frac{q_{22}}{q_{21}}\right| \tag{6.19}$$

Remember that these equations apply in the h.f. range where the lower bounds are neglected. Notes: templates, utilizing the J plants, are obtained of the ratios of the **q**'s in Eqs. (6.18) and (6.19).

Note: (1) t_{21} still needs to be specified even when $f_{21} = 0$; (2) the λ ratio can be anything; and based upon the location of "noisy sensors" might require a trade-off in the phase margin frequency specification for each loop.

Multiplying Eq. (6.18) by Eq. (6.19) yields:

$$|1 + \mathbf{L}_1||1 + \mathbf{L}_2| \geq \left|\frac{\mathbf{q}_{11}\mathbf{q}_{22}}{\mathbf{q}_{12}\mathbf{q}_{21}}\right| = \left|\frac{\mathbf{p}_{12}\mathbf{p}_{21}}{\mathbf{p}_{11}\mathbf{p}_{22}}\right| \tag{6.20}$$

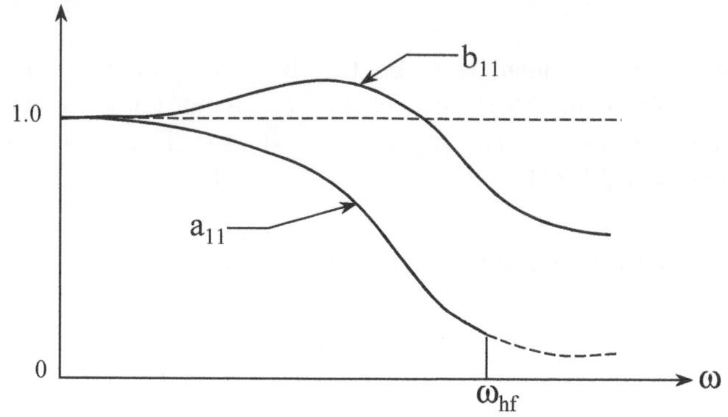

Fig. 6.5 High-frequency (h.f.) cut-off for a_{11}.

For the *2x2* plant, the diagonal dominance condition, for $\omega \to \infty$ (say $\omega > \omega_h$) is given by:

$$\left|\frac{p_{12}\,p_{21}}{p_{11}\,p_{22}}\right| < 1 \tag{6.21}$$

which must be true over the entire range of $P \in \mathcal{P}$. Let

$$\gamma_{12} = \frac{p_{12}\,p_{21}}{p_{11}\,p_{22}} \tag{6.22}$$

Remember that

$$P = \begin{bmatrix} p_{11} & p_{12} \\ p_{21} & p_{22} \end{bmatrix} \to P^{-1} = \frac{1}{\det P}\begin{bmatrix} p_{22} & -p_{12} \\ -p_{21} & p_{11} \end{bmatrix}$$

and

$$q_{11} = \frac{\det P}{p_{22}} = \frac{p_{11}\,p_{22} - p_{12}\,p_{21}}{p_{22}} = p_{11}(1-\gamma) \qquad \text{(a)}$$

(6.23)

$$q_{22} = \frac{\det P}{p_{11}} = p_{22}(1-\gamma) \qquad \text{(b)}$$

Thus, it is the zeros of the q's . not the zeros of the p's which determine if the system is m.p.

6-4 STABILITY ANALYSIS

A $2x2$ plant is used to discuss the stability analysis for a MIMO system. The Nyquist stability criterion[1] should be applied utilizing a polar plot analysis, especially for n.m.p. q's. By substituting the equation for t_{21} into the equation for t_{11} in Eq. (5.55), *with* $f_{ij} = 0$ *for* $i \neq j$, yields:

$$t_{11} = \frac{f_{11}L_1(1+L_2)}{(1+L_1)(1+L_2) - \gamma_{12}} \tag{6.24}$$

Thus, the denominator of Eq. (6.24) yields the characteristic equation:

$$(1+L_1)(1+L_2) - \gamma_{12} = 0 \tag{6.25}$$

At "low" frequencies $|L_1|$ and $|L_2|$ are both $>> |\gamma_{12}|$ and at h.f. $|L_1| = |L_2| \approx 0 \rightarrow 1 - \gamma_{12} = 0$.

Figure 6.6 shows the polar plot of $(1 + L_1)(1 + L_2)$ for all $P \in \mathcal{P}$(shaded area) and the locus of critical points represented by γ_{12}. For a LTI plant, with no parameter uncertainty, whose characteristic equation is $G(s)H(s) + 1 = 0$, the "locus of critical points" is the $-1 + j0$ point. Thus, for this $2x2$ plant the polar plot in Fig. 6.6 can not encircle the γ_{12} locus for $P \in \mathcal{P}$ or $(1 + L_1)(1 + L_2) - \gamma_{12}$ can not encircle the origin. Since the diagonal dominance condition must be satisfied, i.e., as $\omega \rightarrow \infty$ $|p_{11}p_{22}| >> |p_{12}p_{21}|$ then the magnitude $|\gamma_{12}|$ better be < 1 or the γ_{12} locus will be encircled, resulting in an unstable system.

6-5 EQUILIBRIUM AND TRADE-OFFS

There are interactions between the rows of the SFG in Fig. 5.11 via the specifications b_{ij}. Consider the following two examples.

Example 6.1 Trade-off Example #1

The SFG for a $2x2$ plant is shown in Fig. 6.7. For this example assume $|t_{ij}|_{max} \equiv b_{ij}$ and let the ϕ's represent the actual values of the outputs (see Sec. 5-6.2).

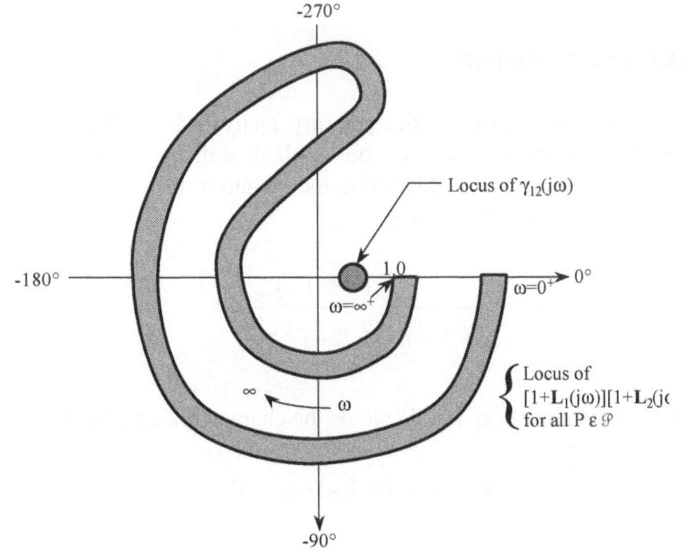

Fig. 6.6 Polar plots of Eq. (6.25).

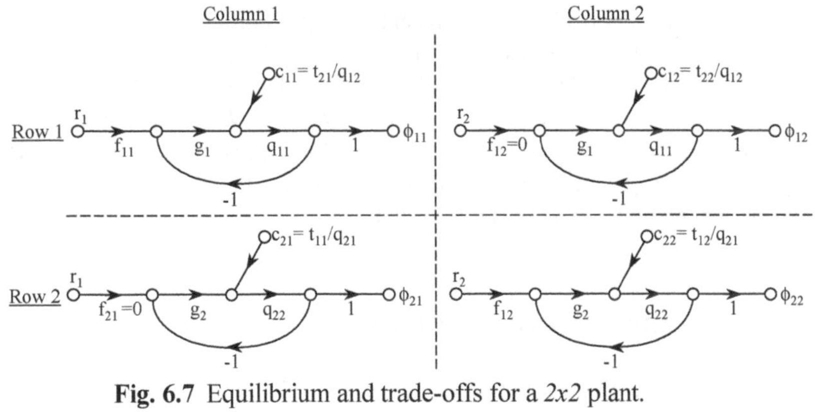

Fig. 6.7 Equilibrium and trade-offs for a *2x2* plant.

(a) **Consider** Suppose, for a fixed value of ω_i:

--- ϕ_{12} "dominates" in the first row of Fig. 6.7, i.e. dominates L_1 and
--- ϕ_{21} "dominates" in the second row, i.e., dominates L_2

then if the value of the specification b_{22} is lowered it helps ϕ_{12} which decreases the demand on L_1.

(b)**Assume** The terms that dominate in each row are in different columns. Thus, it is possible to ease the design by being concerned with one or more b_{ij}'s. For example:

	Column 1	**Column 2**	
	$c_{11} = \dfrac{-b_{21}}{q_{12}}$	$c_{12} = \dfrac{-b_{22}}{q_{12}}$	**Row 1**
	$c_{21} = \dfrac{-b_{11}}{q_{21}}$	$c_{22} = \dfrac{-b_{12}}{q_{21}}$	**Row 2**

If the $1,1$ and $2,2$ terms dominate then it is possible to ease the design by decreasing b_{21} and b_{12}.

Example 6.2 Trade-off Example #2

For Column 1, from the SFG of Fig. 5.11 for a $3x3$ plant:

$$c_{11} = -\left(\frac{b_{21}}{q_{12}} + \frac{b_{31}}{q_{13}}\right), \qquad c_{21} = -\left(\frac{b_{11}}{q_{21}} + \frac{b_{31}}{q_{23}}\right), \qquad c_{31} = -\left(\frac{b_{11}}{q_{31}} + \frac{b_{21}}{q_{32}}\right)$$

The same process, as done in Example 6.1, is used here in doing a "trade-off" on the specifications for one or more b_{ij} in order to hopefully achieve the desired dominance through the analysis of:

$$|1 + \mathbf{L}_i| \geq \frac{|q_{11}|}{b_{11}}\left[\frac{b_{21}}{|q_{12}|} + \frac{b_{31}}{|q_{13}|}\right], \geq \frac{|q_{22}|}{b_{21}}\left[\frac{b_{11}}{|q_{21}|} + \frac{b_{31}}{|q_{23}|}\right],$$

$$\geq \frac{|q_{33}|}{b_{31}}\left[\frac{b_{11}}{|q_{31}|} + \frac{b_{21}}{|q_{32}|}\right] \qquad \textbf{(6.26)}$$

For example, suppose b_{21} is decreased from 0.1 to δ. (Decrease of b_{21} is permissible because the BNIC $|t_{21}| \leq b_{21}$ is required.) This decreases $|c_{11}|$ and $|c_{31}|$ in Fig. 5.11, which eases the burden on L_1 and L_3 for satisfying $y_{11} \in \tau_{11}$ and

$y_{31} \in \tau_{31}$ assuming that y_{11} and y_{31} dominate L_1 and L_3, respectively. But it makes it harder on L_2, because now $|y_{21}| \leq \delta$ instead of 0.1. This does not matter if y_{22} dominates L_2. If so, b_{21} can be decreased until say y_{21} imposes the same burden on L_2 as does y_{22}, denoted by $y_{21}\tilde{\ }y_{22}$. Any further decrease of b_{21} involves "trade-off," i.e., a sacrifice of L_2 for the sake of L_1 and/or L_3. However, it is conceivable that before $y_{21}\tilde{\ }y_{22}$ occurs, either y_{12} (or y_{13})$\tilde{\ }y_{11}$, or y_{32} (or y_{33})$\tilde{\ }y_{31}$ will impose the same burden. There is a bewildering multitude of possibilities and options.

As illustrated by these examples, it may be possible to modify one or more b_{ij} until the dominant terms lie in the same column. To accomplish this may require a "trade-off" in one or more of the system's performance specifications. Is there a simple rule for determining when 'free' easing of burdens is no longer possible, and only "trade-off" is available? This question is partly answered by the following definition and theorem.

DEFINITION

Design equilibrium exists when it is impossible to reduce the burden on any L_i, without increasing it on some other L_j.

THEOREM

A necessary and sufficient condition for design equilibrium is when all the L_i are dominated by the members of the same column, i.e. y_{ij}, $i = 1,2,...,m$ dominate the L_i, with j any fixed column. This does not preclude $y_{ij}, \sim y_{ik} \sim ...$ for one or more i,k. See Ref. 4 for the proof.

Clearly, a number of equilibriums can exist, as any column may be chosen. This variety offers the designer very useful design freedom. He may wish to economize on some L_x, according to the circumstances - for example, if the sensor on y_x is noisier than the others, or q_{xx} has elastic modes in a rather low ω range, so it is important to reduce $|L_x(j\omega)|$ rapidly compared with ω. It is also emphasized that all the above discussion is at a fixed ω_i value. It may be desirable and it is certainly possible to have different columns dominating at different ω_i values, and the design technique offers this kind of flexibility. The potential profit to be gained by exploiting column dominance is greater at high frequency because the δ_R tolerances are greater in this range of ω, i.e., $\delta_R = b_{ii} - a_{ii}$ is greater than in the "low frequency" range.

The above analysis led to the low and medium frequency bounds on L_{i_o} shown in Fig. 6.8.

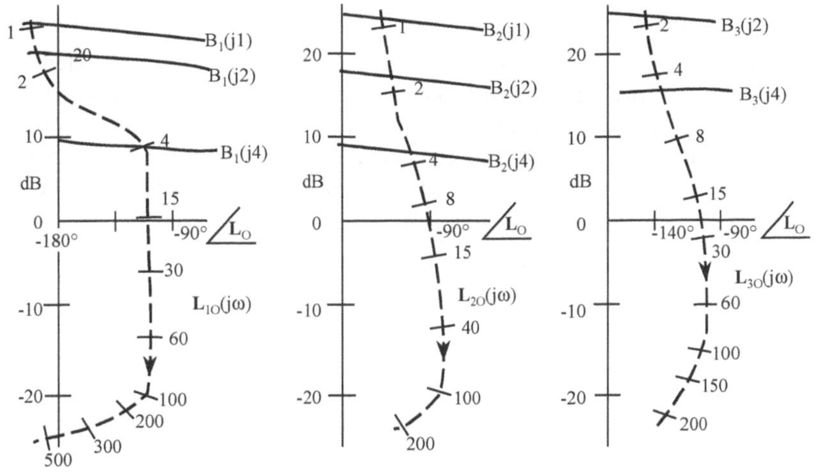

Fig. 6.8 Bounds in the Nichols Chart on $L_{i_O}(j\omega)$, $i=1,2,3$, and L_i designed.

TRADE-OFF

After equilibrium has been reached, it is still possible and may be desirable to do "trade-off" – that is, sacrifice one or more L_{i_o} for others, e.g., in the situations noted above. Thus, say column *1* dominates all L_i but it is highly desirable to reduce L_1. In Fig. 5.11 one or both of b_{21}, b_{31} may be reduced thereby decreasing $|c_{11}|$ but making the specifications harder on y_{21}, y_{31}. Assuming that $y_{11} \cong y_{12}$ then it is also necessary to reduce one or both of b_{22}, b_{32}. At low ω values, the amount of such improvement is quite limited, because $b_{ij} > a_{ij}$ (in the interacting y_{ij}), and the difference between b_{ij} and a_{ij} is small at low ω (see Fig. 6.5). Also, as $b_{ij} \to a_{ij}$, the sacrifice of $L_i \to \infty$, for the sake of very small improvement of some other L. In serious situations, the designer can then consider modifying the (a_{ij}, b_{ij}) tolerances. This is another of the valuable insights offered by the design technique. The trade-off situation is radically different in the high frequency range.

6.5.1 TRADE-OFF IN HIGH FREQUENCY RANGE[4]

At large ω the $a_{ij}(j\omega)$ of the interacting t_{ij} of Figs.6.1a and b become so small that they can be made zero. Hence, the A_{ij} in Table 6.1 can be ignored and each $\tau_{c_{ij}} = b_{ij}$, so all the constraints in the table are now of the D type, i.e., replace the B_{ij} by D_{ij}. There is now no limit on the permissible reduction of the b_{ij} and the problems of "equilibrium" and "trade-off" become so simplified, as to permit analytical derivation of the bounds $\boldsymbol{B}_i(j\omega)$ on $\boldsymbol{L}_{i_O}(j\omega)$. optimally First the claim that the b_{ij} are chosen so that $D_{1i} \sim D_{1j}, \ldots, D_{mi} \sim D_{mj}$, for all i,j is easily proven. Suppose, (with no loss of generality), that D_{11} dominates L_1. Reduce b_{12} (recall it has replaced τ_{c12}) until D_{12}~D_{11}. This certainly doesn't hurt L_1 and helps D_{i2}, $i \neq 1$. Hence, if any D_{i2} dominates L_i this reduction of b_{12} helps L_i. If not, a single D_{i2} does so, then there is at least no harm in this reduction of b_{12}. Similarly, reduce b_{13}, until D_{13}~D_{12}~D_{11}. Using exactly the same argument, one can deduce that D_{23}~D_{22}~D_{21}, etc., for each row.

The above result applies even if the $|q_{11}/q_{12}|_{max}$ are different in the elements of row 1, etc. However, if they are the same, as they are in this case then let $v_{ij} \triangleq |q_{ii}/q_{ij}|_{max}$ which appear in row i of Table 6.1

The above give

$$\frac{b_{11}}{v_{12}\, b_{21}+v_{13}\, b_{31}}=\frac{b_{12}}{v_{12}\, b_{22}+v_{13}\, b_{32}}=\frac{b_{13}}{v_{12}\, b_{23}+v_{13}\, b_{33}} \triangleq \lambda_1 \tag{6.27}$$

Consider the m equations

$$\frac{b_{k1}}{\sum\limits_{i\neq k} v_{ki} b_{i2}}=\frac{b_{k2}}{\sum\limits_{i\neq k} v_{ki} b_{i2}}=\ldots \triangleq \lambda_k \tag{6.28}$$

$$\frac{b_{k1}}{\sum\limits_{i\neq k} v_{ki} b_{i1}}=\lambda_k \qquad for\ k=1,2,\ldots,m \tag{6.29}$$

which constitute a linear homogeneous set in the m b_{k1}. For a solution to exist the determinant of the coefficient matrix must be zero. This matrix, denoted by ϕ has v_{ij} for its off-diagonal and λ_I as its diagonal elements. Exactly the same matrix results by taking any α value in the set of m equations

$$\frac{b_{kx}}{\sum v_{ki}\, b_{i\alpha}}=\lambda_k, \quad k=1,2,\ldots,m \tag{6.30}$$

The choice of the λ_k is up to the designer but he or she can choose only $m-1$ of them. The condition $det\ \phi = 0$, determines the last λ. In this way, the designer can deliberately sacrifice some loops in order to help others.

For any fixed j, the above set of m equations in b_{kj} ($k = 1,\ldots, m$) is homogeneous, so the ratios b_{kj}/b_{1j} ($k = 2,\ldots$,m) are determined by the choice of the λ_k. Even if the b_{kj}/b_{1j} emerge very large, one can always make the b_{1j} small enough so that the b_{kj} satisfy the tolerances. Hence, it is not necessary to solve the equations for the b_{kj}/b_{1j}. Of course, this is true only because all $a_{kj} = 0$. If the above approach is used at a lower ω, where some or all $a_{kj} \neq 0$, the b_{kj}/b_{1j} should be calculated, in order to be certain that the tolerances are not violated.

Example 6.3 Trade-off Example #3

Equations (5.39) - (5.56) for the *2x2* plant, where $f_{ij} = 0 \ (i \neq j)$, in the high frequency range simplify to:

	Column 1	**Column 2**
Row 1	$\left\| t_{11} \approx \dfrac{c_{11} q_{11}}{1 + L_1} \right\| \leq b_{11} \ (a)$	$\left\| t_{12} \approx \dfrac{c_{12} q_{11}}{1 + L_1} \right\| \leq b_{12} \ (b)$
Row 2	$\left\| t_{21} \approx \dfrac{c_{21} q_{22}}{1 + L_2} \right\| \leq b_{21} \ (c)$	$\left\| t_{22} \approx \dfrac{c_{22} q_{22}}{1 + L_2} \right\| \leq b_{22} \ (d)$

(6.31)

Substitute into these equations the expressions for c_{ij}'s from Example 6.1 to obtain the following set of equations:

	Column 1	**Column 2**
Row 1	$\left\| \dfrac{-b_{21} q_{11} / q_{12}}{1 + L_1} \right\| \leq b_{11} \ (a)$	$\left\| \dfrac{-b_{22} q_{11} / q_{12}}{1 + L_1} \right\| \leq b_{12} \ (b)$
Row 2	$\left\| \dfrac{-b_{11} q_{22} / q_{21}}{1 + L_2} \right\| \leq b_{21} \ (c)$	$\left\| \dfrac{-b_{12} q_{22} / q_{21}}{1 + L_2} \right\| \leq b_{22} \ (d)$

(6.32)

The equations in Eq. (6.32) are rearranged as follows, in order to perform a frequency domain analysis:

	Column 1		Column 2	
Row 1	$\lvert 1+L_1 \rvert \geq \left\lvert \frac{q_{11}}{q_{12}} \right\rvert \left\lvert \frac{-b_{21}}{b_{11}} \right\rvert$ (a)		$\lvert 1+L_1 \rvert \geq \left\lvert \frac{q_{11}}{q_{12}} \right\rvert \left\lvert \frac{-b_{22}}{b_{12}} \right\rvert$ (b)	**(6.33)**
Row 2	$\lvert 1+L_2 \rvert \geq \left\lvert \frac{q_{22}}{q_{21}} \right\rvert \left\lvert \frac{-b_{11}}{b_{21}} \right\rvert$ (c)		$\lvert 1+L_2 \rvert \geq \left\lvert \frac{q_{22}}{q_{21}} \right\rvert \left\lvert \frac{-b_{12}}{b_{22}} \right\rvert$ (d)	

Suppose that

$$\lvert \lambda_1 \rvert = \left\lvert \frac{b_{21}}{b_{11}} \right\rvert > \left\lvert \frac{b_{22}}{b_{12}} \right\rvert = \lvert \lambda_2 \rvert \tag{6.34}$$

and

$$\lvert 1/\lambda_1 \rvert = \left\lvert \frac{b_{11}}{b_{21}} \right\rvert < \left\lvert \frac{b_{12}}{b_{22}} \right\rvert = \lvert 1/\lambda_2 \rvert \tag{6.35}$$

then the *1,1* term [Eq. (6.33a)] dominates over the *1,2* term [Eq. (6.33b)] and the *2,2* term [Eq. (6.33d)] dominates over the *2,1* term [Eq. (6.33c)]. Thus a diagonal dominance exists. The *objective* is to balance the dominance across all columns. For example, in this case, b_{22} is increased until both terms match, i.e.,

$$\lambda_1 = -b_{21}/b_{11} = -b_{22}/b_{12} = \lambda_2 = \lambda$$

then

$$\lvert 1+L_1 \rvert \geq \left\lvert \frac{q_{11}}{q_{12}} \right\rvert \lambda$$

for the *1,1* and *1,2* terms of Eq. (6.33) and

$$\lvert 1+L_2 \rvert \geq \left\lvert \frac{q_{22}}{q_{21}} \right\rvert \frac{1}{\lambda}$$

for the *2,1* and *2,2* terms of Eq. (6.33). Thus the dominance is now "balanced" across all columns for this *2x2* plant.

Note that in the high frequency range the problems of equilibrium and trade-off become simplified. This simplification permits an analytical derivation of the bounds $\boldsymbol{B}_i(j\omega)$ on $\boldsymbol{L}_{i_o}(j\omega)$.

6-5.2 SOME UNIVERSAL DESIGN FEATURES

For a *3x3* system the bounds $\boldsymbol{B}_i(j\omega)$ on $\boldsymbol{L}_{3_o}(j\omega)$ tend to be significantly larger (more stringent) than those on $\boldsymbol{L}_{1_o}$, $\boldsymbol{L}_{2_o}$ in the low and middle ω range, where the trade-off opportunities are limited because of the interacting $a_{ij} \neq 0$. In the higher ω range, when Eqs. (6.27) and (6.28) apply, there is much greater scope for such trade-off. This was utilized to help L_3 in the higher ω range by setting $\lambda_1 = \lambda_2 = \lambda$, and solving $det\ \boldsymbol{\Phi} = 0$ for λ_3. As the v_{ij} are functions of ω, $det\ \boldsymbol{\phi} = 0$ gives a relation between the λ_i which is also a function of ω. The result is shown in Fig. 6.9 for $\omega > 100$, for which the $v_{ij}(j\omega)$ are fairly constant with ω. Given $|1 + \boldsymbol{L}_i|^{-1} \leq \lambda_i$ and $\boldsymbol{L}_i = \boldsymbol{L}_{i_o}\boldsymbol{q}_{ii}/\boldsymbol{q}_{ii_o}$, one can find the bounds on $\boldsymbol{L}_{i_o}$, which of course depend on the set $Q = \{q_{ii}\}$. The results are shown in Fig. 6.10, where: Fig. 6.10a corresponds to $\lambda_1 = \lambda_2 = 1$, $\lambda_3 = 2.3$ dB; Fig. 6.10b corresponds to $\lambda_1 = \lambda_2 = 0.7$, $\lambda_3 = 2.7$ dB; and Fig.6.10*c* corresponds to $\lambda_1 = \lambda_2 = 0.5$, $\lambda_3 = 3$ dB. The $\boldsymbol{q}_{ii}(j\omega)$ are almost constant with ω for $\omega_i > 100$, so these bounds apply for all $\omega_i > 100$. As expected, the larger λ becomes, the easier it is on $\boldsymbol{L}_{1_o}$, $\boldsymbol{L}_{2_o}$ and the harder it is on $\boldsymbol{L}_{3_o}$. The big difference in the bounds in Fig. 6.10 is due to the small λ compared with the large λ_3 that were deliberately used. For this reason the actual $\boldsymbol{L}_{3_o}(j\omega)$ was able to reach its final asymptotic slope at $\omega \sim 250$ (see Fig. 6.11) compared with $\sim 450, 230$ for $\boldsymbol{L}_{1_o}$, $\boldsymbol{L}_{2_o}$ even though $|\boldsymbol{L}_{3_o}|$ is much more than $|\boldsymbol{L}_{1_o}|$, $|\boldsymbol{L}_{2_o}|$ at low and medium ω. The $\boldsymbol{L}_{i_o}$ chosen to satisfy the bounds are shown in Fig. 6.11. Design simulation results were highly satisfactory and are shown in Fig. 6.12 for extreme plant parameter combinations.

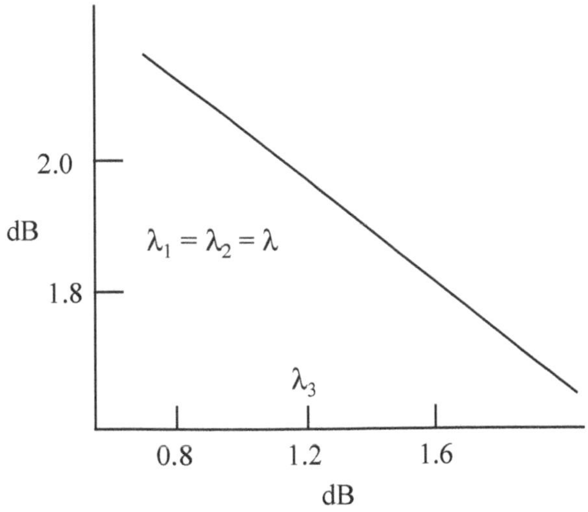

Fig. 6.9 Optimum relation between $\lambda_1 = \lambda_2$ and λ_3 for $\omega > 100$.

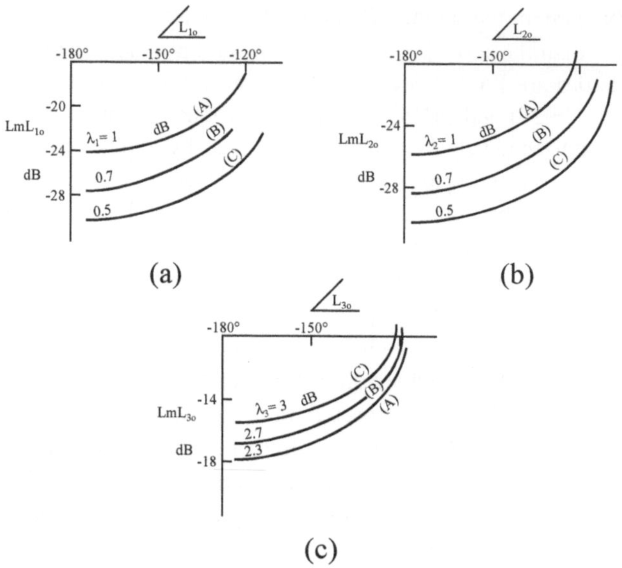

Fig. 6.10 The resulting bounds on L_{1_o} for various λ values for $\omega > 100$.

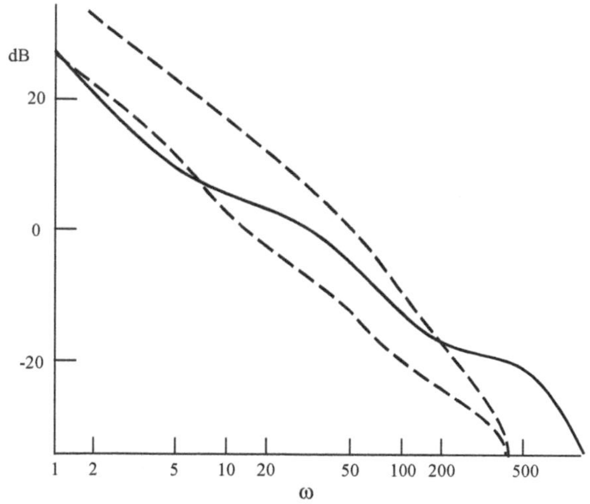

Fig. 6.11 Bode plots of $L_{i_o}(j\omega)$.

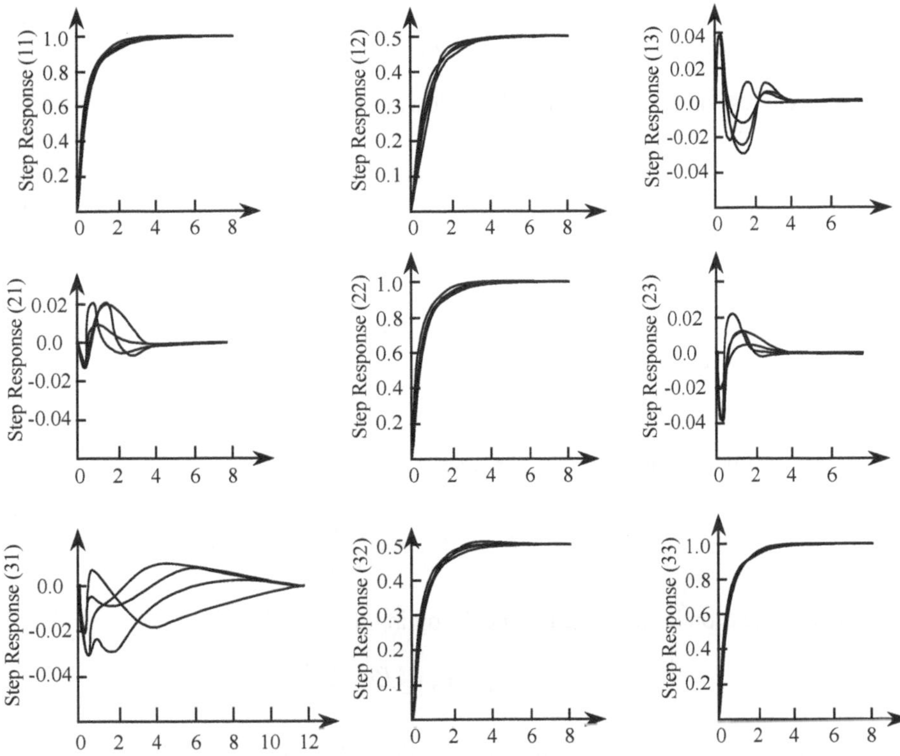

Fig. 6.12 Simulation results for representative step responses.

6-5.3 EXAMPLES – BOUNDS DETERMINATION

Example 6.4 2x2 Plant

Consider the *1,1* MISO equivalent of Fig. 5.11, with $r_1 \neq 0$. Thus:

$$t_{11} = t_{r_{11}} + t_{c_{11}} \quad where$$

$$t_{r11} = \frac{L_1 f_{11}}{1+L_1},\ t_{c11} = \frac{q_{11} c_{11}}{1+L_1},\ c_{11} = \frac{-t_{21}}{q_{12}}$$

The *1,1* cross-coupling expression for the bound determination becomes:

$$t_{c_{11}} = \left| \frac{q_{11} c_{11}}{1 + L_1} \right| \leq \tau_{c_{11}} \qquad \textbf{(6.36)}$$

This expression is rearranged to the following format:

$$| 1 + L_1 | \geq \left| \frac{q_{11} c_{11}}{\tau_{c_{11}}} \right| = \left| \frac{q_{11}}{q_{12}} \right| \frac{b_{21}}{\tau_{c_{11}}} \qquad \textbf{(6.37)}$$

where $|\tau_{21}| = b_{21}$.

Next, consider the $2,1$ MISO equivalent of Fig. 5.11, with $f_{21} = 0$. For this example, since

$$t_{21} \approx t_{c_{21}}$$

Then the upper bound for $t_{c_{21}}$ is b_{21}, i.e.,

$$\tau_{c_{21}} = b_{21} \geq \left| \frac{c_{21} q_{22}}{1 + L_2} \right|$$

This expression is rearranged to the following format:

$$| 1 + L_2 | \geq \frac{b_{11}}{b_{21}} \left| \frac{q_{22}}{q_{21}} \right| \qquad \textbf{(6.38)}$$

where the maximum magnitude b_{11}, from Fig. 6.1c is used for t_{11}.

Equations (6.37) and (6.38) are inverted, respectively, to yield the following equations:

$$\left| \frac{1}{1 + {}_{1}} \right| \leq \left| \frac{{}_{12}}{q_{11}} \right| \frac{\tau_{c_{11}}}{b_{21}} \leq M_{m1} \qquad \text{(a)}$$

$$\textbf{(6.39)}$$

$$\left| \frac{1}{1 + L_2} \right| \leq \left| \frac{{}_{21}}{{}_{22}} \right| \frac{b_{21}}{b_{11}} \leq M_{m2} \qquad \text{(b)}$$

By letting $\ell_1 = 1/L_1$ and $\ell_2 = 1/L_2$, , these equations become

$$\left| \frac{\ell_1}{1 + \ell_1} \right| \leq M_{m_1} \quad \text{(a)} \qquad \left| \frac{\ell_2}{1 + \ell_2} \right| \leq M_{m_2} \quad \text{(b)} \qquad \textbf{(6.40)}$$

which are *now of the mathematical format required in order to use the NC.* Note that:

$$\angle \ell = -\angle \mathbf{L} \tag{6.41}$$

Thus, based on Eqs. (6.40) and (6.41), the *rotated or inverse NC* must be used in conjunction with the templates to determine the cross-coupling effect bounds $\boldsymbol{B}_{c_1}(j\omega_i)$ and $\boldsymbol{B}_{c_2}(j\omega_i)$ for ℓ_1 and ℓ_2, respectively (see Chapter 3). These bounds are transposed to the regular NC to become bounds for $\boldsymbol{L}_1$ and $\boldsymbol{L}_2$, respectively.

Example 6.5 Boundary Determination for Example 6.4

Consider for Eqs. (6.39*b*) and (6.40*b*) that:

$$Lm\left(\frac{b_{21}}{b_{11}}\right) = -40 \text{ dB} \qquad Lm\left(\frac{\boldsymbol{q}_{21}}{\boldsymbol{q}_{22}}\right) = \text{Lm}\,\boldsymbol{q}_{21} - \text{Lm}\,\boldsymbol{q}_{22} \qquad \text{Lm}\,\boldsymbol{q}_{21} = 4 \text{ dB}$$

Thus

$$Lm\left[\frac{1}{1+\boldsymbol{L}_2}\right] = \text{Lm}\left[\frac{\ell_2}{1+\ell_2}\right] \le -40 \text{ dB} + 4 \text{ dB} - \text{Lm}\,\boldsymbol{q}_{22}$$

$$= -36 \text{ dB} - \text{Lm}\,\boldsymbol{q}_{22} \le \text{Lm}\,M_{m2} \tag{6.42}$$

For this example, $\boldsymbol{B}_{c_2}$ is determined in the high frequency range where the templates are essentially a straight line. *Assume* for this example that q_{21} is independent of q_{22} (not true in flight control). The graphical technique for determining this boundary is illustrated in Fig. 6.13 as follows:

Trial 1: From Eq. (6.42)

First trial value for $\boldsymbol{q}_{22}$	M_L	*Actual value of* $\boldsymbol{q}_{22}$
↓	↓	↓

$$\text{Lm}\,\boldsymbol{q}_{22_h} = -36 \text{ dB} - (-18 \text{ dB}) = -18 \text{ dB} \ne \text{Lm}\,\boldsymbol{q}_{22_h} = +6 \text{ dB}$$

$$\text{Lm}\,\boldsymbol{q}_{22_l} = -36 \text{ dB} - (-1 \text{ dB}) = -35 \text{ dB} \ne \text{Lm}\,\boldsymbol{q}_{22_l} = -20 \text{ dB}$$

↓ ↓ ↓

Trial 2:

↓ ↓ ↓

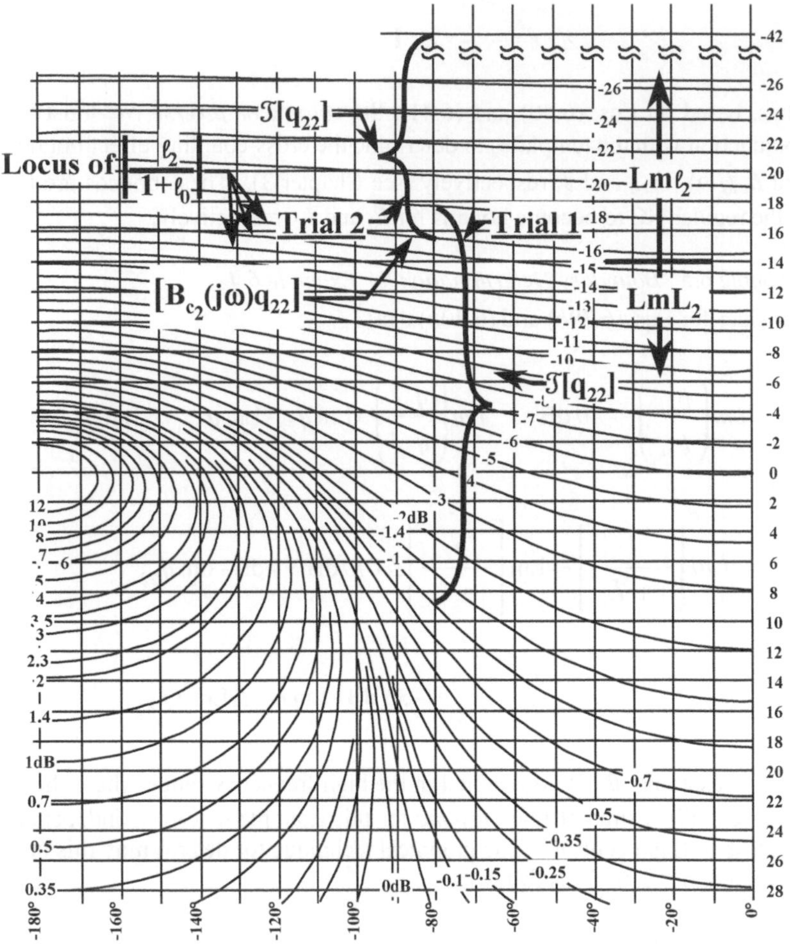

Fig. 6.13 Cross-coupling bound determination for LmL_2 on the inverse Nichols Chart.

$$\text{Lm } \boldsymbol{q}_{22_h} = -36 \text{ dB} - (-42 \text{ dB}) = +6 \text{ dB} = \text{Lm } \boldsymbol{q}_{22_h} = +6 \text{ dB}$$

$$\text{Lm } \boldsymbol{q}_{22_l} = -36 \text{ dB} - (-16 \text{ dB}) = -20 \text{ dB} = \text{Lm } \boldsymbol{q}_{22_l} = -20 \text{ dB}$$

In Trial 2 the trial and actual values agree resulting in obtaining a point on $\boldsymbol{B}_{c_2}$. In a similar manner other points for this boundary can be obtained. This procedure is automated in the MIMO QFT CAD package of Appendix C.

Usually the dominating case(s) on a grid line are quickly found. Thus, the optimal $\boldsymbol{L}_{2_o}$ must lie on or below the boundary

$$-\mathbf{B}_{c_2}(j\omega_i) \text{ vs. } \angle\ell_2$$

on the inverted NC where the values

$$-\mathbf{B}_{c_2}(j\omega_i) \text{ vs. } \angle\ell_2$$

for various values of $-\angle\ell_2 = \angle \boldsymbol{L}_2$, are plotted on the regular NC to obtain the boundary

$$\left[\boldsymbol{B}_{c_2}(j\omega_i)\right]_{q22REG}$$

then $\boldsymbol{L}_{2_o}$ must lie above this boundary.

For Prob. 6.1, in trying to satisfy the requirements for both $\boldsymbol{L}_1$ and $\boldsymbol{L}_2$, Eqs. (6.37) and (6.38), respectively, assume that the specifications are met for $\boldsymbol{L}_1$ and not for $\boldsymbol{L}_2$. Thus, for this situation reducing b_{11} allows a reduction in b_{21} proportionally to maintain the requirement of Eq. (6.37) and in turn satisfy Eq. (6.38) for a satisfactory L_2 since for this example it is stipulated that the *q's* are independent. When the *q's* are not independent then it is necessary to obtain the templates

$$\Im\left(\frac{q_{12}(j\omega_i)}{q_{11}(j\omega_i)}\right) \qquad \Im\left(\frac{q_{21}(j\omega_i)}{q_{22}(j\omega_i)}\right)$$

in order to determine the bounds.

Example 6.6 Cross-coupling Boundary Determination for a 3x3 Plant

Consider the *3x3* MIMO control system of Fig. 5.11. From Eq. (5.28) obtain

$$\begin{bmatrix} \dfrac{1}{q_{11}}+g_1 & \dfrac{1}{q_{12}} & \dfrac{1}{q_{13}} \\ \dfrac{1}{q_{21}} & \dfrac{1}{q_{22}}+g_2 & \dfrac{1}{q_{23}} \\ \dfrac{1}{q_{31}} & \dfrac{1}{q_{32}} & \dfrac{1}{q_{33}}+g_3 \end{bmatrix} \begin{bmatrix} t_{11} & t_{12} & t_{13} \\ t_{21} & t_{22} & t_{23} \\ t_{31} & t_{32} & t_{33} \end{bmatrix} = \begin{bmatrix} f_{11}g_1 & f_{12}g_1 & f_{13}g_1 \\ f_{21}g_2 & f_{22}g_2 & f_{23}g_2 \\ f_{31}g_3 & f_{32}g_3 & f_{33}g_3 \end{bmatrix} \qquad (6.43)$$

The first row of Eq. (6.43) yields:

$$t_{11} = \frac{f_{11}L_1}{1+L_1} + \frac{-q_{11}\left[\frac{t_{21}}{q_{12}} + \frac{t_{31}}{q_{13}}\right]}{1+L_1} = t_{r1} + \frac{q_{11}c_{11}}{1+L_1} = t_{r1} + t_{c11} \qquad (6.44)$$

In a similar manner the expressions for the *1,2* and *1,3* expressions for row one of Fig. 5.11 are obtained.

Given that $|t_{ij}|_{max} \equiv b_{ij}$ and assume, for $|t_{11}|_{max} = b_{11}$, that ϕ_{11} (the actual control ratio) is satisfied. Thus, for this situation it might be difficult to satisfy ϕ_{22}, ϕ_{33}, etc. If this is the case, then the designer needs to do a "trade-off" in trying to achieve the best performance possible. Performing the design in the high frequency range simplifies the task as discussed in Example 6.5. The equations for the determination of the cross-coupling bounds are:

For the first row of Fig. 5.11:

$$|1+\mathbf{L}_1| \geq \frac{|\mathbf{q}_{11}|}{\mathrm{b}_{11}}\left[\frac{\mathrm{b}_{21}}{|\mathbf{q}_{12}|} + \frac{\mathrm{b}_{31}}{|\mathbf{q}_{13}|}\right], \geq \frac{|\mathbf{q}_{11}|}{\mathrm{b}_{12}}\left[\frac{\mathrm{b}_{22}}{|\mathbf{q}_{12}|} + \frac{\mathrm{b}_{32}}{|\mathbf{q}_{13}|}\right], \geq \frac{|\mathbf{q}_{11}|}{\mathrm{b}_{13}}\left[\frac{\mathrm{b}_{23}}{|\mathbf{q}_{12}|} + \frac{\mathrm{b}_{33}}{|\mathbf{q}_{13}|}\right]$$

(a) (b) (c)

(6.45)

For the second row:

$$|1+L_2| \geq \frac{|q_{22}|}{b_{21}}\left[\frac{b_{11}}{|q_{21}|} + \frac{b_{31}}{|q_{23}|}\right], \geq \frac{|q_{22}|}{b_{22}}\left[\frac{b_{12}}{|q_{21}|} + \frac{b_{32}}{|q_{23}|}\right], \geq \frac{|q_{22}|}{b_{23}}[\quad] \qquad (6.46)$$

(a) (b) (c)

For the third row:

$$|1+\mathbf{L}_3| \geq \frac{|\mathbf{q}_{33}|}{\mathrm{b}_{31}}\left[\frac{\mathrm{b}_{11}}{|\mathbf{q}_{31}|} + \frac{\mathrm{b}_{21}}{|\mathbf{q}_{32}|}\right], \geq (\quad), \geq (\quad) \qquad (6.47)$$

(a) (b) (c)

Consider the interactions between the rows of Fig. 5.11 via the b_{ij} specifications. For example, suppose b_{21} is decreased from *0.1* to a δ value. The decrease is acceptable because for BNIC $|t_{21}| \leq b_{21}$ is required. This decrease in turn decreases $|c_{11}|$ and $|c_{31}|$ which eases the burden on L_1 and L_3 for satisfying $y_{11} \in \Im_{11}$ and $y_{31} \in \Im_{31}$. This is based upon the fact that y_{11} and y_{31} dominate L_1 and L_3, respectively. Thus, the design for L_2 is more difficult because $|y_{21}| \leq \delta$ instead of *0.1*. This may not matter if y_{22} dominates L_2. If so, b_{21} can be decreased until, for example, y_{21} imposes the same burden on L_2 as does y_{22} ($y_{21} \sim y_{22}$). Any further decrease in b_{21} may involve a "trade-off," i.e., requiring a sacrifice on the specifications on L_2 for the sake of L_1 and/or L_3. However, it is conceivable that before $y_{21} \sim y_{22}$ occurs, either y_{12}(or y_{13})$\sim y_{11}$ or y_{32}(or y_{33})$\sim y_{31}$. There are other trade-offs that may be possible to investigate.

In summary, for the h.f. range, the main factors to keep in mind in trying to achieve column dominancy are:

1. Adjust the b_{ij}'s until one column dominates.
2. Once Item 1 is accomplished, try to make the non-dominant elements in each row equal the dominant ones by reducing the b_{ij}'s in each denominator. This will make all three columns equal and allows one to use any column as the constraining column. This reduces the *3x3* matrix of constraints to a single constraint on each loop, i.e.,

$$|1+\mathbf{L}_1| \geq A, \quad |1+\mathbf{L}_2| \geq B, \quad |1+\mathbf{L}_3| \geq C$$

3. From Sec. 5-7, with respect to Condition 2 (e.g. a *2x2* plant) if

$$\left|\frac{p_{12}\,p_{21}}{p_{11}\,p_{22}}\right| < 1$$

is not true then renumber the output terminals. For example:

$$\phi_1 \rightarrow \phi_2' \qquad \phi_2 \rightarrow \phi_1'$$

Thus, for the original *2x2* plant:

$$\mathbf{\Phi} = \begin{bmatrix} \phi_1 \\ \phi_2 \end{bmatrix} = [\,p_{ij}\,] \begin{bmatrix} u_1 \\ u_2 \end{bmatrix} = \begin{bmatrix} p_{11}\,u_1 + p_{12}\,u_2 \\ p_{21}\,u_1 + p_{22}\,u_2 \end{bmatrix}$$

For the renumbered system:

$$\Phi = \begin{bmatrix} \phi_1' \\ \phi_2' \end{bmatrix} = \begin{bmatrix} p_{21}' u_1 + p_{22}' u_2 \\ p_{11}' u_1 + p_{12}' u_2 \end{bmatrix} = \begin{bmatrix} p_{21}' & p_{22}' \\ p_{11}' & p_{12}' \end{bmatrix} \begin{bmatrix} u_1 \\ u_2 \end{bmatrix}$$

which results in Condition 2 being satisfied, i.e.,

$$\left| \frac{\mathbf{p}_{22}' \mathbf{p}_{11}'}{\mathbf{p}_{12}' \mathbf{p}_{21}'} \right| > 1$$

4. Items 1-3 apply to an *mxm* plant as well.

6-6 TEMPLATES: SPECIAL CASE

Up to now it is assumed that all p_{ij} elements of $\boldsymbol{P}$ have the same number of excess poles (n) over zeros (w), that is, $\lambda = (n - w) = \lambda$ for all J plants. Thus, in taking

$$\lim_{\omega \to \infty} (p_{ij})_\iota \to \frac{k_{ij}}{(j\omega)^\lambda}$$

the limiting angle for all plants is $\lambda(-90^\circ)$ and the templates become a straight line in the limit. In the event that λ_ι is not the same for all LTI plant cases then in taking

$$\lim_{\omega \to \infty} (p_{ij})_\iota \to \frac{(k_{ij})_\iota}{(j\omega)^{\lambda_\iota}}$$

the limiting angle $\lambda_\iota(-90^\circ)$ for each LTI plant case do not all have the same value. Therefore, the template as $\omega \to \infty$ will not be a straight line.

6-7 SUMMARY

The specifics of QFT Design Method 1 are presented in this chapter. In addition, system sensitivity, stability, and performance trade-off analyses are discussed in detail. The case of plants having RHP zeros and/or poles are also discussed in this chapter. It should be noted that if each p_{ij} of $\boldsymbol{P}$ does not have the same value of excess of poles over zeros then as $\omega \to \infty$ the templates may not be straight lines. The reader is urged, in performing a QFT design, to constantly have in mind the material given in Chapter 9.

7

MIMO SYSTEM DESIGN METHOD 2–MODIFIED SINGLE-LOOP EQUIVALENTS[18,20,31]

7-1 INTRODUCTION

The design technique of Chapter 6 inherently involves some over-design, as seen from Eq. (6.l), in which t_{21}, t_{31} can be any members of their acceptable sets τ_{21}, τ_{31} and q_{12}, q_{13} any member of their uncertainty sets. As noted in Sec. 6-2, it is therefore necessary to use the worst case values which *leads to over-design.* Actually, in the *real world,* there is a correlation between the t_{21}, t_{31} and the q_{11}, q_{13}, etc. For example, in Eq. (5.55) it is possible that q_{12} is large when t_{21} is large in the expression for c_{11}. Such a correlation can only help make c_{11} smaller. Thus, for Method 1, it is not possible to use this correlation, and so one must take the largest t_{21}, the smallest q_{12}, etc. This is the price paid for converting the MIMO problem into the much simpler MISO problems, and avoiding having to work with the horrendous denominator in Eq. (5.25).

Another disadvantage of Method 1 is that there emerges a certain inequality (see Sec. 5-7, the diagonal dominance condition) which must be satisfied by the plant elements; e.g., for $m = 2$, as $\omega \to \infty$ it is:

$$|\mathbf{p}_{11}(j\omega)\mathbf{p}_{22}(j\omega)| > |\mathbf{p}_{12}(j\omega)\mathbf{p}_{21}(j\omega)| \text{ for all } \mathrm{p} \,\varepsilon\, \wp \tag{7.1}$$

or vice versa. Rosenbrock's[61] dominance condition is tougher. It requires this inequality to be satisfied over the entire frequency range, not just as $s \to \infty$. His method can therefore not be used for the *3x3* example of Chapter 6. The equivalent of Eq. (7.1) for $m = 3$, as $\omega \to \infty$, is

$$|\mathbf{p}^*_{123}| = |\mathbf{p}^*_{11}\mathbf{p}^*_{22}\mathbf{p}^*_{33}| > |\mathbf{p}^*_{11}\mathbf{p}^*_{23}\mathbf{p}^*_{32}| + |\mathbf{p}^*_{12}\mathbf{p}^*_{21}\mathbf{p}^*_{33}| + |\mathbf{p}^*_{12}\mathbf{p}^*_{23}\mathbf{p}^*_{31}| + |\mathbf{p}^*_{13}\mathbf{p}^*_{22}\mathbf{p}^*_{31}| + |\mathbf{p}^*_{13}\mathbf{p}^*_{21}\mathbf{p}^*_{32}| \quad (7.2)$$

It turned out that Eq. (7.2) was not satisfied by the FY-16CV *3x3* lateral system for *any* $p \in \mathcal{P}$. Thus, it was necessary to seek a modification of the technique that succeeds in avoiding the condition of Eq. (7.2), and involves less over design than Method 1. This modification lead to the development of Method 2[20]. The necessary constraints for Method 2 are discussed in Sec. 7-6 and in Appendix F.

In summary, Method 1 is used when the diagonal dominance condition can be satisfied and when the BW constraints can not be satisfied by Method 2 (this BW constraint is discussed in a later section). Method 2 is used when the diagonal dominance condition can not be satisfied and/or where over-design needs to be minimized. The MIMO QFT CAD package (Appendix C) contains both design methods.

7-2 DESIGN EQUATIONS FOR THE *2x2* SYSTEM

In Method 2 the same design equations as before are used for t_{11}, t_{12} or alternatively t_{21}, t_{22} (as in Fig 5.11 for $m = 2$). Thus, for Fig. 7.1 and for an impulse input, Eqs. (5.36) through (5.39) yield:

$$t_{1j} = y_{1j} = \frac{f_{1j} L_1 + c_{1j} q_{11}}{1 + L_1} \quad \text{(a)} \qquad L_i = g_i q_{ii} \quad \text{(b)}$$

$$c_{1j} = \frac{-t_{2j}}{q_{12}} \quad \text{(c)} \qquad t_{2j} \in \tau_{2j} \quad \text{(d)} \qquad P^{-1} = [p^*_{ji}] = [1/q_{ji}] \quad \text{(e)} \qquad (7.3)$$

The objective is to choose $f_{11}(s)$, $f_{12}(s)$, and $g_1(s)$ such that the $y_{1j}(s)$ has no RHP poles, and satisfy the tolerances on $|t_{ij}(j\omega)|$, $\forall P \in \mathcal{P}$ and $\forall t_{2j} \in \tau_{2j}$ appearing in c_{1j}. These are precisely the MISO design problems of Sec. 5-6 through 5-8.

When t_{1j} is substituted for y_{1j} in Eq. (7.3), the resulting equations are exact. Hence, if indeed $y_{1j} \in \tau_{1j}$ for all t_{2j} (of c_{1j}) $\in \tau_{2j}$, $P \in \mathcal{P}$, then the design objectives for t_{1j} have been achieved by the f_{11}, f_{12}, and g_1 but only if actually $t_{2j} \in \tau_{2j}$.

The final step, in the *2x2* system, is to choose f_{21}, f_{22}, and g_2 to ensure that the $t_{2v}(s)$ have no RHP poles and that $t_{2v} \in \tau_{2v}$ and $P \in \mathcal{P}$. In Secs. 5-6 and 5-7 the design equations are again Eq. (7.3) with *2* replacing *1* and the c_{2v} containing t_{1v}.

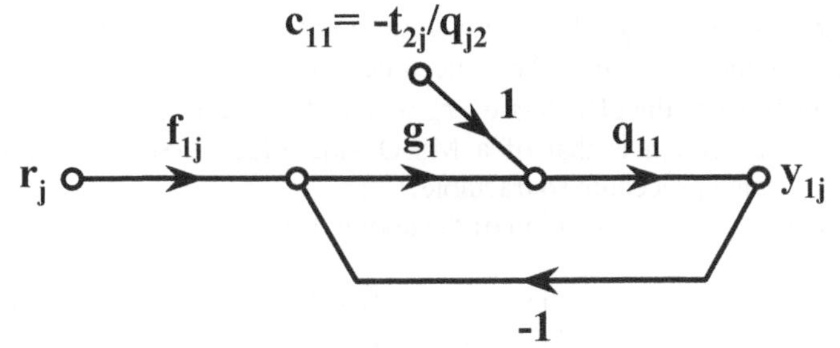

Fig. 7.1 MISO structure for t_{ij} ($j=1,2$), Eq. (7.3).

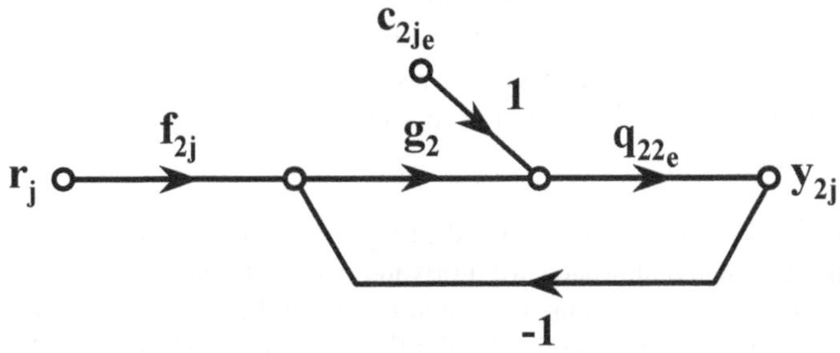

Fig. 7.2 MISO structure for t_{2j} ($j=1,2$), Eq. (7.4).

Instead, here, equations are used that are independent of the t_{1v}, by simply finding t_{2j} from $\boldsymbol{T} = [\boldsymbol{I} + \boldsymbol{PG}]^{-1}\boldsymbol{PGF}$. Thus,

$$t_{2j} = \frac{f_{2j} L_{2e} + c_{2j}}{1 + L_{2e}} \quad \text{(a)} \qquad L_{2e} = g_2 q_{22e} = \frac{g_2 q_{22}(1 + L_1)}{1 - \gamma_{12} + L_1} \quad \text{(b)}$$

$$\gamma_{ji} = \frac{p_{ij} p_{ji}}{p_{ii} p_{jj}} \quad \text{(c)} \qquad c_{2j} = \frac{g_1 f_{1j} p_{21}(1 - \gamma_{12})}{1 - \gamma_{12} + L_1} \quad \text{(d)} \qquad \textbf{(7.4)}$$

$$\boldsymbol{P} = [p_{ji}] \quad \text{(e)}$$

or the MISO structures of Fig. 7.2. This design is done *after* the design of Eq. (6.17a) has been completed by means of Eqs. (7.3a-d), so that L_1, f_{11}, and f_{12} are known (Use L_1 not L_{1_o} -- see Appendix F). It is then necessary to find g_2, f_{21}, and

f_{22} so that in Fig. 5.2 the outputs y_{21} and y_{22} are stable and satisfy the tolerances on $|y_{21}|$ and $|t_{22}|$, respectively. These are single-loop problems similar to Fig. 7.1, except that only the uncertainty $\boldsymbol{P} \in \mathcal{P}$ need be considered, as the c_{2j} in Eqs. (7.4a-e) are not functions of the elements of τ_{ij}, which they are in Eqs. (7.3*a-d*). At each step, design execution is that of a MISO single-loop system -- which is what makes this design procedure so tractable.

The theoretical justification of the above design procedure is as follows:

1. The design specifications are satisfied for t_{2v} of Fig. 7.2, by proper choice of the f_{2v} and g_2 for the given f_{1v} and g_1.
2. Thus, the design is satisfactory for t_{21} and t_{22} because Eqs. (7.4a-b) are exactly the expressions for t_{21} and t_{22}, (even if the specifications for t_{12} and t_{11} are not satisfied).
3. Now Fig. 7.1 has been designed [via Eqs. (7.3a-d)] so that y_{11} and y_{12} are stable and satisfy the specifications on t_{11} and t_{12} $\forall \boldsymbol{P} \in \mathcal{P}$, *if the* t_{2j} *appearing in* c_{1j} *are in* τ_{2j} *(which they are)*.
4. The equations for y_{11} and y_{12} correspond precisely to those for t_{11} and t_{12}, respectively.

Thus, no fixed point theory is needed to rigorously justify this design procedure, although the idea and approach are motivated by the fixed point method in Method 1. There, the design equations for t_{21} and t_{22} are of the same form as Eqs. (7.3a-d), with the c_{2j} functions of t_{1j}, so fixed point theory is required to justify Method 1.

7-3 DESIGN GUIDELINES

The following items are intended to provide the reader with a heuristic insight to the design process.

1. **Designation of the order of loop shaping** -- The order in which the loops are to be designated and designed, i.e., L_1, L_2, etc. in order to achieve *arbitrarily small* (a.s.) sensitivity[20], is based upon the following:

   (a) Choose the loop that satisfies

$$q_{ii} = \frac{\det \boldsymbol{P}}{(\mathrm{Adj}\boldsymbol{P})_{ii}} \quad \text{be m.p.} \tag{7.5}$$

   as *Loop 1* $\rightarrow q_{ii} = q_{11}$. Note: for non-square plants the use of a weighting matrix $\boldsymbol{W} = \{w_{ij}\}$ is required in order to achieve an *effective square plant matrix* $\boldsymbol{P}_e = \boldsymbol{PW}$ since a square plant is required for a QFT design (see Sec. 7-10). If Eq. (7.5) is not satisfied, see Constraint 1 in Sec. 7-8.1, then an adjustment of the "*weighting*" factors w_{ij} may result in satisfying this requirement. Even if this requirement is not achievable a satisfactory design may still be possible depending upon the application (see Chapters 9 and Appendix G).

(b) When applying Method 2, if q_{11} is m.p. then q_{22e}, etc. will all be m.p.

(c) When applying Method 2, in general, if more than one loop can be m.p. choose the loop that has the toughest or most stringent specifications and uncertainty as Loop 1.

2. **Template width** -- If $\exists\omega_h \in$ for all $\omega > \omega_h$ so that the width of $\Im q_{11}(j\omega)$ does not exceed $180^\circ - \gamma$, where γ is the desired phase margin angle, then it is possible to achieve a.s. tolerances. If this condition is not satisfied then it is impossible to achieve a.s. tolerances for $\omega > \omega_h$. This prevents a.s. sensitivity if

$$q_{ii} = \frac{k\prod(s + z_v)}{\prod(s + p_u)} \tag{7.6}$$

with gain k (a $\pm$ uncertainty value) is independent of the signs of z_v and/or p_u. Numerator or denominator factors of the form $(1 + Ts)$, where T has a $\pm$ uncertainty value and is independent of other parameters, must be excluded. See Constraint 2 in Sec. 7-8.1.

3. **For the *2x2* plant if**

(a) γ is "small" then $q_{22e} \cong q_{22}$.

(b) γ_{ji} is "large," i.e., $|\gamma_{ji}| \cong 1$ then q_{22e} can be unstable. This may be acceptable but may lead to a wide BW for L_2.

4. **Transmission zeros** -- If there are transmission zeros in the RHP it only indicates that q_{ij} *may* be n.m.p. or the *det* $\boldsymbol{P}$ *may* have zeros in the RHP only.

7-4 REDUCED OVER-DESIGN[18]

Figures 7.1 and 7.2 are the same as Fig. 5.11 (for *2x2* system)[7] in which t_{21} and t_{22} appear in the cross-coupling c_{11} and c_{2j_e}, respectively. There is inherent over-design in Fig. 7.1 because in reality there is a correlation between the t_{21} and t_{22} and the q_{ij} of $\boldsymbol{P}^{-1}$. This correlation is not being exploited. The uncertainties in $t_{21} \in \tau_{21}$ and $t_{22} \in \tau_{22}$ are assumed independent of $\boldsymbol{P} \in \mathcal{P}$ in Figs. 7.1 and 7.2. But such over-design does not exist in Fig. 7.2 because c_{21} and c_{22} are not functions of the elements of any τ_{ij}. In Fig. 5.11, for the second row of the *2x2* system, the design for t_{21} and t_{22}, by Method 1 involved over-design precisely as in the first row of this figure and in Fig. 7.l. The above procedure can, of course be reversed with y_{21} and y_{22} using Eq. (7.3) with *1* replaced by *2* and with t_{11} and t_{12} using Eqs. (7.4) by exchanging numbers *1,2*.

7-5 *3x3* DESIGN EQUATIONS

Let y_{1j}, $j = 1,2,3$, be the same as in Secs. 5-6 and 5-7, that is:

$$y_{1j} = \frac{f_{1j} L_1 + c_{1j} q_{11}}{1 + L_1}, \quad L_1 = g_1 q_{11}, \quad c_{1j} = -\Sigma t_{kj} / q_{1k} \tag{7.7}$$

giving MISO problems. L_1 and the three f_{1j} are chosen so that y_{1j} are stable and satisfy the tolerances on t_{1j} for all $t_{kj} \in \tau_{kj}$ appearing in c_{ij} in Eq. (7.5c) and for all $\boldsymbol{P} \in \mathcal{P}$. The equations for y_{2j} are obtained from the second row of Fig. 5.12 in which t_{1j} and t_{3j} appear, but the t_{1j} are replaced by the y_{1j} of Eq. (7.7), giving for $j = 1,2,3$

$$y_{2j} = \frac{f_{2j} L_{2_e} + c_{2j}}{1 + L_{2_e}} \quad \text{(a)} \qquad L_{2_e} = \frac{g_2 q_{22}}{1 - \dfrac{\gamma_{12}}{1 + L_1}} \quad \text{(b)}$$

$$p_{21}^{e} = \frac{L_1 p_{21}^{*}}{1 + L_1} \quad \text{(c)} \qquad L_{2_e} = g_2 q_{22_e} \quad \text{(d)}$$

$$c_{2j} = \frac{L_{2e}}{g_2} \left(t_{3j} \left[\frac{p_{21}^{*} p_{13}^{*}}{p_{11}^{*}(1 + L_1)} - p_{23}^{*} \right] - f_{1j} p_{21}^{*} \right) \quad \text{(e)} \tag{7.8}$$

$$\boldsymbol{P}^{-1} = [p_{ij}^{*}] \quad \text{(f)}$$

again resulting in MISO problems. Note that γ_{12} is defined by Eq. (7.4c) where $j = 1$ and $i = 2$. L_1 and the f_{1j} are known from the designs of Eqs. (7.7). In Eqs. (7.8) the f_{2j} and L_{2e} are chosen so that y_{2j} are stable for all $t_{3j} \in \tau_{3j}$ appearing in c_{2j} and, of course, for all $\boldsymbol{P} \in \mathcal{P}$. Although the forms for c_{2j} and L_{2e} in Eqs. (7.8) are different from those in Eqs. (7.7), they are otherwise identical in form, so the design techniques for both are basically the same, as detailed in Ref. 20.

Finally, MISO design equations for t_{3j} are obtained by finding t_{3j} from $\boldsymbol{T} = [\boldsymbol{I} + \boldsymbol{PG}]^{-1}\boldsymbol{PGF}$, or from the third row of Fig. 5.11 for $i = 3$ and eliminating t_{1i} and t_{2i} by means of Eqs. (7.7) and (7.8). Thus, the resulting equations are:

$$t_{3j} = \frac{f_{3j}L_{3e}}{1 + L_{3e}} \quad \text{(a)}, \qquad L_{3e} = \frac{L_3\varsigma}{\varsigma - \Lambda} \quad \text{(b)}, \qquad L_3 = g_3 q_{33} \quad \text{(c)},$$

$$\varsigma = (1 + L_1)(1 + L_{2_e}) - \gamma_{12} \quad \text{(d)},$$

$$\Lambda = \gamma_{23}(1 + L_1) + \gamma_{13}(1 + L_{2e}) - (\gamma_{12}\mu_2 + \gamma_{13}\mu_3) \quad \text{(e)},$$

$$\mu_2 = \frac{p^*_{23}p^*_{31}}{p^*_{21}p^*_{33}} \quad \text{(f)}, \qquad \mu_3 = \frac{p^*_{32}p^*_{21}}{p^*_{31}p^*_{22}} \quad \text{(g)}, \qquad \textbf{(7.9)}$$

$$c_{3j} = \frac{f_{1j}L_1 q_{33}\eta_1 + f_{2j}L_2 q_{33}\eta_2}{\varsigma - \Lambda} \quad \text{(h)},$$

$$\eta_1 = q_{22}p^*_{21}p^*_{32} - p^*_{31}(1 + L_{2_e}) \quad \text{(i)},$$

$$\eta_2 = q_{11}p^*_{12}p^*_{31} - p^*_{32}(1 + L_1) \quad \text{(j)}$$

Note: See Eq. (5.93) for the expressions for the γ_{ij} terms of Eq. (7.9).

Since L_1, L_2, f_{1j}, and f_{2j} are known, the only unknowns in the Eq. (7.9) are the f_{3j} and g_3. These equations constitute single-loop uncertainty problems, for which the technique of Chaps. 3 and 4 apply, i.e. they are chosen so that the t_{3j} are stable and satisfy the tolerances τ_{3j}. Note, again, that at each step, design execution is that of MISO single-loop systems.

The justification of the above design approach is the same as for the *2x2* case. Suppose the nine f_{ij} and three g_i in Eqs. (7.9a-j) are such that the t_{3j} are stable and their tolerances are satisfied (which is so by definition here). Now the design based on Eqs. (7.8a-e) guarantees that the t_{2j} tolerances are satisfied, providing the t_{3j} appearing in c_{2j} *are* $\in \tau_{3j}$ which is the case here. Hence, the t_{2j} tolerances are also satisfied. Finally, the design based on Eqs. (7.7a-c) guarantees that the t_{1j} tolerances are satisfied, providing that the t_{2j} and t_{3j}, appearing in c_{1j} are $\in \tau_{2j}$ and τ_{3j}, respectively, which has been established. In the above, there is some over design in Eqs. (7.7a-c) because the $t_{ij} \in \tau_{kj}$, ($k = 2,3$ and $j = 1,2,3$) appear as cross-coupling effects uncorrelated to the plant uncertainty. In Eqs. (78*a-f*) there is less over-design because only the t_{3j} appears, while there is no such over-design in Eqs. (7.9a-j). Of course, the order can be changed and equations of the form of Eqs. (7.7a-c) are used for the second or third channel, etc.

7-6 EXAMPLE – *3x3* SYSTEM DESIGN EQUATIONS

Consider the design of a *3x3* control system for which $r_1(t) \neq 0$ and $r_2(f) = r_3(t) = 0$ (thus, $f_{i2} = f_{i3} = 0$). This example entails four parts: part (1) -- the set-up of the pertinent equations, and parts (2)-(4) -- the design approach. It is assumed that the loop to be designed first is row 1 of Fig. 5.11.

Part (1) The pertinent equations are (refer to Figs. 5.11 and 5.13):

Row 1

$$t_{11} = \frac{f_{11}L_1 - q_{11}\left(\frac{t_{21}}{q_{12}} + \frac{t_{31}}{q_{13}}\right)}{1+L_1} \text{ (a)}, \qquad t_{12} = \frac{c_{12}q_{11}}{1+L_1} \text{ (b)},$$

$$t_{13} = \frac{c_{13}q_{11}}{1+L_1} \text{ (c)} \tag{7.10}$$

Row 2

$$t_{21} = \frac{f_{21}L_2 - q_{22}\left(\frac{t_{11}}{q_{21}} + \frac{t_{31}}{q_{23}}\right)}{1+L_2} \text{ (a)}, \qquad t_{22} = \frac{c_{22}q_{22}}{1+L_2} \text{ (b)}, \tag{7.11}$$

$$t_{23} = \frac{c_{23}q_{22}}{1+L_2} \text{ (c)}$$

Row 3

$$t_{31} = \frac{f_{31}L_3 - q_{33}\left(\frac{t_{21}}{q_{32}} + \frac{t_{11}}{q_{31}}\right)}{1+L_3} \text{ (a)}, \qquad t_{32} = \frac{c_{32}q_{33}}{1+L_3} \text{ (b)}, \tag{7.12}$$

$$t_{33} = \frac{c_{33}q_{33}}{1+L_3} \text{ (c)}$$

Part (2) Utilizing Method 1, the b_{ij}'s are substituted into Eq. (7.10) in order to yield the f_{11} and g_1 that satisfy the design specifications. In general, it is necessary that L_1 and f_{1j} be designed so that the y_{1_j}'s are stable and satisfy the specifications on t_{1j} for all $t_{kj} \in \tau_{kj}$ in c_{ij} and for all $\boldsymbol{P} \in \mathcal{P}$.
Part (3) Substitute t_{11} obtained in Part (2) into Eqs. (7.11) and (7.12). Apply Method 2 to design f_{21} and g_2 [see Eq. (7.8)]. In general, design f_{2j} and L_{2e} so that the y_{2j}'s are stable for all $t_{3j} \in \tau_{3j}$ in c_{2j} and for all $\boldsymbol{P} \in \mathcal{P}$.
Part (4) Substitute t_{11} and t_{21} obtained in Parts (2) and (3) into Eq. (7.12). Apply Method 2 to design f_{31} and g_3 [see Eq. (7.9)]. Thus, it is now possible to obtain t_{31}, that hopefully meets the specifications, in terms of only the parameters, that is, the b_{ij}'s are not involved in the design of f_{31} and g_3. This is the concept of Method 2 which results in Eq. (7.12) having only two unknowns: f_{31} and g_3.

Note: one may initially start the design with a diagonal pre-filter $\boldsymbol{F}$ matrix in order to simplify the design process. If the design specifications cannot be achieved with a diagonal pre-filter then it will be necessary to utilize a non-diagonal pre-filter matrix.

7-7 *mxm* SYSTEM: $m > 3$

The procedure for generating the design equations for *mxm* MIMO systems with $m > 3$, should be clear from the preceding sections. One uses for any channel (say the first) design equations in which all the t_{ij} ($i \neq 1$) appear as cross-coupling effects. Denote these as Eqs. A. These equations can be derived or obtained from Ref. 7, Eqs. (4a,b) with $u = 1$. For the next chosen channel (say the second), the design starts with Eqs. (4a,b) of Ref. 7, with $u = 2$, in which all the t_{1j} and t_{2j} are eliminated by means of Eqs. A. Denote the resulting design equations as Eqs. B. For the next chosen channel (say the third), start again with Eqs. (4a,b) of Ref. 7, with $u = 3$, but eliminate all the t_{1j} and t_{2j} by means of Eqs. A and B. The process continues until the end, and the theoretical justification is the same as given previously for $m = 2,3$.

Mixtures of the first and second techniques may also be used. For example, for the *3x3* system, Eq. (7.7) type equations are used for the first two rows of Fig. 5.11 for *both* channels 1, 2 and Eqs. (V.9a-j) for channel 3. The theoretical justification is now as follows: the design for channel 3 is correct by definition then the fixed point theory, precisely as in Ref. 7, is used to justify the designs for channels (1,2). This method was used for the *3x3 FY16-CCV* lateral design modes.[30]

The two sets of design equations for t_{1j} and t_{2j} are taken as the mappings on the acceptable sets τ_{ij} and the third set of mappings is simply $t_{3j} \in \tau_{3j}$. The nine f_{ij} and the three g_i have been chosen so that these mappings map τ_{ij} into themselves, etc., so a fixed point exists, etc., as in (Ref. 7). For larger *m*, it is clear that a larger variety of mixtures is possible, giving the designer useful flexibility. However, the designer must understand MISO design theory used in the design execution, which reveals the cost of feedback and the available trade-offs among the loops, in order to be able to exploit this flexibility to its fullest extent.

There are additional important advantages in Method 2 since the diagonal dominance conditions given by Eqs (7.1) and (7.2) are no longer necessary. Instead the principal condition to be satisfied [to achieve "*arbitrarily small (a.s.) sensitivity*"] is that *det* $\boldsymbol{P}$ has no RHP zeros. The reader is referred to Ref. 20 for details.

7-8 CONDITIONS FOR EXISTENCE OF A SOLUTION

This section considers the conditions required for the applicability of the QFT design technique. Also, it considers the inherent, irreducible conditions applicable for LTI compensators in general, and compares the two sets of conditions. This is done for a.s. sensitivity which is defined as the BW achievement of *a.s.* sensitivity of the t_{ij} over *arbitrarily large (a.l.)* BW. Such a.s. sensitivity also achieves attenuation over a.l. BW of *external* disturbances acting on the plant. This problem has also been studied in an abstract setting by Zames and Bensousan.[63]

In Fig. 7.1, it is required that $y_{11} \in \tau_{11}$ and $y_{12} \in \tau_{12}$ for all $\boldsymbol{P} \in \mathcal{P}$, $t_{21} \in \tau_{21}$ and $t_{22} \in \tau_{22}$. In the general *m* case, the cross-coupling component in, for example, y_{11} is:

$$y_{c11} = \frac{c_{1j} q_{11}}{1 + g_1 q_{11}} = - \sum_{k \neq 1} \left[t_{k1} \left(\frac{(Adj\,\boldsymbol{P})_{1k} / (\mathrm{Adj}\boldsymbol{P})_{11}}{1 + g_1 \dfrac{\det \boldsymbol{P}}{(\mathrm{Adj}\boldsymbol{P})_{11}}} \right) \right] \tag{7.13}$$

RHP poles of $(Adj\ \boldsymbol{P})_{1k}$ are normally cancelled by similar poles in *det* $\boldsymbol{P}$, since $(Adj\ \boldsymbol{P})_{1k}$ is a term in the expansion of *det* $\boldsymbol{P}$. RHP zeros of $(Adj\ \boldsymbol{P})_{11}$ are, of course, normally cancelled by similar ones in the denominator. There may be exceptional cases when in *det* $\boldsymbol{P}$, for example, a RHP pole of $(Adj\ \boldsymbol{P})_{1k}$ is cancelled by an identical zero of p^*_{ik} and does not appear in the other terms of *det* $\boldsymbol{P}$. Such cases are excluded

7-8.1 CONDITIONS FOR "a.s. SENSITIVITY" IN SINGLE-LOOP DESIGN

In Eqs. (7.3), it is seen that a.s. tolerances over a.l. BW (i.e. "a.s. sensitivity") for t_{11}, t_{12} are achievable if $L_1 = g_1 q_{11}$ can be made a.l. over a.l. BW. Indeed this is at least theoretically possible, if q_{11} satisfies certain constraints. These have been detailed in Appendix 1 of Ref. 7 so are only qualitatively described here by means of Fig. 7.3.

Figure 7.3 is the extended logarithmic complex plane (NC). Since $\boldsymbol{P}$ ranges over $\mathcal{P}$, the set $\{L_1 = \boldsymbol{g}_1(j\omega)\boldsymbol{q}_{11}(j\omega)\}$ is not a single complex number (at any fixed ω) in the NC but a region, denoted as $\Im_p[\boldsymbol{L}(j\omega)]$, the template of $\boldsymbol{L}_1$, which is the same as $\Im_p[\boldsymbol{q}_{11}(j\omega)]$ but translated vertically by $20log[\boldsymbol{g}_1(j\omega)]$ dB and horizontally by $\angle \boldsymbol{g}_1(j\omega)$ degrees, because $L_1 = g_1 q_{11}$. In a design with significant plant uncertainty $\Im_p[\boldsymbol{L}_1(j\omega)]$ must lie relatively high up, above the zero dB line as shown in Fig. 7.3 for ω_l. This is so over the important ω range of t_{11} and t_{12} (their bandwidths generally), in order to achieve the desired sensitivity reduction. Such large values for $\Im_p[\boldsymbol{L}_1(j\omega)]$ can be maintained theoretically for any finite ω range, if q_{11} is m.p. For those ω for which $\Im_p[\boldsymbol{L}_1(j\omega)]$ is so located, the uncertainty in the magnitude and phase of $\boldsymbol{q}_{11}(j\omega)$ (i.e., the area of $\Im_p[\boldsymbol{q}_{11}(j\omega)]$, can be a.l. Note, however, that in order to maintain $\Im_p[\boldsymbol{L}_1(j\omega)]$ above the zero dB line, any zeros of $\boldsymbol{q}_{11}(j\omega)$ on the $j\omega$ axis must be known and finite in number in order for $g_1(s)$ to be assigned poles there. (Obviously, transcendental compensation can be used for special countable cases.) If the range of such zeros on the $j\omega$ axis is uncertain, then the specifications

$$a_{ij}(j\omega) \leq |t_{ij}(j\omega)| \leq b_{ij}(j\omega), \quad \forall \boldsymbol{P} \in \mathcal{P} \tag{7.14}$$

$$\tau_{ij} = t_{ij}(j\omega) = set\ of\ acceptable\ t_{ij}(j\omega) \tag{7.15}$$

must be modified to permit such zeros of t_{ij}.

Eventually $\boldsymbol{L}_1(j\omega)$ must decrease and $\rightarrow 0$ as $\omega \rightarrow \infty$. Stability over the range of $\mathcal{P}$ requires that $\Im P[\boldsymbol{L}_1(j\omega)]$ move downward in between the vertical lines V_1 and V_2 without the points ... $0_{-1} 0_1 0_2$, ... lying in any of the $\Im P[\boldsymbol{L}_1(j\omega)]$. This appears to allow $(360^\circ - 2\gamma)$ degrees phase width for $\Im P[\boldsymbol{q}_{11}(j\omega)]$, with phase margin angle γ. However, $\angle \boldsymbol{L}_1(j\omega)$ must be negative, on the average, in order for $|\boldsymbol{L}_1(j\omega)|$ to decrease. In practice only the $(180^\circ - \gamma)$ phase width is tolerable in this range. As ω increases and the $\Im P[\boldsymbol{L}_1(j\omega)]$ descend lower on the chart below the zero dB line, clearly their width may increase again, but it is essential that the points ... $0_1, 0_2$, ... never be a part of any $\Im P[\boldsymbol{L}_1(j\omega)]$. Unstable $\boldsymbol{q}_{11}$ are included in the above discussion and do not require separate treatment. It follows from the above that

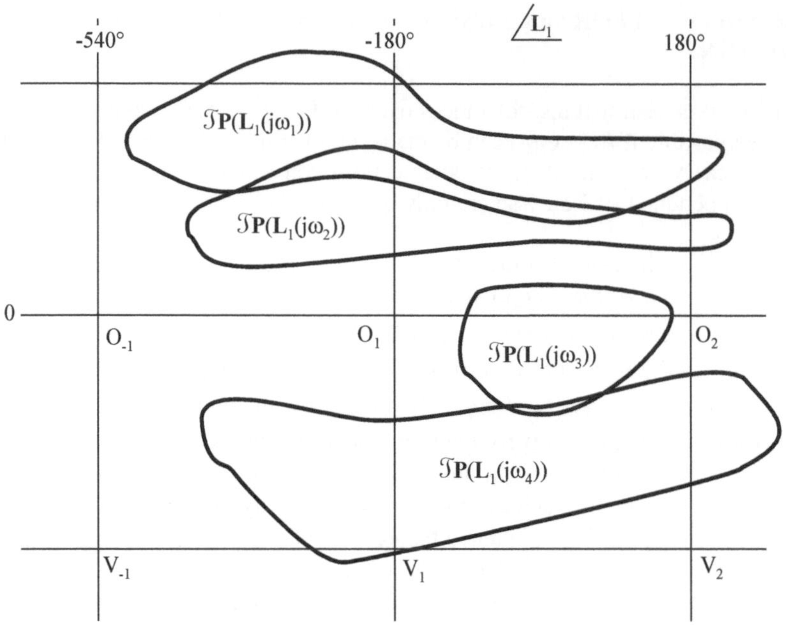

Fig. 7.3 Templates of $L_1(j\omega)$ on logarithmic complex plane (Nichols Chart).

the manageable uncertainties depend on the assigned t_{ij} tolerances, but two important constraints (also see Sec. 7-3), for $m \leq 3$, are stated here for the case of "a.s. sensitivity:"

Constraint 1

$$q_{11} = \det \boldsymbol{P}/(\text{Adj. } \boldsymbol{P})_{11} \text{ must be m.p.}$$

If this constraint is not satisfied for Method 2 the theoretically attainable benefits of feedback are limited. The benefits which do arise may nevertheless suffice for the specific system being designed (see Refs. 26, 27, and 29). Note: if q_{11} is m.p then q_{22_e}, etc. are also m.p.

Constraint 2

Suppose $\exists\omega_h \ni$ for all $\omega > \omega_h$, the width of $\Im p[\boldsymbol{q}_{11}(j\omega)]$ exceeds $(180^\circ - \gamma)$, γ being a desired phase margin angle, then it is impossible to achieve a.s. tolerances for $\omega > \omega_h$. This prevents "*a.s. sensitivity*" if

$$q_{ii} = \frac{k\prod(s+z_v)}{\prod(s+p_u)}$$

with the k uncertainty including a sign change which is independent of the signs of the z_v and p_u. Also excluded is a factor $(1 + \tau s)$ in the numerator or denominator of q_{ii}, with the uncertainty in τ including a sign change which is independent of other parameters.

A good way to apply this constraint, for Method 2, is to obtain:

$$\sum_{\omega \to \infty} \left[\frac{\det \boldsymbol{P}}{\det \boldsymbol{P}_o} \right]$$

where $\boldsymbol{P}_o$ represents the nominal plant matrix. This ratio must not change sign over the range of uncertainty, i.e., $\boldsymbol{P} \in \mathcal{P}$. This automatically takes care of all loops. If this condition is not satisfied then it is impossible, by the usual LTI design techniques, to achieve significant sensitivity reduction.

7-8.2 APPLICATIONS OF SEC. 7-8.1 TO DESIGN METHOD 2[20,52]

Constraints 1 and 2 therefore apply to the $q_{ii} = \det \boldsymbol{P}/(Adj\ \boldsymbol{P})_{ii}$ *of the first channel i, used in the design technique of Secs. 7-2 and 7-5.* So from Constraint 1, q_{ii} must be m.p. $\forall \boldsymbol{P} \in \mathcal{P}$ (RHP poles are tolerable), for "a.s. sensitivity" design. Suppose $\boldsymbol{P} = [p_{ij}]$ has each $p_{ij} \to k_{ij}/s^{\lambda}$ as $s \to \infty$. For $m = 2$:

$$q_{ii} = \frac{p_{11}p_{22} - p_{12}p_{21}}{p_{jj}} \to \frac{k_{11}k_{22} - k_{12}k_{21}}{k_{jj}s^{\lambda}} \tag{7.16}$$

Let $\boldsymbol{K} = [k_{ij}]$, Constraint 2 states that there may be no change in the sign of $\det \boldsymbol{K}/k_{jj}$ as $\boldsymbol{P}$ ranges over $\mathcal{P}$. In this chapter it is assumed that all $k_{ij} > 0$ for all $\boldsymbol{P} \in \mathcal{P}$, Constraint 2 gives $k_{11}k_{22} > k_{12}k_{21}$ $\forall \boldsymbol{P} \in \mathcal{P}$, or vice versa. To remove the ambiguity it is also assumed that the plant terminals are numbered so that for at least one $\boldsymbol{P} \in \mathcal{P}$, $k_{11}k_{22} > k_{12}k_{21}$, so Constraint 2 yields

$$k_{11}k_{22} > k_{12}k_{21} \quad \forall\, \boldsymbol{P} \in \mathcal{P} \tag{7.17}$$

This is a diagonal dominance condition as $s \to \infty$ *which applies only to the first channel being designed when applying Method 2.* Zames and Bensousan[63] have defined a diagonal condition as $s \to \infty$, in more abstract form.

The above, for example, applies to channel no. 1, for which Eqs. (7.3) are used. Equations (7.4) are used for the second channel. The m.p. condition of Constraint 1 therefore applies to $q_{22}(1 + L_1)$, most of which is not new because m.p. $(1 + L_1)$ and $\det \boldsymbol{P}$ are already required. As for Constraint 2, "a.s. sensitivity" can be achieved by L_1 BW << that of L_2 [denoted by BW(L_1) << BW(L_2)], and then Eq. (7.4b) implies Constraint 2 applies to p_{22} of $\boldsymbol{P}$. That is:

$$p_{ii} = \frac{k\prod(s+z_v)}{\prod(s+p_u)}$$

For the condition assumed with Eqs. (7.16) and (7.17), with no sign changes in the k_{ij}, the results are the same. It may also be so in the general case but this would require consideration of simultaneous sign changes among the k_{ij}, which is not done here.

For $m = 3$ the application of the constraints to the first channel makes it applicable to q_{11}. Application to the second, Eqs. (7.8), gives the same results as for the $m = 2$ case, because L_{2e} has the same form in both cases [compare to Eqs. (7.4b) and (7.8b)]. If "a.s. sensitivity" is achieved (as it may be) by BW(L_1) << BW(L_2) << BW(L_3), the result is that the constraints of Sec. 7-8.1 apply to q_{11}, $q_{22}/(1-\gamma_{12})$, and

$$\frac{q_{33}(1-\gamma_{12})}{1-(\gamma_{12}+\gamma_{23}+\gamma_{13})+(\gamma_{12}\mu_2+\gamma_{13}\mu_3)}$$

It has not been ascertained whether these two sets of constraints are identical. However, Sec. 7-8.3 shows that the constraints of Sec. 7-8.1 must always apply to each q_{ii}, $i = 1$ to m. Constraints for $m > 3$ may be similarly developed.

When BW is a specification it is important that one template for each loop be obtained at their respective loop BW specification.

7-8.3 INHERENT CONSTRAINTS[20]

It is important to determine whether the constraints in Sec. 7-8.2 are due to the specific design technique or are inherent in the problem itself. For this purpose examine Eq. (7.4a) for t_{22}. How can "a.s. sensitivity" of t_{22} be achieved despite large uncertainty in $\boldsymbol{P}$? Clearly by large L_{2e}, the usual feedback method. Large L_{2e} is achieved by large g_2q_{22} because large L_1 (needed likewise for small t_{11} sensitivity) gives $L_{2e} = g_2q_{22}$. The latter also attenuates c_{22}, which may not be small because of g_1 in its numerator. This same principle applies to all t_{ij}, and is basically the same as that derived from examination of Eq. (7.3), i.e., there is need for a.l. L_2 and L_1 over a.l. BW in order to achieve "a.s. sensitivity." But do the constraints of Sec. 7-8.1 apply to q_{22} and q_{11}?

This is indeed so, and proven by Eq. (7.3a) and its analog for t_{2j} (by interchanging $1,2$), by simply asking whether a stable t_{11} is possible if $1 + L_1$ has RHP zeros? For if not, and since a.l. L_1 over an a.l. BW is needed, it follows that L_1 (and L_2) must satisfy the constraints. Suppose $(1 + L_1)$ has RHP zeros. These are RHP poles of t_{11}, unless in Eq. (7.3a), for $i = 1$, the numerator of t_{11} has these same zeros. Suppose it has them, and there is a small change in f_{21}. Since $\boldsymbol{F}$ is outside the feedback loops, system stability is unaffected. The zeros of $1 + g_1 q_{11}$ in Eq. (7.3a) are thereby unaffected, so neither should the zeros of the numerator of Eq. (7.3a), for $i = 1$. The term $f_{11} g_1 q_{11}$ is unaffected, but t_{21} is affected [see Eqs. (7.4) with $i = 1$]. Hence, the hypothesis $(1 + L_1)$ has RHP zeros is untenable, so the constraints apply to q_{11} and q_{22}.

From Sec. 7-4, it is clear that the best method is achieved with Eqs. (7.3) by maximum use of the correlation which exists between the $\mathcal{P}$ uncertainty and the $t_{2j} \in \tau_{2j}$ in Eqs. (7.3). A suggestion for this purpose has been given in Ref. 7. That is, one subdivides the plant set into subsets $\mathcal{P}_j$ which are correlated with the subsets τ_{iju} of τ_{ij}. Equations (7.3), similarly Eqs. (7.7) and (7.8), are now applied to these pairs $\mathcal{P}_u$, τ_{iju} separately for each u. To the knowledge of the authors of Ref. 7 this approach has not as yet been attempted in any numerical problem. It is worth noting that the constraints of the diagonal dominance type as $s \to \infty$, also appear in the design technique of Ref. 7. However, they are *always* present there, even if "a.s. sensitivity" is not attempted. In design Method 2 the constraints are in effect only for "a.s. sensitivity." Hence, it is possible that a specific synthesis problem with given τ_{ij}, $\mathcal{P}$ sets may not be solvable by Method 1, but is solvable by Method 2. This is the case in Ref. 7.

7-9 NON-DIAGONAL *G*

The constraints on $\boldsymbol{P}$ in Sec. 7-8 are deduced on the assumption that $\boldsymbol{G}$ is diagonal. Are these constraints eased if a non-diagonal $\boldsymbol{G}$ is used? To answer this question, let $\boldsymbol{H}$ be a fixed LTI pre-compensator matrix inserted ahead of the plant and let $\boldsymbol{V} = \boldsymbol{PH}$ be the new effective plant in the set $\upsilon = \{\boldsymbol{PH}, \boldsymbol{P} \in \mathcal{P}\}$. The design techniques with diagonal $\boldsymbol{G}$, are now applied to set υ instead of set $\mathcal{P}$. If $\boldsymbol{H}$ is helpful in overcoming some constraint, then it is necessary that the constraint violated by $\mathcal{P}$, is not violated by υ. The constraint *det* $\boldsymbol{P}$ is m.p. $\forall \boldsymbol{P} \in \mathcal{P}$ is not eased at all, because *det* $\boldsymbol{V}$ = (*det* $\boldsymbol{P}$)(*det* $\boldsymbol{H}$), and obviously cancellation of RHP zeros of *det* $\boldsymbol{P}$ by *det* $\boldsymbol{H}$ cannot be done for many reasons. The other important constraint involving diagonal dominance as $s \to \infty$, is also not eased, because it applies to the sign of *det* $\boldsymbol{V}$ not changing, as $s \to \infty$. Thus, the constraints on $\mathcal{P}$ for "*a.s. sensitivity*" are not eased by a non-diagonal $\boldsymbol{G}$.[52]

However, $\boldsymbol{H}$ may be very helpful in reducing the amount of feedback needed to achieve specified tolerance sets given by Eqs. (7.14) and (7.15), for a given plant set $\mathcal{P}$, so that a design unachievable by diagonal $\boldsymbol{G}$ (say, because of n.m.p. $\boldsymbol{P}$ or sensor noise problems) may be achievable via $\boldsymbol{H}$. For example in Eq. (7.4a), L_{2_e} must handle the uncertainties due to L_{2_e} itself and attenuate the effective cross-coupling set $\{c_{2j}\}$. For basically non-interacting tolerances on t_{kj} $(k \neq i)$, f_{ij}

is made zero, so only the latter need exist. It may be possible to considerably reduce $|c_{2j}|_{max}$ by means of $\boldsymbol{H}$, by making $\boldsymbol{V} = \boldsymbol{PH}$ quasi-diagonal, even though $\boldsymbol{P}$ has large non-diagonal components.

Off-diagonal plant elements appear in all the design Eqs. (5.40) through (5.46), Eq. (7.4), and Eqs. (7.7) through (7.9) in the 'cross-coupling' components, so their reduction via $\boldsymbol{H}$ is desirable. How is this systematically done in the case of significant $\boldsymbol{P}$ uncertainty? For $m = 2$, let the normalized $\boldsymbol{H}$ have 1 for its diagonal elements and $h_{12} = \mu$, $h_{21} = \upsilon$. Then $v_{12} = \mu p_{11} + p_{12}$ and $v_{21} = p_{21} + \upsilon p_{22}$. The objective is to minimize over $\mathcal{P}$ $\max|v_{12}|,|v_{21}|$ at each ω. Sketches of the sets

$$\left(\frac{p_{12}(j\omega)}{p_{11}(j\omega)}\right),\left(\frac{p_{22}(j\omega)}{p_{21}(j\omega)}\right)$$

in the complex plane, are clearly very helpful in choosing $\mu(j\omega), \upsilon(j\omega)$. However, one should check the effect on the resulting sets of $v_{11} = p_{11} + \upsilon p_{12}$, $v_{22} = \mu p_{21} + p_{22}$, because of the requirements on the loop transmissions due to their uncertainties. The final choice depends on the relative importance of the two terms in the numerators of y_{1j}, y_{2j} in Eqs. (7.3) and (7.4). (See Ref. 7, Secs. 3.2 and 4 for discussion relevant to this topic.)

If the elements of $\boldsymbol{P}$ have a RHP pole in common, i.e., $\boldsymbol{P} = \boldsymbol{P}_l/(s-p)$, then one should not try a diagonal $\boldsymbol{P}$ by means of $\boldsymbol{PH} = \boldsymbol{\Lambda}$ diagonal, because in practice $\boldsymbol{H} = \boldsymbol{P}_a^{-1}\boldsymbol{\Lambda}$ with $\boldsymbol{P}_a^{-1} \neq \boldsymbol{P}^{-1}$ exactly, giving $\boldsymbol{PH} = \boldsymbol{PP}_a^{-1}\boldsymbol{\Lambda}$ with RHP dipoles. Instead, one tries a diagonal $\boldsymbol{P}_l$ by means of $\boldsymbol{P}_l\boldsymbol{H} = \boldsymbol{\Lambda}$, giving $\boldsymbol{H} = \boldsymbol{P}_{la}^{-1}\boldsymbol{\Lambda}$, and $\boldsymbol{PH}$ and $\boldsymbol{P}_l\boldsymbol{P}_a^{-1}\boldsymbol{\Lambda}/(s-p)$.

Boje, University of Natal, Durban, South Africa, utilizes the Perron root interaction measure for decoupling plants (control authority allocation) in multivariable QFT robust control system design[81]. The use of the off-diagonal components in multivariable QFT can ease the diagonal controller design problem. The difficulty is how to design the off-diagonal elements, especially if one must consider engineering factors such as cost-benefit trade-offs of using cross-feeds, strong structure in the plant uncertainty, system integrity and plant input signal levels. Garcia-Sanz, Universidad Publica de Navarra, Pamplona, Spain, has also developed a non-diagonal $\boldsymbol{G}$ decoupling design approach[82] (see Chapter 10) based upon a sequential design methodology.

7-10 ACHIEVABILITY OF A m.p. EFFECTIVE PLANT *det* P_e

For some control systems (e.g., flight control) there are often more control inputs (ℓ) than outputs (m) available. The inputs and outputs of an aircraft may be selected in such a manner as to yield the *mxm* aircraft plant matrix $\boldsymbol{P}(s)$. If the constraint *det* $\boldsymbol{P}(s)$ be m.p. is not satisfied then it may be possible to achieve a m.p. *det* $\boldsymbol{P}_e(s)$ for an effective *mxm* plant $\boldsymbol{P}_e(s)$ by augmenting the basic $mx\ell$ plant $\boldsymbol{P}_b$ by a *ℓxm weighting matrix* $\boldsymbol{W}$, i.e., a gain matrix $\boldsymbol{W} = [w_{ij}]$ or a frequency sensitivity matrix $\boldsymbol{W} = [w_{ij}(j\omega)]$ as shown in Fig. 7.4. The expression for the output $\boldsymbol{y}(s)$ may be obtained by two different approaches:

Case A From the differential equations describing the system of Fig. 7.4 obtain

$$\boldsymbol{D}(s)\boldsymbol{y}(s) = \boldsymbol{N}(s)\boldsymbol{u}(s) \tag{7.18}$$

where $\boldsymbol{D}(s)$ and $\boldsymbol{N}(s)$ are *mxm* matrices of polynomials and the output $\boldsymbol{y}(s)$ and input $\boldsymbol{u}(s)$ are *mx1* vectors. Equation (7.18) is manipulated to yield

$$\boldsymbol{y}(s) = \boldsymbol{D}^{-1}(s)\boldsymbol{N}(s)\boldsymbol{u}(s) = \boldsymbol{P}_e(s)\boldsymbol{u}(s) \tag{7.19}$$

where the *effective plant matrix* is given by

$$\boldsymbol{P}_e(s) = \boldsymbol{D}^{-1}(s)\boldsymbol{N}(s) \tag{7.20}$$

Thus from Eq. (7.20) the following expressions are obtained

$$\boldsymbol{D}(s)\boldsymbol{P}_e(s) = \boldsymbol{N}(s) \tag{7.21}$$

$$\det \boldsymbol{D}(s) \det \boldsymbol{P}_e(\mathrm{s}) = \det \boldsymbol{N}(s) \tag{7.22}$$

$$\det \boldsymbol{P}_e(\mathrm{s}) = \frac{n_e(s)}{d_e(s)} \tag{7.23}$$

where $n_e(s)$ = *det* $\boldsymbol{N}(s)$, $d_e(s)$ = *det* $\boldsymbol{D}(s)$, and $n_e(\mathrm{s})$ and $d_e(s)$ are polynomials. Thus, $n_e(s)$ must not have any RHP zeros in order for Eq. (7.23) to be m.p. and satisfy the constraint that *det* $\boldsymbol{P}_e(s)$ be m.p.

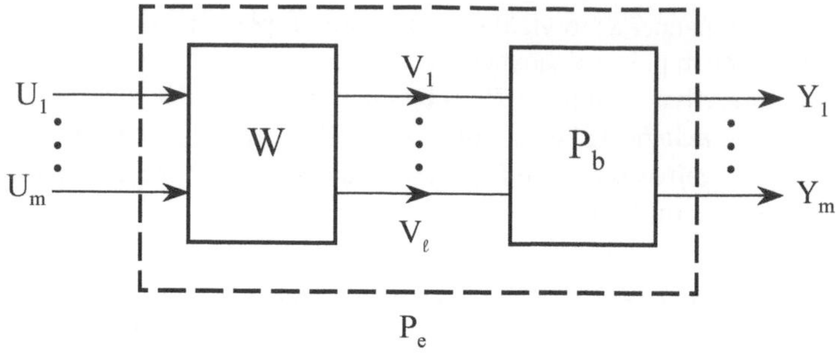

Fig. 7.4 An *mxm* effective plant $\boldsymbol{P}_e(s)$.

Case B From the block diagram of Fig. 7.4, where $\boldsymbol{v}(s)$ is a $\ell x1$ vector and $\boldsymbol{W}$ is only a gain matrix the following expressions are obtained:

$$\boldsymbol{y}(s) = \boldsymbol{P}_b(s)\boldsymbol{v}(s) \qquad \textbf{(7.24)}$$

$$\boldsymbol{v}(s) = \boldsymbol{W}\boldsymbol{u}(s) \qquad \textbf{(7.25)}$$

$$\boldsymbol{y}(s) = \boldsymbol{P}_b(s)\boldsymbol{W}\boldsymbol{u}(s) \qquad \textbf{(7.26)}$$

From Eqs. (7.19) and (7.26) it is seen that the effective plant matrix can also be expressed by

$$\mathbf{P}_e(s) = \mathbf{P}_b(s)\mathbf{W} \qquad \textbf{(7.27)}$$

where $\boldsymbol{P}_b(s) = [p_{ij}(s)]$, $p_{ij}(s) = n_{ij}(s)/d(s)$, $\boldsymbol{W} = [w_{ij}]$, $n_{ij}(s)$ and $d(s)$ are polynomials, $m \le \ell$, and w_{ij} is a gain to be determined in order to try to achieve a m.p. *det* $\boldsymbol{P}_e(s)$. Note that although in this section all the $p_{ij}(s)$ are considered to have a common denominator, i.e., $d(s) = d_{ij}(s)$, in general the p_{ij} elements of $\boldsymbol{P}_b(s)$ can have different denominators

By use of the Binet-Cauchy formula[64] Eq. (7.27) may be expressed as follows:

$$\det \boldsymbol{P}_e(s) = \det[\boldsymbol{P}_b(s)\boldsymbol{W}] = \sum_i [p_i(s)w_i] \qquad \textbf{(7.28)}$$

where $p_i(s)$ and w_i are the determinants of appropriate *mxm* sub-matrices of $\boldsymbol{P}_b(s)$ and $\boldsymbol{W}(s)$, respectively. A sufficient condition for det $[\boldsymbol{P}_b(s)\boldsymbol{W}]$ to be m.p. is that at least one $p_i(s)$ be m.p. A short proof of this fact is as follows: let $p_i(s)$ be the determinant of the *mxm* sub-matrix of $\boldsymbol{P}_b(s)$ formed by the columns indexed by $i_1,\ldots, i_m$. Assume that this $p_i(s)$ is m.p. Choose $\boldsymbol{W}$ so that the *mxm* sub-matrix formed by the rows indexed by $i_1,\ldots, i_m$, is the identity matrix while all remaining rows are zero. Then, clearly det[$\boldsymbol{P}_b(s)\boldsymbol{W}$] = $p_i(s)$. The number of terms in the summation of Eq. (7.28) is given by

$$\alpha_z = \frac{\ell!}{(\ell - m)!\, m!} \tag{7.29}$$

Let the sub-matrices of $\boldsymbol{P}_b$ and $\boldsymbol{W}$, of Eq. (7.28), be represented, respectively, by $\boldsymbol{P}_b$ and $\boldsymbol{W}_{bw}$ where $w = 1, 2, \ldots, \alpha_z$. The *det* $\boldsymbol{P}_e(s)$ obtained from Eq. (7.28) is

$$\det \mathbf{P}_e(s) = \det \mathbf{P}_{b_1}(s) \det \mathbf{W}_{b_1} + \det \mathbf{P}_{b_2}(s) \det \mathbf{W}_{b_2} + \cdots$$
$$+ \det \mathbf{P}_{b_w}(s) \det \mathbf{W}_{b_w} + \cdots + \det \mathbf{P}_{b\alpha_z}(s) \det \mathbf{W}_{b\alpha_z} \tag{7.30}$$

where

$$\boldsymbol{P}_{b_w} = \left[\frac{n_{ij}(s)_w}{d(s)} \right] \tag{7.31}$$

and

$$\boldsymbol{W}_{bw} = [w_{ij_w}] \tag{7.32}$$

Thus each term in Eq. (7.30) can be expressed as

$$\det \boldsymbol{P}_{b_w}(s) \det \boldsymbol{W}_{b_w} = \frac{1}{d^m} \det|\, n_{ij}(s)_w \,| \det|\, w_{ij_w} \,| \tag{7.33}$$

which permits Eq. (7.30) to be written as follows:

$$\det \boldsymbol{P}_e(s) = \frac{1}{d^m(s)} \Sigma(\det|\, n_{ij}(s)_1 \,| \det|\, w_{ij_1} \,| +$$
$$\det|\, n_{ij}(s)_2 \,| \det|\, w_{ij_2} \,| + \cdots + \det|\, n_{ij\alpha_z} \det|\, w_{ij\alpha_z} \,|) = \frac{n_e'}{d^m(s)} \tag{7.34}$$

where n_e' is a polynomial. Since Eq. (7.23) and (7.34) represent the same open-loop system of Fig. 7.4 then

$$\frac{n_e(s)}{d_e(s)} = \frac{n_e'(s)}{d^m(s)} \tag{7.35}$$

which results in $d_e(s) = d(s)$ and

$$n_e(s) = \frac{n_e'(s)}{d^{m-1}(s)} \tag{7.36}$$

Based upon Eqs. (7.34) through (7.36) $d^{m-1}(s)$ must be a factor of $n_{ij}(s)_w$.

The following development illustrates that a sufficient condition for the existence of a m.p. $det\ \boldsymbol{P}_e(s)$ is that at least one $det\ \boldsymbol{P}_b(s)$ must be m.p. Assume

$$\det|\boldsymbol{n}_{ij}(s)_1|\det|\boldsymbol{w}_{ij_1}| = \det|\boldsymbol{n}_{ij}(s)_1|\, k_1 \tag{7.37}$$

in Eq. (7.34) is m.p. where $k_1 = det\ |w_{ij_1}|$ is a scalar. Factor out $d^{m-1}(s)$ from every term in Eq. (7.36) and let

$$k_1 \frac{\det|\boldsymbol{n}_{ij}(s)_1|}{d^{m-1}(s)} = k_1 N_1(s) \tag{7.38}$$

Also, let the summation of the remaining terms, after factoring out $d^{m-1}(s)$ from each term, be expressed as

$$k_2 N_2(s) = \frac{1}{d^{m-1}(s)}[\det|\boldsymbol{n}_{i_j}(s)_2|\det|\boldsymbol{w}_{ij_2}| + \cdots + \det|\boldsymbol{n}_{i_j}(s)_{\alpha z}|\det|\boldsymbol{w}_{ij_{\alpha z}}|] \tag{7.39}$$

where k_2 is a scalar. Note: k_1, k_2, and $N_2(s)$ are now functions of w_{ij} which are to be selected in order to try to achieve a m.p. $\boldsymbol{P}_e(s)$. Based upon Eqs. (7.38) and (7.39), Eq. (7.29) can be expressed as follows:

$$det\ \boldsymbol{P}_e(s) = \frac{1}{d(s)}[k_1 N_1(s) + k_2 N_2(s)] \tag{7.40}$$

In order for $det\ \boldsymbol{P}_e(s)$ to be m.p. then the zeros of the polynomial

$$k_1 N_1(s) + k_2 N_2(s) \tag{7.41}$$

must all be in the LHP. Equation (7.41) is manipulated to the mathematical format of

$$\left(\frac{k_1}{k_2}\right)\left[\frac{N_1(s)}{N_2(s)}\right] = -1 \tag{7.42}$$

which permits a root-locus analysis of Eq. (7.40). Since the zeros of Eq. (7.42) are in the LHP then the weighting factors w_{ij} are selected in hopes that all roots of Eq. (7.41) lie in the LHP for all $\boldsymbol{P}_b \in \boldsymbol{P}$. This assumes that throughout the region of plant parameter uncertainty, the initially chosen m.p. sub-matrix in Eq. (7.34) is m.p. for all $\boldsymbol{P}_b \in \boldsymbol{P}$ and be expressed by Eq. (7.38). To enhance the achievability of m.p. *det* $\boldsymbol{P}_e(s)$ the following guidelines may be used:

1. Determine the number α_y sub-matrices of $\boldsymbol{P}_{b_w}$ of Eq. (7.34) that are m.p.
2. Select one of the m.p. sub-matrices to be identified as Eq. (7.38).
3. The values of w_{ij} of $\boldsymbol{W}_{b_w}$ associated with

   (a) The remaining $\alpha_y - 1$ m.p. $\boldsymbol{P}_{b_w}$ be altered in such a manner as to increase the values of their corresponding *det* $\boldsymbol{W}_{b_w}$.

   (b) The $\alpha_z - \alpha_y$ non m.p. $\boldsymbol{P}_{b_w}$ be altered in such a manner as to decrease the value of their corresponding *det* $\boldsymbol{W}_{b_w}$.

By changing the values of the gains w_{ij} in the manner described will result in the $\alpha_y - 1$ m.p. terms of Eq. (7.39) to dominate in the resulting expression for $N_2(s)$. This dominance of the $\alpha_y - 1$ terms in $N_2(s)$ may result in $N_2(s)$ being m.p. Also, it may enhance the achievability of a m.p. $\boldsymbol{P}_e(s)$ by the analysis of Eq. (7.42).

Once the $n_{ij}(s)$ and w_{ij} are specified, in order to simplify the root-locus computational effort, it is necessary to factor out $d(s)$ from the numerator polynomials of Eqs. (7.38) and (7.39). Depending upon the CAD package that is utilized in determining $d(s)$ and the numerator polynomials of these equations, perfect factoring may not exist.

Now consider the most general case, i.e.,

$$\det \boldsymbol{P}_{b_w}(s) = \det \left| \frac{\boldsymbol{n}_{ij}(s)_w}{\boldsymbol{d}_{ij}(s)_w} \right| = \frac{n_w(s)}{d_w(s)} \tag{7.43}$$

and

$$\det \boldsymbol{W}_{b_w} = \det \left| \frac{\boldsymbol{w}_{ij}(s)_w}{\boldsymbol{h}_{ij}(s)_w} \right| = \frac{w_w(s)}{h_w(s)} \tag{7.44}$$

where the $d_{ij}(s)$ and $h_{ij}(s)$ are not all the same and where $n_w(s)$, $d_w(s)$, $w_w(s)$ and $h_w(s)$ are scalar polynomials. Hence

$$\det \boldsymbol{P}_e(s) = \sum_{\alpha_z} \left[\frac{n_w(s)\, w_w(s)}{d_w(s)\, h_w(s)} \right] \tag{7.45}$$

For each value of w, $n_w(s)/d_w(s)$ has a range of uncertainty which may be correlated with that of the remaining $\alpha_z - 1$ sub-matrices of $\boldsymbol{P}(s)$. Thus, for the general case some or all of the $n_w(s)/d_w(s)$ may have RHP zeros and/or poles. The problem now becomes one of trying to choose fixed $w_w(s)$ and $h_w(s)$ polynomials so that the $det\ \boldsymbol{P}_e(s)$ has no RHP zeros over the entire range, or failing that has them as relatively "far-off"; as possible. This problem has been studied by several researchers and the resulting techniques may be used for this purpose.[16,23]

Thus, the Binet-Cauchy formula permits the determination if an m.p. effective plant $det\ \boldsymbol{P}_e(s)$ is achievable over the region of plant uncertainty.

7-11 SUMMARY

This chapter has presented synthesis techniques for highly uncertain *mxm* MIMO LTI feedback systems with output feedback, with the following features:

(a) There is detailed control over the m^2 individual system transfer functions.
(b) The MIMO uncertainty problem is rigorously converted into a number of MISO uncertainty problems. Solutions of the latter are guaranteed to be satisfactory for the former. Relatively simple MISO single loop feedback techniques can be used to solve the MISO problems.
(c) For "arbitrary small sensitivity" over arbitrary large bandwidth (BW), the technique in Secs. 7.2 through 7.5, give constraints on the plant which are inherent and irreducible, i.e., every LTI compensation technique has these constraints.
(d) Part of the constraints (at infinite s) in (c) were always present in the previously developed MISO equivalent technique,[7] i.e., even if "a.s. sensitivity" was not required. They are present in the new techniques only for a.s. sensitivity. Also, fixed point theory is not required for justification of the new technique.
(e) The over-design inherent in the fixed point techniques,[7] has been reduced, but some over-design is still present.
(f) These techniques are applicable to the design of a re-configurable aircraft with surface failures.[22]

The reader is urged, in performing a QFT design, to constantly refer to Chapter 9 and to Appendices F and G. Also, the reader is referred to Ref. 130 for further discussion on multivariable feedback control system analysis and design.

8

MIMO SYSTEM WITH EXTERNAL DISTURBANCE INPUTS[12]

8-1 INTRODUCTION

Previous chapters have dealt with MIMO tracking control systems with no external disturbances being applied to the plant. This chapter considers the analysis and design of a MIMO external disturbance rejection control system. Therefore, the first portion develops the necessary equations that take into account the MIMO system's external disturbance inputs. The remaining portion applies this development to a real world application problem.[12] Although this chapter addresses the specific aerial refueling problem, the design procedures can be applied to other MIMO external disturbance and tracking/external disturbance problems.

8-2 MIMO QFT WITH EXTERNAL (INPUT) DISTURBANCE

Output disturbance rejection is the primary design criterion in this chapter. Previous discussions of MIMO QFT did not consider external input disturbance in the calculation of cross-coupling rejection bounds. The following development quantifies external uncertain disturbances. Figure 8.1 represents an *mxm* MIMO closed-loop system in which $\boldsymbol{G}(s)$, $\boldsymbol{P}(s)$, and $\boldsymbol{P}_d(s)$ are *mxm* matrices. $\mathcal{P}(s) = \{\boldsymbol{P}(s)\}$ and $\mathcal{P}_d(s) = \{\boldsymbol{P}_d(s)\}$ are sets of matrices due to plant and disturbance uncertainties respectively. The objective is to find a suitable mapping that permits the analysis and synthesis of a MIMO control system by a set of equivalent MISO control systems.

From Fig. 8.1, the following equations are written

$$\boldsymbol{y}(s) = \boldsymbol{P}(s)\boldsymbol{u}(s) + \boldsymbol{P}_d(s)\boldsymbol{d}_{ext}(s) \quad \boldsymbol{u}(s) = \boldsymbol{G}(s)\boldsymbol{v}(s) \quad \boldsymbol{v}(s) = \boldsymbol{r}(s) - \boldsymbol{y}(s) \tag{8.1}$$

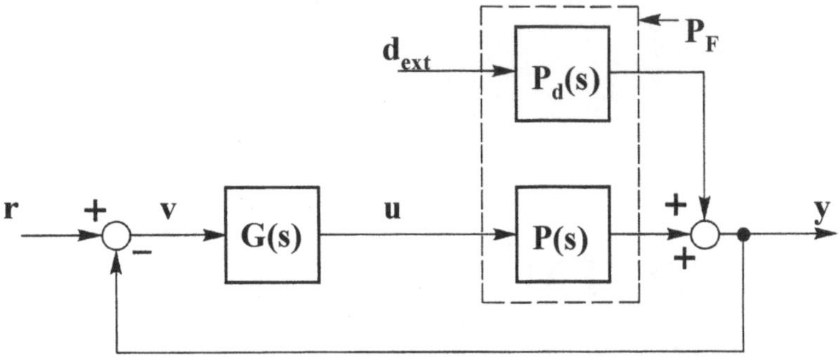

Fig. 8.1 QFT Compensator with output external disturbance.

where for the regulator case with zero tracking input

$$\mathbf{r}(t) = \left[\,0, 0, 0\,\right]^{T} \tag{8.2}$$

From Eqs. (8.1) and (8.2) where henceforth the (s) is dropped in the continuing development,

$$v = -\,y \qquad u = -\,Gy \tag{8.3}$$

which yields

$$y = -\,PGy + P_d d_{ext} \tag{8.4}$$

Equation (8.4) is rearranged to yield:

$$\mathbf{y} = \left[\,\mathbf{I} + \mathbf{PG}\,\right]^{-1} \mathbf{P}_d \mathbf{d}_{ext} \tag{8.5}$$

Based upon unit impulse disturbance inputs for d_{ext}, the system control ratio relating d_{ext} to y is

$$\mathbf{T}_d = \left[\,\mathbf{I} + \mathbf{PG}\,\right]^{-1} \mathbf{P}_d \tag{8.6}$$

Pre-multiply Eq. (8.6) by $[I + PG]$ yields

$$[\mathbf{I} + \mathbf{PG}]\mathbf{T}_d = \mathbf{P}_d \tag{8.7}$$

Pre-multiplying both sides of Eq. (8.7) by $\boldsymbol{P}^{-1}$ results in

$$\left[\boldsymbol{P}^{-1} + \boldsymbol{G}\right]\boldsymbol{T}_d = \boldsymbol{P}^{-1}\boldsymbol{P}_d \quad \textbf{(8.8)}$$

Let

$$\mathbf{P}^{-1} = \begin{bmatrix} p_{11}^{*} & p_{12}^{*} & \cdots & p_{1m}^{*} \\ p_{21}^{*} & p_{22}^{*} & \cdots & p_{2m}^{*} \\ \vdots & \vdots & & \vdots \\ p_{m1}^{*} & p_{m2}^{*} & \cdots & p_{mm}^{*} \end{bmatrix} \quad \textbf{(8.9)}$$

The m^2 effective plant transfer functions are formed as

$$q_{ij} \equiv \frac{1}{p_{ij}^{*}} = \frac{\det \boldsymbol{P}}{\operatorname{adj} \boldsymbol{P}_{ij}} \quad \textbf{(8.10)}$$

the $\boldsymbol{Q}$ matrix is then formed as

$$\boldsymbol{Q} = \begin{bmatrix} q_{11} & q_{12} & \cdots & q_{1m} \\ q_{21} & q_{22} & \cdots & q_{2m} \\ \vdots & \vdots & & \vdots \\ q_{m1} & q_{m2} & \cdots & q_{mm} \end{bmatrix} = \begin{bmatrix} 1/p_{11}^{*} & 1/p_{12}^{*} & \cdots & 1/p_{1m}^{*} \\ 1/p_{21}^{*} & 1/p_{22}^{*} & \cdots & 1/p_{2m}^{*} \\ \vdots & \vdots & & \vdots \\ 1/p_{m1}^{*} & 1/p_{m2}^{*} & \cdots & 1/p_{mm}^{*} \end{bmatrix} \quad \textbf{(8.11)}$$

where $\boldsymbol{P} = [p_{ij}]$, $\boldsymbol{P}^{-1} = [p_{ij}^{*}] = [1/q_{ij}]$, and $\boldsymbol{Q} = [q_{ij}] = [1/p_{ij}^{*}]$. The $\boldsymbol{P}^{-1}$ matrix is partitioned as follows:

$$\boldsymbol{P}^{-1} = [\, p_{ij}^{*} \,] = [\, 1/q_{ij} \,] = \boldsymbol{\Lambda} + \boldsymbol{B} \quad \textbf{(8.12)}$$

where $\boldsymbol{\Lambda}$ is the diagonal part of $\boldsymbol{P}^{-1}$ and $\boldsymbol{B}$ is the balance of $\boldsymbol{P}^{-1}$. Thus $\lambda_{ii} = 1/q_{ii} = p_{ii}^{*}$, $b_{ii} = 0$, and $b_{ij} = 1/q_{ij} = p_{ij}^{*}$ for $i \neq j$. Substituting Eq. (8.12) into Eq. (8.8) with $\boldsymbol{G}$ diagonal, results in

$$[\, \boldsymbol{\Lambda} + \boldsymbol{B} + \boldsymbol{G} \,]\boldsymbol{T}_d = [\, \boldsymbol{\Lambda} + \boldsymbol{B} \,]\boldsymbol{P}_d \quad \textbf{(8.13)}$$

Rearranging Eq. (8.13) produces

$$T_d = [\Lambda + G]^{-1}[\Lambda P_d + BP_d - BT_d] = \{t_{d_{ij}}\} \tag{8.14}$$

Rearranging, for a unit impulse external disturbance inputs, Eq. (8.14) yields:

$$Y(T_d) = [\Lambda + G]^{-1}[\Lambda P_d + BP_d - BT_d] \tag{8.15}$$

This equation defines the desired fixed point mapping, where each of the m^2 matrix elements on the right side of Eq. (8.14) are interpreted as MISO problems. Proof of the fact that the design of each MISO system yields a satisfactory MIMO design is based on the Schauder fixed point theorem.[7] The theorem defines a mapping $Y(T_d)$ where each member of T_d is from the acceptable set $\mathfrak{I}_d$ (see Fig. 5.10). If this mapping has a fixed point, i.e., $T_d \in \mathfrak{I}_d$, then this T_d is a solution of Eq. (8.14).

Figure 8.2 shows the effective MISO loops resulting from a *3x3* system. Since Λ and G in Eq. (8.14) are diagonal, the *(1,1)* element on the right side of Eq. (8.15) for the *3x3* case, for a *unit impulse input*, provides the output

$$y_{d_{11}} = \frac{q_{11}}{1 + g_1 q_{11}} \left[\frac{p_{d11}}{q_{11}} + \frac{p_{d21}}{q_{12}} + \frac{p_{d31}}{q_{13}} - \left(\frac{t_{c21}}{q_{12}} + \frac{t_{c31}}{q_{13}} \right) \right] \tag{8.16}$$

Equation (8.16) corresponds precisely to the first structure in Fig. 8.2.

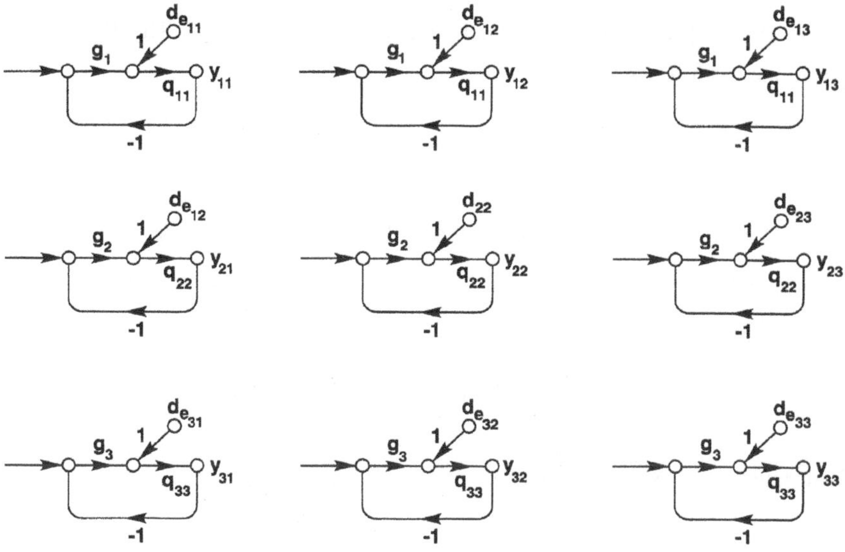

Fig. 8.2 *3x3* MISO equivalent loops for external output disturbance.

Similarly, each of the nine structures in this figure corresponds to one of the elements of $\boldsymbol{Y}(\boldsymbol{T}_d)$ of Eq. (8.15). The control ratios for the external disturbance inputs d_{ext_i} and the corresponding outputs y_i for each feedback loop of Eq. (8.15) have the form

$$y_{ii} = w_{ii}(d_{e_{ij}}) \qquad \textbf{(8.17)}$$

where $w_{ii} = q_{ii}/(1 + g_i q_i)$ and

$$\begin{aligned} d_{e_{ij}} &= (d_{ext})_{ii} - c_{ij} \\ &= \sum_{k=1}^{x} \frac{p_{d_{kj}}}{q_{ik}} - \sum_{k \neq i}^{m} \frac{t_{c_{kj}}}{q_{ik}} \qquad \begin{array}{l} x = \text{number of } \textit{disturbance} \text{ inputs} \\ \text{m} = \text{dimension of square MIMO system} \end{array} \end{aligned}$$

(8.18)

Thus, the *interaction term*, Eq. (8.18), not only contains the cross-coupling interaction but also the external disturbances, i.e.:

$$(d_{ext})_{ij} = \sum_{k=1}^{x} \frac{p_{d_{kj}}}{q_{ik}}$$

(8.19)

$$c_{ij} = \sum_{k \neq i}^{m} \frac{t_{c_{ij}}}{q_{ik}}$$

where $(d_{ext})_{ij}$ represents the external disturbance effects and c_{ij} represents the cross-coupling effects.

Additional equations, quantifying both the external disturbance $(d_{ext})_{ij}$ and the internal cross-coupling effects c_{ij}, are derived to utilize the improved method (Method 2) QFT design technique. These equations are used to define the disturbance bounds for subsequent loops based on the completed design of the first loop. For this development, the equations for the case of a *2x2* MIMO system are presented.

From Eq. (8.17) and for the *1-2* loop case, which is the output of *loop 1* due to *disturbance input 2*, including the cross-coupling terms from *loop 2*, yields, for *unit impulse inputs*, the following control ratio:

$$t_{d_{12}} = y_{12} = w_{11}(d_{e_{12}}) = \frac{q_{11}}{1 + L_1}\left[\frac{p_{d12}}{q_{11}} + \frac{p_{d_{22}}}{q_{12}} - \frac{t_{d22}}{q_{12}}\right] \qquad \textbf{(8.20)}$$

Substituting in for t_{d22} yields:

$$t_{d_{12}} = \frac{q_{11}}{1 + L_1}\left[\frac{p_{d12}}{q_{11}} + \frac{p_{d22}}{q_{12}} - \frac{q_{22} d_{e22}}{(1 + L_2) q_{12}}\right] \qquad \textbf{(8.21)}$$

$$t_{d_{12}} = \frac{q_{11}}{1+L_1}\left[\frac{p_{d_{12}}(1+L_2)q_{12} + p_{d_{22}}(1+L_2)q_{11} - q_{22}q_{11}d_{e_{22}}}{(1+L_2)q_{11}q_{12}}\right] \quad \textbf{(8.22)}$$

$$t_{d_{12}} = \frac{(1+L_2)(p_{d_{12}}q_{12} + p_{d_{22}}q_{11}) - q_{22}q_{11}\left(\frac{p_{d_{12}}}{q_{21}} + \frac{p_{d22}}{q_{22}} - \frac{t_{c12}}{q_{21}}\right)}{(1+L_1)(1+L_2)q_{12}} \quad \textbf{(8.23)}$$

Substituting in for d_{e22} and rearranging yields

$$t_{d_{12}} = \frac{(1+L_2)(p_{d_{12}}q_{12} + p_{d_{22}}q_{11}) - \frac{q_{22}q_{11}p_{d_{12}}}{q_{21}} - q_{11}p_{d_{22}} + \frac{q_{22}q_{11}t_{d_{12}}}{q_{21}}}{(1+L_1)(1+L_2)q_{12}} \quad \textbf{(8.24)}$$

$$t_{d_{12}} = \frac{(1+L_2)(p_{d_{12}}q_{12} + p_{d_{22}}q_{11})q_{21} - q_{22}q_{11}p_{d_{12}} - q_{11}q_{21}p_{d_{22}} + q_{22}q_{11}t_{d_{12}}}{(1+L_1)(1+L_2)q_{12}q_{21}}$$

(8.25)

where $\gamma_{12} = q_{11}q_{22}/q_{21}q_{12}$. Solving for t_{d12} yields

$$t_{d_{12}}[(1+L_1)(1+L_2)] = \frac{(1+L_2)(p_{d_{12}}q_{12} + p_{d_{22}}q_{11})}{q_{12}} - \frac{q_{11}p_{d_{22}}}{q_{12}} - \gamma_{12}p_{d_{12}} + \gamma_{12}t_{d_{12}} \quad \textbf{(8.26)}$$

$$t_{d_{12}}[(1+L_1)(1+L_2) - \gamma_{12}] =$$

(8.27)

$$(1+L_2)\frac{q_{11}}{q_{12}}p_{d_{22}} + (1+L_2)p_{d_{12}} - \frac{q_{11}}{q_{12}}p_{d_{22}} - \gamma_{12}p_{d_{12}}$$

$$t_{d_{12}} = \frac{L_2 \dfrac{q_{11}}{q_{12}} p_{d_{22}} + (1 + L_2 - \gamma_{12}) p_{d_{12}}}{L_1(1 + L_2) + 1 + L_2 - \gamma_{12}} \tag{8.28}$$

Equation (8.28) is rearranged as follows:

$$t_{d_{12}} = \frac{\dfrac{L_2 \dfrac{q_{11}}{q_{12}} p_{d_{22}}}{1 + L_2 - \gamma_{12}} + p_{d_{12}}}{1 + \dfrac{L_1(1 + L_2)}{1 + L_2 - \gamma_{12}}} \tag{8.29}$$

From Eq. (8.29) the effective plant is defined as:

$$\mathrm{q}_{11_e} \equiv \frac{\mathrm{q}_{11}(1 + \mathrm{L}_2)}{1 + \mathrm{L}_2 - \gamma_{12}} \tag{8.30}$$

Substituting Eq. (8.30) into Eq. (8.29), yields:

$$t_{d_{12}} = \frac{\dfrac{L_2 p_{d_{22}} q_{11_e}}{q_{12}(1 + L_2)} + p_{d_{12}}}{1 + g_1 q_{11_e}} \tag{8.31}$$

Thus, in general, for the *2x2* case, the improved method control ratio of the jth interaction input to the ith system output is:

$$t_{d_{ij}} = \frac{\dfrac{L_k p_{d_{kj}} q_{ii_e}}{q_{ik}(1 + L_k)} + p_{d_{ij}}}{1 + g_i q_{ii_e}} \quad \text{where } i = 1, 2 \text{ and } k \neq i \tag{8.32}$$

The interaction bounds (the optimal bounds for a pure regulator control system), representing the cross-coupling and the external disturbance effects, are calculated at a given frequency to satisfy

$$(B_{d_e})_{ij} \geq \left| t_{d_{ij}} \right| = \left| \frac{\dfrac{L_k\, p_{d_{kj}}\, q_{ii_e}}{q_{ik}\,(1+L_k)} + p_{d_{ij}}}{1+g_i q_{ii_e}} \right| \quad \text{where } i=1,2 \text{ and } k \neq i \qquad \textbf{(8.33)}$$

or

$$\left| 1+g_i q_{ii_e} \right| \geq \frac{1}{(B_{d_e})_{ij}} \left| \frac{L_k\, p_{d_{kj}}\, q_{ii_e}}{q_{ik}\,(1+L_k)} + p_{d_{ij}} \right| \quad \text{where } i=1,2 \text{ and } k \neq i \qquad \textbf{(8.34)}$$

The improved QFT method uses these equations to reduce the over-design inherent in the original design process. The order in which loops are designed is important as pointed out in Chapter 7. Any order can be used, but some orders produce less over-design (less bandwidth) than others. The last loop designed has the least amount of over-design, therefore the most constrained loop is done first by Method 1. Then the design is continued through the remaining loop by use of Method 2. As a second iteration, the first loop is then redesigned using Method 2.

At this point it is important to point out that when the interaction term specification is considered, the designer must decide how much is to be allocated for the cross-coupling effects and how much for the external disturbance effects. In other words, the designer can "tune" the external disturbance rejection specification depending on the nature of the interaction term for a particular loop. For example, if one loop is only affected by external disturbance, the interaction term specification would consider external disturbance effects only. But if the loop interaction term is a mix of cross-coupling and external disturbance, the designer must then "tune" the interaction term specification accordingly. Since each loop may not exhibit the same interaction characteristics, interaction term specification tuning provides flexibility in the QFT design process.

8-3 AN EXTERNAL DISTURBANCE PROBLEM

The remaining portion of this chapter applies the QFT design technique to a real-world problem. The intent is to present to the reader a design problem, with as much detail as space permits, from the onset of the specification of the control problem to the verification of the design results by means of linear and nonlinear simulations.

8-3.1 AERIAL REFUELING BACKGROUND

The United States Air Force (USAF) maintains a fleet of large cargo/transport aircraft. Refueling these aircraft during flight provides unlimited range of operation for this fleet of aircraft. However, long flights and multiple air-to-air refueling can seriously strain and fatigue the pilot, decreasing flight safety, and extending recovery time between missions. Hence, automatic control of the receiving aircraft during aerial refueling operations is most beneficial.

Cargo/transport aircraft are generally large and have high moments of inertia. Piloting a large, high inertia aircraft during air-to-air refueling requires intense concentration. The pilot must maintain a very precise position relative to the tanker. He/she maintains position visually, applying the appropriate control inputs when changes in position occur. The pilot must compensate for changes in aircraft dynamics due to taking on fuel, specifically, movements in center-of-gravity and changes in the moments of inertia I_{xx} and I_{yy}. Besides dynamic changes, the pilot must contend with maintaining position in the presence of wind gusts. Since these aircraft can take on large amounts of fuel, up to *250,000* pounds, air-to-air refueling can take up to *30* minutes. Compound this over long flights and multiple refueling, and the pilot's fatigue level increases and can reach an unsafe level. This could endanger the flight crew, and possibly impact the pilot's capability to perform his/her mission.

One way to ease the pilot workload is to implement an automatic flight control system (AFCS) for air-to-air refueling. The AFCS needs to be able to maintain a precise position of the receiving aircraft (receiver) relative to the tanker in the presence of such disturbances as wind gusts, and in the changes of mass and moments of inertia. This AFCS is designed to precisely regulate position relative to the tanker by applying the MIMO QFT disturbance external rejection design method to address the rejection of the disturbances entering the system at the output.

8-3.2 PROBLEM STATEMENT

During air-to-air refueling, the receiver aircraft will change position relative to the tanker. The pilot must pay close attention and take corrective action to maintain position. Excessive changes in position will disconnect the refueling boom from the receiver. An AFCS must be designed to regulate the receiver's position, thus reducing the pilot workload and fatigue factor. By using MIMO QFT, an AFCS is designed that operates throughout the range of the changing aircraft dynamics and rejects disturbances including those at the output.[25]

8-3.3 ASSUMPTIONS

The following assumptions are made:

- Only the desired outputs are of interest for final performance.

- Position of the receiver aircraft relative to the tanker during air-to-air refueling can be accurately measured.
- The CAD packages used, MIMO/QFT, EASY5x, $MATRIX_x$ and Mathematica are adequate for the design process.

The first assumption is required in applying MIMO QFT. The second assumption is required because no sensors are currently in place to measure the position of the receiver relative to the tanker. The third assumption is concerned with the limits of CAD packages and their numerical robustness.

8-3.4 DESIGN OBJECTIVES

The design objectives are: (1) to utilize the aircraft models, developed in Ref. 12, for the QFT design process using a published document[65] containing C-135 cargo aircraft stability derivatives tables and plots; (2) to present a design for multi-channel control laws using MIMO QFT for several flight conditions with special emphasis on aircraft center-of-gravity and weight changes; (3) to simulate the design for linear and nonlinear performance on $MATRIX_x$, and nonlinear performance on EASY5x; (4) to evaluate the new control law; and (5) to validate the MIMO QFT design with disturbances at the output.

8-3.5 SCOPE

The MIMO QFT external disturbance rejection technique is applied to the design of an AFCS regulator for the automatic maintenance of the three-dimensional separation (x, y, and z) of a receiver aircraft relative to a tanker. The AFCS controls the receiver and is independent of the tanker in as much as the tanker is used as the point of reference. The MIMO/QFT CAD package (Appendix C) was utilized to achieve this design. The AFCS is designed to reject disturbances at the x, y, and z outputs in order to keep the receiver aircraft in a volume specified as the area of boom operation.[65] Models are developed for disturbances due to wind gusts and received fuel. The MIMO QFT plant is the bare-aircraft model augmented by a typical Mach-hold, altitude-hold, wing-leveler autopilot. QFT compensators control the reference signal of the autopilot to maintain "formation" during air-to-air refueling. The control system is simulated for linear performance in $MATRIX_x$. A full six-degree-of-freedom nonlinear simulation is performed in EASY5x.

8-3.6 METHODOLOGY

The design approach requires six steps:

- Generate linear time-invariant (LTI) state-space models of the aircraft for different weights and center of gravity.

- Implement a Mach-hold, altitude-hold, wing leveler autopilot that operates for all aircraft models.
- Model the disturbance due to wind gusts and refueling.
- Design the AFCS using QFT.
- Simulate the design on $MATRIX_x$ and EASY5x to validate the AFCS design.

8-3.7 OVERVIEW OF THE AERIAL REFUELING DESIGN PROBLEM

Section 8-4 discusses how the external disturbances are incorporated into the *6* degree of freedom (DOF) aircraft equations. The MIMO QFT mathematical expressions are reformulated based upon these modified aircraft equations. The air-to-air refueling AFCS concept is discussed in Secs. 8-5 and 8-6. The AFCS QFT design is presented in Sec. 8-7 followed by Sec. 8-8 which presents the linear and nonlinear simulations and illustrate the "goodness of the design."

8-4 AIR-TO-AIR REFUELING FCS DESIGN CONCEPT

The aircraft (A/C) modeled in this chapter is the cargo variant of the C-135 class aircraft (C-135B). This A/C is chosen because of the availability of the aerodynamic data.[65] A Mach-hold, altitude hold, and wing-leveler autopilot is included in the C-135B model.[25] Wind gusts and fuel transfer disturbance models are developed as well as the AFCS concept.

8-4.1 C-135B MODELING

EASY5x is used to develop the state-space *6* DOF bare (uncontrolled) A/C model. EASY5x is a computer aided design (CAD) tool written by Boeing Computer Services used to model, simulate, and analyze dynamic systems. The user need only provide A/C stability derivatives, flight conditions, and desired input/outputs. Sixteen bare A/C plants are developed to account for the uncertainty of the C-135B during air-to-air refueling. The 16 models are based on two different coefficients of lift, $C_L = 0.2,\ 0.6$, for *8* different A/C weights. The *Mach 0.69* at *28,500* feet flight condition is considered. These discrete values are selected based on the availability of data, normal refueling speed and altitude, and represent weights ranging from empty/low fuel to loaded/full fuel A/C. Typically, during air-to-air refueling, the C-135B will have a C_L between *0.27* and *0.45*.[65] Therefore, the *16* plant models envelop the structured uncertainty of the C-135B during air-to-air refueling.

The *6* DOF state-space models, generated by EASY5x, are loaded into $MATRIX_x$. $MATRIX_x$ is used to design the autopilot using root-locus design techniques. The autopilot is designed to control all *16* plant cases. The bare aircraft and autopilot are shown in Fig. 8.3. An autopilot is used for two reasons:

1. Autopilots reduce the high frequency cutoff of the A/C
2. All A/C have autopilots

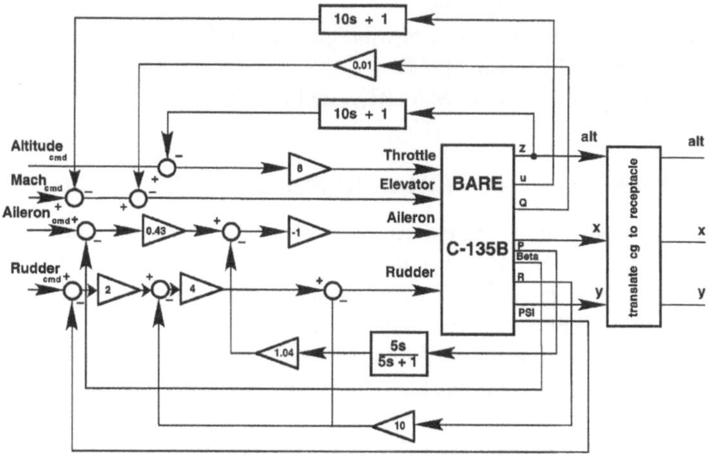

Fig. 8.3 C-135B bare aircraft with autopilot.

Lowering the cutoff frequency of the A/C reduces high frequency parameter uncertainty which in turn reduces the size of the QFT templates. Since autopilots are available, using it in the QFT design eliminates duplication of a control system to provide input to the bare A/C, reducing cost and overhead. The inputs to the autopilot are thrust, elevator, aileron, and rudder commands. The outputs are *z-position* (*altitude*), *x-position*, and *y-position* in a local inertial reference frame where x is positive out the nose of the A/C, y out the right wing, and altitude is positive up. The three outputs frame of reference is translated from the A/C center of gravity (cg) to the approximate location of the air-to-air refueling receptacle on the top of the A/C.

The Mach hold command input is used to control the x position, altitude hold controls altitude, and the rudder command is used to control the y position. Mach and altitude are self evident, rudder is chosen over aileron because the rudder does not roll the A/C. By using the rudder for the QFT controller, good performance is obtained while leaving the aileron controller to handle wing leveling.

8-4.2 DISTURBANCE MODELING

Disturbance models are generated by developing augmented state-space models of the A/C in the presence of wind gusts and fuel transfer inputs. The development in Chapter 3 of Ref. 12, based upon external disturbance inputs represented by the vector $\boldsymbol{d}$, considers three disturbance components: pitch plane wind induced disturbance Γ_{pitch}, lateral channel wind induced disturbance Γ_{lat}, and refueling disturbance Γ_{rf}. Total disturbance modeled is

$$\boldsymbol{\Gamma d} = \Gamma_{pitch} d_{pitch} + \Gamma_{lat} d_{lat} + \Gamma_{rf} d_{rf} \tag{8.35}$$

where Γ_{pitch} and Γ_{rf} are additional inputs that identify pitch plane flight behavior. In the same manner, the input Γ_{lat} identifies the lateral plane flight behavior. The state-space equation now takes on the form

$$\begin{aligned} \dot{x} &= \boldsymbol{Ax} + \boldsymbol{Bu} + \boldsymbol{\Gamma d} \\ y &= \boldsymbol{Cx} \end{aligned} \tag{8.36}$$

Note, as the disturbances are applied to the bare A/C, they enter into the inner most loop, around which the autopilots, and later, the QFT loops are closed.

8-5 PLANT AND DISTURBANCE MATRICES

Based on zero initial conditions, then from Eq. (8.36)

$$s\,\mathbf{x} = \mathbf{A}\,\mathbf{x} + \mathbf{B}\,\mathbf{u} + \Gamma\,\mathbf{d} \tag{8.37}$$

$$\mathbf{x} = [s\,\mathbf{I} - \mathbf{A}]^{-1}\,\mathbf{B}\,\mathbf{u} + [s\,\mathbf{I} - \mathbf{A}]^{-1}\Gamma\mathbf{d} \tag{8.38}$$

$$\begin{aligned} \mathbf{y} &= \mathbf{C}\mathbf{x} = \mathbf{C}\left[s\mathbf{I} - \mathbf{A}\right]^{-1}\mathbf{B}\mathbf{u} + \mathbf{C}\left[s\mathbf{I} - \mathbf{A}\right]^{-1}\Gamma\mathbf{d} \\ &= \mathbf{P}(s)\mathbf{u} + \mathbf{P}_d(s)\Gamma\mathbf{d} \end{aligned} \tag{8.39}$$

where

$$\begin{aligned} \mathbf{P}(s) &= \mathbf{C}[s\,\mathbf{I} - \mathbf{A}]^{-1}\mathbf{B} \\ \mathbf{P}_d(s) &= \mathbf{C}[s\,\mathbf{I} - \mathbf{A}]^{-1} = \{p_{d_{ij}}\} \end{aligned} \tag{8.40}$$

and where the plant model $\boldsymbol{P}_F$ is partitioned into the two matrices $\mathbf{P}(s)$ and $\boldsymbol{P}_d(s)$, $\boldsymbol{P}(s) = \boldsymbol{P}_e(s)$ for a square plant matrix $\boldsymbol{P}(s)$, and the matrix $\boldsymbol{P}_d(s)$ models the transmission from the external disturbance inputs to the output of $\boldsymbol{P}_F$. If $\boldsymbol{P}(s)$ is not a square matrix then a weighting matrix $\boldsymbol{W}(s)$ must be used to yield $\boldsymbol{P}_e(s) = \boldsymbol{P}(s)\boldsymbol{W}$.

Equation (8.39) is represented in Fig. 8.1. Thus, the QFT formulation of Sec. 8-2 is applicable for this problem.

8-6 CONTROL PROBLEM APPROACH

The tanker's position is assumed fixed and hence the receiver aircraft's position is measured from this frame of reference. In this approach the control problem can be viewed as a formation flying problem. The receiver maintains the total obligation of regulating its position. The tanker is free to change course, altitude, and velocity while the receiver compensates for these changes and maintains relative position. Equations are developed that identify perturbations from the set position. These perturbations are viewed as disturbances by the receiver. The perturbations are caused by wind gusts and disturbances due to refueling. Other, un-modeled, disturbances may include the tanker changing course. The control problem's goal is to minimize the perturbations to be within a specified volume of space where the refueling boom can operate. Normal boom operating position and length defines this volume.

The C-135 tanker refueling boom has the following operational constraints: (1) nominal boom operation position is 30° down from horizontal, (2) the boom can move as much as 4° up and down from normal position and continue delivering fuel, (3) it can move as much as 10° up and down from normal position and maintain its connection to the receiver, but cannot deliver fuel, (4) horizontal movement is limited to 10° left and right while maintaining fuel flow, (5) the disconnect limit horizontally is 15° left and right, (6) nominal boom length is *477.5 inches* (*39.8 ft*), (7) it can expand or constrict 13.5 inches and maintain refueling, (8) it can expand or constrict as much as *73.5 inches* and maintain contact but not refueling.[65] These dimensions provide a maximum perturbation from nominal boom position of approximately *2.85 ft* up or down, *7 ft* left or right, in order to maintain fuel flow. In order to maintain connection, the maximum perturbation can be *7.5 ft* up or down, and *11.5 ft* left or right.

Using the tanker as the point of reference, the relationship in Fig. 8.4 can be used to develop the equations required to define the regulation control problem. R is the nominal boom length measured from the boom hinge point on the tanker. Z is the vertical distance between the boom hinge point and receiver aircraft's refueling receptacle. X is the horizontal distance between these same points. Y measures the distance between the center line of the boom hinge point and receiver receptacle.

The following equations are derived

$$R^2 = X^2 + Y^2 + Z^2 \tag{8.41}$$

where

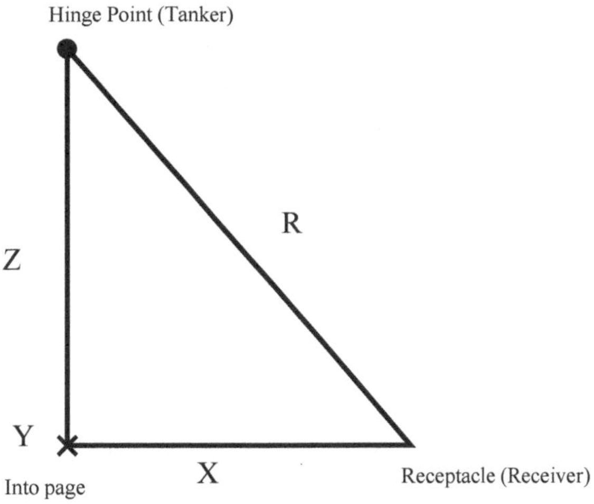

Fig. 8.4 Control problem geometry.

$$
\begin{aligned}
R &= \bar{R} + r \\
X &= \bar{X} + x \\
Y &= \bar{Y} + y \\
Z &= \bar{Z} + z
\end{aligned}
\qquad \textbf{(8.42)}
$$

which are the sums of the nominal positions (over bar terms) and perturbations (lower case terms). Substituting Eq. (8.42) into Eq. (8.41) and squaring the terms, yields

$$\bar{R}^2 + 2r\bar{R} + r^2 = \bar{X}^2 + 2x\bar{X} + x^2 + \bar{X}^2 + 2y\bar{Y} + y^2 + \bar{Z}^2 + 2z\bar{Z} + z^2 \qquad \textbf{(8.43)}$$

Since $r, x, y, z \ll \bar{R}, \bar{X}, \bar{Y}, \bar{Z}$ respectively, Eq. (8.43) is approximated as

$$\bar{R}^2 + 2r\bar{R} = \bar{X}^2 + 2x\bar{X} + \bar{Y}^2 + 2y\bar{Y} + \bar{Z}^2 + 2z\bar{Z} \qquad \textbf{(8.44)}$$

Taking the derivative with respect to time where the over bar terms are constant yields

$$\bar{R}\frac{dr}{dt} = \bar{X}\frac{dx}{dt} + \bar{Y}\frac{dy}{dt} + \bar{Z}\frac{dz}{dt} \qquad \textbf{(8.45)}$$

Integrating, rearranging, and setting $\boldsymbol{r} = \boldsymbol{0}$ yields:

$$r = \frac{\overline{X}}{R}x + \frac{\overline{Y}}{R}y + \frac{\overline{Z}}{R}z = 0 \tag{8.46}$$

defining

$$\begin{aligned} x &= x_T - x_R \\ y &= y_T - y_R \\ z &= z_T - z_R \end{aligned} \qquad \begin{aligned} &T\text{ - Tanker} \\ &R\text{ - Receiver} \end{aligned} \tag{8.47}$$

Thus, $\boldsymbol{r} = \boldsymbol{0}$ if and only if, $x = y = z = 0$. Therefore, the control problem is to design the compensator $\boldsymbol{G}$ of Fig. 8.5 that will satisfy Eqs. (8.46) and (8.47).

8-7 THE QFT DESIGN

The details for the QFT AFCS design are now presented. First the disturbance rejection specification is identified. Next, design of the loop transmissions for all three channels is described. Previous chapters have provided the detailed step-by-step guide for the QFT design process.

8-7.1 DISTURBANCE REJECTION SPECIFICATION

The primary goal in designing the AFCS system is to regulate the position of the A/C receiving fuel relative to the tanker. As discussed in Sec. 8-2, any deviation

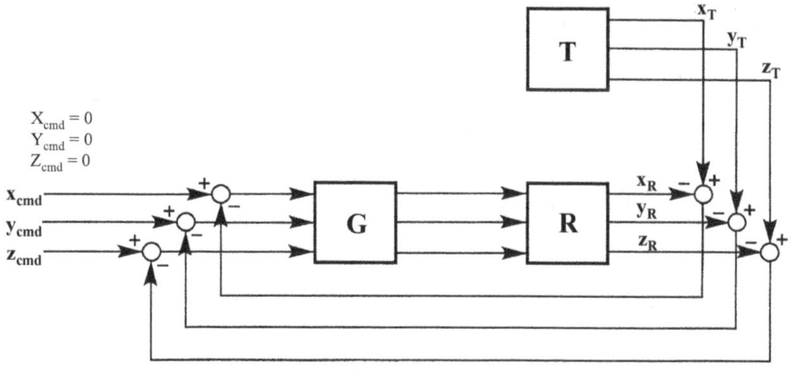

Fig. 8.5 Control problem.

from the nominal set position is considered a disturbance. Hence a disturbance rejection specification is determined based on modeled disturbance inputs and the basic QFT design based upon unit impulse inputs. Since the most severe disturbance is due to wind, the disturbance specification is "tuned" to the wind input of *10 ft/sec*. A maximum deviation from the nominal set position of *2 ft* in any direction is specified which will confine the receiving A/C to a volume that permits continued fuel delivery. Therefore the following disturbance specification is derived. Given an impulse input of magnitude *10 ft/sec*, the system response will deviate no more than *2 ft*. Additionally, the system will attenuate to half the maximum deviation in less than *1 sec*. Equation (8.48) identifies the transfer function for the disturbance rejection specification, and Fig. 8.6 shows the disturbance rejection model response to a *10 ft/sec* impulse input.

$$\text{Generic disturbance rejection model} = \frac{Y(s)}{D_{ext}(s)} = \frac{400s}{(s+1)(s+5)^3} \tag{8.48}$$

An analysis of the *Lm* plot of the disturbance rejection specification, which is superimposed over the *Lm* **P**(*s*) MISO loop plots (see Fig. 8.7), reveals[12] that MISO loops {*2,1*}, {*3,1*}, {*3,2*}, {*1,3*}, and {*2,3*} are below the disturbance specification before compensation is applied.

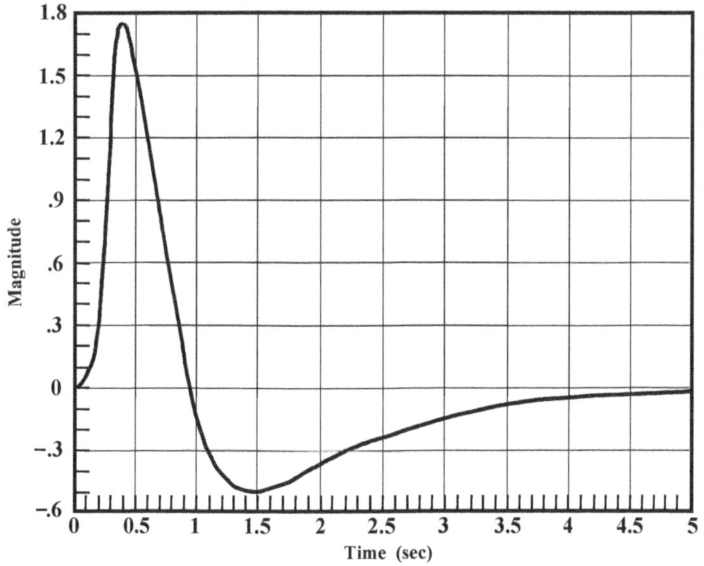

Fig. 8.6 Disturbance rejection model response to *10 ft/sec* impulse.

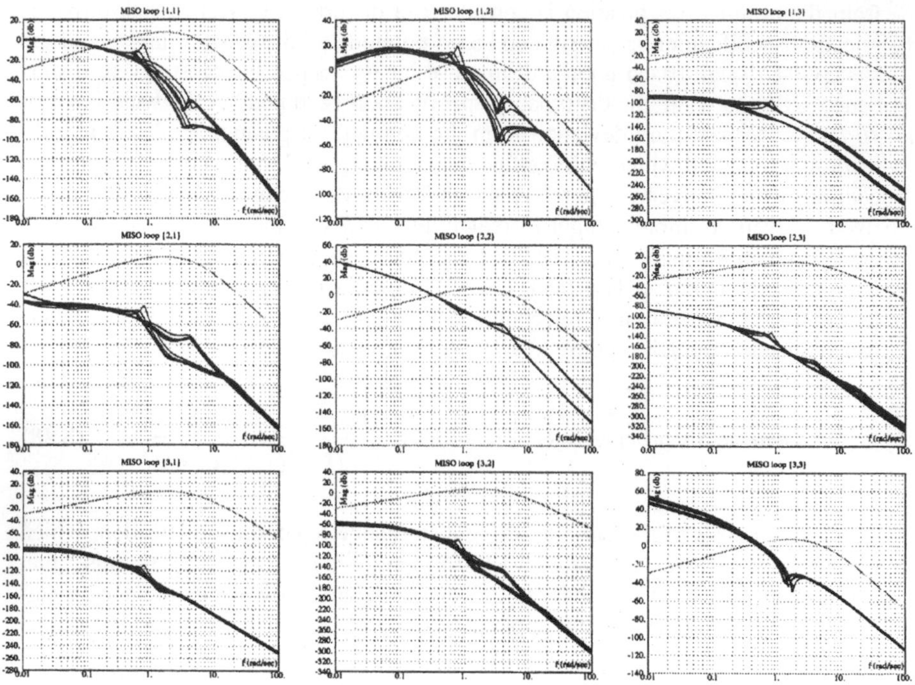

Fig. 8.7 $P(s)$ Log magnitude plot.

8-7.2 LOOP SHAPING

The order of loop shaping is determined by the amount of cross-coupling each MISO loop exerts on each other. Since channel 2 couples strongly into channel 1 it is designed first. The improved method is then applied to utilize the known $\boldsymbol{g}_2$ to recalculate the disturbance bounds for channel 1. Channel 1 is then designed. Channel *3* is designed last since it is completely decoupled and thus a 1x1 SISO system. Note: in practice the loops would have been renumbered in order to be in sequence and "simplify" the performance of the design process.

The band-pass of the plants are relatively low; a benefit of using an autopilot. In shaping the loops the overall system band-pass is designed to remain approximately equal to the plant band-pass. This requirement may require trade-offs on meeting certain higher frequency bounds.

8-7.3 CHANNEL 2 LOOP DESIGN (METHOD 1 "LOOP 1" DESIGN PROCEDURE)

For channel *2* plant case *2* is chosen to be the nominal loop. Plant *2* is chosen because through initial design attempts it proved to be the most difficult to shape around the stability contour. A successful shaping of plant case *2* guarantees stability for all plant cases. Templates, stability and disturbance boundaries are calculated and composite bounds are formed in the MIMO QFT CAD Package. The channel *2* plants are 360° out of phase between the plants derived from the aircraft plant with $C_L = 0.2$ and $C_L = 0.6$. This is evident from the 360° wide templates and the stretching of the bounds over 360°. The phase difference does not present a problem as the MIMO CAD Package is able to accommodate this scenario. Compensator g_2 poles and zeros are added to shape the loop. Channel *2* is relatively easy to shape and proved to have the lowest order compensator g_2.

As shown in Fig. 8.8, the channel *2* loop easily satisfies all QFT loop shaping requirements for composite bound and stability contours, guaranteeing a stable design satisfying the disturbance rejection specification. Figure 8.9 shows the loop shapes on the NC for all *16* plants. From this figure the 360° degree phase difference in some plants is evident. Though there is a phase difference, each plant correctly goes around the stability contour indicating a stable design for all plant cases. The following compensator is designed for this channel:

$$g_2 = \frac{(s+0.25)(s+0.75)(s+1.2)(s+1.3)}{s(s+0.98 \pm j1)(s+10)(s+20)(s+120)} \tag{8.49}$$

8-7.4 CHANNEL 1 LOOP DESIGN (METHOD 2 "LOOP 2" DESIGN PROCEDURE)

After g_2 is designed the improved method (Method *2*) is applied using the equations derived in Sec. 8-2. Utilizing the known structure of g_2 a more accurate calculation of the cross-coupled disturbance from the compensated channel *2* to the uncompensated channel *1* is achieved. The disturbance and hence the composite bounds are generated, based upon Eq. (8.30). These have smaller magnitude compared to those obtained by Method *1*, thus over design is reduced.

For the same reason as in channel 2, the nominal loop for channel *1* is given by plant case *2*. Again, as in channel *2* the templates show a 360° phase difference between the two plant cases of $C_L = 0.2$ and 0.6. But unlike channel *2* there is a magnitude uncertainty evident in the channel *1* templates, see Appendix E in Ref. 12. The magnitude uncertainty arises due to the strong coupling from channel *2* into channel *1*, and also from the difference in the effect of wind disturbance between the *2* classes of A/C plants based on C_L. The plants of $C_L = 0.6$ have a larger wind induced disturbance as shown in Appendix D of Ref. 12. Therefore, these plants have not only more external disturbance, but also larger cross coupling disturbance.

The loop shaping is more difficult for channel *1*. The loop tends to curl at certain frequencies as shown in Fig. 8.10. The curling causes a large change in phase with little or no change in magnitude. This type of behavior makes it difficult to shape a loop that is stable, satisfies the composite bound criteria, and

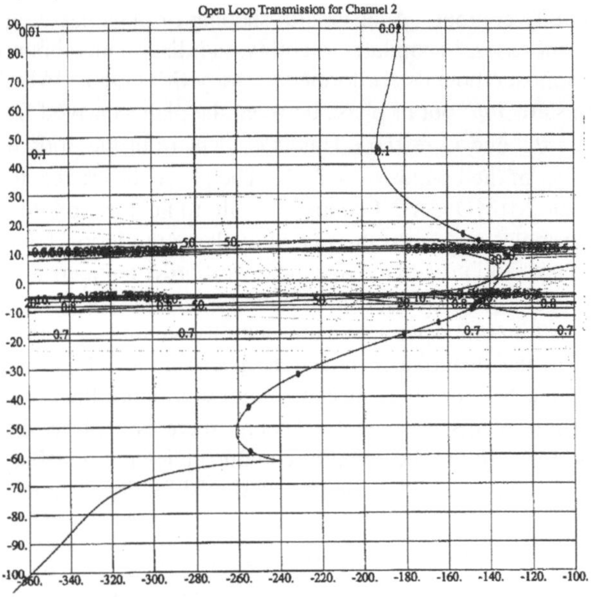

Fig. 8.8 Channel 2 loop shaping; P_o = plant case 2.

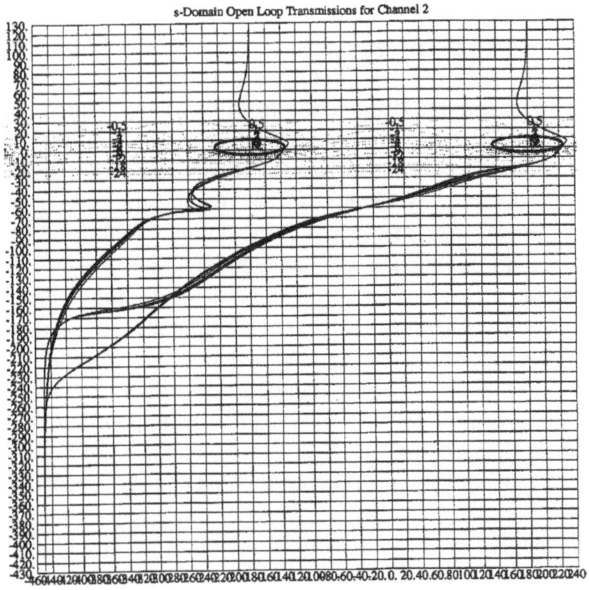

Fig. 8.9 Channel 2 Nichols plot for all plant cases.

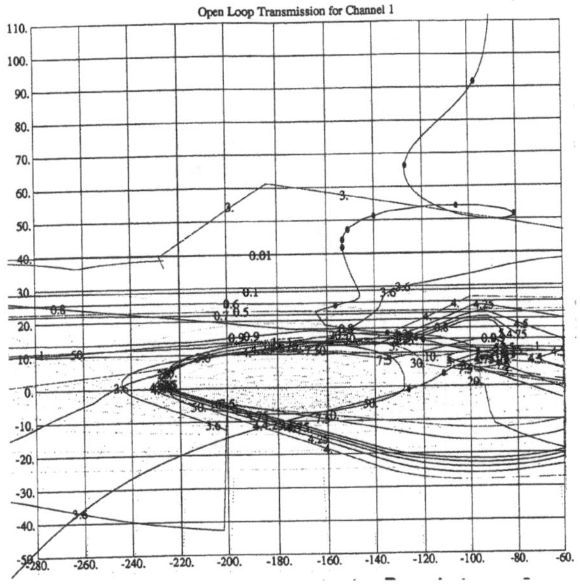

Fig. 8.10 Channel 1 loop shaping; P_o = plant case 2.

maintains a low system band-pass. The loop for channel *1* is shaped with a compromise on the band-pass. A lag-lead compensator is used to "stretch" the low frequency curl. Additional lag-lead compensators are tried to further "stretch" the curl but caused the loop to increase in magnitude as the frequency increased. A loop shape is finally achieved that satisfies the lower frequency bounds, stability, and slightly increases the system band-pass. The channel 1 compensator $\boldsymbol{g}_1$ has a higher order than the channel *2* compensator. This is an indicator of the difficulties in achieving a loop shape that satisfies design criteria.

The Nichols plot of all 16 plants in Fig. 8.11 shows the uncertainty in the low frequency range of the plants. Though there is large phase and magnitude differences between the plants the QFT method is able to achieve a design that satisfies stability and disturbance rejection for all plant cases. The following compensator is designed for channel 1:

$$g_1 = \frac{(s+0.3)(s+0.25\pm j0.433)(s+3)(s+9)(s+1.14\pm j3.747)(s+200)}{s(s+2)(s+0.32\pm j3.184)(s+90)(s+135\pm j65.38)(s+1100)} \quad \textbf{(8.50)}$$

8-7.5 CHANNEL 3 LOOP DESIGN (A SISO DESIGN)

Channel *3* exhibits none of the channel 1 or channel 2 characteristics. The channel 3 templates have relatively small phase and magnitude uncertainty. There is no

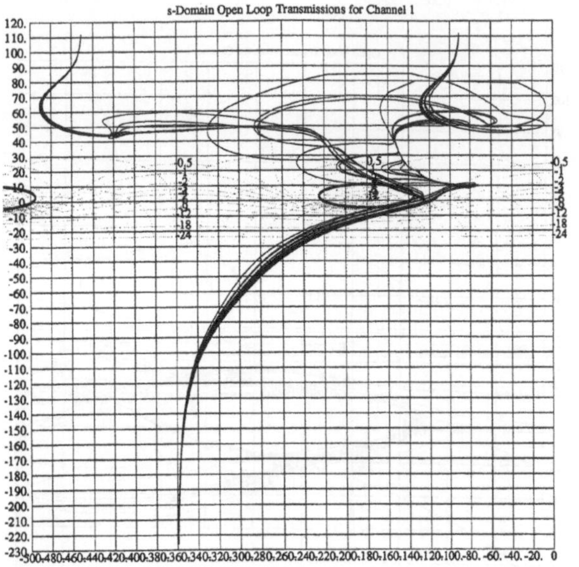

Fig. 8.11 Channel 1 Nichols plot for all plant cases.

coupling from channels 1 or 2 into channel 3. The external disturbances have similar effects on channel 3 for all plant cases.

The lack of cross-coupling disturbance and relatively certain external disturbance is evident in Fig. 8.12 where the bounds collapse around the stability contour. The channel 3 loop has a tendency to curl up as the frequency increases. The main difficulty is to add compensation to shape the loop around the bounds and stability contour at $+180^{\circ}$ and then add further compensation to keep the loop from penetrating the stability region at -180°. To achieve stability the very low frequency bounds are penetrated. This trade-off is considered acceptable since channel 3, y position has the largest margin of disturbance allowed, 7.5 ft, as detailed in Sec. 8-6.

The Nichols plot of Fig. 8.13 shows a very tight grouping of all plant cases. Again, further evidence of relatively small uncertainty in channel 3. Notice the large change in phase with no decrease in magnitude. This is deemed acceptable since it occurs below the *0* dB line at frequencies below the cutoff. The transfer function for the channel 3 compensator is:

$$g_3 = \frac{(s+0.05)(s+0.1)(s+0.2)(s+0.6)(s+1.5\pm j2.6)(s+5)(s+30)}{s(s+0.00025)(s+0.6\pm j1.91)(s+10)(s+35\pm j35.707)(s+37.5\pm j64.95)} \tag{8.51}$$

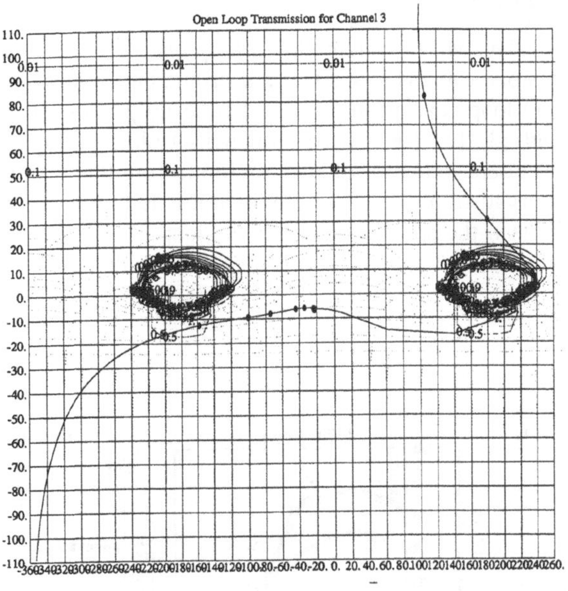

Fig. 8.12 Channel *3* loop shaping; P_o = plant case 2.

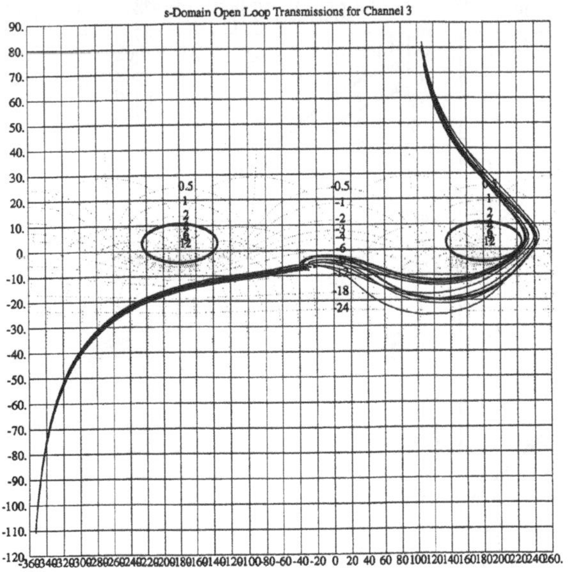

Fig. 8.13 Channel *3* Nichols plot for all plant cases.

8-7.6 CLOSED LOOP *Lm* PLOTS

The overall equivalent MISO system closed loop *Lm* plots are shown in Fig. 8.14. From these plots one can easily see that the disturbance rejection specification is met for all MISO loops except in the low frequency portion of the MISO loops {1,2}, {2,2}, and {3,3}. The closed loop MISO plots of Fig. 8.14, except as noted, are an excellent indicator of success in meeting the design specification. Due to the conservatism of the QFT technique it is left up to the nonlinear simulation to determine if the specifications for the loops {1,2}, {2,2}, and {3,3} are actual met and to attest the goodness of the design for all loops.

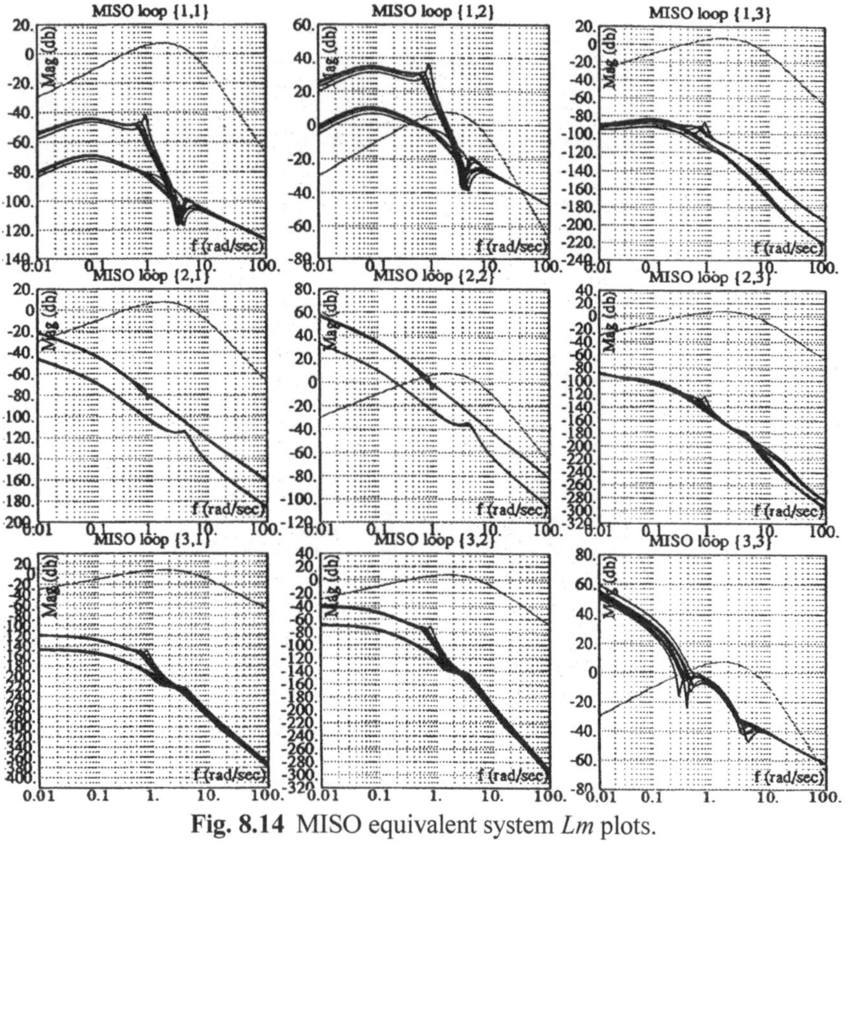

Fig. 8.14 MISO equivalent system *Lm* plots.

8-8 AIR-TO-AIR REFUELING SIMULATIONS

In this section the compensators designed in the previous sections are installed in the AFCS and simulations are run to analyze their performance. Linear simulations are run for all plant cases in $MATRIX_X$. Nonlinear simulations are for two plant cases, one for each $C_L = 0.2$ and 0.6, are performed in EASY5x.

8-8.1 LINEAR SIMULATIONS

Linear simulation are performed in $MATRIX_X$ with the modeled external disturbances forcing the system to deflect from the set point. The simulations are executed in the presence of all external disturbances simultaneously. The results of the linear separation for channel 1 (*Z* separation) demonstrate excellent results with very little perturbation from the set point. Figure 8.15 presents the channel 1 response. The plots demonstrate two distinct responses corresponding to the A/C lift coefficient C_L. The A/C with $C_L = 0.2$ show a maximum perturbation of approximately 0.003 *ft*. Also the response dampens faster for the A/C modeled with $C_L = 0.2$. The aircraft with $C_L = 0.6$ deflected to a maximum value of approximately *0.008 ft* with slower dampening.

The channel 2 (X separation) linear simulation demonstrates similar characteristics for response based on C_L. Again, excellent rejection of external disturbance is achieved as shown in Fig. 8.16. The $C_L = 0.2$ A/C has a maximum de-

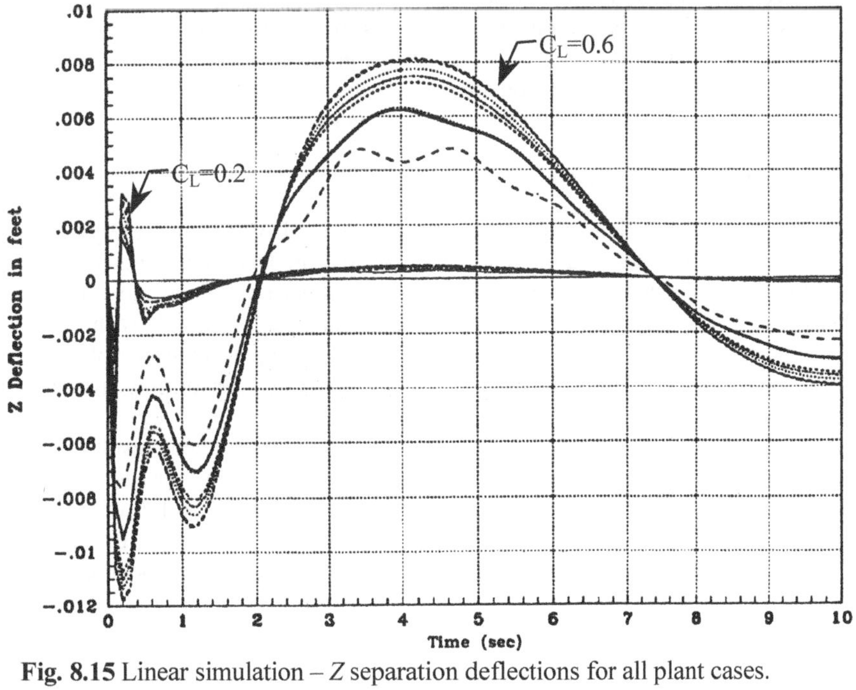

Fig. 8.15 Linear simulation – *Z* separation deflections for all plant cases.

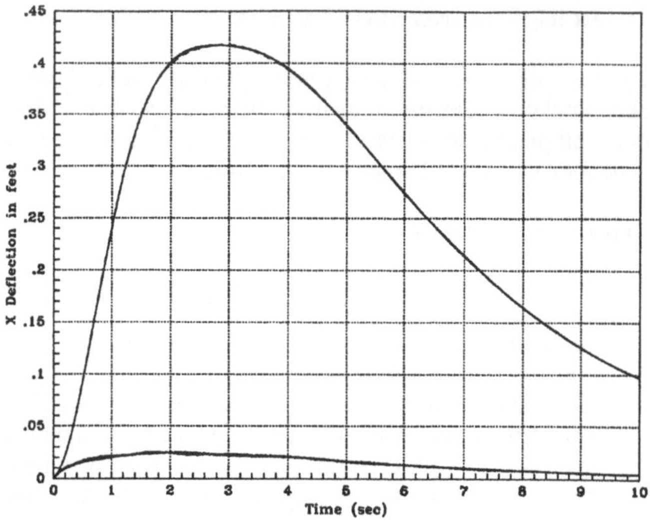

Fig. 8.16 Linear simulation – X position deflection for all plant cases.

flection of approximately 0.025 *ft*, while the C_L = 0.6 A/C deflects approximately 0.425 *ft* from the set point. Recall that the A/C with C_L = 0.6 has a larger uncompensated perturbation due to external wind disturbance.

Channel 3 (Y separation) has the largest perturbation from the set point in the linear simulation, see Fig. 8.17. The maximum perturbation in channel 3 is approximately 1.9 *ft*. Though considerably larger than channels 1 and 2, the channel 3 perturbation remains within the design specification.

8-8.2 NONLINEAR SIMULATIONS

The nonlinear simulation are performed in EASY5x. EASY5x has a Dryden wind gust model preprogrammed in the CAD package. The Dryden wind gust model is used in the nonlinear simulations versus the disturbance model developed in Secs. 8-4.2 and 8-6. Two nonlinear simulations are run: one for an A/C with $C_L = 0.2$ and one for a $C_L = 0.6$. The nonlinear simulations require considerable time to setup and perform, therefore, time limitation prevented performing a nonlinear simulation for each plant case.

The nonlinear simulations demonstrate the same excellent results that are achieved in the linear simulation. The nonlinear results are consistent with the linear results, namely very small perturbations for channels *1* and *2*, with a larger deflection in channel *3*, are recorded as shown in Figs. 8.18 and 8.20. As in the linear simulations, the nonlinear simulations are within the design specifications.

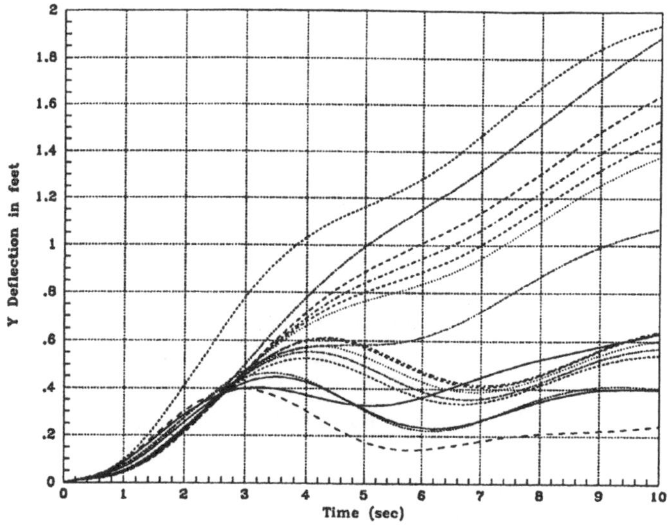

Fig. 8.17 Linear simulation – Y position deflection for all plant cases.

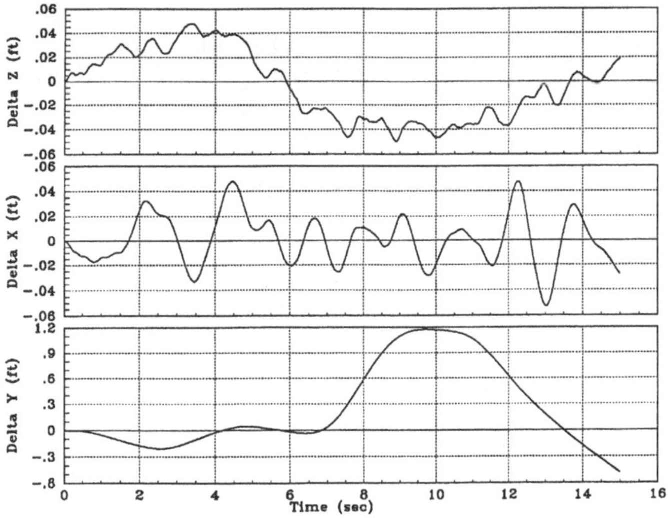

Fig. 8.18 Nonlinear simulation – X, Y, Z, position deflection, plant 1 $C_T = 0.2$.

Also presented in the nonlinear simulation plots, Figs. 8.19 and 8.21, are the control surface and thrust response of the autopilot. The aileron, rudder, and elevator responses are well within the physical capability of these devices. On the other hand the thrust requirements are probably beyond engine response capability. The engine response is most likely due to the autopilot design. The autopilot is a "text book" design and is not very sophisticated. A QFT design using the actual C-135B autopilot can probably achieve similar results without extreme engine response requirements.

8-9 TRACKING/REGULATOR MIMO SYSTEM

A MIMO control system that involves both tracking and external disturbance inputs (see Fig. 8.1) can now be designed by either Method 1 of Chapter 6 or Method *2* of Chapter 7. In Figs 5.10 and 5.11 the c_{ij} terms are replaced by the $d_{e_{ij}}$ terms; that is Fig. 8.2 is modified by including the tracking inputs $r_i \neq 0$. This modified figure now represents the tracking/regulator MIMO control system. The optimal bounds are now a combination of not only the most stringent portion of each of the tracking and cross-coupling effect bounds but also that of the bounds due to the external disturbance $(d_{ext})_{ij}$ of Eq. (8.19).

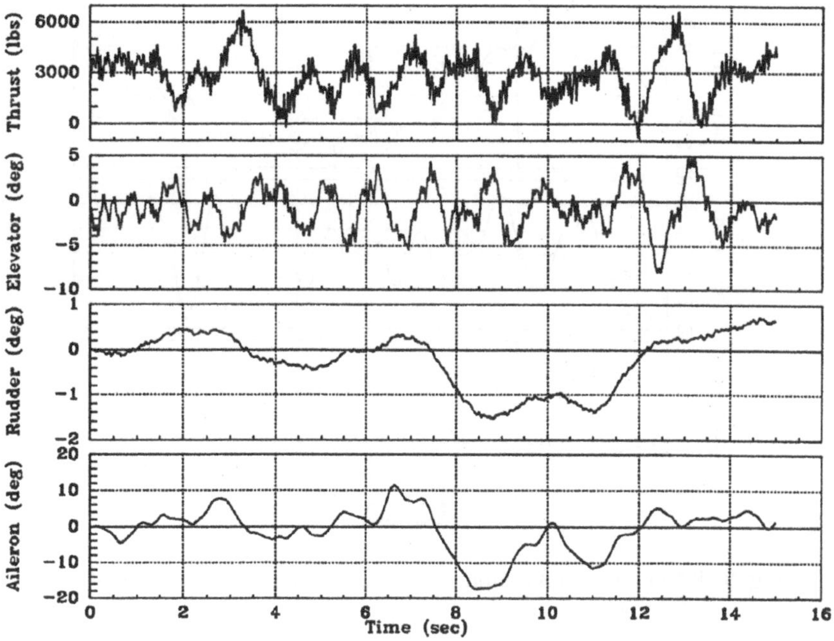

Fig. 8.19 Nonlinear simulation – control surface and throttle response, plant 1.

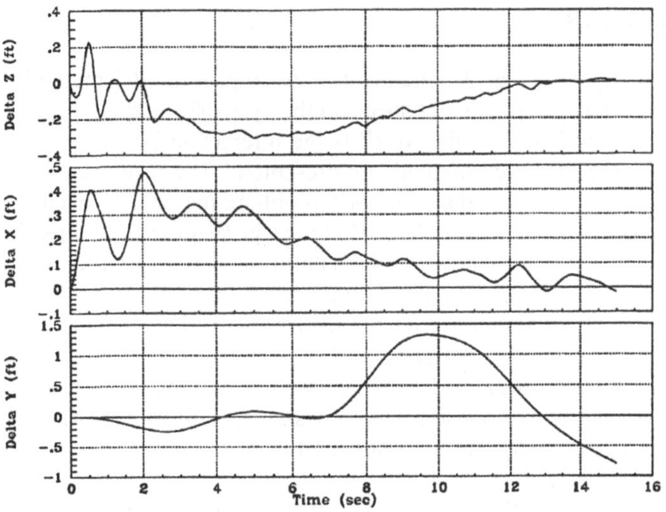

Fig. 8.20 Nonlinear simulation – X, Y, Z position deflection plant 2 $C_L = 0.6$.

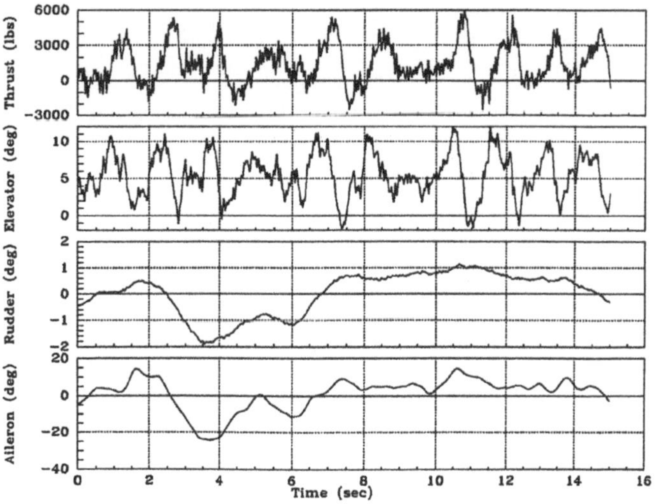

Fig. 8.21 Nonlinear simulation – control surface and throttle position.

8-10 SUMMARY

This chapter presents the development of the improved QFT method to include the effects of external input disturbances on the system's outputs. Equations to calculate the new optimal bounds, which is the interaction bound for a pure regulator system for the improved method, are presented. The results of this development are applied to the design of a real-world problem: the AFCS. Each loop shaping is detailed, covering the particular difficulties in shaping the loops for each channel. Also, the inherent nature of QFT's ability in handling large plant uncertainties is discussed. Finally the *Lm* plots of the closed loop MISO system is shown, indicating a successful design based upon the trade-offs that are required.

The compensators designed in this chapter are integrated into the air-to-air refueling AFCS. Linear and nonlinear simulations are performed with excellent results. The system response is within design specification. The QFT design process worked extremely well in designing the AFCS in the presence of an external output disturbance.

The tracking MIMO QFT CAD package (see Appendix C) was modified to also handle the pure regulation problem.[12] This modified CAD package has been used to design an FCS that involved satisfying both tracking and regulation specifications.[10,73]

9

NOW THE "PRACTICING ENGINEER TAKES OVER"[2,34,37]

9-1 INTRODUCTION

Control system design is an interdisciplinary and multi-stage process and is not a sub-area of mathematics. It is rather an engineering endeavor. While applied mathematics plays a crucial role in facilitating the solution of robust control system design problems, it behooves the engineer to reduce the real-world control task to a tractable mathematical problem. The required simplifying assumptions and the attendant mathematical modeling effort require sound engineering knowledge and judgement. Furthermore, the mathematical solution of the control problem needs to be implemented and the validity and applicability of the modeling assumptions need to be verified by the engineer. Hence, in this chapter the process of applying the QFT robust control system design method is presented from this broad perspective and it is shown that it is uniquely suited to address and solve engineering control system design problems.

The scientific method uses mathematical methods: (1) to gain insights into, (2) to generalize, and (3) to expand the state-of-the-art, in many areas of science and technology. Often this requires the proof of theorems, corollaries, and lemmas. Generally, at this stage, the researcher (1) is not concerned with whether his or her efforts will result in the solution of "real world problems," and (2) applies linear analysis and synthesis techniques most of the time, although most of the real world problems are nonlinear. The scientific method is necessary in order for the researcher to be able "to see the trees from the forest," and thus to be able to achieve positive results.

Once the scientific approach has successfully advanced the state-of-the-art, and where applicable, the engineer must take over and apply the new results to real world problems. The engineer is at the "*interface*" of the real world, and the body of knowledge and theoretical results available in the technical literature.

Thus, the control system design task is a multi-stage process which entails many steps, say from *A* to *Z*. Mathematics is most helpful in taking the engineer through some of these steps, for example, from *P* to *S*. It however behooves the engineer to make the required modeling assumptions, hypotheses, and simplifications that are needed for him or her to proceed from *A* to *O*, so that the mathematical problem is tractable and the existing theory can be applied. Finally, steps *T* to *Z* entail extensive simulations and/or experimental tests where the validity of the model is verified, the implementation issues are addressed, and the design is validated.

Horowitz applied the scientific method in the development of his Quantitative Feedback Theory (QFT) approach to the engineering design of robust control systems.[38] F. Bailey, J. W. Holton, and O. Merino,[39] O. D. I. Nwokah,[66-72] et al (see Ref. 2) have used the scientific method to further enhance the mathematical rigor of the QFT technique, in order to help the engineer to bridge the "*interface*" gap. The goal of these researchers is to establish theorems, corollaries, and lemmas that can tell the designer at the onset whether, given the required performance specifications, and in the face of parametric uncertainty, is a QFT solution feasible. This body of knowledge must be coupled with "the body of engineering knowledge" pertaining to the application, when dealing with nonlinear systems and real world problems. This requires that the engineer have a good understanding of the physical characteristics of the plant to be controlled. This "*bridging the gap*" (the "interface" gap) can be highlighted by the following anonymous quote:

"In **THEORY** (scientist)
There is no difference between theory and practice.

In **PRACTICE** (engineer)
There is a difference between practice and theory."

Thus, an engineer who has a firm understanding of the results of the "scientific method" and has a firm understanding of the nature and characteristics of the plant to be controlled must develop an "engineering method" by developing appropriate "Engineering Rules" (E.R.) that will assist him or her in "*bridging the gap*." Therefore, a successful practical control system design process can be described as an approach that utilizes existing and/or developing new E.R.s by an experienced control system designer in order to bridge the gap between the scientific and the engineering methods. This approach will expedite the engineer's robust control system design and enhance the quality of this design.

In conclusion, the control system design task is a multi-stage process which entails many steps, say from *A to Z*. Mathematics is most helpful in taking the engineer through some of these steps, say from *P to S*. It however behooves the engineer to make the required modeling assumptions, hypotheses, and simplifications that are needed for him or her to proceed from *A to O*, so that the mathematical problem is tractable and the existing theory can be applied. Finally, steps *T to Z* entail extensive simulations where the validity of the model is verified, the implementation issues are addressed, and the design is validated. An example of this multi-stage process is presented in Ref. 33.

9-2 TRANSPARENCY OF QFT

The elements resulting from the application of the scientific method which provide the "transparency" of the QFT design technique and that enhance its ability to solve real world problems are:

(a) **Template** The size of the template (width and height), see Fig. 9.1, tells the engineer at the onset of the design process whether a fixed compensator ***G*** can be synthesized that will yield the desired system performance in the face of the prevailing structured uncertainty. If only the template's height is the problem then the engineer needs to employ straight gain scheduling. When control effector failures need to be accommodated, the width the of template can become excessive and a single successful design might not be possible. In this situation the designer needs to reduce the size of the templates (see E.R. 3 of Sec. 9-3). This can be done by either, reevaluating the design requirements and eliminating individual effector failure cases, or by grouping failure cases so that individual designs can be accomplished and then scheduled.

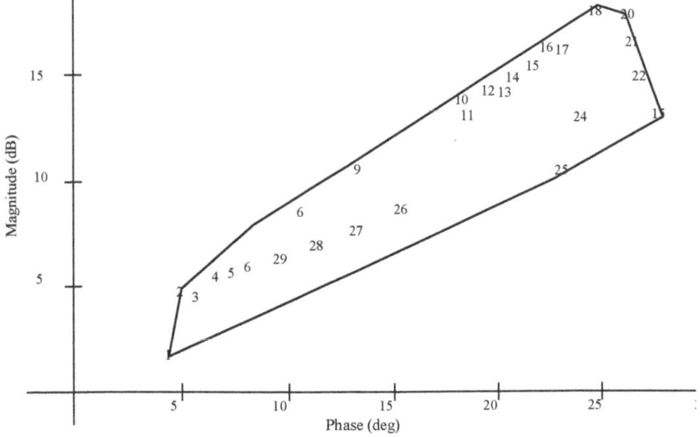

Fig. 9.1 A template representing J LTI plants at $\omega = 30$ rad/sec.

One of the key problems associated with QFT design is choosing LTI plants cases that represent the nonlinear plant in the operational region of interest. Since plants on the interior of the template are guaranteed to meet specifications, the minimum set of LTI plants needed for a QFT design are those that form the boundaries of the templates over the frequency bandwidth of interest. By choosing the minimum number of LTI plant cases required to fully describe the templates of interest, the designer reduces the computation requirements and simplifies validation and verification of the design.

One method used to choose LTI plants is through the use of template expansion. For example (see Chapter 11), in aircraft design the designer knows the proposed operating region of the aircraft, flight envelope (*operating scenario*), and has a nonlinear model of the aircraft for that flight envelope. Nonlinear aircraft models are linearized at operating states, flight conditions. These flight conditions include states such as velocity, altitude, angle of attack, and side-slip angle as well as other parameters such as center of gravity and weight. The problem is to find a minimum set of LTI plant cases that describe the template boundaries utilizing all permutations of the independent parameters that describe the flight envelope. This process is depicted in Fig. 9.2 for two independent parameters.

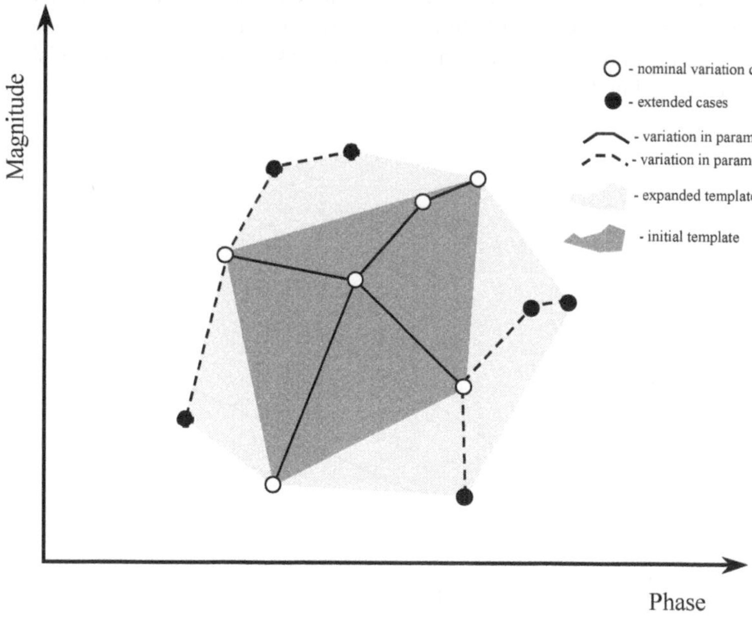

Fig. 9.2 Template expansion procedure.

The first step in the template expansion process is to use engineering knowledge of the aircraft to find the limits of variation of the individual parameters in the flight envelope. The next step is to choose a nominal flight condition as a starting point on the template. Next plot lines on the templates describing the effect of varying each parameter through its limits, in turn, while keeping the remaining parameters fixed at the nominal flight condition. At this point the limits of variation of the parameters describe the boundary of the template. The next step in the process is to expand the boundaries of the template by choosing points on the existing template boundary to be nominal points and then plotting new lines describing parametric variations at each of these nominal points. Variations about these nominal points describe an expanded template boundary. Once the original template boundary has been expanded, the points making up the expanded template boundary are used to expand the boundary again. In this fashion, a template boundary can be developed, graphically, so that all variations are taken into account.

(b) **Phase margin frequency** In order to ensure that the value of the specified phase margin frequency ω_ϕ is not exceeded by any of the plants in the set then, when all J plants in $\mathcal{P}$ are stable, select the nominal plant p_ι to be the plant lying at the "top" vertex of the template (see Fig. 9.1). Furthermore, it is required that this plant always lies, for all template frequencies, at the top of the templates. In the event that one or more plant in the set $\mathcal{P}$ is unstable, the design engineer must then select, as the nominal plant, a plant p_ι that has the highest degree of instability. That is, the plant whose unstable pole lies furthest to the right in the s-plane) as the "worst case plant." This rule for the selection of the nominal plant will facilitate the achievement of the specified value of ω_ϕ. It is advisable that one of the templates, for each loop, be obtained at the respective specified BW(L_i) frequency.

(c) **Signal flow graph (SFG)** For $\boldsymbol{y} = \boldsymbol{P_e u} = \boldsymbol{PWu}$: Fig. 9.3 represents a MIMO QFT control system structure. Having the SFG for the portion of Fig. 9.3 that represents $\boldsymbol{P_e}$ can be helpful in the initial selection of the values of w_{ij} and modifying some of these values during the simulation phase of the design process. The selection of values, etc, are further enhanced by the engineer's firm understanding of the interrelationship of the plant outputs with the inputs to the $\boldsymbol{W}$ matrix.

(d) **Minimum order compensator (controller) (*MOC*) *G*** Why are high-order compensators unacceptable? The implementation of a m-order compensator in a digital flight control system (DFCS) yields an m-order discrete-time dynamical system. The latter is equivalent to a first order dynamical system with m-time delays. Hence, during the first $m-1$ time instants the input has a somewhat limited effect on the output, for the output is partially determined by the m initial conditions. Thus, the control action is delayed.

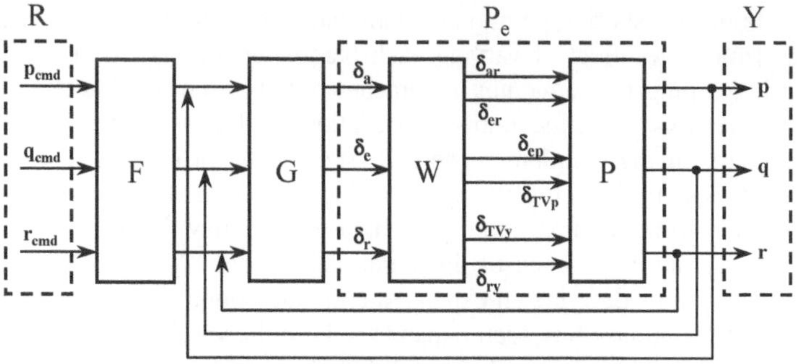

Fig. 9.3 MIMO QFT control structure block diagram.

Example – Consider an *60* Hz FCS sampling rate and a 20th-order H_∞ compensator. The FCS's time delay is $20/60 = 1/3\ s$ which is unacceptable. In general, the time delay caused by an m-order controller is $m/60\ s$.

Another factor that must be considered in maintaining low-order controllers is that on board flight control computers have limited capacity due to other non-control related computing requirements. As a rule, about 30% of the computer capacity is allocated to the DFCS and 70% for non FCS requirements.

To achieve a "low-order" compensator (controller) some designers, or design methods, have recourse to plant $\boldsymbol{P}$, and in-turn $\boldsymbol{P}_e$, "doctoring" or "padding" by inserting additional poles and/or zeros into $\boldsymbol{P}$. This results in an "augmented" plant matrix $\boldsymbol{P}_{eS}$ which these designers base their design on in order to achieve a compensator $\boldsymbol{G}$ whose order equals the order of the augmented plant. Hence, these methods yield high-order compensators.

In QFT, in order to achieve the MOC $\boldsymbol{G}$, during loop shaping, the poles and zeros of the nominal plant q_{ii_o} are put to good use in synthesizing a satisfactory loop shaping transfer function L_{ii_o}. Doing so yields the MOC $g_i = L_{ii_o}/q_{ii_o}$.

Some or all of these elements, or comparable ones, are not available in other optimization based, multivariable control system design techniques. This minimizes their ability to achieve, in a relatively "short design time," a design that meets all the performance specifications that are specified at the *onset of the initial design effort.*

9-3 BODY OF ENGINEERING QFT KNOWLEDGE

Through the many years of applying the QFT robust control design technique to many real world nonlinear problems, the following Engineering Rules have evolved:

9-3.1 E.R.1 Weighting Matrix

An $\ell x m$ weighting matrix $\boldsymbol{W} = \{w_{ij}\}$ is required to achieve a square mxm equivalent plant matrix $\boldsymbol{P}_e$ whenever the plant matrix $\boldsymbol{P}$ is $mx\ell$, i.e.,

$$\boldsymbol{P}_e = \boldsymbol{PW} \tag{9.1}$$

and where

$$\boldsymbol{P}_e = \{p_{ij}\}, \quad \boldsymbol{P}_e^{-1} = \{p_{ij}^*\}, \quad \text{and} \quad \boldsymbol{Q} = \{q_{ij}\} = \{1/p_{ij}^*\} \tag{9.2}$$

Thus, it is desired to know at the onset if it is possible to achieve m.p. q_{ii}'s by the proper selection of the w_{ij} elements. Now, m.p. q_{ii} plants are most desirable for they allow the full exploitation of the "benefits of feedback," i.e., high gain. It turns out that one can apply the Binet-Cauchy theorem[64] (see Chapter 7) to determine if m.p. q_{ii}'s are possible. Also, it may be desirable to obtain complete decoupling for the nominal plant case, i.e.,

$$\boldsymbol{P}_{e_{diag}} = \begin{bmatrix} p_{11} & 0 & \dots & 0 \\ 0 & p_{22} & \dots & 0 \\ \vdots & \vdots & \dots & \vdots \\ 0 & 0 & \dots & p_{mm} \end{bmatrix} = \boldsymbol{PW} \tag{9.3}$$

Although for the non-nominal plants complete decoupling, in general, will not occur, the degree of decoupling will have been enhanced. This greatly facilitates the QFT design process, for less attention needs to be given to cross-coupling effects (c_{ij}) rejection. Method 1 is then more readily applicable, with the additional benefit of reduced closed-loop BW.

9-3.2 E.R.2 n.m.p. q_{ii}'s

For q_{ii}'s that are n.m.p. one must determine if the location of the RHP zero(s) is in a region which will not present a problem for the real-world design problem being considered. For manual flight control systems, if a RHP zero happens to be "close" to the origin (see Fig. 9.4), this is not necessarily deleterious since the pilot inputs a new command before its effect is noticeable; in other words, it is assumed that the unstable pole is outside the closed loop system's lower bandwidth. If this RHP zero is “far out" to the right, it is outside the bandwidth of concern in manual control and it does not present a problem. For these cases a satisfactory QFT design may be achievable.

9-3.3 E.R.3 Templates

The adage “a picture is worth a thousand words” applies to the preliminary task of determining if a robust control solution exists, bearing in mind the need to satisfy tracking specifications, external disturbance and cross-coupling effects rejection, and satisfying the stability bounds. If theorems, corollaries, and/or lemmas pertaining to these bounds, obtained by the scientific method, reveal that no loop shaping solution exists then one must be attuned to stepping back and doing a “trade-off.” In other words, either some specifications need to be relaxed in order to achieve a solution, or one must be willing to live with a degree of gain scheduling. Thus, a graphical analysis of the template shown in Fig. 9.1 can reveal the following:

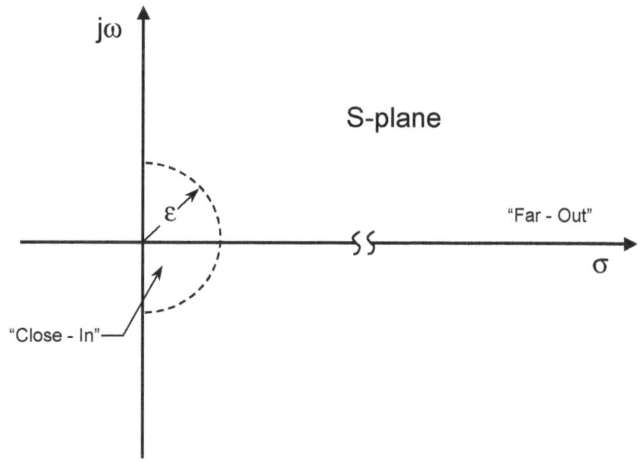

Fig. 9.4 Right-half-plane analysis.

(a) The maximum template height, in dB, is too large thus not enabling the determination of the tracking bounds for a given ω_i. One can then decide if gain scheduling is required and is feasible in order to yield a design that yields the tracking bounds.

(b) The situation where the templates are too "wide" (the magnitude of phase angle width) thus prohibiting a QFT solution or a solution by any other multivariable design technique. This is especially true for real-world control problems that involve control effector failures accommodation.[10,73] In these design problems, generally, the worst failure case is the culprit in generating this large "angle width." Thus, in order to achieve a solution one needs to relax the requirement that the "worst failure case" be accommodated. Naturally, when this situation arises, it is necessary to stipulate for what failure case or cases a successful design is achievable. In determining "reasonable failure cases" that can be accommodated by robust (not adaptive) control – one must consider if 10%, 25%, 50%, 80% failure still permits enough control authority! This degree of failure can only be determined by a person who is knowledgeable of the physical plant to be controlled. In general, knowledge of the plant (application) "is king" when it comes to the design of a feedback compensator or controller for the said plant.

(c) The effects of the structured uncertainty on the template's geometry is now discussed. Thus, in flight control[34] (see Chapter 11), linearized plants that represent different flight conditions in the flight envelope are extracted from a nonlinear truth model. An attempt is made to choose flight conditions in such a way as to fully cover the flight envelope with the templates. To do this, a nominal flight condition for an unmanned research vehicle was chosen to be *50 kts* forward velocity, *1000 ft* altitude, a weight of *205 pounds* and center of gravity at *29.9%* of the mean aerodynamic chord. From this nominal flight condition, each parameter was varied, in steps, through maximum and minimum values. These variations produced an initial set of templates. On these templates, variations caused by each parameter are identified. Each variation, when shown on the template, identified an expanding area of the flight envelope, that required more plants for a better definition, see Fig. 9.2.

If all the p_{ij}'s of $\boldsymbol{P}$ do not have the same value of λ (excess of poles over zeros) then as $\omega \rightarrow \infty$ the templates may not become straight lines. A possible method of reducing the size of the templates is given by E.R.8.

9-3.4 E.R.4 Design Techniques

No matter what design method one uses, performance specifications must be realistic and commensurate with the real world plant being controlled. Situations have occurred where the conclusion was reached that no acceptable design was possible. For these situations when one "stepped back" and asked the pertinent question "was something demanded that this plant physically cannot deliver regardless of the control design technique?", it was determined that some or all of the prescribed performance specifications were unrealistic.

9-3.5 E.R.5 QFT Method 2

Arbitrarily picking the wrong order of the loops to be designed (loop closures) by Method 2, can result in the nonexistence of a solution due to the chosen order of closures (see Sec. 5-6.4). This may occur if the solution process is based on satisfying an upper limit of the phase margin frequency ω_ϕ for each loop upper bound response $\mathbf{T}_{R_U}$. The proper order of the loops to be designed by Method 2, entails picking the loops in the order of increasing values of the desired ω_ϕ; i.e., first close loop 1, then loop 2, etc. where $\omega_{\phi 11} < \omega_{\phi 22} < \omega_{\phi 33} < \ldots$. Indeed, by Method 2 it is known that ω_ϕ of the succeeding designed loop is larger than the previously designed loops.

As an example, consider the necessity of satisfying the desired BW frequency $\omega_{h_{ii}}$ for each loop. Associated with this BW frequency is a corresponding phase margin frequency $\omega_{\varphi_{ii}}$ that can be determined from Fig. 5.14. In reality, for a MIMO system, the determination of the $\omega_{\varphi_{ii}}$ must be determined from Fig. 5.15. Thus, when applying Method 2 for a MIMO system if a constraint is satisfying the desired P.S. on the BW frequency then it is necessary to determine the corresponding $\omega_{\varphi_{ii}}$ for each desired loop BW frequency. The phase margin frequencies are arranged in the manner indicated in the previous paragraph which determines the value of the phase margin frequency that must be achieved during the loop shaping process of each corresponding loop.

9-3.6 E.R.6 Minimum Order Compensator (Controller) (MOC)

In order to ensure the smallest possible order compensator/controller, one starts the loop shaping process by using the loop's nominal plant $L_{o1} = q_{11o}$, and then zeros and poles are successively added in order to obtain the required loop shape, resulting in:

$$L_o(s) = \frac{L_{o1}(s)(s - z_1)\ldots(s - z_w)}{(s - p_1)\ldots(s - p_v)} \tag{9.4}$$

Finally, the compensator is obtained from $g_1 = L_o/q_{11o}$. Thus, the nominal plant's poles and zeros are being used to shape the loop. This insures that the ensuing compensator/controller is of the lowest order, which is highly desirable.

9-3.7 E.R.7 Minimum Compensator Gain

To minimize the effects of noise, saturation, etc., it is desirable to minimize the amount of gain required in each loop *i*, while at the same time meet the performance specifications in the face of the given structured uncertainty. To achieve this goal, a control system designer, with a good understanding of the Nichols Chart and a good interactive QFT CAD package, can use his "engineering talent" to make use of the "dips" in the composite $\boldsymbol{B}_{oi}(j\omega_j)$, see Fig.9.5. The designer by shaping $\boldsymbol{L}_{oi}$ to pass through these dips, *where feasible*, can ensure achieving the minimum compensator gain that is realistically possible. To achieve this by an automatic loop shaping routine may be difficult.

9-3.8 E.R.8 Basic *mxm* Plant *P* Preconditioning

When appropriate, utilize unity feedback loops for the *mxm* MIMO plant $\boldsymbol{P}$ that will yield an *mxm* preconditioned plant matrix $\boldsymbol{P}_P$. In some flight control system design problems an inner loop controller such as that of an angle of attack command has been used to achieve a $\boldsymbol{P}_P$. The templates $\Im\boldsymbol{P}_P(j\omega_i)$, in general can be smaller in size than the templates $\Im\boldsymbol{P}(j\omega_i)$. This template reduction size is predicted by performing a sensitivity analysis (see Sec. 14.2 of Ref. 1). The QFT

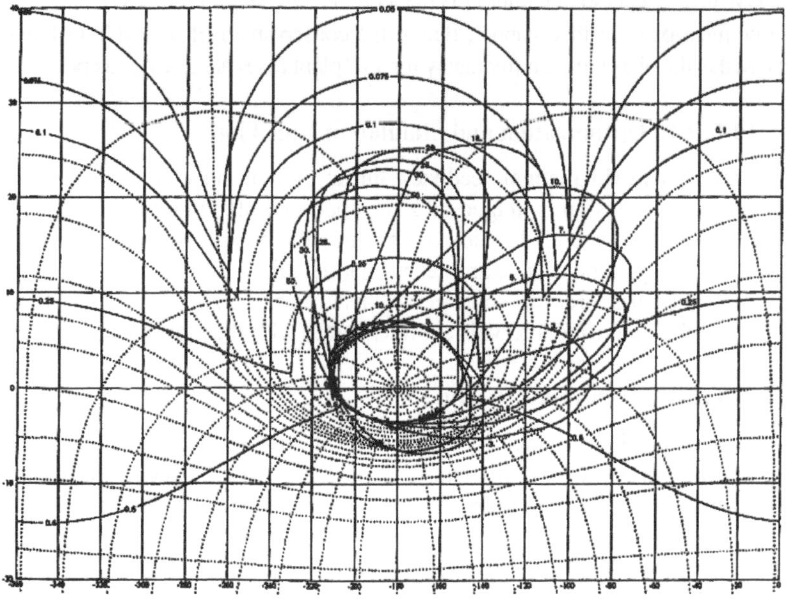

Fig. 9.5 QFT stability and composite bounds.

design is performed utilizing the preconditioned matrix $\boldsymbol{P}_P$. This concept has been used in a number of MIMO QFT designs.[11,12,65] (See Sec. 8-4.1.)

9-3.9 E.R.9 Nominal Plant Determination

It is easy to determine the phase margin angle γ, the gain margin, and the phase margin frequency ω_ϕ of a feedback control system using the NC. Thus, QFT affords the robust establishment of these FOM. Indeed, by choosing the nominal plant:

(a) To correspond to the maximum dB plant, for all templates, on the ω_{ϕ_i} $BW(L_i)$ template *ensures that* $\omega_{\phi_i} \le BW(L_i)$ *for all plants.*

(b) To be the uniformly maximum dB plant, for all templates, ensures the achievement of *a robustly guaranteed gain margin* is easily accomplished.

(c) Whichever is the "left-most" plant on all templates ensures that the desired γ is robustly achieved.

With respect to item (a), if the maximum dB plant is not the same on all templates within the BW then select the maximum dB plant on the template for the desired ω_{ϕ_t}. Selecting the nominal plant in this manner, due to the conservatism of the QFT technique, will ensure that the phase margin frequency for all J plant cases are as close as possible to ω_{ϕ_t}.

9-3.10 E.R.10 Optimization and Simulation Run Time

In many real-world manual feedback linear or nonlinear control problems, the goodness of the design is judged on a pre-specified planning time horizon beyond which the performance is less important since the human operator will inject new inputs to the system. For example, in manual flight control the time horizon is determined by the aircraft's short period dynamics, e.g., *5* seconds and there is no interest in the long time intervals commensurate with the slow phugoid dynamics.

9-3.11 E.R.11 Asymptotic Results

Asymptotic results provided by mathematical analysis are not as useful as they seem to be. Consider the manual control disturbance rejection case where fast disturbance attenuation is more desirable than total disturbance rejection which entails a very long "settling time." Transient performance is sometimes of more interest than the ultimate steady-state output.

9-3.12 E.R.12 Controller Implementation

Tight performance specifications and a high degree of uncertainty require small sampling intervals T. Unfortunately, the smaller the value of T, the greater the degree of accuracy that is required to be maintained. The numerical accuracy is enhanced by a factored representation of the controller and pre-filter.[13] For example, by use of the bilinear transformation the controller $G_z(z)$ of Fig. 9.6b is obtained from the compensator $G_c(s)$ or $G_c(w)$ of Fig. 9.6a. The equivalent cascaded transfer function representation (factored representation) of $G_z(z)$, shown in Fig. 9.6c, is utilized to obtained the algorithm for the software implementation of $G_z(z)$.

9-3.13 E.R.13 Non-Ideal Step Function for Simulation

For mathematical analysis of LTI systems ideal step functions are utilized. In the real world, ideal step functions do not exist. Thus, for a more realistic test in determining a control system's performance through simulation a *ramped-up step function* is used as shown in Fig. 9.7. Use of this type of a realistic forcing function has the tendency to minimize the saturating aspects of the system.

9-4 NONLINEARITIES -- THE ENGINEERING APPROACH

All the current robust control design methods, including QFT, yield linear compensators for linear, but uncertain, plants. Hence, the achieved robust performance applies to "small signals" only. The intrinsic scalability property, which is afforded by linearity, breaks down in the face of non-linearity. The worst offenders are saturation type non-linearities. The latter are encountered in actuators, which, unfortunately, are invariably located at the plant inputs. Both displacement and rate saturation significantly reduce the achievable benefits of (high gain) feedback. Thus, consider the extreme case of zero inputs to linear plants: the output will always be zero, irrespective of the (linear) plant, and so

$$I(s) \rightarrow \boxed{G_c(s) = \frac{N(s)}{D(s)}} \rightarrow O(s) \qquad I(w) \rightarrow \boxed{G_c(w) = \frac{N(w)}{D(w)}} \rightarrow O(w)$$

(a) The design compensators

$$I(z) \rightarrow \boxed{G_c(z) = \frac{N(z)}{D(z)}} \rightarrow O(z)$$

(b) z- domain compensator

$$I(z) \rightarrow \boxed{G_1(z)} \xrightarrow{O_1(z)} \varsigma \cdots \varsigma \rightarrow \boxed{G_j(z)} \xrightarrow{O_j(z) = O(z)}$$

(d) z- domain J cascaded controllers

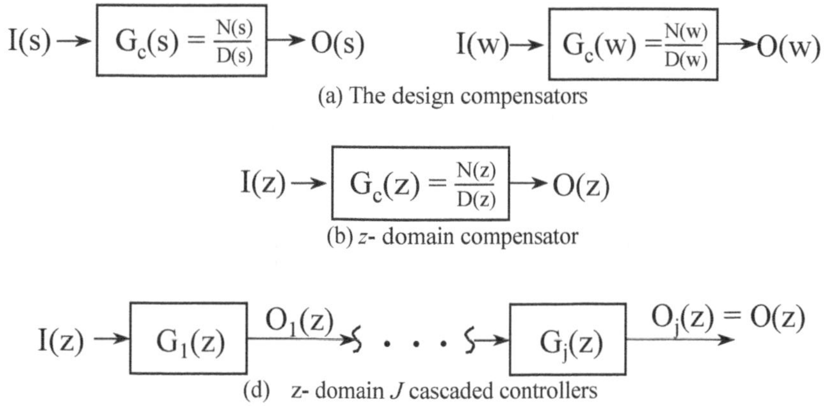

(e)

Fig. 9.6 s- or w-domain to z-domain bilinear transformation: Formulation for implementation of the $G(z)$ controller.

infinite robustness is achieved. The lack of robustness becomes evident when the plant is being driven hard (with a large input signal – it is "slewed") and nonzero inputs are applied. This is also generally true in robust feedback control systems once significant slewing is attempted, and is due to non-linearity and saturation. Furthermore, non-linearity and saturation need to be addressed when feedback control is used to stabilize open-loop unstable plants. Indeed, from a "small signal" point of view saturation is equivalent to opening the feedback loop. This will have catastrophic consequences, for it will cause instability and departure.

9-5 PLANT INVERSION

Given an m-output vector $\boldsymbol{y}$, an m-input vector $\boldsymbol{u}$ for Fig. 9.8, the LTI plant transform equations can be written in the form:

$$\boldsymbol{D}(s)\boldsymbol{y}(s) = \boldsymbol{N}(s)\boldsymbol{u}(s) \tag{9.5}$$

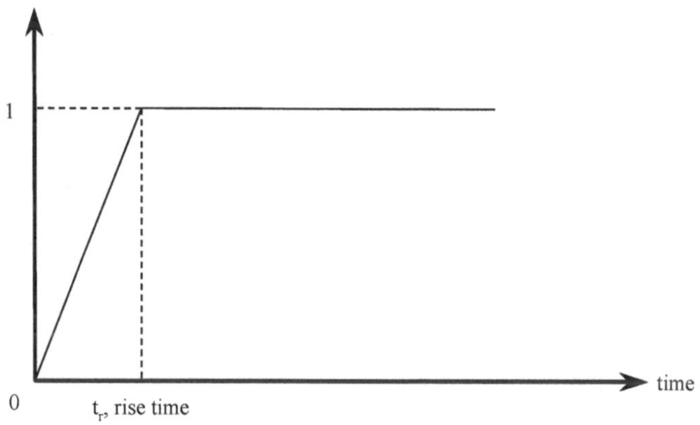

Fig. 9.7 A ramped-up step function.

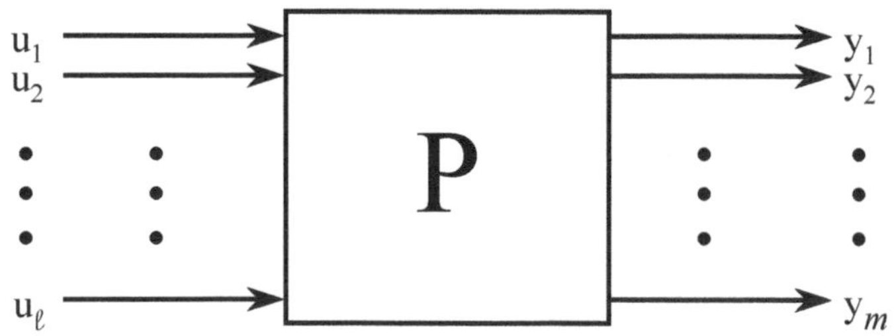

Fig. 9.8 A MIMO plant.

with $\boldsymbol{D}(s) = [d_{ij}(s)]$ and $N(s) = [n_{ij}(s)]$ being *mxm* polynomial matrices in *s*. The resulting plant matrix $\boldsymbol{P}$ ($y = \boldsymbol{P}\boldsymbol{u}$) is

$$\boldsymbol{P} = \boldsymbol{D}^{-1}\boldsymbol{N} = \frac{(Adj\ \boldsymbol{D})\boldsymbol{N}}{det\,\boldsymbol{D}} = [p_{ij}(s)] \quad \textbf{(9.6)}$$

where the elements of $\boldsymbol{P}$ are transfer functions. Suppose that the inverse plant

$$\boldsymbol{P}^{-1} = [p_{ij}^{*}] = [1/q_{ij}] \quad \textbf{(9.7)}$$

is needed, for example, as in some of the QFT techniques. The designer can obtain it either from Eq. (9.5), i.e., since $\boldsymbol{u} = \boldsymbol{P}^{-1}\boldsymbol{y}$,

$$\boldsymbol{P}^{-1} = \boldsymbol{N}^{-1}\boldsymbol{D} = \frac{(Adj\ \boldsymbol{N})\boldsymbol{D}}{det\ \boldsymbol{N}} \quad \textbf{(9.8)}$$

or from the state equations i.e.:

$$\dot{\boldsymbol{x}} = \boldsymbol{A}\boldsymbol{x} + \boldsymbol{B}\boldsymbol{u} \ \ \text{(a)} \qquad \boldsymbol{y} = \boldsymbol{C}\boldsymbol{x} \ \ \text{(b)} \quad \textbf{(9.9)}$$

that describe the nth-order plant, where the matrices $\boldsymbol{A}$, $\boldsymbol{B}$, $\boldsymbol{C}$ are *nxn*, *nxm*, and *mxn*, respectively, and which, in turn, yields the following expressions:

$$\boldsymbol{y} = \boldsymbol{C}[s\boldsymbol{I} - \boldsymbol{A}]^{-1}\boldsymbol{B}\boldsymbol{u} \quad \textbf{(9.10)}$$

In other words,

$$\mathbf{P} = \mathbf{C}[\mathrm{s}\mathbf{I} - \mathbf{A}]^{-1}\mathbf{B} = \left\{\frac{\mathrm{n_{ij}}}{\mathrm{d}}\right\} \quad \textbf{(9.11)}$$

where n_{ij} and d are polynomials of degree ℓ and n, respectively, in s, $\ell \le n$, and d is the characteristic polynomial of $\boldsymbol{A}$.

For the numerical calculation of $\boldsymbol{P}^{-1}$ it is much better to use Eq. (9.8) rather than Eq. (9.11). The reason why is best illustrated by considering a *2x2* ($m = 2$) plant utilizing the second approach. Thus, from Eq. (9.11):

$$\boldsymbol{P}^{-1} = \frac{Adj\left\{\dfrac{n_{ij}}{d}\right\}}{det\,\{n_{ij}\}/\,d^2} = \frac{Adj\,\{n_{ij}\}/d}{det\,\{n_{ij}\}/\,d^2} = \frac{d^2\,Adj\,\{n_{ij}\}}{d\ det\,\{n_{ij}\}} \quad \textbf{(9.12)}$$

In general, for the *mxm* control system:

$$P^{-1} = \frac{d^m \, Adj\,\{n_{ij}\}}{d^{m-1} \, det\,\{n_{ij}\}} \tag{9.13}$$

Thus, if P^{-1} is obtained from Eq. (9.13), rather than directly from Eq. (9.8), then $m-1$ cancellations of polynomials from the numerator in Eq. (9.13) with the $m-1$ polynomials in its denominator is required. The numerical poles/zeros cancellations will of course not be exact, because of the inevitable computer round-off. Note that $m-1$ such cancellations of each zero of the numerator with the $m-1$ poles of the denominator must occur. Thus, in order to recover numerical accuracy one must factor all the numerator and denominator polynomials of P^{-1} and then check out the inevitable inexact cancellations. The following set of roots, each having at least 8 significant digits of accuracy, are entered into the CAD package to create the polynomial:

$$\text{Roots} = \{1.0123456, 2.0123456, \ldots, 50.0123456\}$$

The factored form polynomial is expanded into coefficient form. The roots of the expanded polynomial are then obtained and plotted. . Figure 9.9[36] illustrates the degree of accuracy of factoring this polynomial by various 1992 CAD packages. The MATHEMATICA package yields the most accurate roots which expedites the cancellation of the common factors of Eq. (9.13).

Conclusion Obtain det P and P^{-1} directly from Eq. (9.5). To do this, one needs the plant equations in the form of Eq. (9.5), which may not be readily available from the state-space form. Therefore, the designer is very strongly advised at the very onset of the design process to obtain, if possible, the data in the form of Eq. (9.5).

9-6 INVERTIBILITY

The question is often asked: "How does one know that $P(s)$ is invertible?" Obviously, the plant needs to be a square matrix, i.e., P is mxm. Then, bearing in mind that the entries of P are not real numbers but instead are (rational) functions in the dummy variable s, formal invertibility is almost guaranteed. Indeed, the following holds:

> **Theorem** -- The mxm matrix $P(s)$ in Eq. (9.11) is invertible iff the system (A, B, C) is controllable and observable. Moreover, in the formulation of Eq. (9.12) no "pole/zero" cancellations occur iff the above system is controllable and observable.
>
> **Proof outline** -- If the plant is not controllable and observable then even a formal inversion of P won't be possible because some of the rows and/or columns of P will be linearly dependent over the real field.

Finally, controllability and observability of the control system are "a given" in the real world where over modeling should be avoided at all costs.

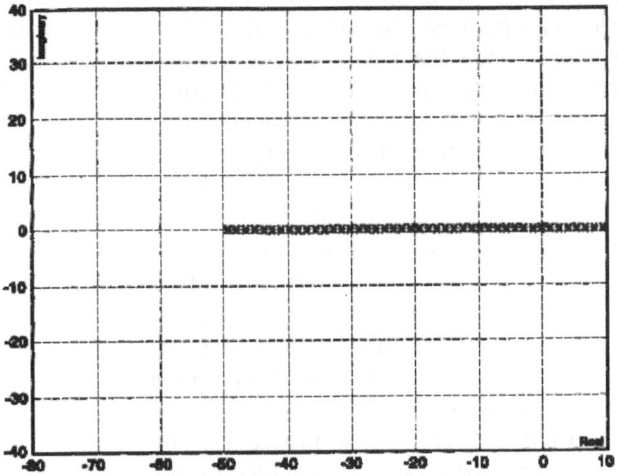

Roots of high precision polynomial (Mathematica).

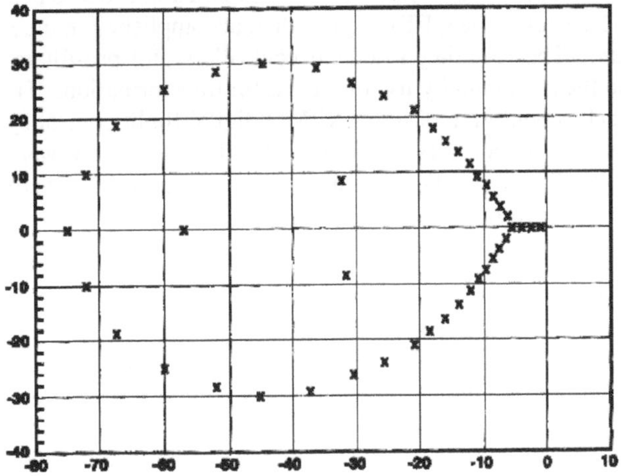

Roots of standard precision polynomial ($MATRIX_X$, Control-C, Matlab, Macsyma).

Fig. 9.9 Figures from Sating's AFIT Thesis.[36]

9-7 PSEUDO-CONTINUOUS-TIME (PCT) SYSTEM

In Chapter 4 the QFT design method applied to MISO sampled-data (S-D) control systems is presented. As pointed out in this chapter, a m.p. *s*-domain plant becomes a n.m.p. plant in the *w*-domain. Also, as is well known, in converting an analog system whose stability, in general, is determined by the value of the open-loop gain *K*, to a S-D system, its stability is now determined not only by the value of *K* but is also a function of the sampling-time *T*. In general, in converting an analog system to a S-D system, as pointed out in Chapter 4, the degree of system stability is decreased. Because, of the *w*-domain n.m.p. characteristic, there is a "restricted" frequency band $\Delta\omega = \omega_K - \omega_i$, $\omega_K > \omega_i$ (or $v_K > v_i$), in which loop shaping must be accomplished while satisfying all the bounds. Such a "restricted" frequency band does not exist for analog systems. Thus, for S-D systems this "restricted" frequency band makes loop shaping a little more difficult to accomplish. Finally, systems that are strictly proper in the *s*-domain are proper but not strictly proper in the *w*-domain

As shown in Chapter 4, a technique for "by-passing" the restricted frequency band problem for a *w*-domain QFT S-D system design is to convert this S-D system to a pseudo-continuous-time (PCT) system. The criteria for converting a given S-D system to a PCT system must be satisfied in order to accomplish a satisfactory QFT design. The QFT design is then accomplished in the *s*-domain for the PCT system. The resulting *s*-domain controllers and pre-filters are then transformed into the z-domain by use of the Tustin transformation. The PCT QFT design approach was used to design a MIMO digital flight control system for an unmanned research vehicle[34] (see Chapter 10). This approach was also used to design a MIMO digital robotic control system.[35] The designs were successful and met all desired performance specifications.

9-8 BODE'S THEOREM[40]

It is well known, based upon Bode's theorem, that putting a unity feedback loop around a plant ***P*** containing parametric uncertainty, in general, will shift the region of uncertainty from one of low frequency to a high frequency range. Now, QFT is good at handling structured (parametric) uncertainty, namely, uncertainty at low frequencies and within the control bandwidth of interest—as opposed to other robust control design methods which address unstructured uncertainty, i.e., un-modelled (e.g., high frequency) dynamics. Hence, applying QFT to an effective plant transfer function ***P*** that includes the unity-feedback loop, may in some cases prevent a successful design. For example, this situation can occur in the design of a flight control system (FCS) where one initially closes an aileron/rudder interconnect feedback loop. The purpose of using an aileron/rudder interconnect is to minimize the cross-coupling effect (also referred to in flight control as *adverse yaw*). When QFT is applied to design a MIMO FCS that already has incorporated an aileron/rudder interconnect, the Bode effect prevents a successful design because the uncertainty has been transferred to an unfavorable high frequency band, as amply reflected in the size of the templates. However, QFT, in effect, essentially does the same job as the use of an aileron/rudder interconnect, for in MIMO QFT the use of high gain will decouple the equivalent set of MISO plants. Hence, the direct application of QFT to the design of the lateral directional channel FCS obviates the need for an aileron/rudder interconnect, for QFT automatically does the job; provided that the inputs and the outputs of the MISO controlled system are properly chosen. For example, in flight control the plant inputs are to the aileron and rudder deflection and the outputs are roll rate and side-slip.

9-9 THE CONTROL DESIGN PROCESS[34]

As demonstrated throughout this chapter – *"In Bridging the Gap"* – there are many important factors that must be considered, both from a theoretical viewpoint and the real world aspects of the control application, during and upon the completion of a successful control system design. The major factors that play a vital role in the design process of a flight control system are discussed in the following chapter and are depicted in Fig. 9.10. This figure shows four major aspects of the control design process: control theory (design techniques) simulation, implementation and flight test. Throughout the text it is stressed that QFT has the ability from the inception of the control problem to the testing and proving of the final designed control system to holistically take into account, *during the design process,* the following items:

(a) The important considerations of control authority allocation
(b) The imbedded desired performance specifications
(c) An awareness of the imperfect real world aspects of the particular control system that is being designed

9-10 SUMMARY

Engineers throughout the world are applying the results of the scientific method to achieving solutions for real world problems. As an example, an aerospace firm has stated to Professor M. Grimble, University of Strathclyde, the following:

> "The QFT approach has the obvious advantage that it is close to engineers' existing experience on classical design methods. However, it provides facilities to deal with uncertainty which are not available in traditional

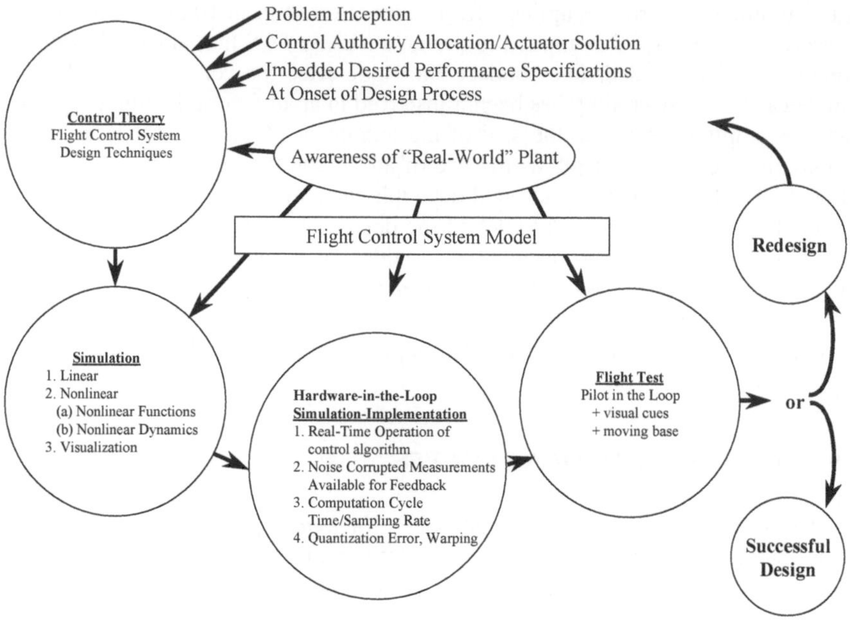

Fig. 9.10 The QFT control system design process.

methods. More recent tools such as H_∞ design also show promise but are very different to the existing procedures used in parts of the aerospace industry. The QFT approach therefore appears to have the attractive features of providing a link with existing techniques whilst at the same time providing many of the advanced features needed for the 90's high performance systems. What might be needed are tools for the future which combine the attractive features of QFT and H_∞ approaches."

In this chapter guidelines are provided to the control engineer on how to interface between the scientific and engineering methods. In conclusion, an attempt is made to bridge the often lamented gap between theory and practice.

10

QUANTITATIVE NON-DIAGONAL COMPENSATOR DESIGN FOR MIMO SYSTEMS [97,98,99,100,101,134,137]

10-1 INTRODUCTION

A fully populated (non-diagonal) matrix compensator allows the designer much more design flexibility to govern MIMO systems than the classical diagonal controller structure. This Chapter introduces a methodology (Method 3) to extend the classical diagonal QFT compensator design (Methods 1 and 2, presented in Chapters 5 to 8), to a fully populated matrix compensator design for MIMO systems with model uncertainty.

In this Chapter three cases are studied: the reference tracking, the external disturbance rejection at the plant input and the external disturbance rejection at plant output. Therefore, the role played by the non-diagonal compensator elements g_{ij} ($i \neq j$) is analyzed to present a QFT design methodology for a fully populated matrix compensator. The definition of three coupling matrices ($\boldsymbol{C}_1$, $\boldsymbol{C}_2$, $\boldsymbol{C}_3$) and a quality function η_{ij} of the non-diagonal elements are used to quantify the amount of loop interaction and to design the non-diagonal compensators respectively. This yields a criterion that proposes a sequential design methodology of the fully populated matrix compensator in the QFT robust control frame that yields n equivalent tracking SISO systems and n equivalent disturbance rejection SISO systems.[97,134,137]

The off-diagonal elements of the compensator matrix reduce (or cancel if there is no uncertainty) the level of coupling between loops. As a result, the diagonal elements g_{kk} of the non-diagonal method need less bandwidth than the diagonal elements of the previous diagonal $\boldsymbol{G}$ methods. The Chapter ends with a real-world example: an industrial SCARA robot manipulator which is controlled by using the non-diagonal MIMO QFT methodology.[98,100]

10-2 THE COUPLING MATRIX

The objective of this section is to define a measurement index (the coupling matrix) that allows one to quantify the loop interaction in MIMO control systems. Following the Horowitz ideas introduced in Chapter 5, consider an nxn linear multivariable system (see Fig. 10.1), composed of a plant $\boldsymbol{P}$, a fully populated matrix compensator $\boldsymbol{G}$, and a pre-filter $\boldsymbol{F}$. These matrices are defined as follows:

$$\boldsymbol{P}=\begin{bmatrix} p_{11} & p_{12} & \cdots & p_{1n} \\ p_{21} & p_{22} & \cdots & p_{2n} \\ \vdots & \vdots & \ddots & \vdots \\ p_{n1} & p_{n2} & \cdots & p_{nn} \end{bmatrix};\quad \boldsymbol{G}=\begin{bmatrix} g_{11} & g_{12} & \cdots & g_{1n} \\ g_{21} & g_{22} & \cdots & g_{2n} \\ \vdots & \vdots & \ddots & \vdots \\ g_{n1} & g_{n2} & \cdots & g_{nn} \end{bmatrix};$$

$$\boldsymbol{F}=\begin{bmatrix} f_{11} & f_{12} & \cdots & f_{1n} \\ f_{21} & f_{22} & \cdots & f_{2n} \\ \vdots & \vdots & \ddots & \vdots \\ f_{n1} & f_{n2} & \cdots & f_{nn} \end{bmatrix} \qquad \textbf{(10.1)}$$

In Fig. 10.1 are shown a plant input disturbance transfer function $\boldsymbol{P}_{di}$, and a plant output disturbance transfer function $\boldsymbol{P}_{do}$, where $\boldsymbol{P} \in \mathcal{TP}$, and $\mathcal{TP}$ is the set of possible plants due to uncertainty. The reference vector $\boldsymbol{r}'$ and the external disturbance vectors at plant input $\boldsymbol{d}_i'$ and plant output $\boldsymbol{d}_o'$ are the inputs of the system. The output vector $\boldsymbol{y}$ represents the variables to be controlled.

The plant inverse $\boldsymbol{P}^{-1}$, denoted by $\boldsymbol{P}^*$ in this Chapter, is presented in the following format:

$$\boldsymbol{P}^{-1}=\boldsymbol{P}^{*}=\left[p^{*}_{ij}\right]=\boldsymbol{\Lambda}+\boldsymbol{B}=\begin{bmatrix} p^{*}_{11} & 0 & 0 \\ 0 & \cdots & 0 \\ 0 & 0 & p^{*}_{nn} \end{bmatrix}+\begin{bmatrix} 0 & \cdots & p^{*}_{1n} \\ \cdots & 0 & \cdots \\ p^{*}_{n1} & \cdots & 0 \end{bmatrix} \qquad \textbf{(10.2)}$$

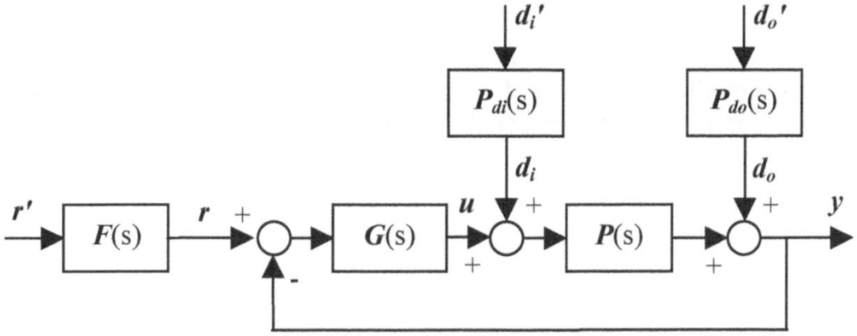

Fig. 10.1 Structure of a 2 Degree of Freedom MIMO System.

and where the compensator matrix is broken up into two parts as follows:

$$G = G_{\mathrm{d}} + G_{\mathrm{b}} = \begin{bmatrix} g_{11} & 0 & 0 \\ 0 & \ldots & 0 \\ 0 & 0 & g_{nn} \end{bmatrix} + \begin{bmatrix} 0 & \ldots & g_{1n} \\ \ldots & 0 & \ldots \\ g_{n1} & \ldots & 0 \end{bmatrix} \tag{10.3}$$

Note that $\boldsymbol{\Lambda}$ is the diagonal part and $\boldsymbol{B}$ is the balance of $\boldsymbol{P}^*$; and that G_{d} is the diagonal part and G_{b} is the balance of $\boldsymbol{G}$.

The following three sub-sections introduce a measurement index to quantify the loop interaction in the three classical cases: reference tracking, external disturbances at the plant input, and the external disturbances at the plant output. In this chapter the measurement index is called the *coupling matrix* $\boldsymbol{C}$ and, depending on the case, shows three different notations: C_1, C_2, C_3, respectively. The use of these coupling matrices enables the achievement of essentially n equivalent tracking SISO systems and n equivalent disturbance rejection SISO systems.

10-2.1 TRACKING

The transfer function matrix of the control system for the reference tracking problem, without any external disturbance (Fig. 10.1), is written as shown in Eq. (10.4),

$$y = \left(I + P\,G\right)^{-1} P\,G r = T_{y/r} r = T_{y/r} F r' \tag{10.4}$$

Using Eq. (10.2) and (10.3), Eq. (10.4) is rewritten as,

$$T_{y/r}r = \left(I + \Lambda^{-1}G_{\mathrm{d}}\right)^{-1}\Lambda^{-1}G_{\mathrm{d}}r + \left(I + \Lambda^{-1}G_{\mathrm{d}}\right)^{-1}\Lambda^{-1}\left[G_{\mathrm{b}}r - \left(\mathbf{B} + G_{\mathrm{b}}\right)T_{y/r}r\right] \tag{10.5}$$

which is another expression to represent the same idea introduced in the previous Chapters 5 to 8 (see Sec. 5-5.1) as,

$$y_{\mathrm{ij}} = t_{\mathrm{ij}}^{\mathrm{y/r}}\, r_{\mathrm{j}} = \left(t_{\mathrm{r_{ii}}} + t_{\mathrm{c_{ij}}}\right) r_{\mathrm{j}} \quad , \quad \mathrm{i,j} = 1,2,\ldots\mathrm{n}$$

where,

$$t_{\mathrm{r_{ii}}} = w_{\mathrm{ii}}\, g_{\mathrm{ii}} \;; \qquad t_{\mathrm{c_{ij}}} = w_{\mathrm{ii}}\, c_{\mathrm{ij}}$$

$$w_{\mathrm{ii}} = \frac{1}{p_{\mathrm{ii}}^{*} + g_{\mathrm{ii}}}$$

$$c_{\mathrm{ij}} = -\sum_{\mathrm{k \neq i}} t_{\mathrm{kj}}\, p_{\mathrm{ik}}^{*} \quad , \quad \mathrm{k} = 1,2,\ldots\mathrm{n}$$

An analysis of Eq. (10.5), the closed-loop transfer function matrix, reveals that it can be broken up into two parts as follows:

i. A diagonal term $T_{\mathrm{y/r\_d}}$ *given by,*

$$T_{\mathrm{y/r\_d}} = \left(I + \Lambda^{-1}G_{\mathrm{d}}\right)^{-1}\Lambda^{-1}G_{\mathrm{d}} \tag{10.6}$$

that represents a pure diagonal structure. Note that it does not depend on the non-diagonal part of the plant inverse $\boldsymbol{B}$ nor on the non-diagonal part of the compensator G_{b}. It is equivalent to n reference tracking SISO systems formed by plants equal to the elements of Λ^{-1} when the n corresponding parts of a diagonal G_{d} control them, as shown in Fig. 10.2a. In this figure $t_i(s)$ represents the closed-loop control ratio.

ii. A non-diagonal term $T_{\mathrm{y/r\_b}}$ *given by,*

$$T_{\mathrm{y/r\_b}} = \left(I + \Lambda^{-1}G_{\mathrm{d}}\right)^{-1}\Lambda^{-1}\left[G_{\mathrm{b}} - \left(\mathbf{B} + G_{\mathrm{b}}\right)T_{y/r}\right] = \left(I + \Lambda^{-1}G_{\mathrm{d}}\right)^{-1}\Lambda^{-1}C_1 \tag{10.7}$$

that represents a non-diagonal structure. It is equivalent to the same n previous systems with cross-coupling (internal) disturbances $c_{1_{ij}}\, r_j$ at the plant input and to n disturbance rejection SISO systems (Fig. 10.2b).

In Eq. (10.7), the matrix C_1 is the only part that depends on the non-diagonal parts of both the plant inverse B and the compensator G_{b}. Hence, this matrix comprises the coupling, and from now on C_1 represents the *coupling*

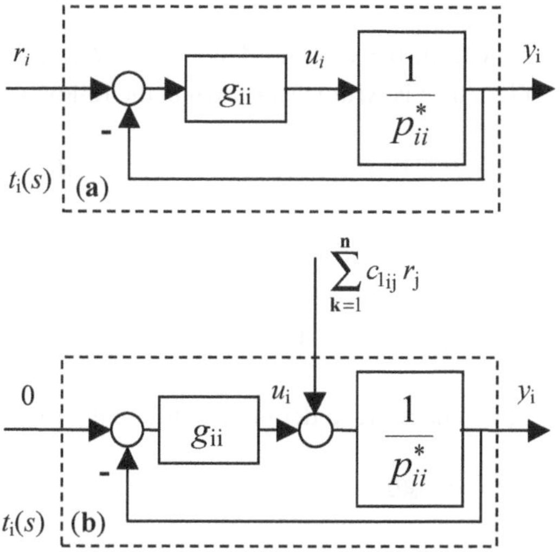

Fig. 10.2 *i*-th equivalent decoupled SISO systems.

matrix $\boldsymbol{C}$ of the equivalent system for reference tracking problems. The bracketed term in Eq. (10.7) represents $\boldsymbol{C}_1$, i.e.,

$$\boldsymbol{C}_1 = \boldsymbol{G}_b - (\mathbf{B} + \boldsymbol{G}_b)\boldsymbol{T}_{y/r} = \begin{bmatrix} 0 & c_{1_{12}} & \cdots & c_{1_{1m}} \\ c_{1_{21}} & 0 & \cdots & c_{1_{2m}} \\ \vdots & \vdots & \ddots & \vdots \\ c_{1_{m1}} & c_{1_{m2}} & \cdots & 0 \end{bmatrix} \quad \textbf{(10.8)}$$

Each element $c_{1_{ij}}$ of this matrix obeys,

$$c_{1_{ij}} = g_{ij}\left(1-\delta_{ij}\right) - \sum_{k=1}^{n}\left(p^*_{ik} + g_{ik}\right)t_{kj}\left(1-\delta_{ik}\right) \quad \textbf{(10.9)}$$

where δ_{ki} is the Kronecker delta that is defined as,

$$\delta_{ki} = \begin{cases} \delta_{ki} = 1 \Leftrightarrow k = i \\ \delta_{ki} = 0 \Leftrightarrow k \neq i \end{cases} \quad \textbf{(10.10)}$$

and which is an extension of the cross-coupling c_{ij} elements introduced in Chapter 5.

10-2.2 DISTURBANCE REJECTION AT PLANT INPUT

The transfer matrix from the external disturbance $\boldsymbol{d}_i^{'}$, at the plant input, to the plant output $\boldsymbol{y}$ (Fig. 10.1) is written as shown in the following equation:

$$\boldsymbol{y} = (\boldsymbol{I} + \boldsymbol{P}\,\boldsymbol{G})^{-1}\,\boldsymbol{P}\,\boldsymbol{d}_i = \boldsymbol{T}_{y/di}\,\boldsymbol{d}_i = \boldsymbol{T}_{y/di}\,\boldsymbol{P}_{di}\,\boldsymbol{d}_i^{'} \qquad \textbf{(10.11)}$$

Using Eqs. (10.12) and (10.3), Eq. (10.11) is rewritten as:

$$\boldsymbol{T}_{y/di}\boldsymbol{d}_i = \left(\boldsymbol{I} + \Lambda^{-1}\boldsymbol{G}_{\mathrm{d}}\right)^{-1}\Lambda^{-1}\boldsymbol{d}_i - \left(\boldsymbol{I} + \Lambda^{-1}\boldsymbol{G}_{\mathrm{d}}\right)^{-1}\Lambda^{-1}\left[(\mathbf{B} + \boldsymbol{G}_{\mathrm{b}})\boldsymbol{T}_{\mathrm{y/di}}\right]\boldsymbol{d}_i \qquad \textbf{(10.12)}$$

From Eq. (10.12) it is possible to define two different terms as follows:

i. A diagonal term $\boldsymbol{T}_{\mathrm{y/di\_d}}$ *given by,*

$$\boldsymbol{T}_{\mathrm{y/di\_d}} = \left(\boldsymbol{I} + \Lambda^{-1}\,\boldsymbol{G}_{\mathrm{d}}\right)^{-1}\Lambda^{-1} \qquad \textbf{10.13)}$$

Again, Eq. (10.13) is equivalent to n regulator SISO systems, as shown in Fig. 10.3a.

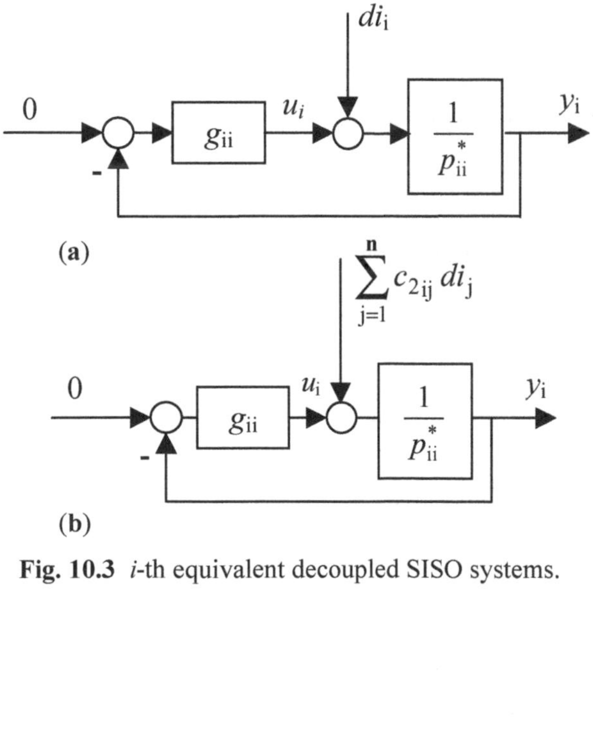

Fig. 10.3 *i*-th equivalent decoupled SISO systems.

ii. A non-diagonal term $\boldsymbol{T}_{\mathrm{y/di\_b}}$ *given by,*

$$\boldsymbol{T}_{\mathrm{y/di\_b}} = \left(\boldsymbol{I} + \boldsymbol{\Lambda}^{-1}\boldsymbol{G}_{\mathrm{d}}\right)^{-1}\boldsymbol{\Lambda}^{-1}\left(\mathbf{B} + \boldsymbol{G}_{\mathrm{b}}\right)\boldsymbol{T}_{y/di} = \left(\boldsymbol{I} + \boldsymbol{\Lambda}^{-1}\boldsymbol{G}_{\mathrm{d}}\right)^{-1}\boldsymbol{\Lambda}^{-1}\boldsymbol{C}_2 \qquad \textbf{(10.14)}$$

represents a non-diagonal structure which is equivalent to the same n previous systems with external disturbances $c_{2_{\mathrm{ij}}}\, di_{\mathrm{j}}$ at the plant input, as shown in Fig. 10.3b.

In Eq. (10.14), the matrix $\boldsymbol{C}_2$ comprises the coupling. Thus, from now on $\boldsymbol{C}_2$ represents the *coupling matrix* of the equivalent system for external disturbance rejection at the plant input problems and is given by

$$\boldsymbol{C}_2 = (\boldsymbol{B} + \boldsymbol{G}_b)\boldsymbol{T}_{y/di} \qquad \textbf{(10.15)}$$

Each element $c_{2_{\mathrm{ij}}}$ of this matrix obeys,

$$c_{2_{\mathrm{ij}}} = \sum_{\mathrm{k=1}}^{\mathrm{n}}(p^{*}_{\mathrm{ik}} + g_{\mathrm{ik}})\ t_{\mathrm{kj}}\ (1 - \delta_{\mathrm{ik}}) \qquad \textbf{(10.16)}$$

where δ_{ki} is the Kronecker delta defined in Eq. (10.10).

10-2.3 DISTURBANCE REJECTION AT PLANT OUTPUT

The transfer matrix from the external disturbance $\boldsymbol{d}_o'$, at the plant output, to the output $\boldsymbol{y}$ (Fig. 10.1) is written as shown in Eq. (10.17),

$$\boldsymbol{y} = (\boldsymbol{I} + \boldsymbol{P}\boldsymbol{G})^{-1}\boldsymbol{d}_o = \boldsymbol{T}_{y/do}\,\boldsymbol{d}_o = \boldsymbol{T}_{y/do}\,\boldsymbol{P}_{do}\,\boldsymbol{d}_o' \qquad \textbf{(10.17)}$$

Using Eqs. (10.2) and (10.3), repeating the procedure of the previous subsections, Eq. (10.17) is rewritten as:

$$\boldsymbol{T}_{y/do}\mathbf{d}_o = \left(\boldsymbol{I} + \boldsymbol{\Lambda}^{-1}\boldsymbol{G}_{\mathrm{d}}\right)^{-1}\mathbf{d}_o + \left(\boldsymbol{I} + \boldsymbol{\Lambda}^{-1}\boldsymbol{G}_{\mathrm{d}}\right)^{-1}\boldsymbol{\Lambda}^{-1}\left[\mathbf{B} - \left(\mathbf{B} + \boldsymbol{G}_{\mathrm{b}}\right)\boldsymbol{T}_{y/do}\right]\mathbf{d}_o \qquad \textbf{(10.18)}$$

From Eq. (10.18) it is also possible to define two terms:

i. A diagonal term $\boldsymbol{T}_{\mathrm{y/do\_d}}$ *given by,*

$$\boldsymbol{T}_{\mathrm{y/do\_d}} = \left(\boldsymbol{I} + \boldsymbol{\Lambda}^{-1}\boldsymbol{G}_{\mathrm{d}}\right)^{-1} \qquad \textbf{(10.19)}$$

Once more, Eq. (10.19) is equivalent to the *n* regulator SISO systems showed in Fig. 10.4a.

ii. *A non-diagonal term* $\boldsymbol{T}_{\text{y/do\_b}}$ *given by,*

$$\boldsymbol{T}_{\text{y/do\_b}} = \left(\boldsymbol{I} + \boldsymbol{\Lambda}^{-1}\boldsymbol{G}_{\text{d}}\right)^{-1}\boldsymbol{\Lambda}^{-1}\left[\boldsymbol{B} - (\boldsymbol{B} + \boldsymbol{G}_{\text{b}})\boldsymbol{T}_{y/do}\right] = \left(\boldsymbol{I} + \boldsymbol{\Lambda}^{-1}\boldsymbol{G}_{\text{d}}\right)^{-1}\boldsymbol{\Lambda}^{-1}\boldsymbol{C}_3 \qquad \textbf{(10.20)}$$

that represents a non-diagonal structure. It is equivalent to the same *n* previous systems with external disturbances $c_{3\text{ij}}\ do_{\text{j}}$ at the plant input, as shown Fig. 10.4b.

In Eq. (10.20), the matrix $\boldsymbol{C}_3$ comprises the coupling. Thus, from now on it represents the *coupling matrix* of the equivalent system for external disturbance rejection for the plant output problems and is given by

$$\boldsymbol{C}_3 = \boldsymbol{B} - (\boldsymbol{B} + \boldsymbol{G}_b)\boldsymbol{T}_{y/do} \qquad \textbf{(10.21)}$$

Each element of the coupling matrix, $c_{3\text{ij}}$ obeys,

$$c_{3\text{ij}} = p^*_{\text{ij}}\ (1 - \delta_{\text{ij}}) - \sum_{k=1}^{n} (p^*_{\text{ik}} + g_{\text{ik}})\ t_{\text{kj}}\ (1 - \delta_{\text{ik}}) \qquad \textbf{(10.22)}$$

where δ_{ki} is the Kronecker delta as defined in Equation (10.10).

10-3 THE COUPLING ELEMENTS

In order to design a MIMO compensator with a low coupling level, it is necessary to study the influence of every non-diagonal element g_{ij} on the

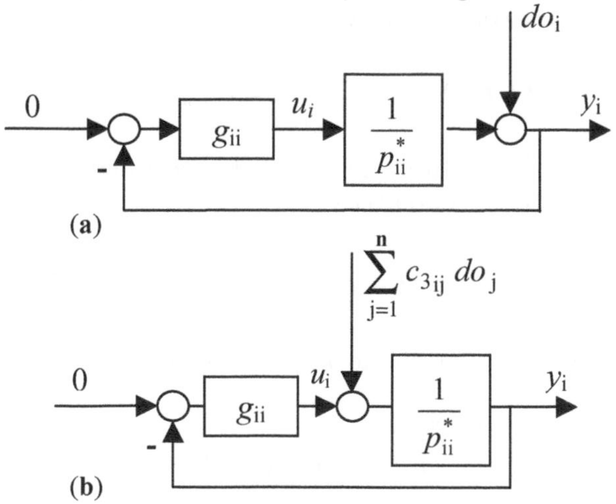

Fig. 10.4 *i*-th equivalent decoupled SISO systems.

coupling elements c_{1ij}, c_{2ij} and c_{3ij} as defined by Eqs. (10.9), (10.16) and (10.22), respectively.

To easily quantify the coupling effects, these elements are simplified by applying the following hypothesis.

Hypothesis H1: suppose that in Eqs. (10.9), (10.16) and (10.22),

$$\left|\left(p_{ij}^{*}+g_{ij}\right)t_{jj}\right| >> \left|\left(p_{ik}^{*}+g_{ik}\right)t_{kj}\right|, \text{ for } k \neq j, \text{ and in the bandwidth of } t_{jj} \quad \textbf{(10.23)}$$

Note that it is not a too bold hypothesis, considering that it is desirable for the diagonal elements t_{jj} to be much larger than the non-diagonal elements t_{kj}, when once the pairing of the most convenient variables have been applied. Thus,

$$\left|t_{jj}\right| >> \left|t_{kj}\right|, \text{ for } k \neq j, \text{ and in the bandwidth of } t_{jj} \quad \textbf{(10.24)}$$

Now, two simplifications are applied to facilitate the quantification of the coupling effects c_{1ij}, c_{2ij}, c_{3ij}.

Simplification S1: Using Hypothesis H1, Eqs. (10.9), (10.16) and (10.22), which describe the coupling elements in the tracking problem, the disturbance rejection at the plant input and disturbance rejection at plant output respectively, are rewritten as,

$$c_{1ij} = g_{ij} - t_{jj}\left(p_{ij}^{*}+g_{ij}\right) \quad ; \quad i \neq j \quad \textbf{(10.25)}$$

$$c_{2ij} = t_{jj}\left(p_{ij}^{*}+g_{ij}\right) \quad ; \quad i \neq j \quad \textbf{(10.26)}$$

$$c_{3ij} = p_{ij}^{*} - t_{jj}\left(p_{ij}^{*}+g_{ij}\right) \quad ; \quad i \neq j \quad \textbf{(10.27)}$$

Simplification S2: The elements t_{jj} are respectively computed for each case from the equivalent system derived from Eqs. (10.6), (10.13) and (10.19), so that,

$$t_{jj} = \frac{g_{jj}\left(p_{ij}^{*}\right)^{-1}}{1+g_{jj}\left(p_{ij}^{*}\right)^{-1}} \quad \textbf{(10.28)}$$

$$t_{jj} = \frac{\left(p_{ij}^{*}\right)^{-1}}{1+g_{jj}\left(p_{ij}^{*}\right)^{-1}} \quad \textbf{(10.29)}$$

$$t_{jj} = \frac{1}{1+g_{jj}\left(p_{ij}^{*}\right)^{-1}} \quad \textbf{(10.30)}$$

Due to Simplifications S1 and S2, the coupling effects c_{1ij}, c_{2ij}, c_{3ij} are computed as,

$$c_{1ij} = g_{ij} - \frac{g_{jj}\left(p_{ij}^{*} + g_{ij}\right)}{\left(p_{jj}^{*} + g_{jj}\right)} \quad ; \quad i \neq j \tag{10.31}$$

$$c_{2ij} = \frac{\left(p_{ij}^{*} + g_{ij}\right)}{\left(p_{jj}^{*} + g_{jj}\right)} \quad ; \quad i \neq j \tag{10.32}$$

$$c_{3ij} = p_{ij}^{*} - \frac{p_{jj}^{*}\left(p_{ij}^{*} + g_{ij}\right)}{\left(p_{jj}^{*} + g_{jj}\right)} \quad ; \quad i \neq j \tag{10.33}$$

10-4 THE OPTIMUM NON-DIAGONAL COMPENSATOR

As stated previously, the purpose of non-diagonal compensators is to reduce the coupling effect in addition to achieving the desired tracking loop performance specifications. The optimum non-diagonal compensators for the three cases (tracking and disturbance rejection at the plant input and output) are obtained making the loop interaction of Eqs. (10.31)-(10.33) equal to zero.

Note that both elements p_{ij}^{*} and p_{jj}^{*} of these three equations are uncertain elements of $\boldsymbol{P}^{*}$. In general every uncertain plant p_{ij}^{*} can be any plant represented by the family:

$$\left\{p_{ij}^{*}\right\} = p_{ij}^{*N}\left(1 + \Delta_{ij}\right) \ , \ 0 \leq \left|\Delta_{ij}\right| \leq \Delta p_{ij}^{*} \ , \ \text{for } i, j = 1, \ldots, n \tag{10.34}$$

where p_{ij}^{*N} is the selected nominal plant for the non-diagonal controller expression and Δp_{ij}^{*} is the maximum of the non-parametric uncertainty radii $\left|\Delta_{ij}\right|$ (see Fig. 10.5b). Note: Δp_{ij}^{*} depends on the selection of p_{ij}^{*N}.

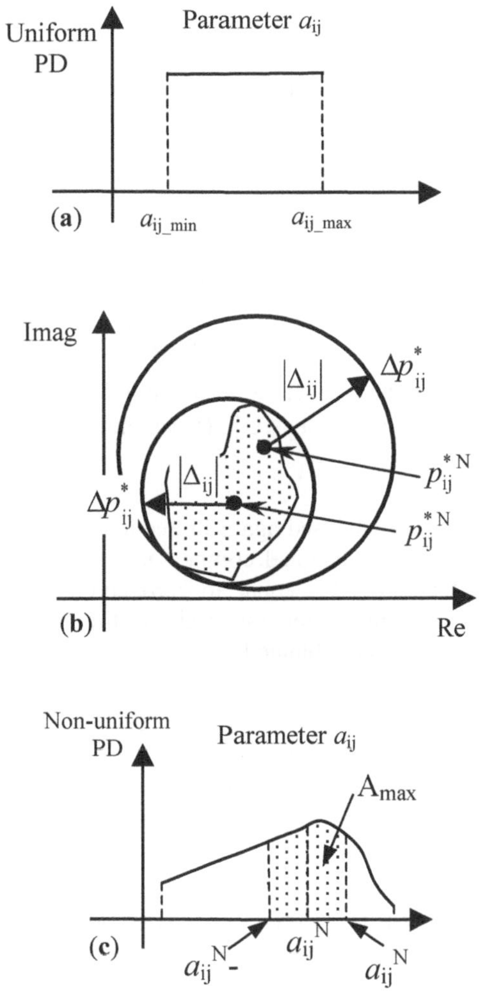

Fig. 10.5 (a) and (c) Probability Distribution of the parameter a_{ij}; (b) two possibilities of the maximum non-parametric uncertainty radii Δp^*_{ij} that comprises the plant templates.

The selected plants p^{*N}_{ij} and p^{*N}_{jj} that are chosen for the optimum non-diagonal compensator must comply with the following rules:

a) If the uncertain parameters of the plants show a uniform Probability Distribution (see Fig. 10.5a), which is typical in the QFT methodology, then the elements p^*_{ij} and p^*_{jj} for the optimum non-diagonal compensator are the plants p^{*N}_{ij} and p^{*N}_{jj}. These plants minimise the maximum of the non-parametric uncertainty radii Δp^*_{ij} and Δp^*_{jj} that comprise the plant templates (see Fig. 10.5b).

b) If the uncertain parameters of the plants show a non-uniform Probability Distribution (see Fig. 10.5c), then the elements p^*_{ij} and p^*_{jj} for the optimum non-diagonal compensator are the plants p^{*N}_{ij} and p^{*N}_{jj}, whose set of parameters maximize the area of the Probability Distribution in the regions $[a_{ij}-\varepsilon, a_{ij}+\varepsilon]$ and $[a_{jj}-\varepsilon, a_{jj}+\varepsilon]$ ($\forall$parameter a_{ij}, b_{ij},..., a_{jj}, b_{jj}...) respectively.

These rules of selection are analysed again in Section 10.5, where the coupling effects with the optimum non-diagonal compensator are computed. By setting Eqs. (10.31), (10.32) and (10.33) equal to zero and using Eq. (10.34), the optimum non-diagonal compensator for each of the following cases [Eqs. (10.35), (10.36) and (10.37)] are obtained.

10-4.1 TRACKING

$$g^{\text{opt}}_{ij} = F_{pd}\left(g_{jj} \frac{p^{*N}_{ij}}{p^{*N}_{jj}} \right), \text{ for } i \neq j \qquad \textbf{(10.35)}$$

10-4.2 DISTURBANCE REJECTION AT PLANT INPUT

$$g^{\text{opt}}_{ij} = F_{pd}\left(- p^{*N}_{ij} \right), \text{ for } i \neq j \qquad \textbf{(10.36)}$$

10-4.3 DISTURBANCE REJECTION AT PLANT OUTPUT

$$g^{\text{opt}}_{ij} = F_{pd}\left(g_{jj} \frac{p^{*N}_{ij}}{p^{*N}_{jj}} \right), \text{ for } i \neq j \qquad \textbf{(10.37)}$$

where the function $F_{pd}(A)$ means in every case a proper stable and minimum phase function made from the dominant poles and zeros of the expression A.

10-5 THE COUPLING EFFECTS

The rules of Sec. 10.4 are utilized for choosing the plants p_{ij}^{*N} and p_{jj}^{*N}. These plants are inserted into Eqs. (10-35) – (10.37) in order to obtain the respective g_{ij}^{opt} which are in turn utilized for determining the minimum achievable coupling effects given by Eqs. (10.38), (10.40), and (10.42). In a similar manner, the maximum coupling effects for the diagonal compensator matrix case, given by Eqs. (10.39), (10.41), and (10.43), are computed by substituting $g_{ij} = 0$ ($i \neq j$) into the coupling expressions of Eqs. (10.31)-(10.33), respectively.

10-5.1 TRACKING

$$\left| c_{1ij} \right|_{g_{ij}=\, g_{ij}^{opt}} = \left| \psi_{ij} \left(\Delta_{jj} - \Delta_{ij} \right) g_{jj} \right| \qquad \textbf{(10.38)}$$

$$\left| c_{1ij} \right|_{g_{ij}=0} = \left| \psi_{ij} \left(1 + \Delta_{ij} \right) g_{jj} \right| \qquad \textbf{(10.39)}$$

10-5.2 DISTURBANCE REJECTION AT PLANT INPUT

$$\left| c_{2ij} \right|_{g_{ij}=g_{ij}^{opt}} = \left| \psi_{ij}\, \Delta_{ij} \right| \qquad \textbf{(10.40)}$$

$$\left| c_{2ij} \right|_{g_{ij}=0} = \left| \psi_{ij} \left(1 + \Delta_{ij} \right) \right| \qquad \textbf{(10.41)}$$

10-5.3 DISTURBANCE REJECTION AT PLANT OUTPUT

$$\left| c_{3ij} \right|_{g_{ij}=g_{ij}^{opt}} = \left| \psi_{ij} \left(\Delta_{ij} - \Delta_{jj} \right) g_{jj} \right| \qquad \textbf{(10.42)}$$

$$\left| c_{3ij} \right|_{g_{ij}=0} = \left| \psi_{ij} \left(1 + \Delta_{ij} \right) g_{jj} \right| \qquad \textbf{(10.43)}$$

where,

$$\psi_{ij} = \frac{p_{ij}^{*N}}{\left(1+\Delta_{jj}\right) p_{jj}^{*N} + g_{jj}} \tag{10.44}$$

and the uncertainty is,

$$0 \le \left|\Delta_{ij}\right| \le \Delta p_{ij}^{*} \;,\; 0 \le \left|\Delta_{jj}\right| \le \Delta p_{jj}^{*} \;,\; \text{for } i, j = 1,\ldots,n$$

The coupling effects, calculated for the pure diagonal compensator cases, result in three expressions (10.39), (10.41) and (10.43) that still present a non-zero value when the selected (p_{ij}^{*N}, p_{jj}^{*N}) - actual plant mismatching due to the uncertainty disappears: $\Delta_{ij} = 0$ and $\Delta_{jj} = 0$. However, the coupling effects obtained with the optimum non-diagonal compensators [see Eqs. (10.38), (10.40) and (10.42)] tend to zero when the mismatching disappears.

10-6 QUALITY FUNCTION OF THE DESIGNED COMPENSATOR

Figure 10.6 shows the appearance of three different coupling bands for a common system. The maximum $|c_{ij}|_{g_{ij}=0}$, computed from Eqs. (10.39), (10.41) or (10.43), and the minimum coupling effects without any non-diagonal compensator g_{ij} limit the first one: the top cross-hatched pair. The second pair of curves (dashed lines) are bounded by the maximum (the upper dashed curve) and the minimum (the lower dashed curve) coupling effects with a non-optimum decoupling element g_{ij}. Finally, the minimum coupling effect $|c_{ij}|_{g_{ij}=g_{ij}^{opt}}$ with the optimum decoupling element g_{ij}^{opt} (the bottom solid curve) presents a maximum value, computed from Eqs. (10.38), (10.40) or (10.42).

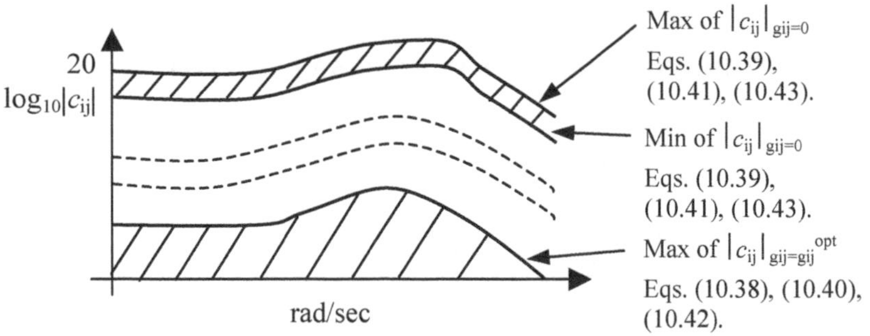

Fig. 10.6 Coupling effect bands with different non-diagonal compensators.

From this analysis a quality function η_{ij} is defined for a non-diagonal compensator g_{ij} (i≠j) so that,

$$\eta_{ij}(\%) = 100 \left\{ \frac{\log_{10}\left[\dfrac{\max\left\{\left|c_{ij}\right|_{g_{ij}=0}\right\}}{\max\left\{\left|c_{ij}\right|_{g_{ij}=g_{ij}}\right\}}\right]}{\log_{10}\left[\dfrac{\max\left\{\left|c_{ij}\right|_{g_{ij}=0}\right\}}{\max\left\{\left|c_{ij}\right|_{g_{ij}=g_{ij}^{opt}}\right\}}\right]} \right\} \tag{10.45}$$

The quality function becomes a proximity measure of the coupling effect c_{ij} to the minimum achievable coupling effect. Thus, the function is useful to quantify the amount of loop interaction and to design the non-diagonal compensators.

If η_{ij} is closed to 100%, then the coupling effect is a minimum and the g_{ij} compensator tends to be the optimum one. A suitable non-diagonal compensator maximizes the quality function of Eq. (10.45).

10-7 DESIGN METHODOLOGY

The compensator design method is a sequential procedure by closing thc loops.[93] It does not assign any special role to the upper and lower triangular elements of the fully populated matrix compensator $\boldsymbol{G}$ but it is only necessary to fulfil Hypothesis H1.

Methodology

Step A. *Input/Output pairing and loop ordering.* First, the methodology begins paring the plant inputs and outputs with the Relative Gain Analysis (RGA) technique.[92] This is followed by arranging the matrix $\boldsymbol{P}^*$ so that $\left(p_{11}^*\right)^{-1}$ has the smallest phase margin frequency, $\left(p_{22}^*\right)^{-1}$ the next smallest phase margin frequency, and so on.[101] The sequential compensator design technique (as described in Fig. 10.7), composed of n stages (n loops), utilizing the following steps (B and C) is repeated for every column $k = 1$ to n.

Step B. *Design of the diagonal compensator elements* g_{kk}. This design of the element g_{kk} is calculated using the standard QFT loop-shaping technique for the inverse of the equivalent plant $\left(p_{kk}^{*e}\right)^{-1}$ in order to achieve robust stability and

$$G = \begin{bmatrix} g_{11} & 0 & \cdots & 0 & \cdots & 0 \\ g_{21} & 0 & \cdots & 0 & \cdots & 0 \\ \cdots & & \cdots & & \cdots & \\ g_{kl} & 0 & \cdots & 0 & \cdots & 0 \\ \cdots & & \cdots & & \cdots & \\ g_{nl} & 0 & \cdots & 0 & \cdots & 0 \end{bmatrix} \Rightarrow \begin{bmatrix} g_{11} & g_{12} & \cdots & 0 & \cdots & 0 \\ g_{21} & g_{22} & \cdots & 0 & \cdots & 0 \\ \cdots & & \cdots & & \cdots & \\ g_{kl} & g_{k2} & \cdots & 0 & \cdots & 0 \\ \cdots & & \cdots & & \cdots & \\ g_{nl} & g_{n2} & \cdots & 0 & \cdots & 0 \end{bmatrix} \Rightarrow \cdots$$

Step 1   Step 2

$$\cdots \Rightarrow \begin{bmatrix} g_{11} & g_{12} & \cdots & g_{1k} & \cdots & g_{1n} \\ g_{21} & g_{22} & \cdots & g_{2k} & \cdots & g_{2n} \\ \cdots & & \cdots & & \cdots & \\ g_{kl} & g_{k2} & \cdots & g_{kk} & \cdots & g_{kn} \\ \cdots & & \cdots & & \cdots & \\ g_{nl} & g_{n2} & \cdots & g_{nk} & \cdots & g_{nn} \end{bmatrix}$$

Step n

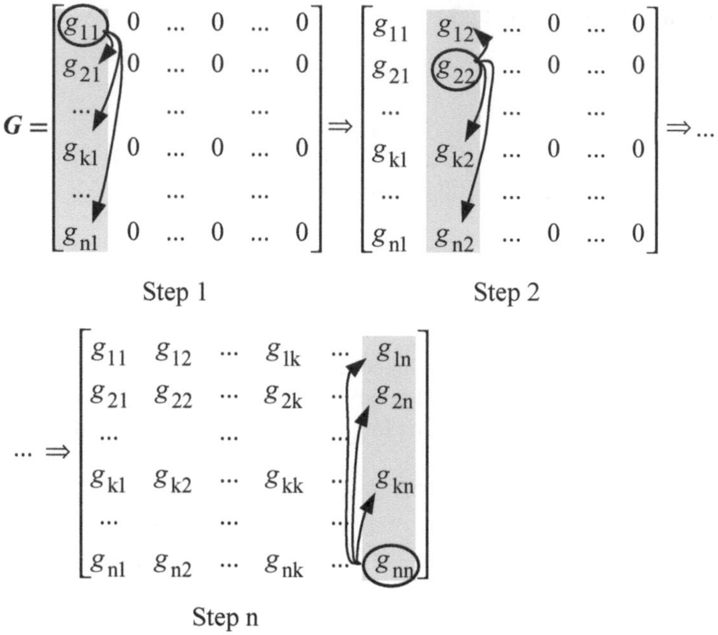

Fig. 10.7 n stages of the sequential compensator design technique.

robust performance specifications. The equivalent plant satisfies the recursive relationship[93] given by Eq. (10.46).

$$\left[p_{ii}^{*e} \right]_k = \left[p_{ii}^* \right]_{k-1} - \frac{\left(\left[p_{i(i-1)}^* \right]_{k-1} + \left[g_{i(i-1)} \right]_{k-1} \right)\left(\left[p_{(i-1)i}^* \right]_{k-1} + \left[g_{(i-1)i} \right]_{k-1} \right)}{\left[p_{(i-1)(i-1)}^* \right]_{k-1} + \left[g_{(i-1)(i-1)} \right]_{k-1}} ; \ i \geq k; \ \left[P^* \right]_{k=1} = P^*$$

(10.46)

This equation is an extension for the non-diagonal case of the recursive expression proposed by Horowitz[20] as the *Improved design technique*, also called Method 2 in Chapter 7.

At this point the design has also to fulfil two stability conditions:[135] a) $L_i(s) = g_{ii}(s) (p_{ii}^{*e})^{-1}$ has to satisfy the Nyquist encirclement condition and b) no RHP pole-zero cancellations have to occur between $g_{ii}(s)$ and $(p_{ii}^{*e})^{-1}$.

If the system requires the tracking specifications as

$$a_{ii}(j\omega) \leq \left| t_{ii}^{y/r}(j\omega) \right| \leq b_{ii}(j\omega)$$

and since

$$t_{ii}^{y/r} = t_{rii} + t_{cii}$$

the tracking bounds b_{ii} and a_{ii} are corrected to take into account the cross-coupling specification $\tau_{c_{ii}}$, so that:

$$b_{ii}^{c} = b_{ii} - \tau_{cii} \ , \ a_{ii}^{c} = a_{ii} + \tau_{cii} \tag{10.47}$$

$$t_{cii} = w_{ii}\, c_{ii} \leq \tau_{cii} \tag{10.48}$$

$$a_{ii}^{c}(j\omega) \leq \left|t_{rii}(j\omega)\right| \leq b_{ii}^{c}(j\omega) \tag{10.49}$$

These are the same corrections proposed by Horowitz[95,20] for the original MIMO QFT Methods 1 and 2 (Chapters 5 to 8).

However, for the non-diagonal compensator these corrections are less demanding. The coupling expression $t_{c_{ii}} = w_{ii}\, c_{ii}$ is now minor as compared to the previous diagonal methods [for instance compare Eqs. (10.38) and (10.39)]. That is the off-diagonal elements g_{ij} ($i \neq j$) of the matrix compensator attenuate or cancel the cross coupling effects. This results in the diagonal elements g_{kk} of the non-diagonal method requiring a smaller bandwidth than the diagonal elements of the diagonal compensator methods.

Step C. *Design of the non-diagonal compensator elements* g_{ij}. The (n–1) non-*diagonal elements* g_{ik} *(*$i \neq k$*,* $i = 1,2,\ldots n$*) of the k-th compensator column are* designed to minimise the cross-coupling terms c_{ik} given by Eqs. (10.31)-(10.33). The optimum compensator elements [see Eqs. (10.35)-(10.37)], are utilized in order to achieve this goal,.

Once the design of $\boldsymbol{G}(s)$ has finished, to ensure that the last non-diagonal elements designed do not destabilize the closed-loop system, the design has also to fulfil two stability conditions:[135] c) no Smith-McMillan pole-zero cancellations have to occur between $\boldsymbol{P}(s)$ and $\boldsymbol{G}(s)$ and d) no Smith-McMillan pole-zero cancellations have to occur in $|\boldsymbol{P}^*(s) + \boldsymbol{G}(s)$.

Remark

Although it is very remote, theoretically there exists the possibility of introducing right-half-plane (RHP) transmission zeros due to the controller design. This undesirable situation can not be detected until the whole multivariable system design is completed. To avoid this problem, the proposed methodology (Steps A, B and C) is introduced in the next procedure.[175]

Stage 1: Design of the controller matrix. First of all, the matrix compensator $\boldsymbol{G}(s)$ is designed conforming to the methodology described in Steps A, B and C.

At the end of this stage we will be able to evaluate the transmission zeros of the whole multivariable system.

Stage 2. Calculation of transmission zeros. The multivariable zeros of $\boldsymbol{P}(s)\ \boldsymbol{G}(s)$ are determined using the Smith-McMillan form over the set of possible plants P due to the uncertainty. If there are not new RHP zeros apart from those that might already be present in $\boldsymbol{P}(s)$, the method concludes. Otherwise, proceed to Stage 3.

Stage 3. Modification of the RHP transmission zero positions. Once we have observed that there exist RHP transmission zeros introduced by the matrix compensator elements, we proceed to rectify this undesirable situation by modifying the non-diagonal elements placed in the last column of the matrix $\boldsymbol{G}(s)$, according to the Smith-McMillan expressions.

Step D. *Pre-filter.* The design of a pre-filter $\boldsymbol{F}(s)$ is necessary in case of reference tracking specifications. Once the full matrix compensator $\boldsymbol{G}(s)$ has been designed the pre-filter does not present any difficulty, because the final $\boldsymbol{T}_{y/r}$ function shows less loop interaction. Therefore, the pre-filter $\boldsymbol{F}$ can be matrix diagonal.

10-8 SOME PRACTICAL ISSUES

The sequential non-diagonal MIMO QFT technique introduced in this Chapter arrives at a robust stable closed-loop system if, for each $\boldsymbol{P} \in \mathcal{TP}$, [135]

- each $L_i(s) = g_{ii}(s)\ (p_{ii}^{*e})^{-1}$, i=1, ..., n, satisfies the Nyquist encirclement condition,
- no RHP pole-zero cancellations occur between $g_{ii}(s)$ and $(p_{ii}^{*e})^{-1}$, i=1, ..., n,
- no Smith-McMillan pole-zero cancellations occur between $\boldsymbol{P}(s)$ and $\boldsymbol{G}(s)$, and
- no Smith-McMillan pole-zero cancellations occur in $|\boldsymbol{P}^*(s) + \boldsymbol{G}(s)|$

On the other hand, the resulting matrix $\boldsymbol{PG}$ should be checked in every step of the methodology to ensure that RHP transmission zeros or unstable modes have not been introduced by the new compensator elements g_{kk} or g_{ik}, which would obviously cause an unnecessary loss of control performance. If these n.m.p. zeros appear due to the designed compensator elements, supplementary constraints in the determinant of $\boldsymbol{PG}$ should be imposed to re-calculate the compensator. There again, if the plant elements, p_{kk} or p_{ik}, are the cause of the introduction of non-minimum phase elements in the equivalent plant $\left(p_{kk}^{*e}\right)^{-1}$, the theory proposed for n.m.p. MISO feedback systems in Chapter 4 can be applied to properly design the compensators in the loop-shaping step.

Incidentally, arbitrarily picking the wrong order of the loops to be designed can result in the nonexistence of a solution. This may occur if the solution process is based on satisfying an upper limit of the phase margin frequency ω_ϕ for each loop. To avoid that potential problem, as it has been introduced in step **A** of the methodology, loop *i* having the smallest phase margin frequency has to be chosen as the first loop to be designed (see E.R.5 in Chapter 9). The loop that has the next smallest phase margin frequency is next, and so on.[101]

Finally, it is important to notice that the calculation of the equivalent plant $\left(p_{kk}^{*e}\right)^{-1}$ usually introduces some exact pole-zero cancellations. This operation can be precisely done by using symbolic mathematic tools, but could be erroneously done when using numerical calculus due to the typical computer round errors.

10.9 EXAMPLE: NON-DIAGONAL MIMO QFT COMPENSATOR DESIGN FOR A SCARA ROBOT TRACKING SPECIFICATIONS

The non-diagonal MIMO QFT compensator design technique is now applied to control a real-world problem: a SCARA robot.[100] Figure 10.8 shows the AdeptOne robot manipulator, and the two joint angles δ_1 and δ_2 that are to be controlled in this example. The plant model, the desired performance specifications, the compensator design, the implementation and the actual experimentation are introduced in the following sub-sections.

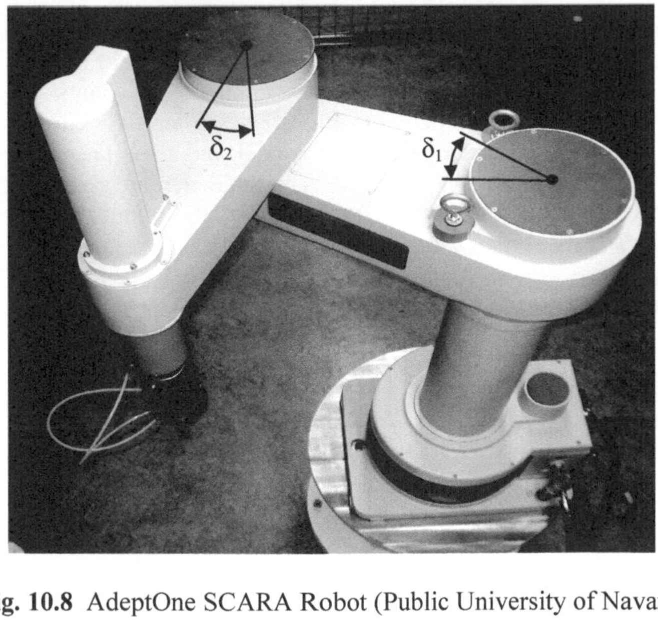

Fig. 10.8 AdeptOne SCARA Robot (Public University of Navarra).

10-9.1 PLANT MODEL

The Lagrange equations' method is used to find the following equations [Eqs. (10.50) and (10.51)], which describe the non-linear dynamic behaviour of the two-link system.[100] The actual inputs are the shaft torques τ_1 and τ_2, applied on joints 1 and 2, respectively, which are developed by power amplifiers (electrical motors) whose amplifier gains are k. Thus, the plant inputs are $u_1 = k\tau_1$ and $u_2 = k\tau_2$ and the plant outputs are the angles δ_1 and δ_2.

$$\left[\alpha_1 + 2\alpha_3 \cos(\delta_2)\right]\ddot{\delta}_1 + \left[\alpha_2 + \alpha_3 \cos(\delta_2)\right]\ddot{\delta}_2 + v_1\dot{\delta}_1 + \mu_1 \mathrm{sgn}(\dot{\delta}_1) = \tau_1 = \frac{u_1}{k} \tag{10.50}$$

$$\left[\alpha_2 + \alpha_3 \cos(\delta_2)\right]\ddot{\delta}_1 + \alpha_2\ddot{\delta}_2 + v_2\dot{\delta}_1 + \mu_2 \mathrm{sgn}(\dot{\delta}_2) = \tau_2 = \frac{u_2}{k} \tag{10.51}$$

where v_i are the coefficients of viscous friction, μ_i the Coulomb friction associated with link i, and

$$\left.\begin{aligned} \alpha_1 &= I_1 + I_2 + m_1 x_1^2 + m_2\left(l_1^2 + x_2^2\right) \\ \alpha_2 &= I_2 + m_2 x_2^2 \\ \alpha_3 &= m_2 l_1 x_2 \end{aligned}\right\} \tag{10.52}$$

are the parameters. In Eq. (10.52) I_i, m_i and x_i are the moment of inertia, mass and position of the i-th link respectively, and l_1 as the length of link 1.

The input signals u_1 and u_2 are computed in counts [ct] and are commanded to the robot motors by the amplifiers. After a system identification technique,[100] the parameters of the robot model and their uncertainty, with a uniform Probability Distribution, are found. The parameters are given in Table 10.1.

Now it is also possible to consider the Coulomb frictions as disturbances and the cosine value of δ_2 as an uncertain parameter h between -1 and +1. Thus, it is easy to find the following linear transfer functions, utilizing Eqs. (10.50) to (10.52), which are the elements of the plant $\boldsymbol{P}$ defined as,

$$\begin{bmatrix}\delta_1 \\ \delta_2\end{bmatrix} = \boldsymbol{P}\begin{bmatrix}u_1 \\ u_2\end{bmatrix} = \begin{bmatrix} p_{11} & p_{12} \\ p_{21} & p_{22}\end{bmatrix}\begin{bmatrix}u_1 \\ u_2\end{bmatrix} \tag{10.53}$$

Table 10.1 Parameters with uncertainty (uniform probability distribution). They are multiplied by a gain of k = 75 [ct/N·m] due to the electronic power amplifier.

	Minimum	Maximum	Nominal
α_1 k [ct s²/rad]	719	813	766
α_2 k [ct s²/rad]	186	200	193
α_3 k [ct s²/rad]	134	230	182
ν_1 k [ct s/rad]	67	381	224
ν_2 k [ct s/rad]	11.6	91.9	51.75
μ_1 k [ct]	344	358	351
μ_2 k [ct]	262	323	292.5

$$p_{11}(s) = \frac{\alpha_2 s + \nu_2}{s\Delta(s)} \frac{1}{k} \tag{10.54}$$

$$p_{12}(s) = \frac{-(\alpha_2 + \alpha_3 h)}{\Delta(s)} \frac{1}{k} \tag{10.55}$$

$$p_{21}(s) = \frac{-(\alpha_2 + \alpha_3 h)}{\Delta(s)} \frac{1}{k} \tag{10.56}$$

$$p_{22}(s) = \frac{(\alpha_1 + 2\alpha_3 h)s + \nu_1}{s\Delta(s)} \frac{1}{k} \tag{10.57}$$

where

$$\Delta = \chi_2 s^2 + \chi_1 s + \chi_0 \tag{10.58}$$

with the following coefficients,

$$\left.\begin{aligned} \chi_2 &= \alpha_2(\alpha_1 + 2\alpha_3 h) - (\alpha_2 + \alpha_3 h)^2 \\ \chi_1 &= \alpha_2 \nu_1 + \nu_2(\alpha_1 + 2\alpha_3 h) \\ \chi_0 &= \nu_1 \nu_2 \end{aligned}\right\} \tag{10.59}$$

10-9.2 PERFORMANCE SPECIFICATIONS

The desired performance specifications for the SCARA robot manipulator are the following, *i.* Robust Stability. $\left|t_{ii}(j\omega)\right| \le 1.2$, for $i = 1,2$, $\forall \omega$, which involves a phase margin of at least 50° and a gain margin of at least 1.833 (5.26 dB).

ii. Control effort constraint. Control signals have to be lower than 32767 [ct] for a disturbance rejection at the plant output of about 20°.

iii. Disturbance rejection at plant input. The maximum allowed error has to be 30° for torque disturbances of 1000 [ct].

iv. Loop Coupling. Reduce the coupling effect as much as possible.

v. Tracking specifications. $\left|t_{ii}^{y/r}(j\omega)\right|$ has to achieve tracking tolerances (Fig. 10.9) defined by,

$$a_{ii}(\omega) \le \left|t_{ii}^{y/r}(j\omega)\right| \le b_{ii}(\omega) \quad \text{for } i = 1,2 \tag{10.60}$$

where,

$$b_{ii}(\omega) = \left| \frac{12.25\left[(j\omega)/30 + 1\right]}{(j\omega)^2 + 5.25(j\omega) + 12.25} \right| \tag{10.61}$$

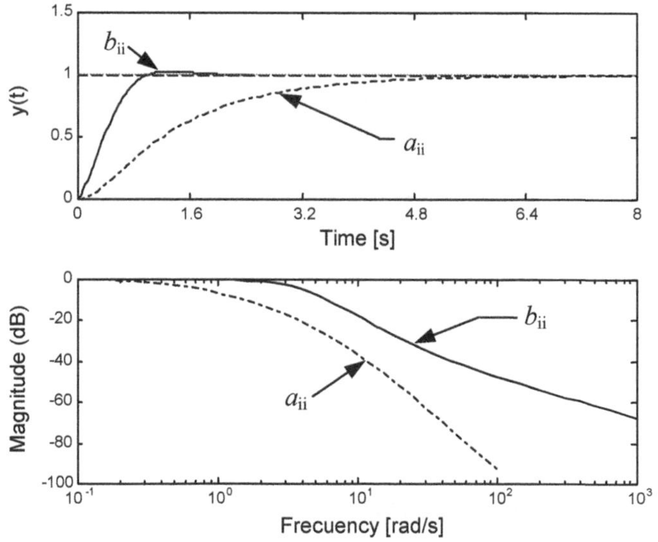

Fig. 10.9 Robust tracking specifications.

$$a_{ii}(\omega)=\left|\frac{2.25}{\left[(j\omega)^2+4.5(j\omega)+2.25\right]\left[(j\omega)/10+1\right]}\right| \tag{10.62}$$

For practical implementation the above specifications are limited by the achieved sampling time of 10 ms.

10-9.3 COMPENSATOR DESIGN

- Step A *Input/Output pairing and loop ordering*

The first step is the Relative Gain Analysis (RGA). In this example the analysis yields a very obvious result: angle δ_1 has to be controlled by motor 1 (u_1), and angle δ_2 by motor 2 (u_2). This pairing produces a multivariable model which is diagonally dominant for all parameter variations within the uncertainty range. Then the diagonal elements t_{ij} are much larger that the non-diagonal ones t_{kj}, fulfilling Hypothesis H1. The first RGA element λ_{11}, plotted in Fig. 10.10 with the model uncertainty, shows that the robot arm presents a very coupled behavior at some frequencies. At low frequencies, below 0.06 rad/s, the coupling is very low, but as the frequency increases the system presents a more pronounced coupled dynamics effect. The required bandwidth of the system derived from tracking specifications lies between approximatcly 2 and 3.5 rad/s. In this frequency range the maximum value of λ_{11} is greater than 4.5.

- Step B.1 *Design of the first diagonal compensator element*, $g_{11}(s)$.

By using the model uncertainty of $\left(p_{11}^{*}\right)^{-1}$ and the desired specifications for Loop 1, the compensator of Eq. (10.63) is found, satisfying all the performance objectives (see Fig. 10.11a). The design also fulfils the first two stability requirements.

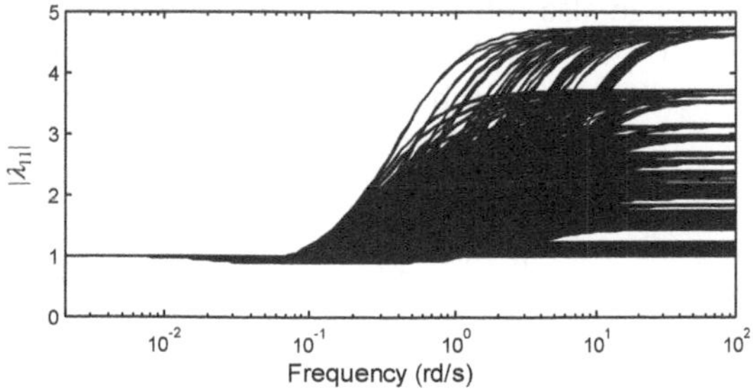

Fig. 10.10 Element λ_{11} of the Relative Gain Analysis.

By using the model of $\left(p_{11}^{*}\right)^{-1}$ and the desired specifications for loop 1, the compensator of Eq. (10.63) is found, satisfying all the performance objectives (see Fig. 10.11a). The design also fulfils the first two stability conditions: that is to say, a) $L_2(s) = g_{22}(s)\,(p_{22}^{*e})^{-1}$ satisfies the Nyquist encirclement condition and b) no RHP pole-zero cancellations occur between $g_{22}(s)$ and $(p_{22}^{*e})^{-1}$.

$$g_{11}(s) = \frac{1.65\,s^2 + 4.3840\,s + 2.6190}{\left(s^2 + 829.2\,s + 1.545\times10^5\right)s}\,10^9 \tag{10.63}$$

- Step C.1 *Design of the non-diagonal compensator element*, $g_{21}(s)$.

Taking into account the optimum compensator for the reference tracking problems of Eq. (10.35), the compensator g_{21} of Eq. (10.64) is designed minimising the coupling effect c_{21}. Figure 10.11b shows the frequency plot of the obtained coupling reduction.

$$g_{21}(s) = \frac{3860\,s^2 + 10300\,s + 6130}{848\times10^{-6}s^3 + 0.00703\,s^2 + 1.31\,s + 0.346} \tag{10.64}$$

- Step B.2. *Design of the second diagonal compensator element*, $g_{22}(s)$.

The following equivalent plant $\left(p_{22}^{*e}\right)^{-1}$ derived from Eq. (10.46) is calculated,

$$\left[p_{22}^{*e}\right]_2 = \left[p_{22}^{*}\right]_1 - \frac{\left(\left[p_{21}^{*}\right]_1 + \left[g_{21}\right]_1\right)\left(\left[p_{12}^{*}\right]_1 + \left[g_{12}\right]_1\right)}{\left[p_{11}^{*}\right]_1 + \left[g_{11}\right]_1} \tag{10.65}$$

The compensator of Eq. (10.66) is determined that satisfies all the performance specifications (see Fig. 10.12a) for the above equivalent plant $(p_{22}^{*})^{-1}$ with uncertainty. The design also fulfils the first two stability conditions: that is: a) $L_2(s) = g_{22}(s)\,(p_{22}^{*e})^{-1}$ satisfies the Nyquist encirclement condition and b) no RHP pole-zero cancellations occur between $g_{22}(s)$ and $(p_{22}^{*e})^{-1}$.

$$g_{22}(s) = \frac{88.1s^2 + 225s + 110}{\left(10^{-6}s^2 + 0.371\times10^{-3}s + 0.0344\right)s} \tag{10.66}$$

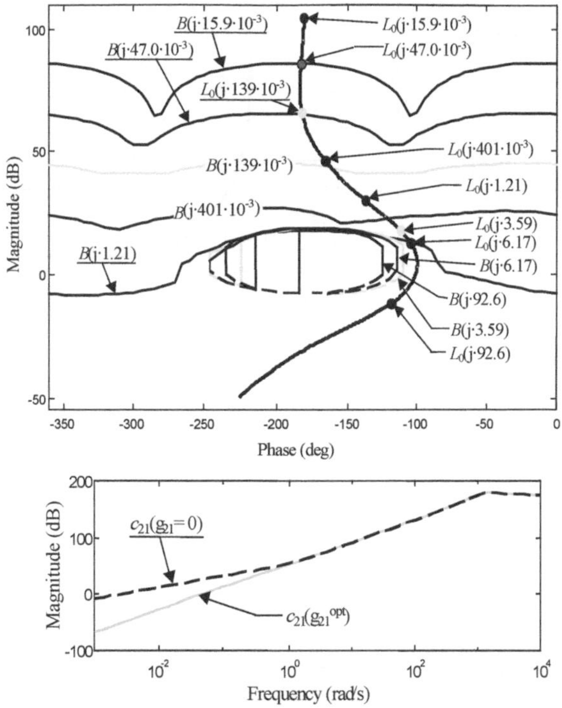

(a)

(b)

Fig. 10.11 QFT MIMO compensator design: (a) Loop-shaping $L_0(s) = g_{11}(s)\left[p_{11}^*(s)\right]^{-1}$; (b) c_{21} coupling reduction.

- Step C.2. *Design of the non-diagonal compensator element*, $g_{12}(s)$.

The compensator g_{12} of Eq. (10.67) is designed minimising the coupling effect c_{12}. Figure. 10.12b shows the frequency plot of the obtained coupling reduction.

$$g_{12}(s) = \frac{20600s^2 + 52600s + 25700}{0.193\times10^{-3}s^3 + 0.0717s^2 + 6.66s + 1.78} \quad \textbf{(10.67)}$$

At this point, the last two stability conditions must be checked to ensure that the non-diagonal element $g_{12}(s)$ designed in the last step does not destabilize the closed-loop system. That is to say: c) no Smith-McMillan pole-zero cancellations occur between $\boldsymbol{P}(s)$ and $\boldsymbol{G}(s)$ and d) no Smith-McMillan pole-zero cancellations pole-zero cancellations occur in $|\boldsymbol{P}^*(s) + \boldsymbol{G}(s)|$. The system is stable. In addition the system has no RHP transmission zeros.

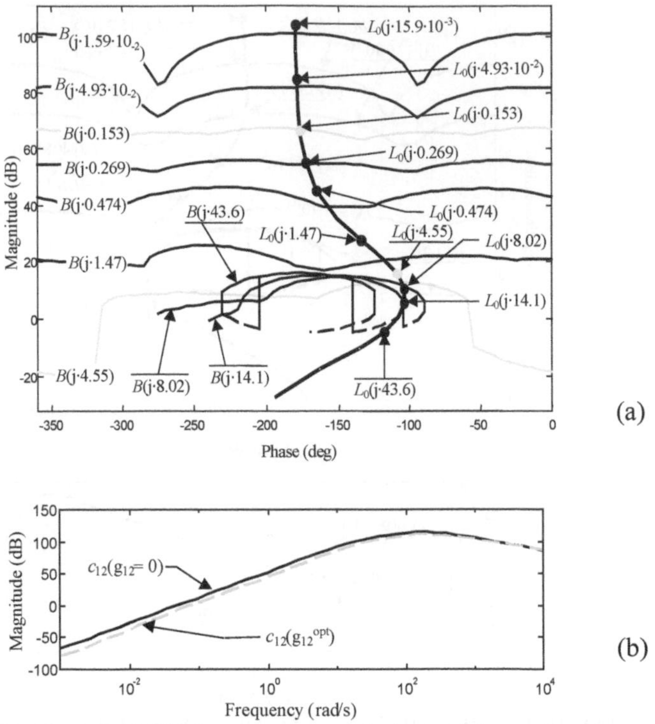

Fig. 10.12 QFT MIMO compensator design: (a) Loop-shaping $L_0(s) = g_{22}(s)\left[p_{22}^*(s)\right]^{-1}$; (b) c_{12} coupling reduction.

- Step D. Pre-filters.

The open loop pre-filters of Eq. (10.68) and Eq. (10.69) are included in order to satisfy time domain specifications for reference tracking.

$$f_{11}(\mathrm{s}) = \frac{14.3}{\mathrm{s}^2 + 7.5620\mathrm{s} + 14.3} \tag{10.68}$$

$$f_{22}(\mathrm{s}) = \frac{3.026}{\mathrm{s}^2 + 6.355\mathrm{s} + 3.026} \tag{10.69}$$

Note that the design of the controller for this example is done in the s-domain. That is, the feedback compensator ***G***(s) and the pre-filter ***F***(s) are first synthesized and then, by use of the Tustin transformation, the digital controller ***G***(z) and pre-filter ***F***(z) are obtained.

10-9.4 EXPERIMENTAL RESULTS

To control the AdeptOne SCARA robot,[126] a digital form of the designed non-diagonal QFT compensator is implemented in a Motorola 68040 microprocessor (25 MHz), with a 8 Mbyte DRAM memory and a VME bus card, and using a sampling time of 10 ms. The two plant outputs, angles δ_1 and δ_2, are measured by encoders, and the two plant inputs, signals u_1 and u_2, are applied to a power amplifier that commands two direct drives that move the arms (see Fig. 10.13).

The obtained results with the non-diagonal QFT controller, when the reference r_1 for angle δ_1 is commanded from 0 up to 45 degrees and while the reference r_2 for angle δ_2 is kept constant (zero degrees), are shown in Fig. 10.14. The same experiment with only the pure diagonal controller is shown in Fig. 10.15.

Similarly, the results obtained with the non-diagonal QFT controller, when the reference r_2 for angle δ_2 is commanded from 0 up to 45 degrees and while the reference r_1 for angle δ_1 is kept constant (0 degrees), are shown in Fig. 10.16. The same experiment with only the pure diagonal controller is shown in Fig. 10.17.

Both real experimental results show how the non-diagonal controller (see Fig. 10.14 and 10.16) reach a significative reduction of the coupling effect with respect to the performance of the pure diagonal controller (see Fig. 10.15 and 10.17).

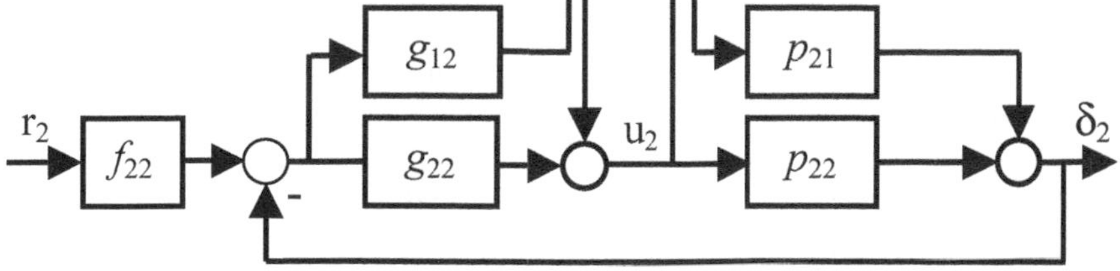

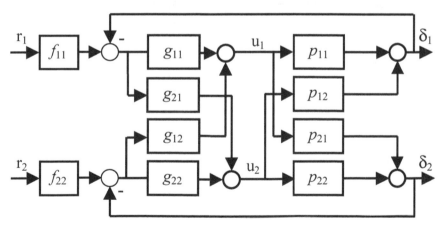

Fig. 10.13 Block diagram of the control system.

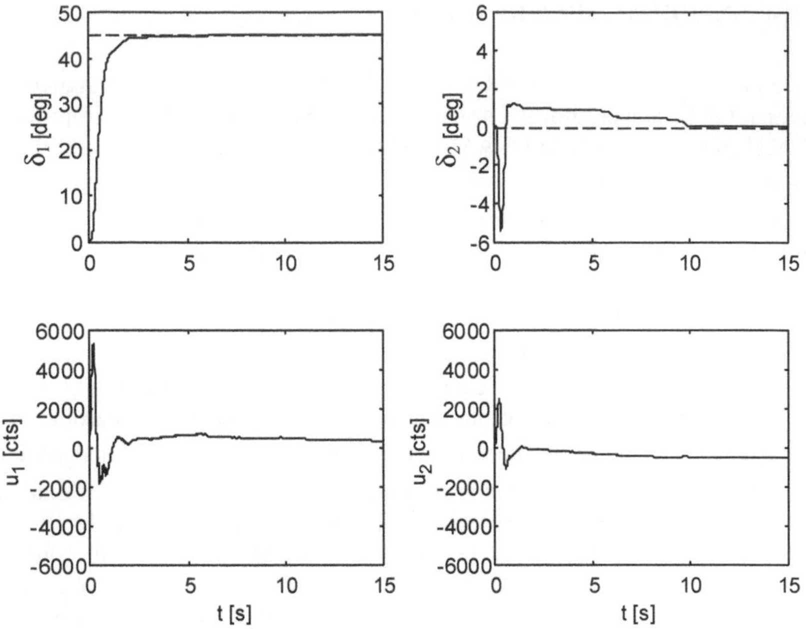

Fig. 10.14 Step input at r_1 with a fully populated (non-diagonal) matrix controller.

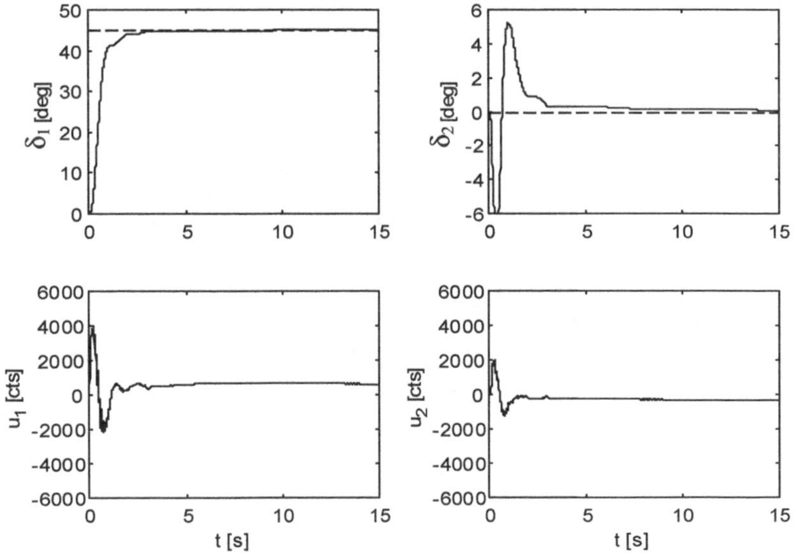

Fig. 10.15 Step input at r_1 with a pure diagonal matrix controller.

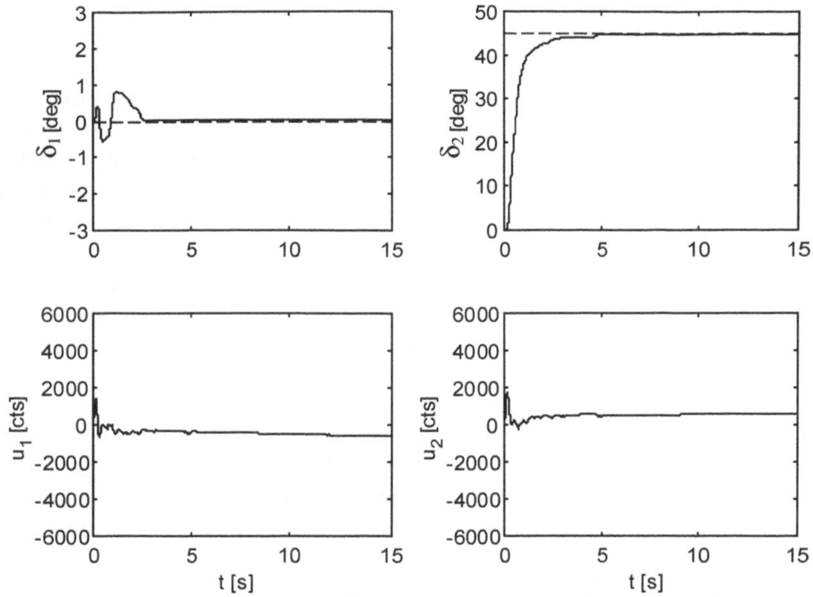

Fig. 10.16 Step input at r_2 with a fully populated (non-diagonal) matrix controller.

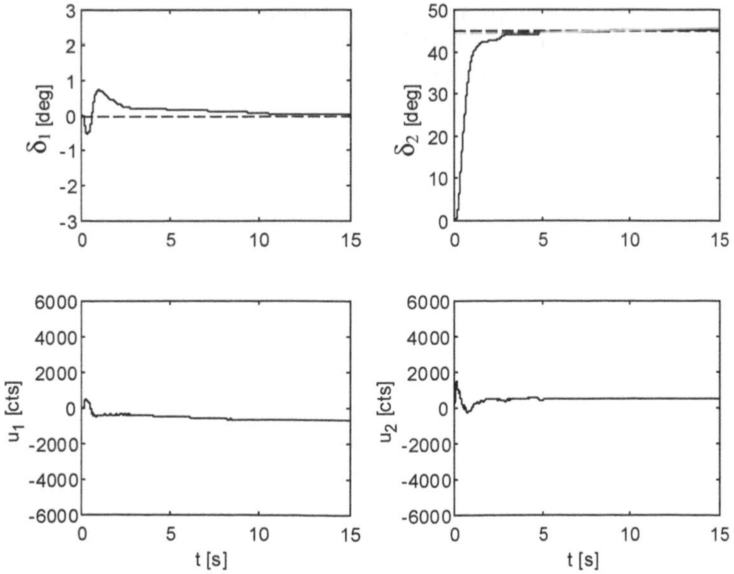

Fig. 10.17 Step input at r_2 with a pure diagonal matrix controller.

10-10 OTHER METHODS[81,109,110,111,132,135,136,176,177]

Some other methods for non-diagonal multivariable QFT robust control system design have been introduced in the last few years. Franchek and Nwokah present a sequential loop frequency approach that utilizes a fully populated matrix controller to meet performance specifications which may include system integrity requirements.[109] Boje and Nwokah utilize the Perron-Frobenius root interaction measure to design a pre-compensator that reduces the level of coupling between loops, before a diagonal QFT controller matrix is attempted.[81,111] Yaniv introduces an approach that emphasizes the bandwidth of a non-diagonal pre-controller multiplied by the classical diagonal controller.[110] De Bedout and Franchek developed sufficient conditions to guarantee closed-loop internal stability for sequentially designed multivariable feedback control systems with unstable Smith-McMillan pole-zero cancellations or with non-minimum phase Smith-McMillan zeros.[135] Kerr, Jayasuriya and Asokanthan also analyze in depth the stability of the sequential and non-sequential QFT design methods for MIMO systems.[136] The reader is referred to the literature for more details.

On the other hand, in the last few years a significant number of papers related to controller fragility have appeared. The fragility problem arises when the controller synthesis techniques and/or the digital implementation tend to produce control laws with a high sensitivity of closed-loop stability to small changes in controller coefficients. As it is claimed in the literature, the controller fragility is mainly produced by using popular robust and optimal control synthesis methods, standard model-based identification techniques, certain parameterizations, and different levels of accuracy in controller implementation. Unfortunately this issue has been neglected in standard textbooks on control and system identification. A novel solution, based on the Quantitative Feedback Theory, was introduced by Garcia-Sanz, Brugarolas and Eguinoa.[132] The method analyses the controller resiliency/fragility to small changes in its coefficients, taking also into account the plant model with parametric and non-parametric uncertainty and the robust stability/performance specifications.

Finally, many real processes have an intrinsically distributed nature (*e.g.* chemical reactors, drainage and sewage water networks, large flexible space structures, electrical networks, etc.). This means that some decisive variables depend not only on time but on spatial distribution and topology as well. These are distributed parameter systems (DPS) and their behavior can be described by partial differential equations (PDE) with some boundary and initial conditions. A novel solution, based on the Quantitative Feedback Theory, was introduced by Garcia-Sanz, Huarte and Asenjo.[176,177] It extends the classical QFT methodology for lumped systems to DPS controlled by a one-point feedback structure. The method considers a spatial distribution of the points where the inputs and the outputs of the control system are applied (actuators, sensors, disturbances and control objectives) and introduces a new set of transfer functions (TF) that describe the relationship between the distributed inputs and outputs of the system. Based on these TFs, the work extends the classical stability and performance specifications to the DPS case and presents a new set of quadratic inequalities to define the QFT bounds. The method can also deal with uncertainty in the spatial distribution of the inputs and the outputs.

10-11 SUMMARY

A QFT methodology to design fully populated matrix (non-diagonal) controllers to solve the MIMO reference tracking and the external disturbances rejection problems at plant input and output, in the presence of model plant uncertainty, is presented in this Chapter.

The definition of both a coupling matrix and a quality function of the non-diagonal elements are used to quantify the amount of loop interaction and to design the non-diagonal controllers respectively. This yields a criterion that makes it possible to propose a sequential design methodology of the fully populated matrix controller in the QFT robust control frame. The use of the coupling matrices $\boldsymbol{C}_i$ essentially enables the achievement of n equivalent tracking SISO systems and n equivalent disturbance rejection SISO systems.

The technique is applied to design a non-diagonal MIMO QFT controller for a SCARA robot manipulator. The actual experiments that have been carried out with an AdeptOne robot show a significant reduction of the MIMO coupling effects when the non-diagonal controller is implemented.

In addition, because the off-diagonal elements of the matrix controller attenuate the cross coupling (or cancel if there is no uncertainty) then the diagonal elements g_{kk} of the non-diagonal method (Method 3) need less bandwidth than the diagonal elements of the previous diagonal methods (Methods 1 and 2).

Summarizing the methodology, Table 10.2 gathers the most important expressions obtained for the three problems studied in this chapter: reference tracking and rejection of external disturbances at plant input and at plant output.

Table 10.2. Summary

	Reference tracking	**External disturbances at plant input**	**External disturbances at plant output**
T	$T_{y/r} = [I + \Lambda^{-1} G_d]^{-1} \Lambda^{-1} \{G_d + [G_b - (B + G_b)\, T_{y/r}]\}$	$T_{y/di} = [I + \Lambda^{-1} G_d]^{-1} \Lambda^{-1} \{I - [(B + G_b) T_{y/di}]\}$	$T_{y/do} = [I + \Lambda^{-1} G_d]^{-1} \{I + \Lambda^{-1}[B - (B + G_b)\, T_{y/do}]\}$
C	$C_1 = G_b - (B + G_b) T_{y/r}$	$C_2 = (B + G_b) T_{y/di}$	$C_3 = B - (B + G_b) T_{y/do}$
c_{ij}	$c_{1ij} = g_{ij} - \dfrac{g_{jj}(p^*_{ij} + g_{ij})}{(p^*_{jj} + g_{jj})}$; $i \neq j$	$c_{2ij} = \dfrac{(p^*_{ij} + g_{ij})}{(p^*_{jj} + g_{jj})}$; $i \neq j$	$c_{3ij} = p^*_{ij} - \dfrac{p^*_{jj}(p^*_{ij} + g_{ij})}{(p^*_{jj} + g_{jj})}$; $i \neq j$
g_{ij}^{opt}	$F_{pd}\left(g_{jj} \dfrac{p_{ij}^{*N}}{p_{jj}^{*N}} \right)$, $i \neq j$	$F_{pd}\left(- p_{ij}^{*N} \right)$, $i \neq j$	$F_{pd}\left(g_{jj} \dfrac{p_{ij}^{*N}}{p_{jj}^{*N}} \right)$, $i \neq j$

11

THE DESIGN AND IMPLEMENTATION PROCESS FOR A ROBUST CONTROL SYSTEM

11-1 INTRODUCTION

Chapters 2 through 8 and Chapter 10 have involved the theoretical development of the QFT technique and the associated control system design process with minimum concern with the real-world aspects of the control problem. As pointed out in Chapter 9, in facing the technological problems of the 21st century, it is necessary that engineers of the future must be able to *bridge the gap* between the scientific and engineering methods. Thus, Chapter 9 presents a set of Engineering Rules (E.R.) as a first step towards achieving this goal. This chapter provides the next step in enhancing this goal: overcoming problems encountered during design and implementation of a QFT control system in the real world. Included are testing of the assumptions that are made in order to design with linear models, perform nonlinear simulations, implement hardware, account for un-modeled effects, etc. Most of the real world implementation problems are the result of assumptions made during the design process.

Control design problems generally involve real world nonlinear plants. In utilizing control system design techniques, which require linear plant models, it is necessary that assumptions be made that allow simplification of these nonlinear plants, i.e., "assume linear behavior" that result in obtaining linear plant models. Thus, it is important for the designer to follow a design and implementation process that allows the testing of the assumptions as early in the process as possible so the control system can be redesigned, for example, to take into account un-modeled effects. As described in Chapter 1 and detailed later in this chapter, the control design process should include simulation of the control system on increasingly realistic models which helps transition to implementation on real world applications.

This chapter begins with an in depth discussion of the control system design process. Following is a discussion of issues the control designer faces in implementing control systems in the real world, such as integrator wind-up, hardware/software interface, and bending modes.

11-2 CONTROL SYSTEM DESIGN PROCESS

In order to design a control system for a real world control problem, the designer must follow a design process such as that shown in Fig. 1.1, redrawn here as Fig. 11.1. This figure represents a design process that moves the designer from the problem definition stage to the successful control system implementation in steps of increasing reality. If the control system does not meet performance specifications at any stage of the process, the control system is redesigned and retested. In general, as the simulations become more realistic, they also become more expensive both in cost and time. Therefore, it is very important to be able to find potential problems early in the design process for the control system. The ovals inside the circle in Fig. 11.1 indicate the features of the QFT technique that assist in the design of control systems and can best meet performance specifications and be implemented on the real world system. The following sections describe the individual stages of the control design and implementation process.

11-2.1 FUNCTIONAL REQUIREMENTS

Before the design process can begin, the designer must have a clear understanding of the problem that needs to be solved. That is, the designer must understand what the controlled system is required to do and what are its operational requirements. The designer must also understand the environment in which the system is required to operate, i.e., the environmental requirements. Together these two requirements make up what is referred to as the *functional requirements*. If the designer does not start with a clear understanding of the functional requirements, costly time can be wasted in the design-test-redesign cycle. If during the design process, it becomes clear that the functional requirements cannot be met, the designer might be called upon to use engineering judgement and the knowledge of the goals of the controlled system to modify these requirements. Note, this is not a step that a control designer normally takes on his own.

11-2.2 PERFORMANCE SPECIFICATIONS

Performance specifications are essentially mathematical models developed from the functional requirements and are utilized during the design process in order to achieve the desired system performance robustness. Since performance specifications are normally only interpretations of the functional requirements, the designer

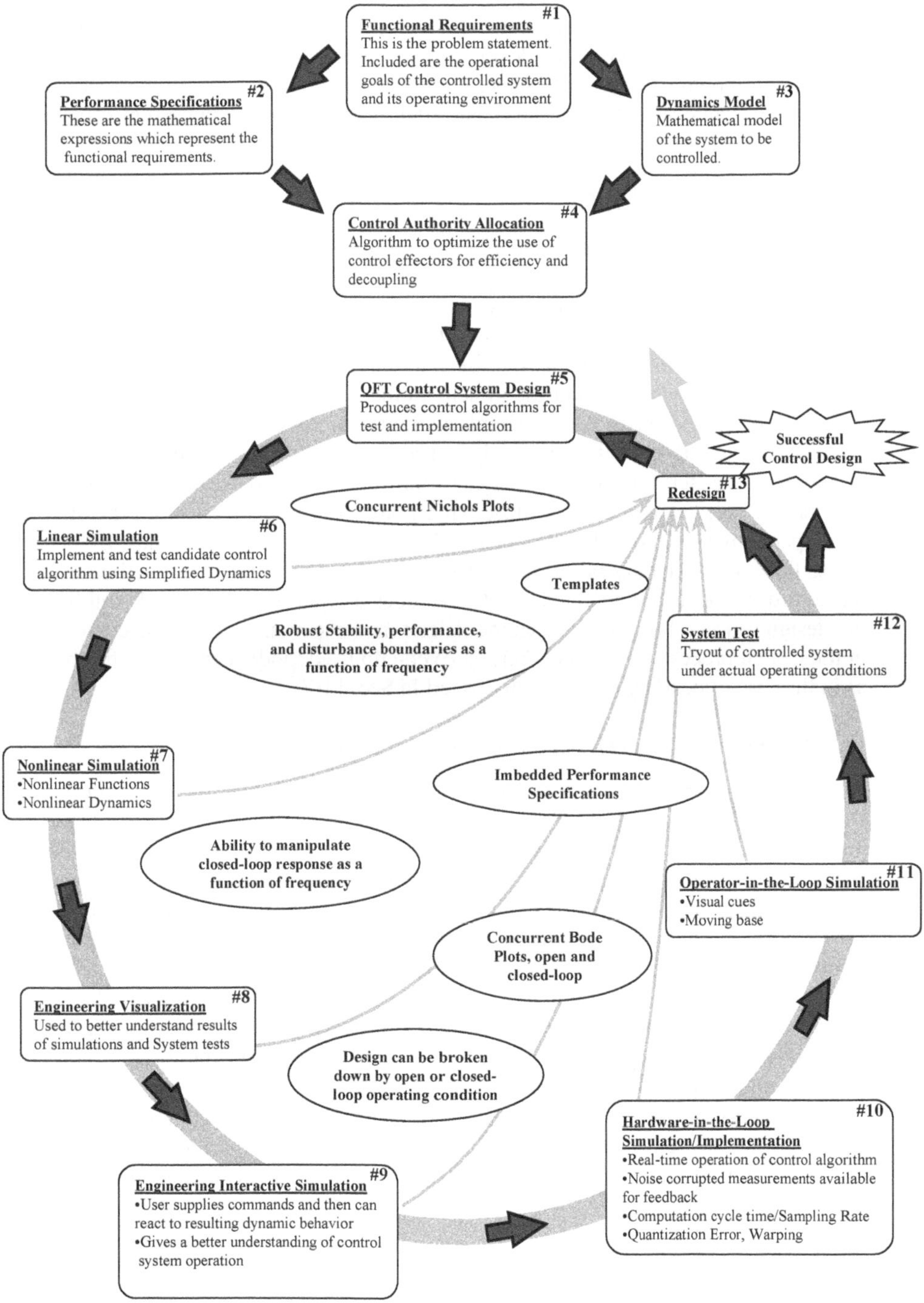

Fig. 11.1 The QFT control system design process: Bridging the Gap.

must be aware of how the specifications and requirements relate and what trade-offs need to be made. During the design process, the designer might need to apply engineering judgment in order to make the necessary modifications to the specifications that, while still meeting the requirements, enables achieving a robust control system design.

11-2.3 DYNAMICS MODEL

The *dynamics model* is a mathematical model of the system to be controlled. The model is developed from a knowledge of the system and its operating requirements. This model can be as simple as a linear-time-invariant (LTI) transfer function or a complicated set of nonlinear differential and algebraic equations with time varying parameters. In many cases, a simplified model of the dynamical system can be used to represent the system in the control design process. In fact, the designer should try to use as simple a model as possible that represents the important system dynamics in the design process. For example, from an analysis of the LTI transfer functions a designer may be able to determine their nondominating poles and zeros, i.e., those which have a negligible effect on the system's performance (those that lie outside the system's bandwidth). Thus, by deleting the nondominating poles and zeros from these LTI transfer functions reduced order models are obtained. Not only does a reduced order model simplify the design process, but also reduces the risk of introducing numerical inaccuracies in the design process. But remember, an oversimplified model can lead to trouble as in the case of bending modes as discussed in Sec. 11-7.

11-2.4 CONTROL AUTHORITY ALLOCATION

An important part of the design process is the *control authority allocation* assigned to each of the control effectors. Depending on the dynamical system, there may be redundant control effectors, i.e. the number of control effectors available to the controller may be greater than the number of controlled variables. Also, the control effectors available may induce cross-coupling in the dynamical system and do not clearly control any one variable. In these cases, judgement must be exercised by the designer, based upon knowledge of the real-world operating characteristics of the plant, in determining the percentage of the control authority that is allocated to the various controlled variables. That is, a method for determining the percentage of control power available from each control effector to each controlled variable must be determined. The optimization of the control effectors' control authority allocation can be used to help decouple the system and assist in achieving the desired robust system performance. This control authority allocation is accomplished by the proper selection of the w_{ij} elements of the weighting matrix $\boldsymbol{W}$.

11-2.5 QFT CONTROL SYSTEM DESIGN

As described in detail in this text, the QFT design process is used to develop mathematical algorithms that can be implemented in order to achieve the desired control system performance. Implementation issues and insights provided by the QFT process to the designer are discussed in the following sections. A QFT design can be accomplished by hand using Nichols and Bode plots, but computer software such as Mathematica, Matlab, and Matrix_x greatly simplify the design process.

11-2.6 LINEAR SIMULATION

Once the control algorithms have been designed, they are implemented along with linear representations of the dynamical system. These systems are simulated and the results are compared to the specifications. Since QFT design involves linearizing nonlinear equations, the control system must be simulated with each LTI transfer function to check the result against the specifications. If some or all of the specifications have not been met, the designer can either redesign the control system or reexamine the requirements. In some cases, the initial specified requirements may not be realistic. For designs that involve control effector damage, the designer must ensure that the assumed percentage of effector damage is realistic with respect to its associated remaining control authority available to satisfy the control system performance requirements; for example, to still be able to fly an aircraft. Also, the designer must ensure that the system performance is close enough to the specifications to meet the overall functional requirements.

11-2.7 NONLINEAR SIMULATION

Once the control system has passed the linear simulation testing phase, the simulation complexity is increased by adding nonlinear components and any other components that are removed to simplify the simulation. As with the linear simulations it may be necessary to accomplish a redesign or a revaluation of the specifications (performance specifications, control authority allocation, and/or the percentage of control effector failure).

11-2.8 ENGINEERING VISUALIZATION

After each of the simulations it is valuable to animate, by a computer simulation, the dynamics data to better understand exactly what occurs during the simulation. Note that the three dimension engineering visualizations integrate all of the dynamics of the simulation. For example, in the case of an aircraft this means that the designer can view the angle of attack, pitch rate, pitch attitude, forward velocity, vertical velocity, and altitude simultaneously. Instead of trying to decipher the position and attitude of the aircraft from six two dimensional plots, the designer can obtain a clearer understanding from watching the computer animation of the maneuver. For more specific details of the maneuver the designer can then return to the data plots.

11-2.9 ENGINEERING INTERACTIVE SIMULATION

When there is an operator involved in the controlled system, for example, a pilot flying an aircraft, it is often useful for the designer to use an interactive simulation in order to obtain a better understanding of the operation of the system. It should be noted that in reality the pilot is a part of the overall flight control system, i.e., he forms the "outer loop" of the control system. Thus, this type of control system is referred as a *manual flight control systems.* An interactive simulation provides the designer with the ability to implement the control system in the same fashion that it will be implemented on the dynamical system. The interactive simulation also gives the designer the ability to test the system continuously throughout the operating environment. In the case of a control system designed for an aircraft, the interactive simulation involving a pilot gives the designer the ability to perform a simulated flight test before the design leaves his/her desk. Such simulations, for a specified aircraft, are often performed by a pilot, for example at the Wright-Patterson AFB Lamars simulator.

11-2.10 HARDWARE-IN-THE-LOOP SIMULATION/IMPLEMENTATION

At this stage of design and implementation the control system algorithms are implemented on the same type of hardware systems as those that control the dynamical system. Other hardware components such as actuators and sensors are also connected to the system. This allows simulation of real-time operation of control algorithm, noise corrupted measurements for feedback, and computation cycle time/sampling rate quantization errors. A hardware-in-the-loop-simulation is also useful to ensure that commands issued from the control system move the effectors in the correct directions and the outputs of the feedback sensors have the correct polarity.

11-2.11 OPERATOR-IN-THE-LOOP SIMULATION

In order to insure the controlled system meets the requirements of the human operator a simulation is set up to allow the operator to interact with a simulation of the system. Many of these simulations surround the operator with visual cues and some inject motion into the simulation. These types of simulations are used to improve the handling qualities of the controlled system by giving the operator a chance to try out the controlled system and then using his or her responses to help shape a redesign.

11-2.12 SYSTEM TEST

The final testing of the control system involves implementation on the dynamical system and operational testing. Once the controlled system has been shown to meet the performance specifications for the operating environment, a successful control design has been achieved.

11-2.13 REDESIGN

At every stage of the control system design and implementation process the designer makes a decision to move to the next stage or to redesign (modify) the control system. Once the control system is modified the simulation testing is repeated.

11-3 DESIGN PROCESS EXAMPLE

The Lambda Unmanned Research Vehicle (URV) shown in Fig. 11.2 is a remotely piloted aircraft with a wingspan of *14 ft* and is operated by the U.S. Air Force for research in flight control technology. The objectives of the project described in this section are as follows:

1. To design robust flight control systems using the QFT design technique
2. To flight test these designs
3. To implement an inner loop FCS on the Lambda URV that would be part of an autonomous flight control system
4. To illustrate some of the real-world problems that are encountered in performing the control system design process are shown in Fig. 11.1.

In accomplishing this design project required four cycles around the control design

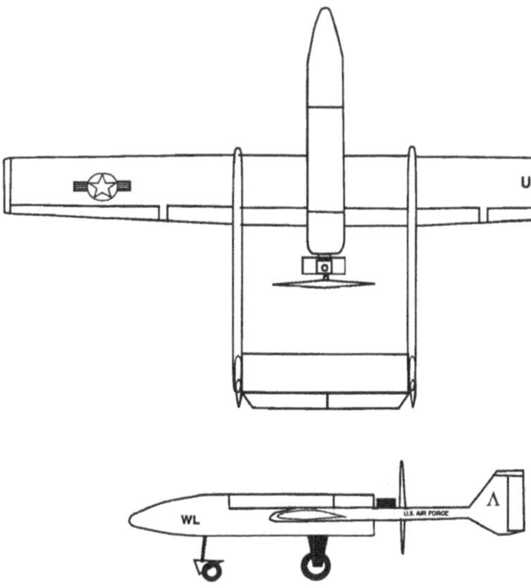

Fig. 11.2 Lambda Unmanned Research Vehicle (URV).

process loop. These four design cycles are:

Cycle 1 – This cycle involved the satisfaction of only the first two of the project objectives.
Cycle 2 – Cycle 1 was repeated but involved the design of an improved integrator wind-up limiter.
Cycle 3 – A redesign of the FCS was accomplished to satisfy requirements 1 through 3.
Cycle 4 – A refinement of the plant model was made in order to take into account a bending mode that was neglected in the previous designs.

Cycles 1 and 3 were unsuccessful and cycles 2 and 4 produced successful flight tests.

The QFT digital flight control system design presented in this chapter is performed as a *pseudo-continuous-time* (*PCT*) design. The resulting *s*-domain compensators and pre-filters are transformed into *z*-domain controllers and pre-filters by use of the Tustin transformation.

11-3.1 FIRST DESIGN CYCLE

Requirements

There were two major design requirements for this project. The first was a desire to develop a robust flight control system using QFT, and take the design through flight test. The second was a need for an inner loop FCS on Lambda that would interface with an autonomous waypoint directed autopilot.

Specifications

The time response specifications were selected base on the open-loop response of Lambda. The pitch rate was an under-damped response that settled fairly quickly. Overshoot and settling time were chosen to be *25%* and *1 sec.*, respectively, for pitch rate response. Roll rate was an over-damped response that settled quickly, and the settling time was chosen to be one second. Yaw rate was also underdamped, but it did not reach steady state as fast as the other two. Yaw rate overshoot and settling time were chosen to be *15%* and *2 secs.*, respectively. These specifications were transformed into LTI transfer functions for use in the QFT design, see Sec 3-4.

Aircraft Model

The aircraft model developmental process began with the use of Digital Datcom, a computer program which predicts stability and control derivatives for aerospace vehicles based on the physical characteristics of the vehicle. Datcom information forms the baseline model of the aircraft. This baseline model was refined by using system identification software to estimate the aerodynamic derivatives from actual flight test data.[78] Maximum likelihood identification was used to identify the natural frequency and damping ratios of the short period and roll modes. This information combined with the Datcom information provided a working model for the flight control system design.

FCS Design

There were two QFT designs accomplished at the Air Force Institute of Technology[65,79] (AFIT). The first was based on the DATCOM model of Lambda alone. The second design was based on the DATCOM model with the refinements made with system identification. This second design used linearized transfer functions to represent Lambda in various flight conditions, covering the entire proposed flight envelope, to accomplish the design and for linear simulations.

Linear Simulations

All FCS designs were simulated using $Matrix_x$ and LTI state space models representing the full flight envelope of Lambda. After successful linear simulations, non-linearities such as control surface travel limits were introduced into the linear simulation.

Nonlinear Simulations

A nonlinear simulation was developed at the Air Force Research Laboratory (formerly the Wright Laboratory) that incorporated a six degree of freedom simulation, automatic trim calculation, air vehicle kinematics, and control surface saturation. While this design produced the desired responses in the linear simulation, when implemented in the nonlinear simulation the original control system exhibited undesirable behavior due to the initial assumptions about allowable gain being incorrect. Thus, the allowable gain was modified to achieve a redesigned controller.

Hardware-in-the-Loop Simulation

Software from the nonlinear simulation were used to develop a hardware-in-the-loop simulation.[80] This simulation allowed the implemented FCS, which is programmed on a EPROM chip, to be tested in the aircraft. When the FCS was implemented in this simulation, it was discovered that the angular rate sensors had high levels of noise, with peak values on the order of *0.5 deg/sec*. The FCS amplified this noise and this effectively masked any control command signal. The noise was recorded and was incorporated into the nonlinear simulation. The MIMO QFT computer-aided-design program, developed by AFIT (see Appendix C), for designing control systems allowed for a rapid redesign. The noise problem was minimized by lowering the loop transmission gain and then testing the resulting FCS in the nonlinear simulation. This remedy was an "engineering decision" in order to obtain a satisfactory design. In the third design cycle a more satisfactory resolution of the noise problem was achieved. Once simulations of the redesign were satisfactory, the FCS was flight tested (Flight Test #1).

Flight Test #1

Two major difficulties caused the first flight test to fail; the first was reversed polarity on an angle sensor and the second was an integrator wind-up limiter scheme that did not work. Since the inner loop FCS was to be implemented as a part of an autonomous system, turn coordination logic was implemented around the inner loop FCS that relied on the roll angle. Post flight analysis of the flight test video and data showed that the polarity of the roll angle sensor was incorrect, thus, when the aircraft was commanded to bank, the rudder was commanded to deflect in the wrong direction. The FCS was thus turned off and the testing involving the lateral control channel was terminated. Later, during the same flight test, when the FCS pitch channel was turned on, the aircraft developed a high pitch rate. This test was also terminated and post analysis revealed that the scheme used to limit integrator wind-up had caused a numerical instability.

11-3.2 SECOND DESIGN CYCLE

Requirements

The requirements for the second design cycle did not change from the original requirements. An additional requirement was incorporated for the second design cycle that involved the design of an improved integrator wind-up limiter.

Specifications and Aircraft Model

The specifications and the aircraft model for the second design cycle did not change from the original requirements.

FCS Design

Since the problems encountered in the first test had nothing to do with the QFT designed FCS, the same QFT FCS designed for the first flight test, was used in the second flight test. During the second flight test, there was no attempt to use a turn coordination algorithm. The insertion of an integrator wind-up limiter involved a different form of the controller implementation for the second design cycle. In this cycle instead of each of the controllers being implemented by a single software algorithm relating their respective outputs to their respective inputs, they were implemented in the manner described by E.R.12 of Chapter 9. That is, the continuous time domain transfer functions were factored into poles and zeros in order to create first order cascaded blocks as illustrated in Fig. 9.6. These transfer functions were individually transformed into the discrete time domain. The individual transfer functions were then implemented, by their own respective software algorithm. This implementation allowed limitations to be placed only on those pieces of the FCS that contained pure integrators and provided the controller accuracy that is indicated by E.R.12.

Linear, Nonlinear, Hardware-in-the-Loop Simulation

All simulations consisted of checking out the new implementation of the FCS. There were no problems encountered during any of these simulations.

Flight Test #2

On 20 Nov 92, the temperature was in the *60°F+* range with winds at *5* to *7 mph*. Lambda was flown in manual mode for take-off, setup, and landing. Due to problems with the first flight test the FCS was engaged only during the test maneuvers. The maneuvers performed consisted of unit step commands in all three axes. This set of maneuvers was first performed with the QFT FCS and then with the open loop aircraft. As shown in Fig. 11.3, the QFT FCS performed as it was designed. The figure shows the responses of Lambda to a step pitch down command. The dotted lines in the plot represent the specified T_{R_U} and T_{R_L}. It is important to note that during this maneuver the aircraft covered a large portion of its dynamics envelope by varying in forward air speed from *75 kts* to *110 kts*.

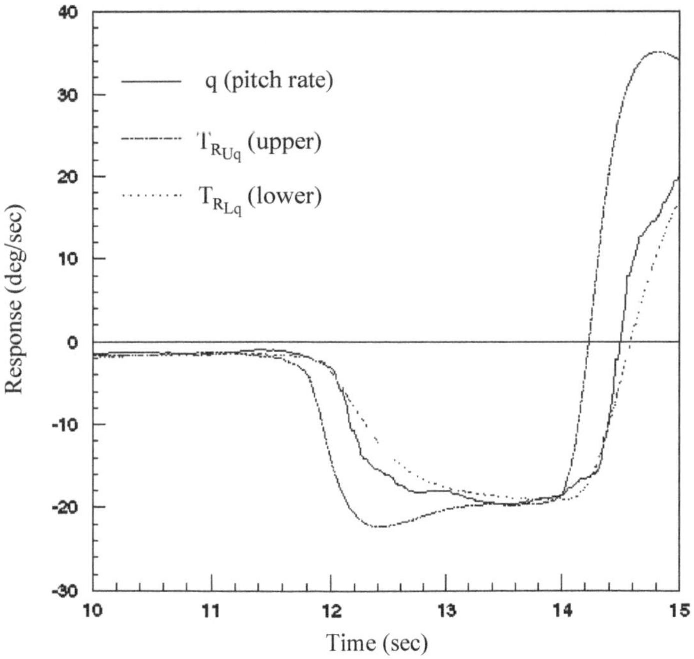

Fig. 11.3 Response to pitch-down command.

11-3.3 THIRD DESIGN CYCLE

Requirements

The requirements for the third design cycle had not changed from the original requirements. This cycle involved the design of an inner loop FCS that had intrinsic turn coordination. Also, the sensor noise problem was reduced by an order of magnitude, to about 0.05°/sec peak to peak, by the addition of a hardware noise filter on the output of the sensors. It was determined that the noise originated from a motor on the sensor was a high frequency being sampled at a lower frequency. Thus, this aliased noise had a relatively high BW. The remedy was to place a filter at the sensor output before the sampler. This allowed a redesign of the FCS to improve the system performance.

Specifications

For this iteration of the design a *sideslip angle command* was incorporated as part of the inner loop controller. Since Lambda has a sideslip sensor, a sideslip command was used to cause the aircraft to intrinsically fly coordinated turns. That is, the goal of turn coordination is to reduce the sideslip angle to zero during a turn by using the proper amount of rudder deflection during the turn. Changing to sideslip command allowed the use of the yaw rate sensor to implement a *yaw damper* to reduced the dutch roll mode oscillations. This yaw damper was implemented by adding a washout filter, designed through the use of a root locus plot. The yaw damper was designed and then incorporated in the aircraft model for a FCS design. During the second flight test the pilot felt that the aircraft's roll rate response was too slow. Therefore, the roll rate response specification was change to match that of the pitch rate. After this change the roll specifications for overshoot and settling time were *25%* and *1 sec*, respectively.

Aircraft Model

The sensor improvement, mentioned above, was included in the nonlinear aircraft model by recording actual noise and inserting it as a block in the model. During the system identification work for the second aircraft model, some of the parameters had been scaled incorrectly. This caused some modeling errors. After the second flight test these errors were corrected through the use of system identification applied to flight test data that resulted in a refinement of the aircraft model.

FCS Design

Matrix_x was used to develop linearized plant models about flight conditions in the flight envelope. An attempt was made to choose flight conditions in such a way as to fully describe the flight envelope with the templates. To do this a template expansion process was developed and is explained in Sec. 11-4.

Linear, Nonlinear, and Hardware-in-the-Loop Simulations

The refined Lambda model was implemented in all three simulations. The FCS was implemented in the cascaded method outlined previously. All simulations produced the desired responses to given stimulus.

Flight Test #3

During the third flight test, when the FCS was engaged, the aircraft exhibited an uncontrolled pitching, or porpoising, behavior. While the post flight test analysis was inconclusive, a longitudinal bending mode at *13.2 rad/sec* seemed to be the likely cause.

11-3.4 FOURTH DESIGN CYCLE

Requirements

The requirements for the fourth flight test had not changed from the original requirements, but involved a refinement in the aircraft model to incorporate effects of the bending mode discovered in Flight Test #3.

Specifications

The specifications for the fourth design cycle were the same as those for the third cycle.

Aircraft Model

A model of the porpoising behavior encountered in the third flight test was identified by assuming that the behavior was caused by an un-modeled effect. Various models were incorporated into the nonlinear model and simulated. This simulation used the identical flight test inputs as simulation inputs and compared the simulated outputs to the flight test data. Using this procedure, see Sec. 11-7, a violation of the gain margin was ruled out by increasing the inner loop gain in the model and observing the response. Instability caused by actuator rate limiting was ruled out by inserting severe rate limited actuator models in the nonlinear simulation. When a bending mode, modeled as a lightly damped pair of poles, was inserted in the model, the simulated responses were very similar to the flight test results.

FCS Design

$Matrix_x$ was used to develop linearized plant models about the given flight conditions and the FCS was redesigned based on the model containing the bending mode. Note, when the FCS from design cycle three, using the aircraft model with the bending mode, there were violations of stability criteria in the frequency domain and, as expected, the porpoising behavior occurred.

Linear, Nonlinear, and Hardware-in-the-Loop Simulation

A fourth design cycle was accomplished using the new model. This design was implemented and all three simulations were run and tested. This FCS design simulation responded within specifications and, as expected, the porpoising effect was eliminated.

Flight Test #4

The fourth flight test occurred in September 1993. The field conditions were a little gusty, but within acceptable limits for the flight test. During the flight the FCS was engaged and then left engaged for the entire series of tests. The FCS performed as designed. The intrinsic turn coordination scheme worked as designed. The pilot was pleased with the handling qualities and felt comfortable flying with the FCS engaged at all times. His one criticism was that the roll rate was too slow. Since the roll rate was limited by the maximum roll rate detectable by the roll rate gyro, the problem was unavoidable. When the data was examined, it was found that all of the *60 Hz* data had been lost, but much of the *10 Hz* data had been captured. Analysis of this data showed that the FCS did cause Lambda to respond within the specified envelope, during onset of the command, but, in some cases, Lambda's response exhibited more overshoot and longer settling time than specified. These problems could be attributable to the gusty conditions, since no gust disturbance was specified during the design process. More flight testing of this FCS will be required to answer this question.

11-4 SELECTION OF DESIGN ENVELOPE

At the onset of a QFT design, the designer must select a set of operating conditions in order to obtain the LTI transfer functions that represent the dynamical system that are required for the design. These LTI plants determine the template contours. The problem is which operating conditions to choose. Only those operating conditions that yield points that lie on the contour of the templates, for all frequencies of interest, are necessary. Choosing too many LTI plants may yield points that lie inside the template contours and can lead to computational problems during the design. Note by applying engineering insights it is readily determined that the template contours and not the LTI plants which lie within the template's contour determine the performance bounds that need to be satisfied by the synthesized functions. Thus, the computational workload and associated problems may be minimized by reducing the number of plants to be utilized in the design process to only those plants that lie on the template contours.

Through engineering knowledge of the problem the designer is able to determine the particular parameters that effect the operating conditions and the physical limits of these parameters. In the case of Lambda the parameters that were varied to set the operating conditions were airspeed, altitude, weight, and center of gravity. Gross limits were set for these values from knowledge of the aircraft and the possible flight envelope. Next, the template expansion process was used to find the set of operating conditions that fully described the flight envelope.

The template expansion process, shown in Fig. 11.4, is a graphical process that tracks the effect of variations of the parameters which are involved in selecting the operating conditions and determine the resulting LTI plants. The process is as follows:

1. Determine the important parameters that describe the operating condition and their minimum, maximum, and nominal values.
2. Choose a template frequency for the expansion process. This frequency should be representative of the dynamic system in the bandwidth of interest. At the end of the process, other template frequencies should be checked to insure that a complete set of operating conditions have been chosen.
3. For the template frequency of step 2, plot the dB vs. phase values of the nominal operating condition.
4. On this same graph, plot the results of varying each parameter through its maximum and minimum while holding the rest of the parameters at their nominal values. This forms an initial template.
5. Identify the variations caused by each parameter. This can be accomplished by connecting the points on the template due to each parameter variation.

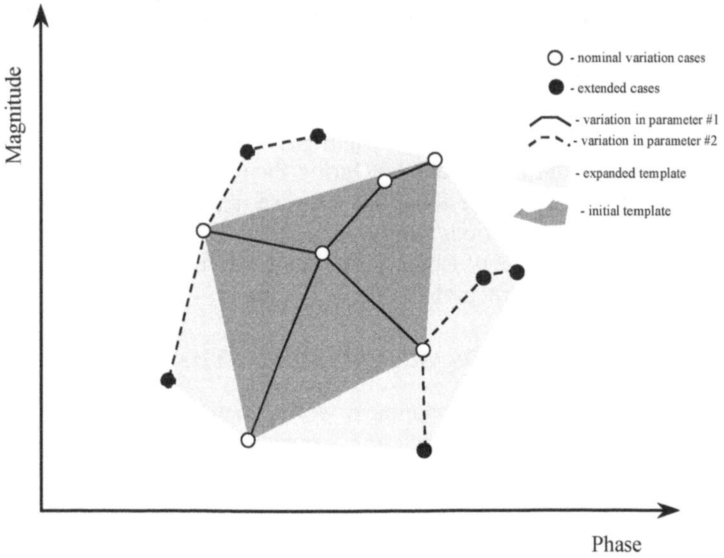

Fig. 11.4 Template expansion process.

6. Choose the two parameters that cause the largest variations and use these to expand the template. This is accomplished by holding the remaining parameters at their nominal values and plotting the four points of the templates resulting from the extremes of the two parameters.
7. Use the outside points, on this expanded template, as nominal points for further expansion with other parameters.
8. Choose other frequencies in the bandwidth of interest to ensure that the operating envelope is completely defined.

For Lambda, a nominal flight condition was chosen to be *50 kts* velocity, *1,000 ft* altitude, a weight of *205 lbs*, and center of gravity at *29.9%* of the mean aerodynamic chord. From this nominal trim flight condition, each parameter was varied, in steps, through maximum and minimum values, while holding the other parameters at their nominal trim values. These variations produced an initial set of templates. On these templates, the variation corresponding to each parameter was identified. Each variation when translated, on the template, identified an expanded template area of the flight envelope that required more plants for better definition.

11-5 CONTROL SYSTEM IMPLEMENTATION ISSUES

An implementation problem that can cause stability and performance problems is integrator wind-up. This is the situation that occurs when the controlled system can not respond quickly enough to the commands from the controller and the commanded values keep increasing due to integrator action. A situation like this occurs when a control effector has reach its limits. The longer the system is in this state the more the commanded value increases. The problem occurs when the controller tries to reverse the command, the commanded value must be "integrated" back down to the operational range before it becomes effective. In order to prevent integrator wind-up, anti-windup algorithms must be applied to integrators during implementation. During the QFT design process the controller is in the form of transfer functions that can be of any order. For implementation, these transfer functions can be separated into first and second order transfer functions (see E.R.12 of Chapter 9). With the transfer functions separated in this manner individual integrators can be limited.

11-6 HARDWARE/SOFTWARE CONSIDERATION

During the modeling and development of the control system, assumptions were made as to the polarity of feedback and command signals. During implementation these assumptions must be tested. This is one of the reasons to use a hardware-in-the-loop simulation. With this type of simulation the control algorithms can be implemented and the control effectors can be monitored during simulated operation. Feedback signals can be checked by moving sensors by hand, if possible.

The other phenomena that a hardware-in-the-loop simulation can identify is the effects of feedback noise on the controlled system. If the feedback noise is within the bandwidth of the control system, and the noise has not been included in the modeling or simulation, the controller may need to be redesigned to account for the noise. This might result in a trade off between performance and noise rejection. Sometimes it is possible to implement a hardware filter after the sensor, but before the sampler to reduce the noise in the bandwidth of interest.

11-7 BENDING MODES

During the design of a control system, the effects of higher frequency modes on stability and performance must be considered. In aircraft, one source of higher frequency modes is structural bending. A control system that excites a bending mode in a flying aircraft can produce disastrous consequences. During the modeling process it is very important to include the effects of these higher frequency modes so they can be minimized during the design process. In the case of Lambda, the existence of a bending mode was discovered during a flight test.

11-7.1 LAMBDA BENDING EXAMPLE

Following the initial flights, the aircraft operators decided that they would prefer a different feedback structure in the FCS that included turn compensation. Thus, to implement turn compensation, a sideslip angle command was incorporated as part of the inner loop controller. The goal of turn coordination is to keep the sideslip angle at zero during turns by using the proper amount of rudder deflection. Since Lambda has a sideslip sensor, sideslip feedback was used to cause the aircraft to fly coordinated turns.

Changing to sideslip command also allowed the use of the yaw rate sensor to implement a yaw damper to reduce the dutch roll mode oscillations. This yaw damper was implemented by adding a washout filter, designed through the use of a root locus plot. The yaw damper was designed and then incorporated in the aircraft model for a FCS design.

When this design was finally flight tested, a porpoising behavior was observed. To ensure flight safety, Lambda was flown to a safe altitude by the pilot before the QFT FCS was engaged. The pilot had Lambda flying in level flight when the longitudinal portion of the QFT FCS was engaged. At this point Lambda began oscillations in the pitch axis and the QFT FCS was disengaged immediately. In order to collect sensor data on this behavior, Lambda was flown back to level flight, the longitudinal portion of the QFT FCS was engaged and the sensor data was recorded for further analysis.

Pitch attitude data from this flight is shown in Fig. 11.5 whose high resolution data was at a 60 Hz sample rate.

11-7.2 UN-MODELED BEHAVIOR

A model of the porpoising behavior was identified by assuming that the behavior was caused by an un-modeled effect. Various proposed models were incorporated

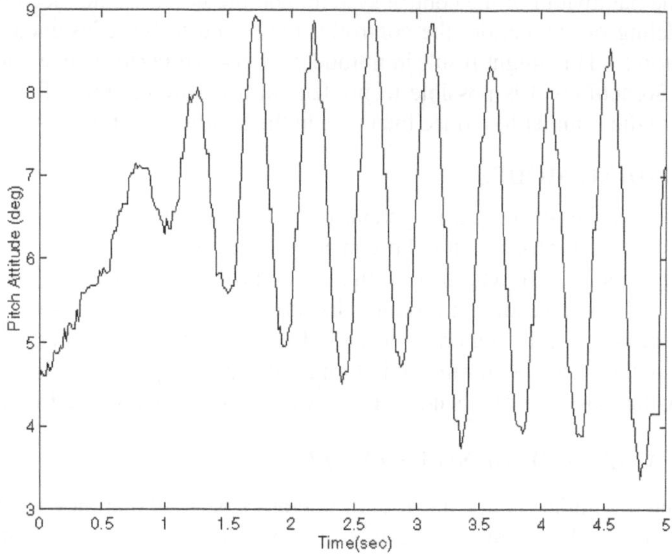

Fig. 11.5 Pitch resonance during flight.

into a nonlinear model of Lambda and simulated. This simulation used the actual flight test inputs as simulation inputs and compared the simulated outputs to the flight test data. Using this procedure, a violation of the gain margin was ruled out by increasing the inner loop gain in the model and observing the response. Instability caused by actuator rate limiting was ruled out by inserting severe rate limited actuator models in the nonlinear simulation.

Upon reviewing the video record of the flight, it was suggested that the aircraft appeared to have a second-order bending mode in the longitudinal axis. It was possible to excite and observe such a mode by tapping rhythmically on the tail of the aircraft.

A bending mode modeled as a lightly damped pair of poles at ω_n = *13.2 rad/sec*, just within the bandwidth of the FCS, was inserted in the nonlinear simulation as shown in Fig. 11.6. This model generated a pitch acceleration signal from elevator deflection which was passed through the second order filter:

$$\dot{q} = \frac{-20}{s^2 + 5.28s + 174.2}\delta_{elev}$$

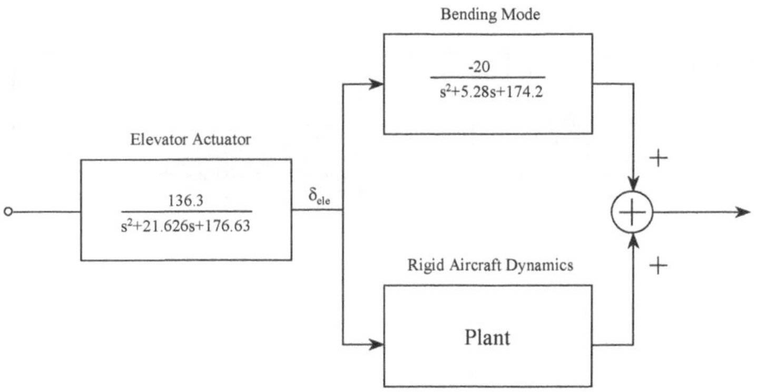

Fig. 11.6 Lambda bending model structure.

The simulated response was very similar to the flight test results. Matrix_X was used subsequently to develop new linearized plant models containing the bending mode about the given flight conditions. The Bode plots of these models are shown in Fig. 11.7.

The new plant models were entered into the MIMO QFT CAD software. The FCS was redesigned based on the new models using the FCS from the previous design cycle as a baseline. The previous controller was:

$$g_{11}(s) = \frac{1093(s+8.5)(s+11)(s+3.9 \pm j2)}{s(s+2)(s+80)(s+36 \pm j48)}$$

The MIMO QFT CAD software showed that, with the old controllers, there were violations of stability criteria on the Nichols chart.

The standard method of design would be to add a notch filter to keep the mode from becoming excited. The bending mode is close enough in frequency to the performance bandwidth of Lambda that care needs to be taken to design a controller that will be able to take advantage of the available bandwidth to deliver performance, stability, and disturbance rejection without exciting the bending mode.

A standard notch filter would not take advantage of any beneficial dynamics at frequencies near the bending mode. It would also increase the order of the compensator. As an alternative, the inner loop filter was revised to compensate for the new information. It was also possible to design a fourth-order compensator to replace the earlier fifth-order design, lowering the complexity of the controller instead of increasing it.

The new controller was determined to be:

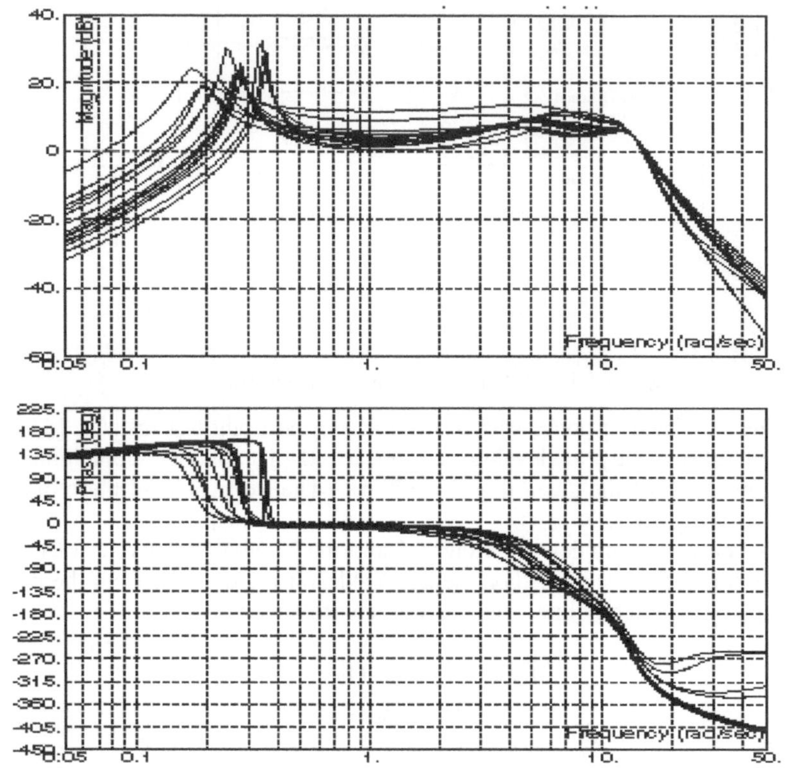

Fig. 11.7 Lambda bending models.

$$g_{11}(s) = \frac{125(s+1)(s+2.5 \pm j9.4)}{s(s+10)(s+35 \pm j35.7)}$$

A characteristic of a bilinear transformation is that, in general, it transforms an unequal-order transfer function ($n_s \neq w_s$) in the s-domain into one for which the order of the numerator is equal to the order of its denominator ($n_z = w_z$) in the z-domain. This characteristic must be kept in mind when synthesizing $g_i(s)$ and $f_{ii}(s)$. Therefore, a non-dominating s-domain zero is inserted into f_{11} at –150.

With the MIMO QFT CAD program it was possible to shape the loop so that at *5 rad/sec* the loop intersected a point on the Nichols Chart where the stabil-

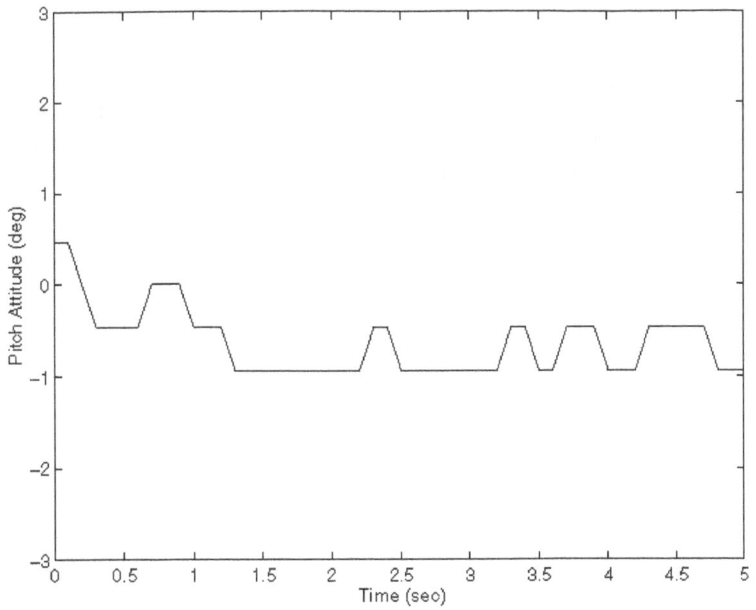

Fig. 11.8. Pitch attitude during flight.

ity boundary and the performance boundary met. This was an optimal point for the loop to pass through given Lambda's performance bandwidth.

The new aircraft model was implemented in the nonlinear simulations and tested with both filters. As expected, the resonance occurred with the FCS that was designed in Design Cycle #3. The FCS resulting from Design Cycle #4 responded within specifications. The new FCS passed a hardware-in-the-loop simulation and was scheduled for a flight test.

During the next flight test, the field conditions were gusty, but within acceptable limits for the experiment. The QFT FCS was engaged and there was no noticeable oscillation. The pilot was very pleased with the handling qualities and felt comfortable flying with the FCS engaged for the entire series of tests. The only problems encountered were some roll performance problems which could be attributed to the windy conditions.

Pitch response during this flight is shown in Fig. 11.8. Unfortunately, the test data recording function failed during the flight so that the only data available is low resolution data ($\pm 0.5°$) recorded at *10 Hz*.

11-8 SUMMARY

Control design and implementation in the real world is an iterative process. Initial steps are performed with linear models that have been formulated with simplifying assumptions. After successful testing of the designed control system, based upon these simplified models, it is tested on increasingly realistic (nonlinear) models. At any point in the design process, if the control system does not meet performance and stability specifications, the control system must be redesigned and retested on the simplified models. This redesign is followed, once again, by testing on the nonlinear model (see Fig. 11.1). At every point of the design process the designer must be aware of test assumptions so engineering judgement can be used to help guide the design to a successful implementation and operation. The bottom line is that the controlled system must meet the requirements set out at the beginning of the process.

12

TIME DELAY SYSTEMS WITH UNCERTAINTY: QFT DESIGN INVOLVING SMITH PREDICTORS[124,125]

12-1 INTRODUCTION

It is well-known that systems with a dominant (long) time delay with respect to plant dynamics have an inherent bandwidth limitation that greatly increases the difficulty of achieving a satisfactory performance of feedback controllers. For example chemical and biological processes, hydraulic and pneumatic systems, spacecrafts in deep space, etc. usually involve transport and communication time delays. In these cases a particular effort has to be applied in the compensator design to achieve the desired performance specifications, within the physical limits impose by the time delay.

For open-loop stable systems, the tracking problem can be improved substantially by introducing dead-time compensation. One of the most and effective long time delay compensator in use today is the Smith Predictor (*SP*), introduced by O.J.M. Smith in 1957.[141] It employs, in an inner loop, a model of the plant characterized by a rational linear transfer function $P_q(s)$ and a time delay θ_q to cancel the real plant output $Y(s)$. Then it uses the un-delayed output $Y^*(s)$ as the feedback signal for control calculation. Figure 12.1 shows the control scheme of the Smith Predictor. Its transfer function can be written as,

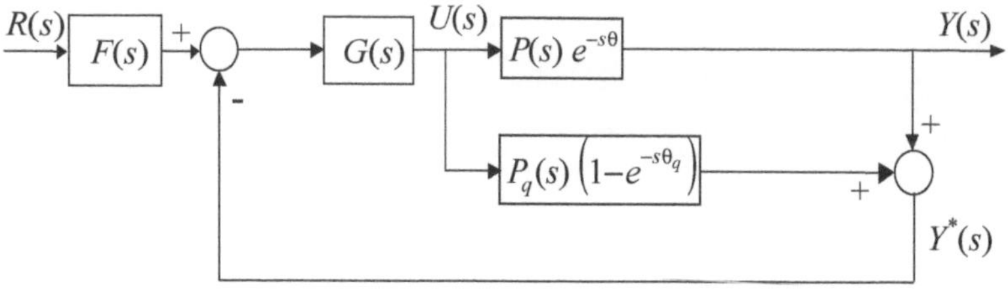

Fig. 12.1 Smith Predictor compensator structure.

$$\frac{Y(s)}{R(s)}=\frac{G(s)P(s)\,e^{-s\theta}}{1+G(s)P_q(s)+G(s)\left[P(s)e^{-s\theta}-P_q(s)e^{-s\theta_q}\right]}F(s) \tag{12.1}$$

where $P(s)\ e^{-s\theta}$ is the plant to be controlled, $P_q(s)\ e^{-s\theta_q}$ the plant model used by the *SP* and $G(s)$ the compensator.

If the model matches the real plant, i.e., the rational linear transfer function is $P_q(s) = P(s)$ and the time delay is $\theta_q = \theta$, then Eq. (12.1) can be simplified to,

$$\frac{Y(s)}{R(s)}=\frac{G(s)P(s)e^{-s\theta}}{1+G(s)P_q(s)}F(s) \tag{12.2}$$

This simplification removes the time delay from the control loop and converts the corresponding control design in a delay free problem. In addition the simplicity of the method allows the *SP* compensator to be implemented by using low cost digital micro-controllers. These reasons make the *SP* one of the most popular methods for compensating systems with time-delay. Indeed, the *SP* is available as a standard algorithm in many commercial devices.

However, although the method has the capability of transforming a time-delay control design to a delay free problem, it still presents two important difficulties to be solved. First of all the original *SP* structure cannot reject load disturbance for processes with integration. In this context different modifications have been proposed to deal with that problem. See for instance: Astrom *et al.* 1994, Matausek and Micic 1996, Normey-Rico and Camacho 1999, etc.[142,143,144]

On the second hand it is well known that the *SP* technique is very sensitive to model-plant mismatch (plant uncertainty), either in the time delay or in the rational part of the model. Under those circumstances considerable stability margins may not guarantee stability even with small modeling errors.

See for example: Ioannides *et al.* 1979, Palmor 1980, Horowitz 1983, Yamanaka and Shimemura 1987.[145,146,147,148]

Accordingly with these difficulties in mind, this chapter presents a method introduced by Garcia-Sanz and Guillen[124,125] to tune any *SP* compensator structure (the original one or the modified ones) when the system presents model uncertainty in both, the rational part and the time delay. The method is composed of two steps. The first one is based on bandwidth frequency considerations and the second one introduces an algorithm to improve the design by using the QFT technique.

12-2 METHODOLOGY OF DESIGN

The *SP* diagram shown in Fig. 12.1 is rearranged to an equivalent structure, as shown in Fig. 12.2, where the expressions of the blocks $P_{eq}(s)$ and $Q(s)$ are shown in the Eqs. (12.3) through (12.5), and where $P(s)$ and θ present uncertainty.

$$H(s)=\left(1-e^{-s\theta_q}\right)\left(\frac{P_q(s)}{P(s)}\right)+e^{-s\theta} \tag{12.3}$$

$$Q(s)=\frac{e^{-s\theta}}{H(s)} \tag{12.4}$$

$$P_{eq}(s)=P(s)H(s)=\left(1-e^{-s\theta_q}\right)P_q(s)+P(s)e^{-s\theta_q} \tag{12.5}$$

An analysis of the equivalent *SP* structure shows that, if there is no uncertainty in the model, that is to say $H(s)$ = 1, the time delay is eliminated from $Y^*(s)$. However, if there is some amount of model-plant mismatch, then the expression $H(s)$ is different from one. As a consequence, the control system is affected by the uncertainty through $H(s)$, and particularly in the blocks $P_{eq}(s)$ and $Q(s)$.

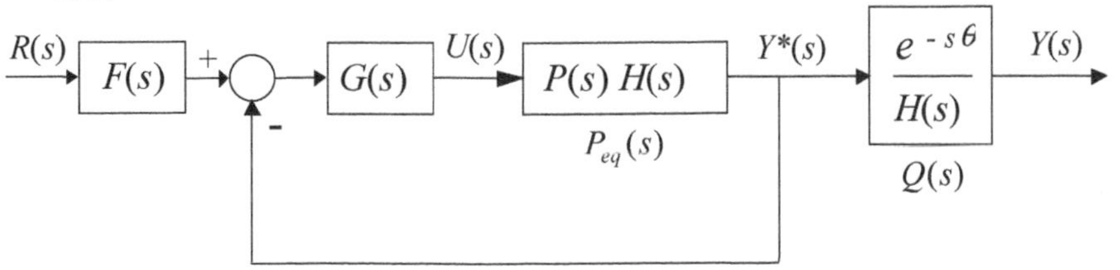

Fig. 12.2 Equivalent diagram of the Smith Predictor.

In this case, the selection of the model $P_q(s)\, e^{-s\theta_q}$ has a great impact on the set of values that $H(s)$ adopts over the space of uncertainty. For this reason, based upon an analysis a method is developed in selecting the model for the *SP* taking into account on how that choice affects the control system performance.

The methodology proposed in this chapter breaks down the study into two complementary steps. The first one analyses the influence of the selected *SP* model through the block $Q(s)$ in the final performance. Afterwards, using the QFT technique, the second step studies how the selection of the *SP* model affects the compensator design for the system $P_{eq}(s)$. [124,125]

First Step

The $H(s)$ term, which appears in the $Q(s)$ block as a post-filter of the control loop (Fig. 12.2), may be responsible for deterioration of the system performance if resonant peaks appear at frequencies below the system bandwidth. In that case, $Y^*(s)$ is distorted when passing through the block. Some years ago, a brief paper written by Santacesaria and Scattolini[149] introduced this problem and proposed a graphic solution for the simple case of model-plant mismatch in the time delay,

$$\theta_q \neq \theta \ \text{ and } \ P_q(s) = P(s) \tag{12.6}$$

This chapter extends these ideas to the complete case of model-plant mismatch, in both time delay and rational part,

$$\theta_q \neq \theta \ \text{ and } \ P_q(s) \neq P(s) \tag{12.7}$$

The study is based on the analysis of the magnitude of $\boldsymbol{Q}(j\omega)$ over the frequency range of interest and for the whole space of parameter uncertainty. Any increase in the model-plant mismatch moves the resonant peaks to lower frequencies, thus limiting achievable closed-loop bandwidth. For a specified bandwidth specification (BW), this can be stated as:

<u>*Criterion* 1</u>. A *SP* model must be selected such that the resulting $\boldsymbol{Q}(j\omega)$ does not distort $Y^*(s)$ for frequencies up to BW and for every possible plant $P_i \in \mathcal{P}$, where i = 1, 2, ...n.

That is to say, the *SP* model must satisfy,

$$\left|\, 20\log_{10}\left|\boldsymbol{Q}(j\omega)\right| \,\right| \leq 3\,\text{dB} \quad , \quad 0 \leq \omega \leq \text{BW} \quad , \quad \forall \left[\boldsymbol{P}_q(j\omega) e^{-j\omega\theta_q} \right] \in \mathcal{P} \tag{12.8}$$

From Eq. (12.4) and taking into account that $\log_{10}|1/\mathbf{H}| = -\log_{10}|\mathbf{H}|$, Eq. (12.8) is equivalent to,

$$\left|20\log_{10}|\boldsymbol{H}(j\omega)|\right| \leq 3\,\mathrm{dB} \quad , \quad 0 \leq \omega \leq \mathrm{BW} \quad , \quad \forall \left[\boldsymbol{P}_q(j\omega)\, e^{-j\omega\theta_q}\right] \in \mathcal{P} \qquad \textbf{(12.9)}$$

The algorithm that implements the *First Step* is outlined in Table 12.1. It is a general procedure that can be used for any plant model with any sort of rational part and any time delay, for any kind of model-plant mismatch and for any *SP* structure. The procedure finds the subset of plant models that could be used by the *SP* without the *Q(s)* block causing too much distortion of the output.

Table 12.1. Outline for the *First Step* algorithm

Item	
1	Define a grid over the uncertainty of $P_i(s)\, e^{-s\theta i} \in \mathcal{P}$. This allows the algorithm to treat with a finite set of possible plants.
2	Fix the desired closed-loop bandwidth specification BW.
3	Select a plant $[P_q(s)\, e^{-s\theta_q}] \in \mathcal{P}$ as the *SP* model.
4	Compute, for the *SP* model, the magnitude of $\boldsymbol{Q}(j\omega)$ for every plant $\in \mathcal{P}$.
5	If any magnitude of $\boldsymbol{Q}(j\omega)$ exhibits some amplification greater than 3 dB at frequencies lower than the desired bandwidth, then that *SP* model can deteriorate the system performance. Hence it is rejected. Otherwise, the selected *SP* model passes the step.
6	Repeat from Item 3 to Item 5 for every possible *SP* model in $\mathcal{P}$.
7	Finally, the set of admissible *SP* models for the plant is obtained.

Second Step

If there is a collection of possible *SP* models satisfying the *First Step*, then an additional degree of freedom is still available in the selection of a model for the *SP*. This additional degree of freedom utilizes the amount of change suffered by the equivalent-plant templates as a second criterion to guide a new *SP* model selection, by using the QFT methodology.

Two equivalent-plant QFT templates calculated for the same frequency and corresponding to two different *SP* models $[P_q(s)\, e^{-s\theta_q}]$ could differ in shape, producing different QFT bounds and hence different compensators. This fact introduces the question about which is the best model of the plant that has to be chosen, from among those satisfying the *First Step,* to obtain the least demanding templates (in height and width) and ease the loop-shaping of *G*(s).

According to QFT, the smaller the area of the templates, the easier the compensator design becomes, and the better performance can be achieved.

Let $\Im \boldsymbol{T}(j\omega)$ represent the actual-plant templates and $\Im \boldsymbol{T}_{eq}(j\omega)$ represent the equivalent-plant templates when plant $[P_q(s)\ e^{-s\theta_q}]$ has been selected as the *SP* model. Also, let $A(\cdot)$ represent the area of a template on the Nichols Chart; and let Ω represent the (discrete) set of frequencies of interest, with n_ω frequencies. The following cost function is proposed as a measure of suitability of a specific *SP* model:

$$I = \frac{1}{n_\omega} \sum_{\omega \in \Omega} \frac{W(\omega)\ A\left[\boldsymbol{T}_{eq}(j\omega)\right]}{A[\boldsymbol{T}(j\omega)]} \tag{12.10}$$

where $W(\omega)$ represents weights that can be used to emphasize critical frequencies (cross over frequency, resonant modes, etc.). This cost function is a weighted sum of normalized areas. A *SP* plant model that leaves invariant every actual-plant template area is assigned a cost $I = 1$ if unity weights are used.

Criterion 2. From the possible *SP* plants obtained by Criterion 1, select the *SP* plant that presents the minimum cost function I [see Eq. (12.10)].

The algorithm that implements the *Second Step* out is outlined in Table 12.2. Again, it is a general procedure that can be used (a) for any plant model with any sort of rational part and any time delay, (b) for any kind of model-plant mismatch and (c) for any *SP* structure. The model of the plant selected for the *SP* structure with the proposed methodology avoids distortion within the operating bandwidth (*First Step*) and presents the least restrictive templates to the compensator design stage (*Second Step*). Finally, though the development is made for continuous systems, the results obtained remain valid for digital control systems incorporating digital *SP*s.

Table 12.2. Outline for the *Second Step* algorithm

Item	
1	Select a plant $P_q(s)\ e^{-s\theta_q}$ as the *SP* model from those that have successfully passed the *First Step* criterion.
2	Compute the equivalent-plant templates of $P_{eq}(s)$ [see Eq. (12.5)] over the frequency range of interest.
3	Calculate the area of the templates, and then the cost function I. [see Eq. (12.10)]
4	Repeat from Items 1 to 3 for every *SP* model that successfully passes the *First Step* criterion.
5	Select the model that results in the minimum cost function I.

12-3 A SYNTHESIS EXAMPLE

For a simple example to clarify the above ideas, consider the plant,

$$P(s)e^{-s\theta} = \frac{K}{\tau s+1}e^{-s\theta} \tag{12.11}$$

This plant captures the essential dynamics of many chemical, biological and industrial processes. Obviously the mathematical *SP* model that describes the behavior of that plant is the first order model with time delay given by,

$$P_q(s)e^{-s\theta_q} = \frac{\hat{K}}{\hat{\tau} s+1}e^{-s\hat{\theta}} \tag{12.12}$$

For this simple example, the parameters of the plant adopt values that belong to the space of uncertainty given by,

$$K \in [1,2],\ \tau \in [1,2],\text{and } \theta \in [1,2] \tag{12.13}$$

12-3.1 APPLYING FIRST STEP

As is mentioned above, the *First Step* of the methodology is based on the analysis of the magnitude of the frequency response of ***Q(jω)*** over the frequency range of interest and for the complete set of parameter uncertainty. For the proposed plant and *SP* model, the magnitude of the frequency response of ***Q(jω)*** is written as,

$$|\boldsymbol{Q}(j\omega)| = \frac{1}{\sqrt{(1+M\,x)^2+(M\,y)^2}} \tag{12.14}$$

where,

$$M = \left(\frac{\hat{K}}{K}\right)\left(\frac{1}{1+\omega^2\hat{\tau}^2}\right) \tag{12.15}$$

$$x = (1+\omega^2\tau\hat{\tau})\left[\cos\omega\theta - \cos\omega(\hat{\theta}-\theta)\right] - \omega(\tau-\hat{\tau})\left[\sin\omega(\hat{\theta}-\theta)+\sin\omega\theta\right] \tag{12.16}$$

$$y = (1+\omega^2\tau\hat{\tau})\left[\sin\omega(\hat{\theta}-\theta)+\sin\omega\theta\right] + \omega(\tau-\hat{\tau})\left[\cos\omega\theta - \cos\omega(\hat{\theta}-\theta)\right] \tag{12.17}$$

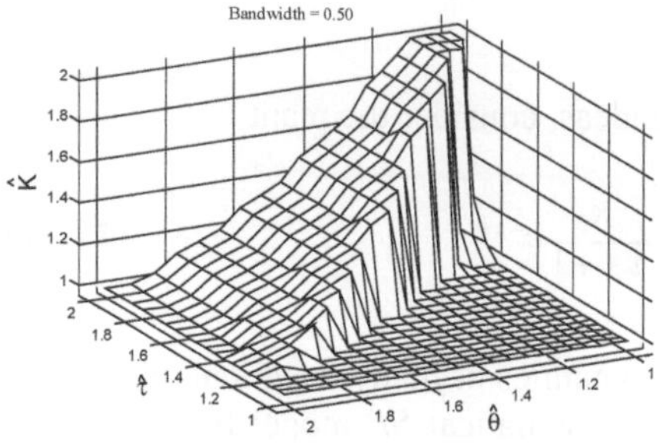

Fig. 12.3 Parameters. First order model with delay. (BW = 0.50 rad/sec).

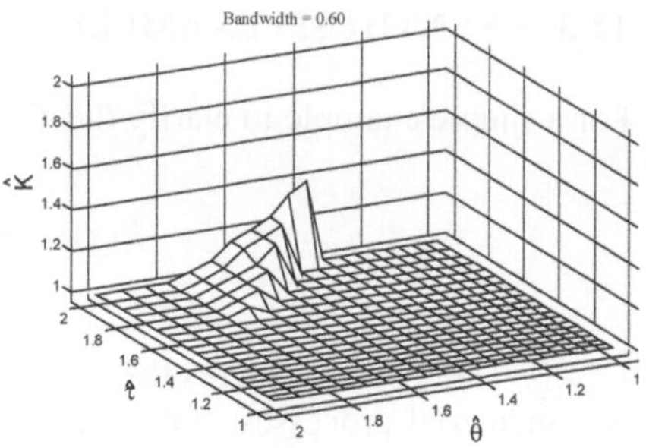

Fig. 12.4 Parameters. First order model with delay. (BW = 0.60 rad/sec).

By fixing the desired closed-loop bandwidth to BW = 0.5 rad/s (Item 2 of Table 12.1), the set of admissible *SP* models which fulfill the bandwidth specification are easily found by following Items 3 through 7 of the *First Step* procedure. The results of this example are plotted as a 3D object, where each of the three axes represent the uncertainty of every parameter (see Fig. 12.3). The models the *SP* could adopt are those located inside the 3D figure. To obtain Fig. 12.3, a grid of 20 values for each parameter is used. Similarly Fig. 12.4 shows the 3D object obtained for the bandwidth specifications of 0.60 rad/s. As might have been expected, the larger the bandwidth specification the smaller the set of possible models. As illustrated by Figs. 12.3 and 12.4, the usual selection of a 'mean' *SP* model (with the mean parameters) does not seem to be appropriate when using Criterion 1. In the limit, if the bandwidth specification is larger enough, it is presumed that there is only one possible model.

12-3.2 APPLYING SECOND STEP

Given the bandwidth specification BW = 0.5 rad/s (*Step one*, Fig. 12.3), there is an additional degree of freedom still available in selecting the *SP* model within the 3D object. As an example, select three models inside this 3D figure, so that,

$$\begin{aligned}&\text{Case A}: \hat{K}=1\,;\ \hat{\tau}=1.1157\,;\ \hat{\theta}=2\\&\text{Case B}: \hat{K}=1\,;\ \hat{\tau}=2\,;\ \hat{\theta}=1.9284\\&\text{Case C}: \hat{K}=1\,;\ \hat{\tau}=2\,;\ \hat{\theta}=1\end{aligned} \tag{12.18}$$

Figure 12.5 shows the contour of the template of the original first order plant with delay (Eq. 12.11) for the whole space of parameter uncertainty (Eq. 12.13), for a particular frequency of $\omega = 1$ rad/sec and without the *SP* structure.

To study how the insertion of the *SP* block affects the system, three individual models, defined by three different set of parameters (see Eq. 12.18) are utilized to study the effect of utilizing a *SP*. The location of these three models, within and on the boundary of the template, at the frequency value of $\omega = 1$ rad/sec are shown in Fig. 12.5 and are labeled as A, B, and C.

Figure 12.6 presents the shape variation of the templates of $P_{eq}(s)$ (Eq. 12.5) for the frequency value of $\omega = 1$ rad/sec, when each of the models A, B and C are individually inserted into the *SP* structure. Notice that the shape of the templates depends strongly on the selected *SP* model.

Applying the *Second Step* (see Table 12.2) for these admissible three models, and using the value $W(\omega) = 1\ \forall\omega$ in Eq. (12.10), the best-cost value of $I = 0.9437$ is found for case A. The worst case, given here for comparison purposes, corresponds to case B with a cost of $I = 1.0740$. Case C presents a cost of $I = 0.9964$.

In order to illustrate the ideas proposed in the second criterion, select models A and C. Figures 12.7 and 12.8 show the templates for both cases over the frequency range of $\omega = [\,0.1 \quad 0.4 \quad 0.7 \quad 1\,]$ rad/sec.

Following the *Second Step* algorithm, it is noted that the set of templates for case A present a smaller area than the set of templates for case C. For this reason and looking for the best compensator design, select case A instead of case C.

To validate this choice, a complete QFT design of the compensator $G(s)$, for both cases A and C, is now presented. Equations (12.19) through (12.21) exhibit the desired stability and performance specifications for both cases.

Stability specification:

$$\left|\frac{\boldsymbol{G}(j\omega)[\boldsymbol{P}(j\omega)\boldsymbol{H}(j\omega)]}{1+\boldsymbol{G}(j\omega)[\boldsymbol{P}(j\omega)\boldsymbol{H}(j\omega)]}\right| \le M_m, \quad \text{where } M_m = 1.2$$

(12.19)

$$\text{for } \omega = [0.01 \;\; 0.1 \;\; 0.3 \;\; 0.5 \;\; 1] \text{ rad/sec}$$

Tracking specifications:

$$T_{R_L}(\omega) \le \left|\frac{G(j\omega)[P(j\omega)H(j\omega)]}{1+G(j\omega)[P(j\omega)H(j\omega)]}\right| \le T_{R_U}(\omega) \qquad \textbf{(12.20)}$$

where :

$$T_{R_L}(\omega) = \left| \frac{0.08}{(j\omega)^3 + 2.4(j\omega)^2 + 0.84(j\omega) + 0.08} \right| \text{ the lower bound}$$

$$T_{R_U}(\omega) = \left| \frac{0.0040\left[(j\omega)^2 + 22(j\omega) + 40\right]}{(j\omega)^2 + 0.36(j\omega) + 0.16} \right| \text{ the upper bound}$$

for $\omega = \begin{bmatrix} 0.01 & 0.1 & 0.3 & 0.5 \end{bmatrix}$ rad/sec

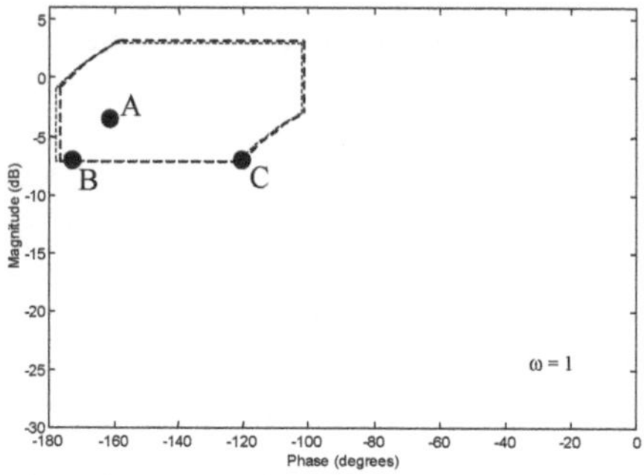

Fig. 12.5 Template for the first order model with delay, without the SP. (ω = 1 rad/sec).

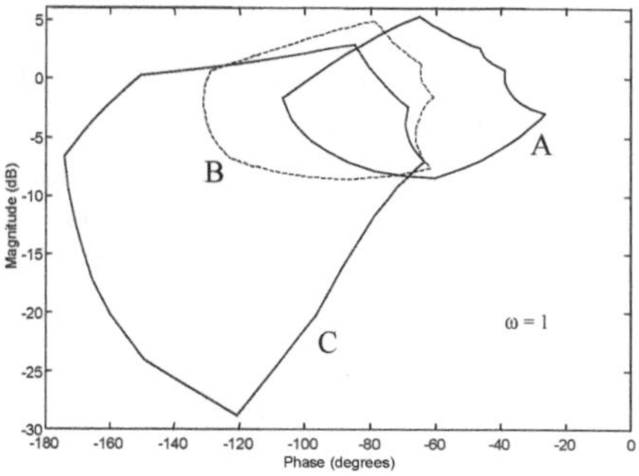

Fig. 12.6 Templates for the first order model with delay and Smith Predictor A, B or C.

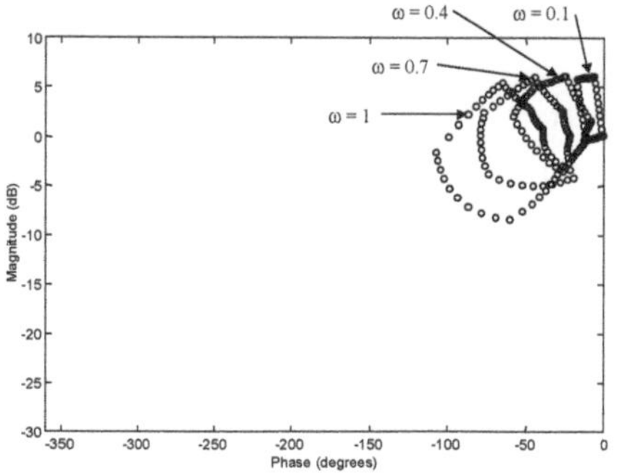

Fig. 12.7 Templates for Case A.

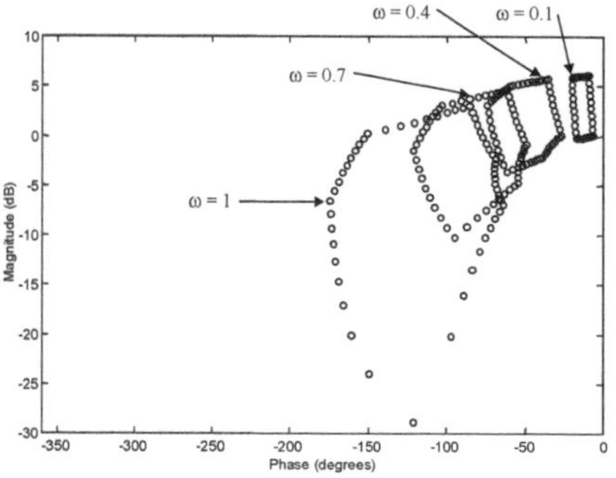

Fig. 12.8 Templates for Case C.

The QFT bounds, calculated taking into account the parameter uncertainty and the above performance specifications, and the loop shaping of $L(s) = G(s)\ P_{eq}(s)$ with the simple compensator structure of $G(s) = k/s$ are illustrated in Figs. 12.9 and 12.10, for cases A and C, respectively.

Figure 12.9 shows the loop shaping of *L(s)* when model A is selected for the *SP* structure. The best-designed compensator for this model is given by:

$$G(s) = \frac{0.1389}{s} \tag{12.21}$$

In addition, Fig. 12.10 shows the loop shaping of *L(s)* when model C is selected for the *SP* structure. The best-designed compensator for this model is given by:

$$G(s) = \frac{0.1671}{s} \tag{12.22}$$

Looking at the loop-shape obtained by both compensators, one concludes that case A fulfils the bounds constraints, while case C does not. This means that the selected model of the plant, used in the *SP,* not only modifies the templates and the loop shaping stage, but also affects the compensator structure itself.

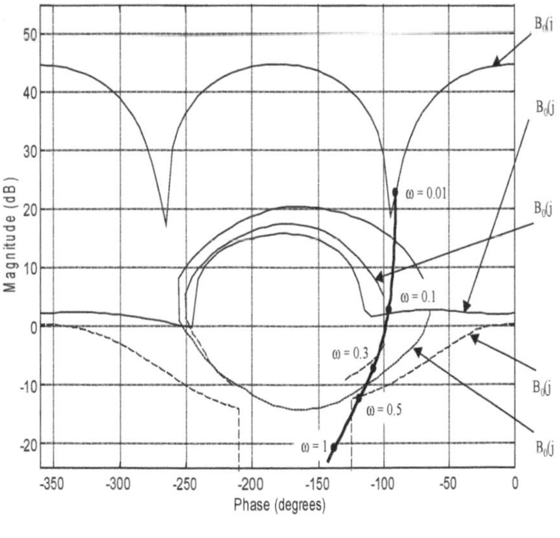

Fig. 12.9 Loop-shaping for Case A.

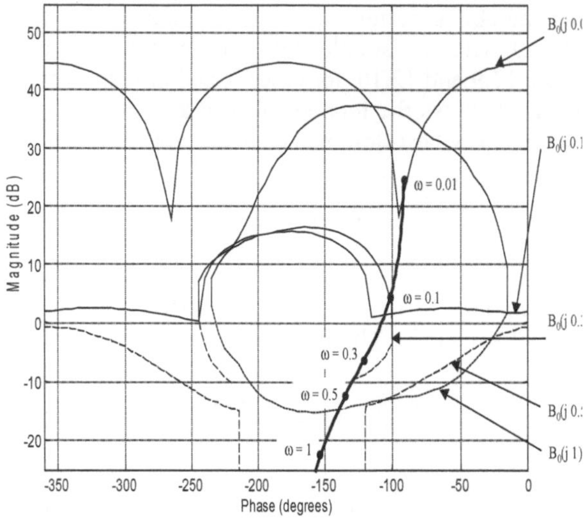

Fig. 12.10 Loop-shaping for Case C.

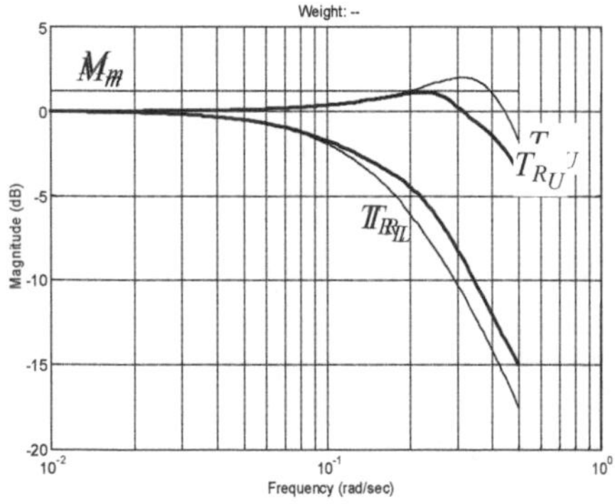

Fig. 12.11 Analysis for Case A.

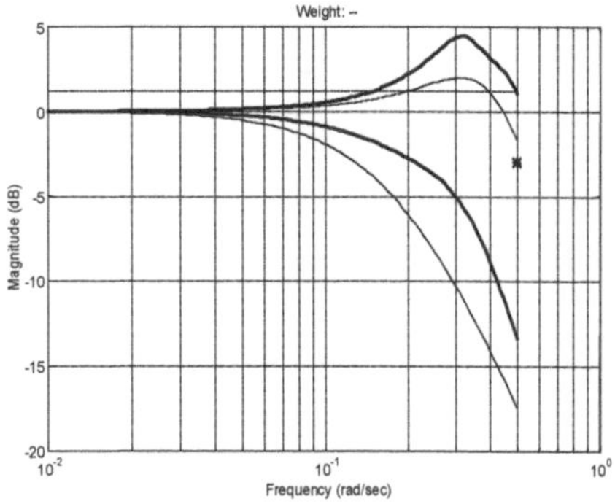

Fig. 12.12 Analysis for Case C.

Finally, the plots in Figs. 12.11 and 12.12 show the closed loop response, in dB, versus the specifications (the upper and lower performance bounds), for the worst (over all uncertainty) cases and for models A and C respectively. The dashed lines represent the performance specifications M_m, T_{R_U} and T_{R_L} and the solid lines show the response of the control system for the worst case. As one might expect, the first trial (model A) presents a good performance, i.e., all performances lie within the bounds (see Fig. 12.11), while the response of the control system with model C does not fulfill the required specifications (see Fig. 12.12).

12-4 APPLICATION TO A HTST PASTEURIZATION PLANT

To validate the proposed methodology in an actual non-linear process with a variable time delay, the new algorithms are applied to a High Temperature Short Time (HTST) pasteurization plant.

12-4.1 MODEL OF THE PASTEURIZATION PLANT

The plant PCT23, manufactured by Armfield[127], is a miniature version (1.2 m × 0.6 m × 0.6 m) of a real industrial HTST pasteurization process (see Fig. 12.13). It consists of a bench mounted process unit, connected to a dedicated control console to provide access to various signals associated with measurements and control.

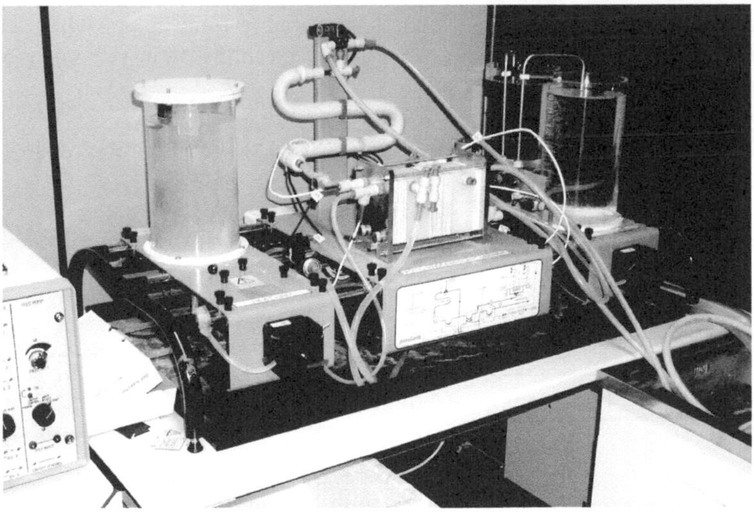

Fig. 12.13 HTST Pasteurization Process (Armfield). Public University of Navarra.

A diagram of the plant is shown in Fig. 12.14. The raw product is stored in a tank from where it is pumped to a heat exchanger of counter-current parallel flows. The product is heated to the pasteurization temperature using a hot water flow coming from a closed circuit with a heater. The product is kept at the pasteurization temperature for a few seconds (holding time), while it crosses through a thermally insulated tube. The following considerations clarify the control problem faced in this chapter:

- Raw product temperature T_r is assumed constant (about 21°C).

- Pasteurisation temperature T_p is the control variable. Temperature requirements strongly depend upon product conditions. On the other hand, the pasteurizer parameters are shown to be nearly temperature independent. With this in mind, pasteurisation temperatures about 40°C are demanded herein. While this may seem rather low for pasteurization, it allows experiments to be speeded up without compromising reliability.

- Hot water temperature T_w is held constant at 49°C. This is achieved by means of an autonomous controller which manipulates the power supplied to the heater.

- The disturbance input, product-fluid flow F_p is allowed to vary within 180 and 400 ml/min. This is a source of variable time delay, which can be computed by dividing the volume of the holding tube (82 cm^3) by the product flow.

- Hot-fluid flow F_h is the manipulated variable, which is in the range 100 to 500 ml/min.

With a sampling of 4 seconds, the model of the pasteurization plant is identified as,

$$\frac{T_p(z^{-1})}{F_h(z^{-1})} = \frac{K\,z^{-d}}{(1-0.8825\,z^{-1})(1-0.8249\,z^{-1})}; \quad K \in [0.056,\, 0.118],\; d \in [3, 7] \tag{12.23}$$

where the nonlinearities of the process (from varying flows) are incorporated as parameter uncertainty.[112]

The aim of this experimental example is to investigate the proposed *SP* design so as to ensure the temperature at the output of the holding tube is maintained within a specified range around the set-point value, despite disturbances and model uncertainty.

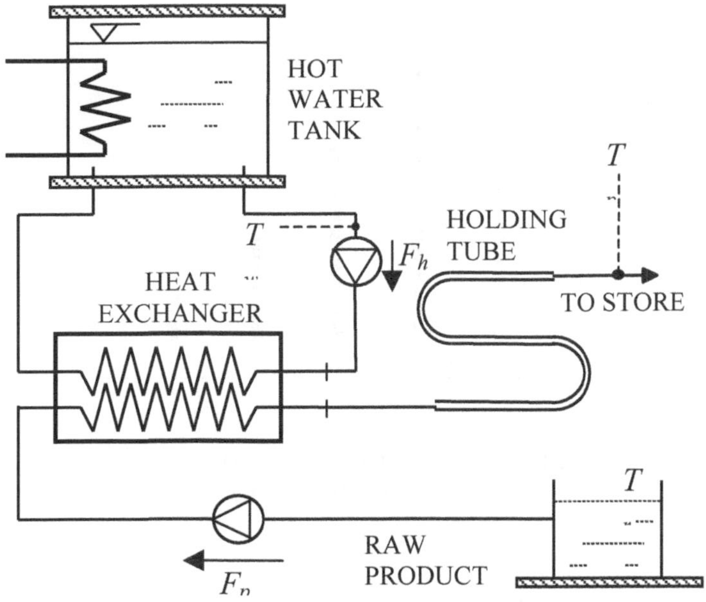

Fig. 12.14 Diagram of the HTST pasteurization process.

12-4.2 CONTROL OF THE PASTEURIZATION PLANT

Figures 12.15 and 12.16 show the *SP* models satisfying the *First Step* for two different bandwidths: 0.030 and 0.037 rad/s respectively. Integer values for the delay and 50 values for *K* are used to obtain the figures.

Now, specifying a desired bandwidth of BW = 0.037 rad/s (see Fig. 12.16), the best and worst models according to Criterion 2, with cost values of 1.0263 and 2.3619 are, respectively:

Case α: best model ($\hat{K} = 0.0642$; $\hat{d} = 6$)

Case β: worst model ($\hat{K} = 0.0704$; $\hat{d} = 3$)

Figure 12.17 shows the equivalent-plant templates corresponding to both cases for a frequency of 0.04 rad/s. The actual-plant template for this frequency is also depicted. Although only integer delays are considered when computing actual-plant templates, they are proved to adequately describe partial delays, following the results of de Paor and O'Malley,[150] for the frequency range

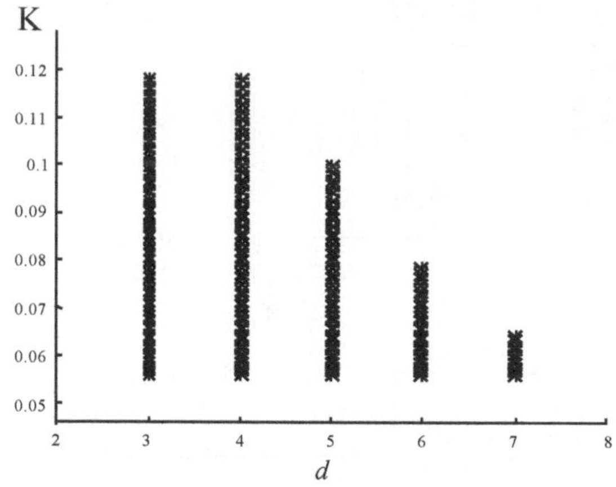

Fig. 12.15 *SP* models that satisfy First Step criterion. (BW = 0.03 rad/s).

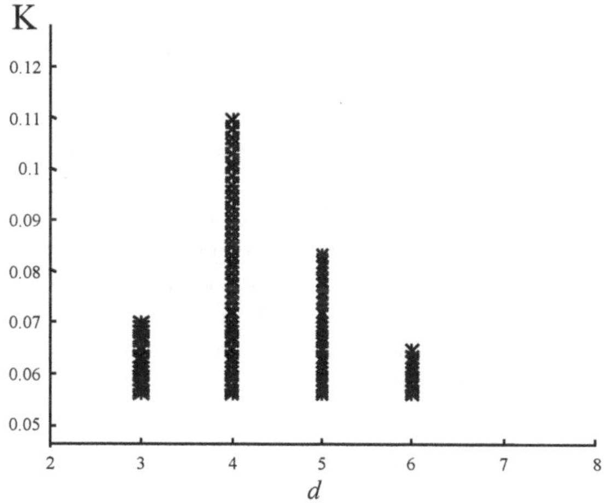

Fig. 12.16 *SP* models that satisfy First Step criterion. (BW = 0.037 rad/s).

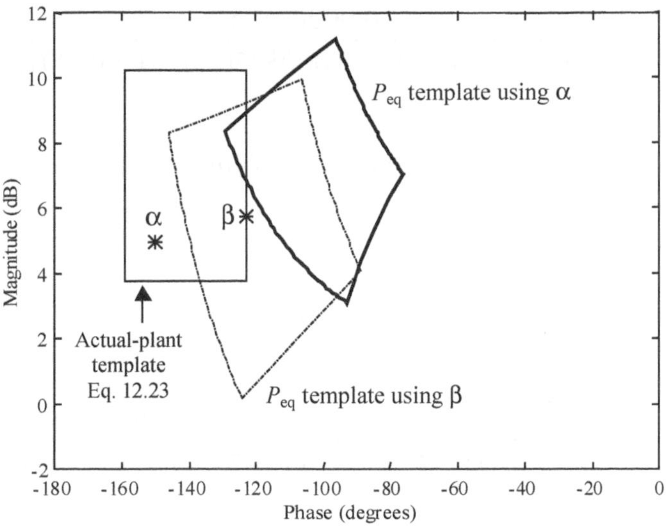

Fig. 12.17 Templates for cases original, α and β (ω = 0.04 rad/s).

of interest. It should be mentioned, however, that template computation for the general case of digital control systems with partial delays does not seem to be addressed yet in the literature. The required robust stability and tracking specifications for the closed loop system are the following.

Robust stability specifications:

$$\left| \frac{G(z)\,[P(z)\;H(z)]}{1+G(z)\,[P(z)\;H(z)]} \right| \le 1.15$$

for $z = e^{j\omega T}$, $\omega = [0.001,\; 0.005,\; 0.01,\; 0.03,\; 0.05,\; 0.07]$ rad/s

(12.24)

Tracking specifications:

$$T_{R_L}(\omega) \le \left| \frac{F(z)G(z)[P(z)H(z)]}{1+G(z)[P(z)H(z)]} \right| \le T_{R_U}(\omega)$$

(12.25)

for $z = e^{j\omega T}$ and $\omega = [0.001,\ 0.005,\ 0.01,\ 0.03,\ 0.05,\ 0.07]$ rads/sec.

$$T_{R_L}(\omega) = \left| \frac{0.00000125}{(j\omega)^3 + 0.06(j\omega)^2 + 0.000525(j\omega) + 0.00000125} \right|$$

$$T_{R_U}(\omega) = \left| \frac{0.025\ \left[(j\omega)^2 + 0.55(j\omega) + 0.025\right]}{(j\omega)^2 + 0.025(j\omega) + 0.000625} \right| \tag{12.26}$$

The loop controller $G(z)$ and a pre-filter $F(z)$ for both cases α and β are shown in Eqs. (12.27) through (12.30).

$$G_\alpha(z) = \frac{1.75(1-0.87z^{-1})(1-0.84z^{-1})(1-0.83z^{-1})(1-0.79z^{-1})(1-0.02z^{-1})}{(1-z^{-1})(1-0.40z^{-1})(1-0.40z^{-1})(1-0.35z^{-1})(1-0.23z^{-1})(1+0.37z^{-1})} \tag{12.27}$$

$$F_\alpha(z) = \frac{0.2565(1-0.9625z^{-1})}{(1-0.9131z^{-1})(1-0.8893z^{-1})} \tag{12.28}$$

$$G_\beta(z) = \frac{4.70(1-0.83z^{-1})(1-0.82z^{-1})(1-0.82z^{-1})(1-0.80z^{-1})(1-0.76z^{-1})}{(1-z^{-1})(1-0.41z^{-1})(1-0.30z^{-1})(1-0.29z^{-1})(1-0.27z^{-1})(1+0.19z^{-1})} \tag{12.29}$$

$$F_\beta(z) = \frac{0.3419(1-0.9622z^{-1})}{(1-0.8914z^{-1})(1-0.8812z^{-1})} \tag{12.30}$$

Figures 12.18 and 12.19 show the Nichols plots of both designs. Though all performance specifications are met for both cases, high-frequency gain is lower for case α. This is a result of the less demanding templates.

Figures 12.20(a) (pasteurization temperatures) and Figs. 12.20(b) and 12.20(c) (hot-fluid flows) compare the experimental result of both designs when applied to the real pasteurization plant. The experiment was scheduled as follows: initial set-point was 36°C, and initial product flow was 6.7 10^{-3} litr./sec. At $t = 1200$ sec. the set-point was changed to 38°C. At $t = 2400$ sec. the product flow was set to 3 10^{-3} litr./sec. Finally, at $t = 3600$ sec. the set-point was raised to 40°C. As expected, the first design (α) clearly outperforms the second one (β), obtaining better set-point tracking and disturbance rejection, and featuring a less oscillatory control signal.

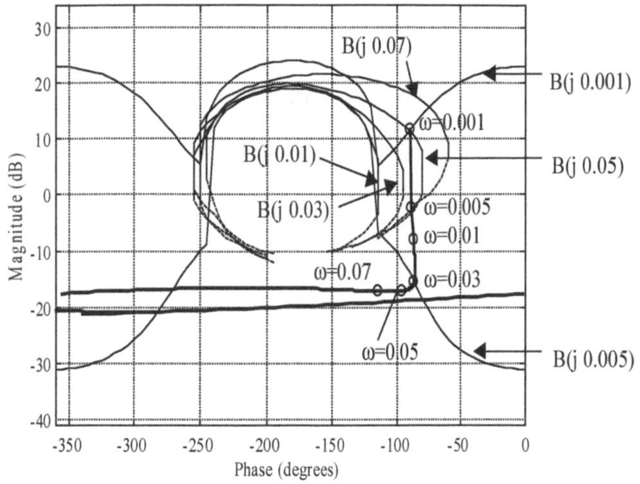

Fig. 12.18 Nichols Chart for Case α.

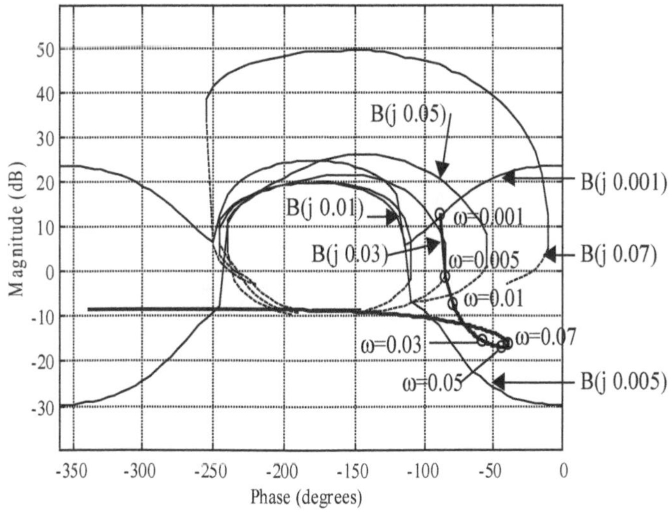

Fig. 12.19 Nichols Chart for Case β.

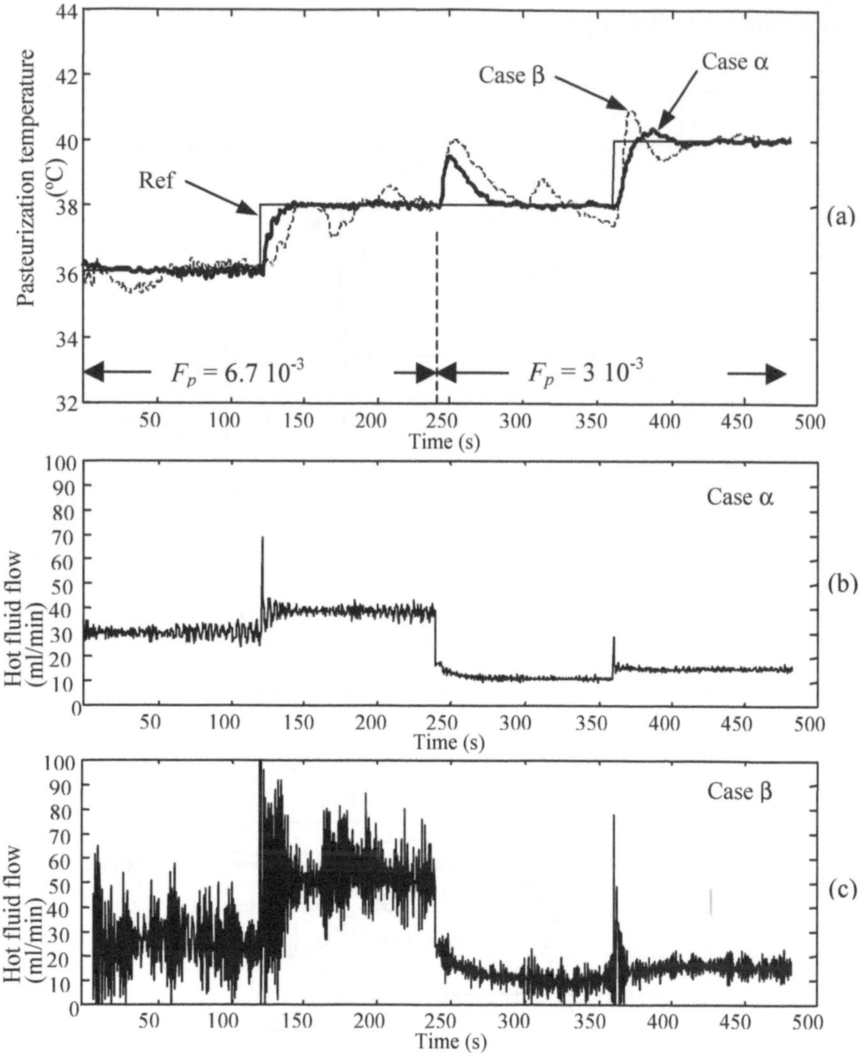

Fig. 12.20 Response of the pasteurisation plant: (a) T_p obtained for case α (solid line) and case β (dashed line); (b) control signal case α; (c) control signal case β.

12-5 SUMMARY

It is well known that the Smith Predictor (*SP*) may be very sensitive to model-plant mismatch, resulting in a poor performance when model uncertainty is present. This chapter introduces a method to design *SPs* when the plant is not precisely known. The method includes two criteria. The first criterion (*First Step*) is based on bandwidth frequency considerations. It finds the set of models of the plant so that, if the *SP* adopts one of them, the desired bandwidth BW is not reduced by the effect of the parameter uncertainty. The second criterion (*Second Step*) introduces some guidelines to improve the design of the *SP* by using the QFT technique.

13

QFT DESIGN TECHNIQUES APPLIED TO REAL-WORLD INDUSTRIAL SYSTEMS[103,104,133,165]

13-1 INTRODUCTION

This chapter presents two challenging real-world examples where QFT has been successfully applied. The first section proposes a QFT robust control strategy to improve the performance of an activated sludge wastewater treatment plant (WWTP). The main control objective is to minimize the effluent nitrogen compounds (Ammonia and Nitrates), rejecting the plant disturbances and insuring robust stability. The strategy incorporates in the controller design the multivariable and nonlinear model n° 1 of the International Water Association (IWA) for aeration tank reactions, calibrated with real data of the WWTP of Crispijana (Spain). The results obtained with the QFT controllers show a tighter control of the effluent nitrogen compounds and a significant reduction (energy saving) of the dissolved oxygen demand.[103,104]

The second part of the chapter presents the large variable-speed direct-drive multi-pole wind turbine TWT1650 designed by M.Torres. The section summarizes some experimental results of the applied QFT controllers. The first TWT1650 prototype started working at Cabanillas Wind Farm (Spain) in August 2001. Since then more than 10 wind turbines have been installed and a large amount of experimental data has been collected. The controller design was made using advanced QFT robust control strategies based on both, mathematical modeling and analysis of the experimental data. This portion of the chapter introduces the main advantages of the multi-pole system and shows some of the most representative experimental results of the TWT1650 with the QFT controllers under medium and extreme wind conditions.[133,165]

13-2 QFT BASED CONTROL OF A BIOLOGICAL WASTEWATER TREATMENT PLANT[103,104]

The need for improving the quality and controlling of the world-wide water source is based upon:

1. More than 70% of the Earth's surface is covered by water.
2. Every living thing (humans, animals, plants, etc.) consists mostly of water.
3. However, in the last few decades the human impact on water resources has been very important.
4. Industrial and domestic facilities consume and pollute more and more of the drinkable water available.

In that context, the requirements for wastewater treatment have become increasingly more stringent (European Union Council Directive 91/271/EEC, Federal Water Pollution Control Act of 1972, 77 and 87, USA). One of the best tools to protect the water environment from negative effects produced by pollutants are the Wastewater Treatment Plants (WWTP), which control and reduce the concentration of pernicious substances in water.

This section presents the design of a 2×2 closed-loop control system, based on the Quantitative Feedback Theory, and their implementation on the regulation of a WWTP. The control strategy incorporates the multivariable and nonlinear International Water Association (IWA) model[151] for aeration tank reactions, calibrated with real data of the WWTP of Crispijana. The main control objective is to minimize the effluent nitrogen compounds (Ammonia -*SNH*- and Nitrates -*SNO*- simultaneously), rejecting the plant disturbances and insuring robust stability.

13-2.1 THE WASTEWATER TREATMENT PLANT

The process considered is an Activated Sludge WWTP with Nitrification and Denitrification (see the D-N configuration shown in Fig. 13.1).

Nitrogen is present in many forms in wastewater, e.g. ammonium, nitrate, nitrite, etc. Nitrogen is an essential nutrient for biological growth and is one of the main constituents in all-living organisms. The presence of nitrogen in effluent wastewater, however, presents many problems and can be very toxic to aquatic organisms.[152]

When untreated wastewater arrives to the WWTP most nitrogen is present in the form of ammonium NH_4^+. Nitrogen can be removed by a three-step procedure: Anoxic zone (D), oxic zone (N) and settler (see Fig. 13.1).

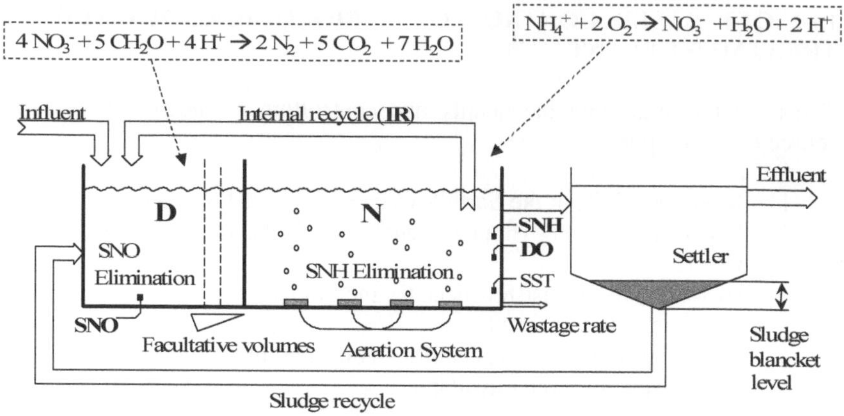

Fig. 13.1 Wastewater plant diagram (D-N configuration).

In the N-zone, ammonium is oxidized to nitrate NO_3^- with an aeration system. This process is called *Nitrification* and can be summarized by the formula:

$$NH_4^+ + 2\,O_2 \rightarrow NO_3^- + H_2O + 2\,H^+$$

In the D-zone, the nitrate previously produced under aerobic conditions is transformed to nitrogen gas N_2 by a process called *Denitrification.* This reaction takes place in an anaerobic environment where the bacteria responsible for denitrification respire with nitrate instead of oxygen. The process can be summarized by the formula:

$$4\,NO_3^- + 5\,CH_2O + 4\,H^+ \rightarrow 2\,N_2 + 5\,CO_2 + 7\,H_2O$$

The plant configuration needs an internal recycle to support the denitrification process. The recycling process supplies the nitrates produced in the N-zone to the D-zone.

13-2.2 INPUT AND OUTPUT VARIABLES

The control objective of the process is to minimize the effluent nitrogen compounds, that is to say, the ammonium *(SNH)* and the nitrate *(SNO)* concentrations, according to the standard regulations and rejecting the plant disturbances. The input variables (actuators) used in the process are the Dissolved Oxygen *(DO)* level in the oxic reactor (N-zone), supplied by an aeration system, and the Internal Recycle *(IR)* flow rate, which provides nitrates to the anoxic zone of the plant (D-zone).

The *DO* level has been the most studied variable in the operation of WWTPs. The *DO* lower limit sets the process requirements and the maintenance of the suspension of solids (*SST*). On the other hand, an excess of ventilation is very expensive and does not raise substantially the level of the *DO* because of the oxygen saturation in the water. Under classical operation, the desired *DO* level in the oxic reactor is fixed at a specific value (open-loop control). This level is usually the reference of a PI controller that evaluates the required airflow.

Proper operation of the *IR* is based on the following trade-off. On the one hand, its flow rate must be high enough to supply the required nitrates to the anoxic zone to allow the denitrification. On the other hand, an excessive pumping can also inhibit the denitrification because of the contribution of oxygen from the oxic zone. Under classical operation, the *IR* is set to a fixed value too (open-loop control).

Alternatively, the control strategy presented here modifies continuously the values of the *DO* level and the *IR* flow to improve the performance of the plant and the energy saving (closed-loop control). In this manner, the controller considers the plant as a multivariable *2x2* system, where the output variables are the *SNH* and the *SNO* concentrations at the outflow, and the input variables are the *DO* level and the *IR* flow. As in the previous chapters, the Relative Gain Analysis (RGA)[92] is used to select the input-output pairs to minimize the amount of interaction among the resulting loops. Figure 13.2 shows the first element λ_{11} of the Relative Gain Matrix $\mathbf{\Lambda}$, which represents the relative gain between the output *SNH* and the input *DO,* calculated over the full range of the plant operation.

Accordingly to the value of λ_{11}, around 1, the two loops are decoupled: the *DO* level as the input variable to control the *SNH* concentration at the outflow (first loop) and the *IR* flow rate as the input variable to control the *SNO* concentration at the outflow (second loop). The system is showed in Eqs. (13.1) and Fig. 13.3.

$$\mathbf{P}'(z) = \begin{bmatrix} P'_{11}(z) & P'_{12}(z) \\ P'_{21}(z) & P'_{22}(z) \end{bmatrix} ; \qquad \mathbf{D}(z) = \begin{bmatrix} D_1(z) \\ D_2(z) \end{bmatrix}$$

$$\mathbf{Y}(z) = \begin{bmatrix} Y_1(z) \\ Y_2(z) \end{bmatrix} ; \qquad \mathbf{U}(z) = \begin{bmatrix} U_1(z) \\ U_2(z) \end{bmatrix} \tag{13.1}$$

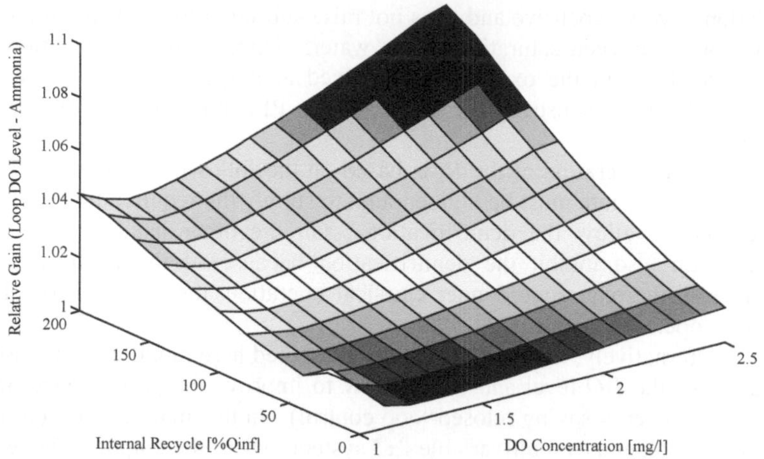

Fig. 13.2 *SNH - DO* element (λ_{11}) of the Relative Gain Matrix.[92]

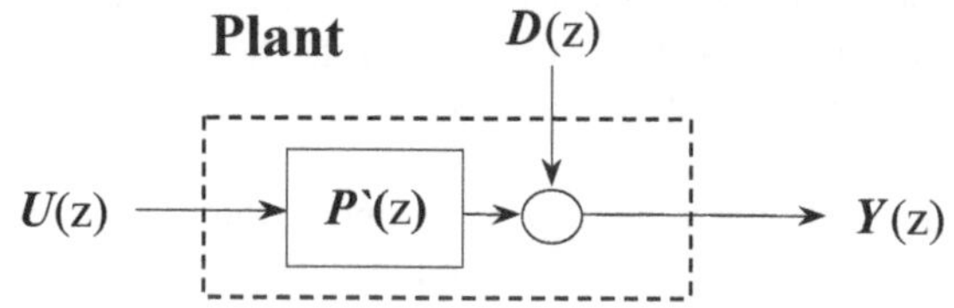

Fig. 13.3 Block-diagram of the open-loop system.

where $U_1(z)$ represents the *DO* level and $U_2(z)$ the *IR* flow rate (the plant inputs); $Y_1(z)$ represents the instantaneous effluent *SNH* concentration and $Y_2(z)$ the instantaneous effluent *SNO* concentration (the plant outputs); and $D_1(z)$ and $D_2(z)$ the effect of the instantaneous ammonia and nitrates load inflows (disturbances) at the outputs of the plant $Y_1(z)$ and $Y_2(z)$.

13-2.3 MODEL OF THE PROCESS

There are two models of the WWTP used in this chapter. The first one *(Truth-model)*, used in the simulations, is based on the general model of an activated sludge process developed by the IWA[151] and the second model of the dynamic response of the Settler developed by the CEIT.[153] It results in a multivariable non-linear system with 34 states and is calibrated with experimental data of the WWTP of Crispijana (Spain).

The second model *(Simplified-model)*, used to design the QFT controller, represents the diagonal elements of the plant (see Eq. 13.2). It exhibits a simple digital structure with one zero and two poles and describes the dynamics of the non-linear process considering the complexity as parameter uncertainty.

$$\boldsymbol{P}'(z)=\left\{P'_{ij}(z)\right\}=\frac{c_{ij}+d_{ij}z^{-1}}{1+a_{ij}z^{-1}+b_{ij}z^{-2}}\;;\quad i=j=1,2 \qquad \textbf{(13.2)}$$

The two models and the controller are implemented using *Simulink, Matlab*, and *C/C*$^{++}$.

13-2.4 PARAMETER ESTIMATION

The parameters (a_{ij} , b_{ij} , c_{ij} , d_{ij} , $i = j = 1, 2$) of the *Simplified-model* are estimated[154] from computer simulations of the *Truth-model* with seasonal temperature variations, changes in the load inflow and in the influent flow rate of the plant (±50%), and variations in the *DO* level and the *IR* flow rate. In this manner, the parameters of the *Simplified-model* are estimated using a collection of 10,000 models that describe the complete set of operating conditions of the plant.

13-2.5 THE CONTROL STRATEGY

This section is divided into two different parts: the structure and the design of the controller. The control specifications, the templates generation, the bounds and the controller design are analysed.

Structure of the Controller

The standard requirements for nitrates and ammonia concentrations of the plant effluent are daily averages and not instantaneous data. Moreover, the influent concentrations during the day can have very different values. The aim of the controller is to reject the effect of those load inflow variations in the daily average of the effluent concentration. Thus, the real objective is to control the average $\overline{\boldsymbol{Y}}(z)$, where,

$$\overline{\mathbf{Y}}(z) = \frac{Tz\left(1 - z^{-n}\right)}{z - 1}\mathbf{Y}(z) \tag{13.3}$$

and where T is the sampling time, T = 1/72 day = 20 minutes, and $n\ T = 1$ day.[103] In this manner, the proposed control structure is shown in Fig. 13.4, where

$$\mathbf{G}'(z) = \begin{bmatrix} G'_{11}(z) & 0 \\ 0 & G'_{22}(z) \end{bmatrix}$$

is the controller.

The analysis of the frequency response of the closed loop system shows that the term $1 - z^{-n}$ introduces changes in magnitude and phase. From Fig. 13.4 obtain:

$$\mathbf{U}(z) = \mathbf{G}'(z)[\,\overline{\mathbf{R}} - \overline{\mathbf{Y}}(z)] \tag{13.4}$$

Where

$$\mathbf{G}'(z) = \frac{\mathbf{G}(z)}{\left(1 - z^{-n}\right)} \quad ; \quad \text{and} \quad \mathbf{P}'(z) = \mathbf{P}(z)\frac{z-1}{T\,z}$$

and where

$$\mathbf{u}(t) = \mathbf{u}(t - nT) + \mathcal{Z}^{-1}\left\{\mathbf{G}(z)\left[\overline{\mathbf{R}}(z) - \overline{\mathbf{Y}}(z)\right]\right\} \tag{13.5}$$

Note that $\mathcal{Z}^{-1}$ represents the inverse $\mathcal{Z}$-transform. So the control output works in an incremental manner. This mode of operation leads to the expression

$$\overline{\mathbf{Y}}(z) = \left[\mathbf{I} + \mathbf{P}(z)\mathbf{G}(z)\right]^{-1}\left[\mathbf{P}(z)\mathbf{G}(z)\overline{\mathbf{R}}(z) + \overline{\mathbf{D}}(z)\right] \tag{13.6}$$

where $\overline{\boldsymbol{D}}(z)$ represents the daily average of $\boldsymbol{D}(z)$.[103]

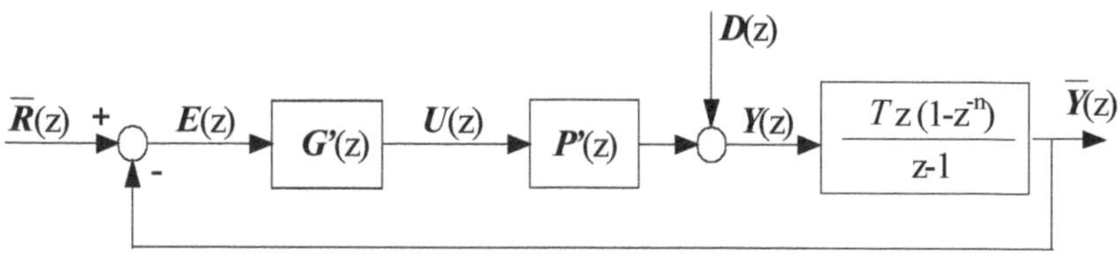

Fig. 13.4 Proposed control system structure.

Design of the Controller

Specifications

The most important performance specifications of the control policy of a WWTP are in the following order: to maintain the plant operative, to comply with the requirements of effluent quality, and to operate the process with maximum efficiency and minimum running costs.

In the context of water pollution standards, the main control objective of the process is to keep the daily average of the effluent ammonia concentration $\overline{SNH}$ around 1.5 mg/l (± 15%), and the daily average of the effluent nitrates concentration $\overline{SNO}$ around 1.8 mg/l (± 10%), rejecting the plant disturbances and insuring simultaneously robust stability.

The capability of the ammonia and nitrates concentration sensors is limited to a 20 minutes sampling time (T = 1200 sec.) which results in a Nyquist frequency of about ω_N = 0.002618 rad/sec. Thus, the sampling time is sufficient for the control objectives of the system.

Looking at the typical behavior of the process, one of the most important disturbances of the plant is the continuous variation of the daily average of the influent ammonia and nitrates loads. That phenomenon affects the process as a plant output disturbance, as shown in $\boldsymbol{D}(z)$ (Figure 13.4), and has a natural frequency of about $1.16\ 10^{-6}$ rad/sec. However, in the real process it is not possible to move the set-point of the aeration system (first loop) very fast. So, the bandwidth of the aeration system is limited to frequencies around $5.79\ 10^{-6}$ rad/sec for energy saving purposes. Similarly, the variation of the *IR* flow rate (second loop) is low-filtered to frequencies around $34.72\ 10^{-6}$ rad/sec. In this manner the controller does not try to reject disturbances at high frequency.

Templates Generation

Figures 13.5 and 13.6 show the contour of the templates of the plant, $P_{11}(z)$ and $P_{22}(z)$ respectively. They have been generated from the *Simplified-model* estimated at the 10,000 operating conditions previously presented, and at representative frequencies. The set of frequencies selected for the analysis and design is ω = [0.1 0.3 1 3 10 30 60 100 150 226.2]/86400 rad/sec.

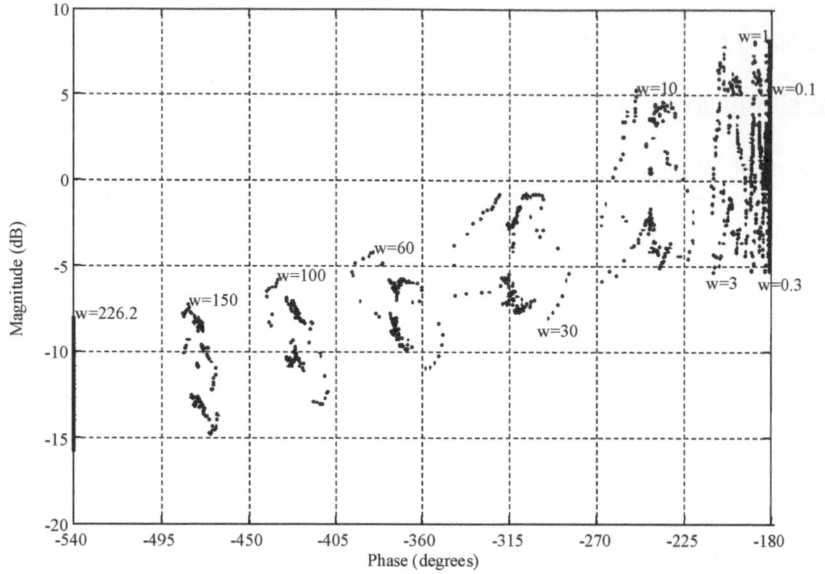

Fig. 13.5 $P_{11}(z)$ plant templates (*SNH*). ω multiplied by 86400.

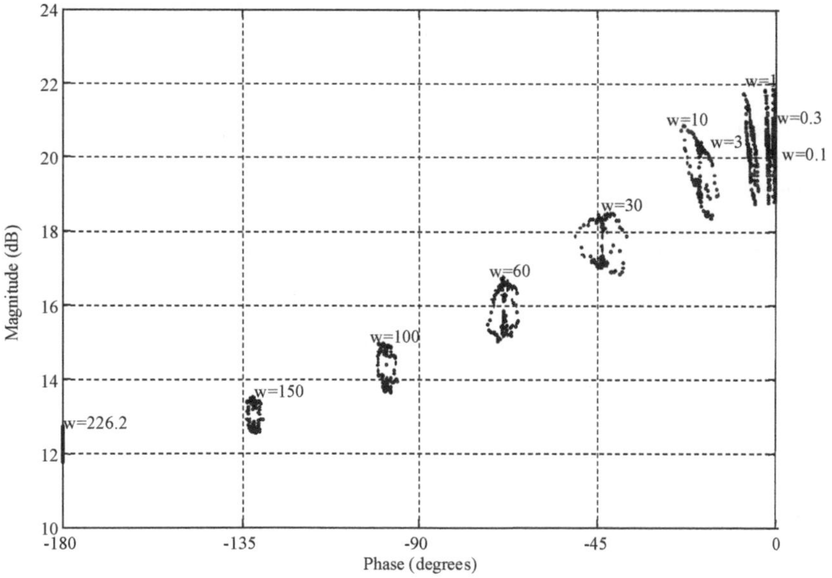

Fig. 13.6 $P_{22}(z)$ plant templates (*SNO*). ω multiplied by 86400.

Robust Stability and Performance Bounds

Following the environmental standards, the main feedback specification is to design a controller, $\boldsymbol{G}(z)$, such that the closed loop system presents a robust disturbance rejection for the plant output disturbances $\boldsymbol{D}(z)$. Remembering the specifications defined before, the output/disturbance transfer function constraints are,

$$\left|\frac{Y_1(z)}{D_1(z)}\right|_{z=e^{j\omega T}} \leq \left|\frac{4s}{s+3}\right|_{s=j\omega} ; \forall P_{11} \in \mathcal{P}, \quad \omega \in [0,\ 0.002618] \text{ rad/sec} \quad \textbf{(13.7)}$$

$$\left|\frac{Y_2(z)}{D_2(z)}\right|_{z=e^{j\omega T}} \leq \left|\frac{2s}{s+5}\right|_{s=j\omega} ; \forall P_{22} \in \mathcal{P}, \quad \omega \in [0,\ 0.002618] \text{ rad/sec} \quad \textbf{(13.8)}$$

In addition, the closed loop system must be robust stable. In order to achieve a conservative design, the stability specification are given by,

$$\left|\frac{P_{11}(z)G_{11}(z)}{1+P_{11}(z)G_{11}(z)}\right|_{z=e^{j\omega T}} \leq 1.3; \quad \forall P_{11} \in \mathcal{P}, \quad \omega \in [0,\ 0.002618] \text{ rad/sec} \quad \textbf{(13.9)}$$

$$\left|\frac{P_{22}(z)G_{22}(z)}{1+P_{22}(z)G_{22}(z)}\right|_{z=e^{j\omega T}} \leq 1.3; \quad \forall P_{22} \in \mathcal{P}, \quad \omega \in [0,\ 0.002618] \text{ rad/sec}$$

(13.10)

which implies at least 45° lower phase margin and 3.17 dB lower gain margin.

Controller Design

After the loop transmission QFT bounds are determined on the Nichols Chart, then the loop shaping process for $L_{ij}(z) = G_{ij}(z)\ P_{oij}(z)$ is performed. Figures 13.7 and 13.8 show $L_{11}(z)$ and $L_{22}(z)$ respectively, and the QFT bounds at the frequencies of interest. The resulting controller has the form,

$$\mathbf{G}(z) = \begin{bmatrix} G_{11}(z) & 0 \\ 0 & G_{22}(z) \end{bmatrix} \quad \textbf{(13.11)}$$

where the Ammonia and the Nitrates controllers are respectively,

$$G_{11}(z) = -2.6\,\frac{(z-0.9565)}{(z-0.9228)} \tag{13.12}$$

$$G_{22}(z) = 0.12\,\frac{z}{(z-0.6)} \tag{13.13}$$

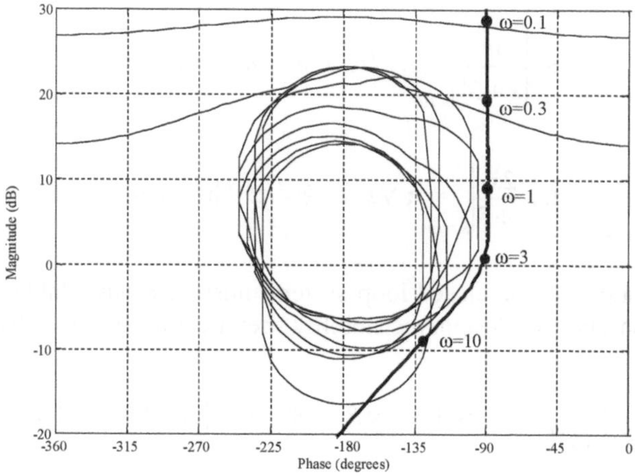

Fig. 13.7 Loop transmission function $L_{11}(z)$ *for SNH.* ω multiplied by 86400.

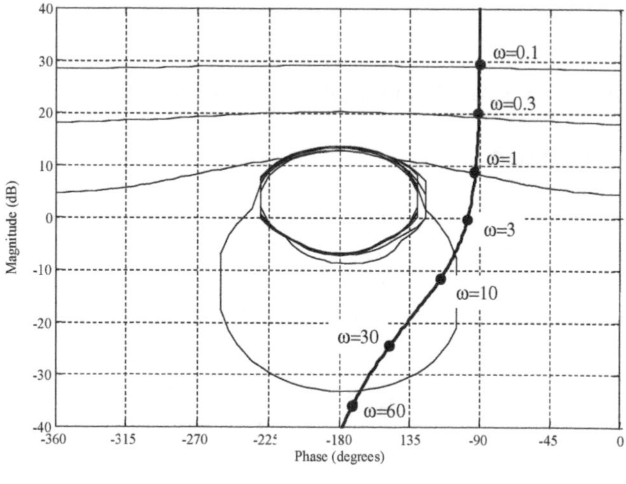

Fig. 13.8 Loop transmission function $L_{22}(z)$ *for SNO.* ω multiplied by 86400.

13-2.6 EXPERIMENTAL RESULTS

Controller Analysis I

First the designed control system is evaluated using the *Simplified-model* with the complete set of uncertainty. Figures 13.9 and 13.10 show the worst (over all uncertainty cases) closed loop response magnitude (solid line) versus the specifications (dashed line). The robust stability analysis is shown in Figs. 13.9a and 13.10a. The robust disturbance rejection analysis is shown in Figs. 13.9b and 13.10b. Both loops fulfill the specifications over the uncertainty range.

Controller Analysis II

To validate the controllers, the final algorithms are also implemented in a computer simulation of the multivariable and non-linear *Truth-model,* calibrated with real data of the WWTP of Crispijana.

The most relevant characteristics and operational variables of the real plant, used in the simulations, are:

- Influent flow rates (Q_{in}): 120,000 m^3/day
- Oxic reactor volume: 25,270 m^3
- Anoxic reactor volume: 10,830 m^3
- Settler volume: 24,200 m^3
- Sludge age (SRT): 12 days
- Nominal Sludge recycle flow rate: 1.0 Q_{in}
- Nominal Internal recycle flow rate: 1.0 Q_{in}
- Nominal Dissolved Oxygen concentration : 2 mg/l
- Design temperature: 13 °C

Figure 13.11a shows the daily average of the influent ammonia load, normalized with respect to the design value of the plant, in percentage. Figure 13.11b shows the daily average of the temperature in Celsius degrees. Both operational variables were obtained by the Crispijana WWTP laboratory (see Fig. 13.12) during 40 days in spring '96.

The main control objective of the first loop is to guarantee the standard requirements of ammonia concentration in the plant effluent, fixed on a daily average $\overline{SNH}$ lower than 2 mg/l. Accordingly, the set point is fixed to 1.3 mg/l. The maximum allowed value (saturation limit) of the *DO* variable is 2 mg/l. In Fig. 13.13 the daily average of the effluent ammonia concentration $\overline{SNH}(t)$ is shown as a dashed line and the control input *DO(t)* is shown as a solid line, both in mg/l, which are obtained with the $G_{11}(z)$ controller, under the influent ammonia load and the temperature conditions described in Fig. 13.11.

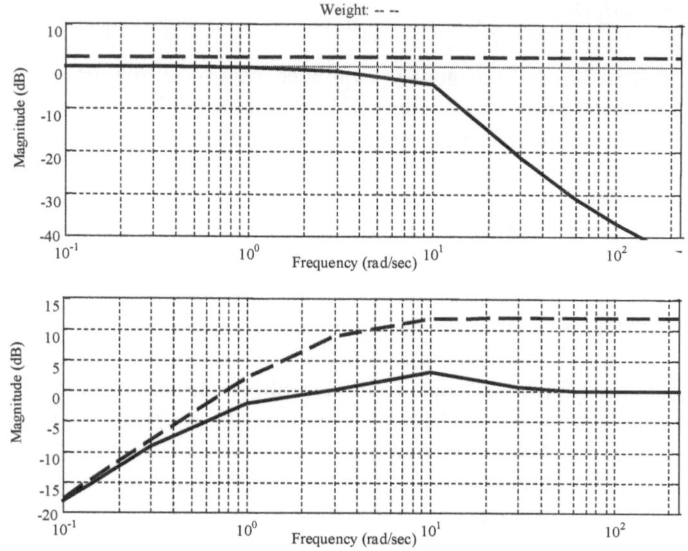

Fig. 13.9 *SNH/DO* loop: (a) analysis of robust stability and (b) analysis of robust output disturbance rejection. ω multiplied by 86400.

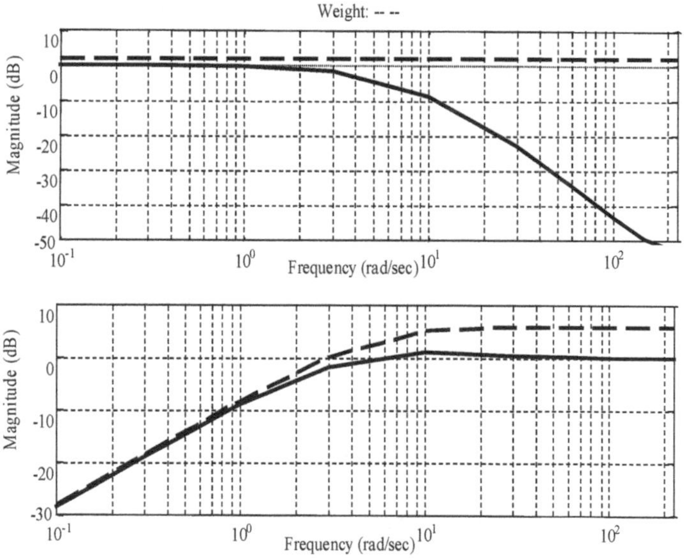

Fig. 13.10 *SNO/IR* loop: (a) analysis of robust stability and (b) analysis of robust output disturbance rejection. ω multiplied by 86400.

Simultaneously, the main control objective of the second loop is to guarantee the standard requirements of nitrates concentration in the plant effluent, fixed on a daily average $\overline{SNO}$ lower than 2 mg/l. For this reason, the controller tries to strengthen the denitrification process in the D tank to minimize the nitrates concentration in the plant effluent. Thus, in the present experiment the set-point of the nitrates concentration in the D tank is fixed to 0.5 mg/l. The maximum allowed value of the *IR* variable is 200% of the design value of the influent flow rate. In Fig. 13.14 the daily average of the nitrates concentration $\overline{SNO}(t)$ is shown in the D tank as a dashed line [mg/l] and the control input $IR(t)$ is shown as a solid line [per unit of the influent flow rate] with the $G_{22}(z)$ controller, under the influent ammonia load and the temperature conditions described in Fig. 13.11.

Figures 13.13 and 13.14 show how the system is able to regulate within the required specifications for both loops, except in the intervals when external disturbances saturate the control inputs:

- First Loop. *DO* saturated: [7→12, 13→14, 16→17, 34→35] days.
- Second Loop. *IR* saturated: [19→24, 29→34] days.

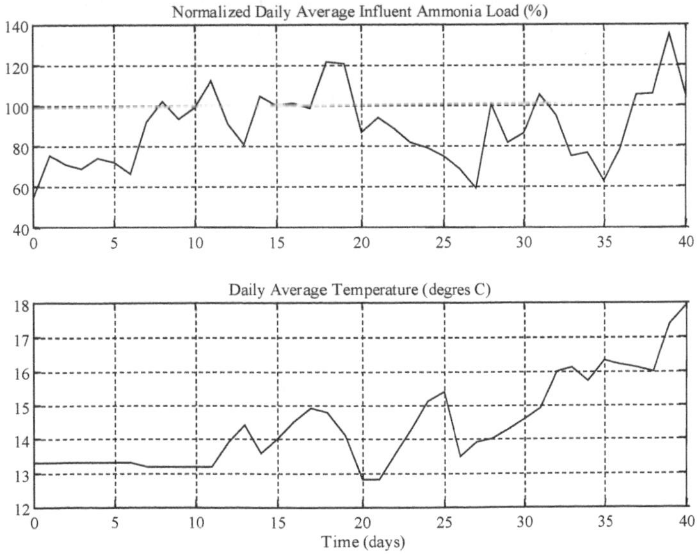

Fig. 13.11 (a) Daily average of the Influent Ammonia Load (%) and (b) daily average of the Temperature (°C)

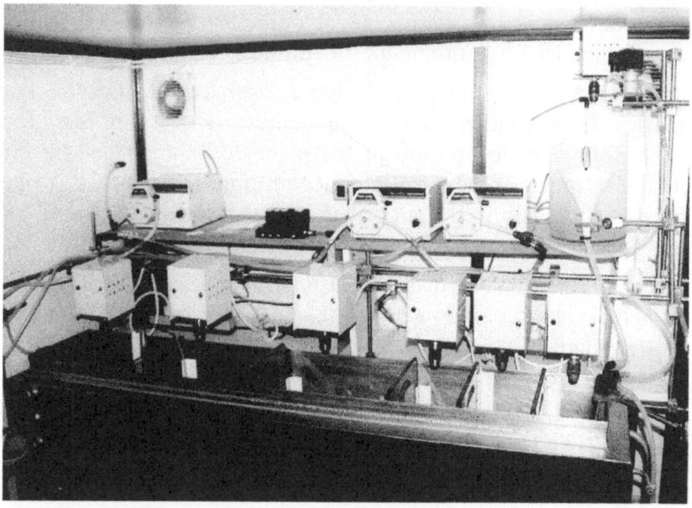

Fig. 13.12 Biological reactors of the Laboratory of Crispijana, Spain. (Courtesy of AMVISA).

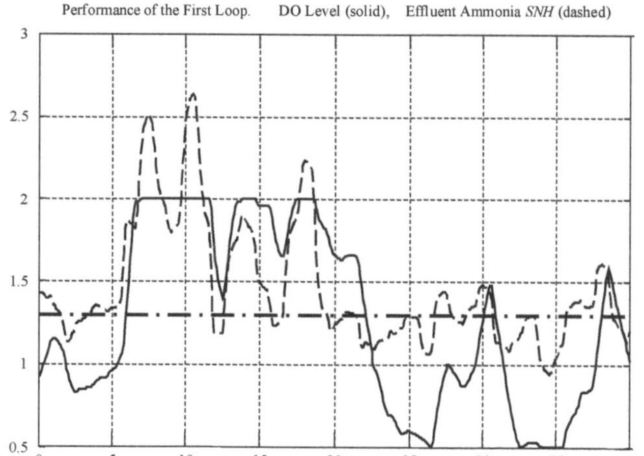

Fig. 13.13 First loop performance with $G_{11}(z)$.

In these cases, the system runs in the best possible conditions, obtaining a good performance when the saturation disappears. The control system achieves satisfactory performance of the effluent ammonia and nitrates concentrations over the whole range of operational conditions. It also obtains a notable reduction of the running costs, minimizing the oxygen supplied by the aeration system.

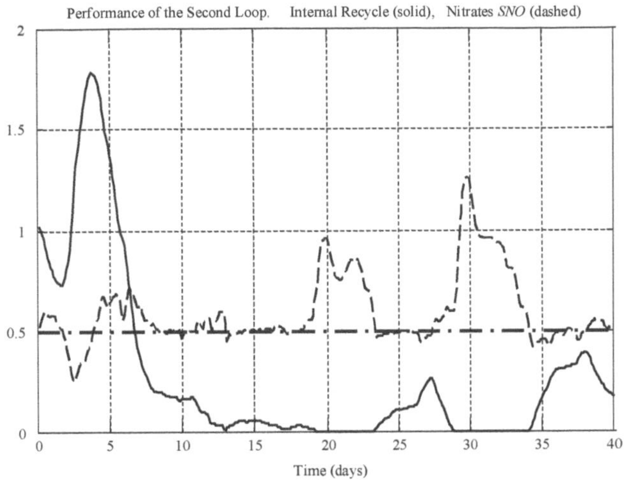

Fig. 13.14 Second loop performance with $G_{22}(z)$.

Fig. 13.15 Wastewater Treatment Plant of Crispijana, Spain. (Courtesy of AMVISA)

13-2.7 CONCLUSIONS PART I.

A QFT robust control strategy was proposed to improve the performance of an activated sludge wastewater treatment plant (WWTP). The main control objective was to minimize the effluent nitrogen compounds (Ammonia and Nitrates simultaneously), rejecting plant disturbances and insuring robust stability. The control strategy incorporated the model No. 1 of the International Water Association (IWA) for aeration tank reactions (*Truth-model*). It resulted in a multivariable and non-linear system with 34 states which was calibrated with experimental data of the WWTP of Crispijana (see Fig. 13.15).

A second model (*Simplified-model*) with a simple digital structure (having one zero and two poles), calibrated over 10,000 operating conditions, was used to design the controller. This linear model describes the dynamics of the process considering the complexity as parameter uncertainty. The development of this model allowed the generation of the templates, to obtain the robust stability and disturbance rejection bounds, and to perform the loop-shaping process in order to determine the QFT controller. The design was made keeping in mind the energy saving in the plant operation.

The behavior of the final algorithms was investigated implementing them in computer simulations of the multivariable and non-linear *Truth-model.* The results obtained with the QFT controllers show a tighter control of the effluent nitrogen compounds and a significant reduction (energy saving) of the dissolved oxygen demand.

13-3 QFT BASED CONTROL OF A VARIABLE-SPEED DIRECT-DRIVE MULTI-POLE SYNCHRONOUS WIND TURBINE: TWT1650[133,165]

The 2004 report of the European Wind Energy Association (*Wind Force* 12, 2004)[157] states that by 2020 wind power will meet 12% of world energy consumption, generating an output in the range of 3000 TWh/year with 1200 GW of capacity. Over the last five years the average annual growth rate of cumulative wind turbines installed has been almost 32%. The average size of new turbines being installed is expected to grow over the next decade from today's figure of 1.2 MW to 1.5 MW in 2007 and 2.0 MW in 2013. The total investment needed to reach a level of approximately 1200 GW capacity by 2020, from the current 40.3 GW (2003), is estimated at 692 billion euro over the whole period. The cost per unit of wind-powered electricity will decrease from the current 804 euro per installed kW and 3.79 cents of euro per kWh to 512 euro per installed kW and 2.45 cents of euro per kWh in 2020, which is a reduction of 36%. The annual saving of carbon dioxide will be approximately 1832 million tons by 2020.

In this growing field, this chapter introduces some general ideas about the control of variable-speed wind turbines and presents the design and experimental results of the TWT1650 wind turbine, designed by M.Torres and controlled with QFT strategies.

13-3.1 VARIABLE-SPEED WIND TURBINE CONTROL

The integration of control systems into a wind turbine (WT) can be easily understood by using the simplified model of Fig. 13.16. It shows a horizontal drive shaft that connects a large rotor inertia at one end (blades) with a generator at the other end. The wind applies an aerodynamic torque T_a on the rotor. The grid or the power electronics applies an electrical torque T_e on the generator. The acceleration or deceleration of the machine depends on the positive or negative sign of the difference ($T_a - T_e$) respectively.

The aerodynamic torque T_a applied on the rotor depends on the rotor tip speed ratio, blade geometry, wind speed and yaw angle, so that,

$$T_a = \frac{\rho \pi R^2 C_p V^3}{2\Omega} \cos(\varphi) \tag{13.14}$$

where C_p is the rotor power coefficient, ρ is the air density, R the rotor radius, V the wind speed, β the pitch angle, Ω the rotor speed and φ the yaw angle.

The aerodynamic rotor power coefficient C_p is a family of nonlinear functions that depends on the rotor tip speed ratio $\lambda = \Omega R/V$, for every β, according to Eq. (13.15) and Fig. 13.17.

$$C_p = f_{Non-linear}\{[V,\Omega],\beta\} \tag{13.15}$$

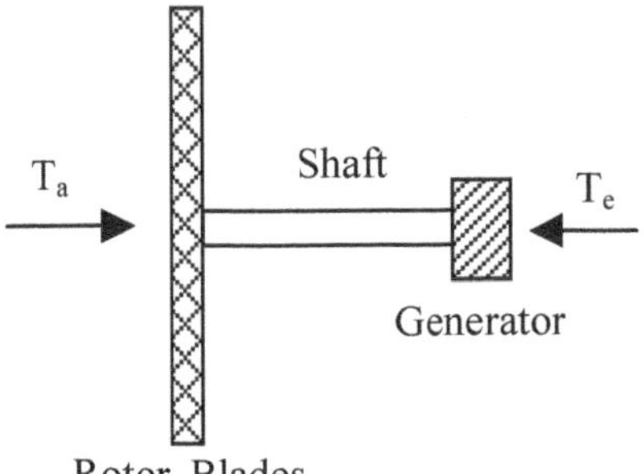

Fig. 13.16 Simplified wind turbine model.

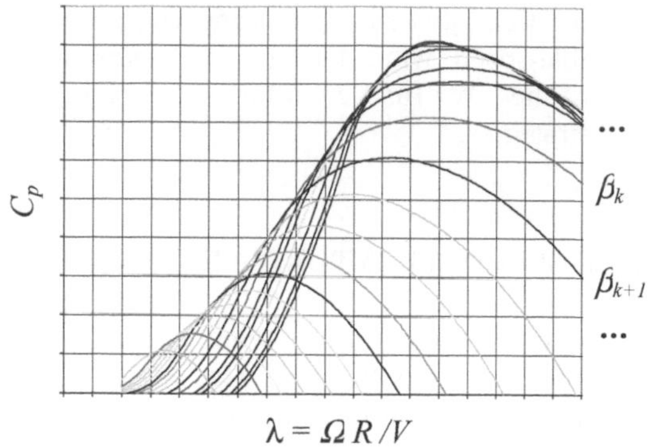

Fig. 13.17 Aerodynamic rotor power coefficient C_p.

On the other hand, the electrical torque T_e depends on the generator's connection to the grid. If the generator is connected directly, T_e depends on the aerodynamic torque and the system dynamics. Then the machine is a constant speed WT.

However, if there is an AC-AC power converter between the generator and the grid, then T_e depends on the control system and the machine is a variable speed WT. In this case, before rated power is reached (see Fig. 13.18, Zone 1), the machine can obtain a better efficiency C_p than for the constant speed case. For every pitch angle β, when the wind speed V changes and moves the rotor from the maximum rotor power coefficient C_p (see Fig. 13.17), the control system can change the electrical torque accordingly, changing afterwards the rotor speed Ω and coming back to the maximum C_p.

WT can also be arranged according to conventional and multi-pole machines. The former has a gearbox that multiplies the mechanical rotor speed before it affects the generator. The latter, however, are direct-drive, avoiding the mechanical gearbox by increasing the number of poles of the generator.

Respect to the control system design, one of the most complex issues is the mathematical modeling of the WT. In the last decade several authors have studied the problem. See for instance the excellent reports by Leithead and Rogers (1996),[158,159] Bongers and Engelen (1987),[160] and Sheinman and Rosen (1991).[161,162] In addition, specific control strategies to govern WTs can be found in the survey written by La Salle et al. (1980).[163] Gain-scheduling controllers were applied by Leith and Leithead (1997)[164] and QFT robust controllers by Torres and Garcia-Sanz (2004).[165]

For simulation purposes the software package Bladed, developed by Garrad Hassan (1997, 2003),[166] is one of the most powerful tools in the field. It is also recognized by the certification WT offices as a standard to simulate mechanical loads, etc. On the other hand, the main standard rules and design's regulations for WTs can be found in Germanischer Lloyd (1994, 2004)[167] and

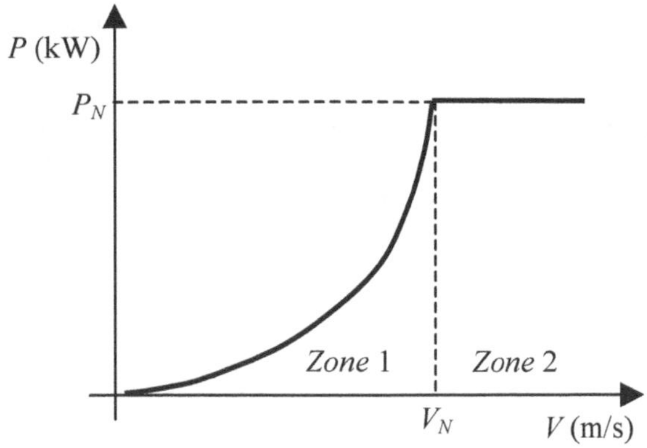

Fig. 13.18 Active power / Wind speed plot.

the European Standard IEC 61400.[168] For experimental measurements, the European network MEASNET (1996)[169] established a general metrology procedure to find experimentally the value of every physical quantity related with wind turbines. For more information about WTs read the excellent books written by Freris (1990)[170], Burton *et al.* (2001)[171] and Manwell *et al.* (2002).[172] These texts present a complete view of the Wind Energy field.

Finally, international companies that design variable-speed multi-pole WTs nowadays are but a few, due to the novelty and technological complexity of the system, even knowing the more capabilities and better performance of these machines. Among others are Enercon (Germany), M.Torres (Spain) and Lagerwey (Netherlands). In this context, the next sections introduce the variable-speed multi-pole synchronous wind turbine TWT1650 and the experimental results obtained when applying QFT controllers.

13-3.2 TWT DESCRIPTION

The TWT (Torres Wind Turbine) machine is a gearless variable-speed pitch-controlled direct-drive multi-pole synchronous generator, with a large rotor diameter (72 m), 1650 kW of power and an advanced control system based on QFT robust strategies (see Fig. 13.19).

The design of the TWT was made keeping in mind four objectives: high reliability at every work condition, optimum energy efficiency at every wind speed, maximum electrical power quality at the output and low cost maintenance.

Taking advantage of the multidisciplinary characteristic of the M.Torres company, the design avoided the classical way where every engineering team (mechanical, aerodynamic, electrical, electronics and control engineers) worked independently in a sequential manner. On the contrary, the design of the TWT was carried out according to an integrated design philosophy, where the engineering teams work together in a simultaneous and concurrent manner. This integrated view of the design requires that control engineers play a central role from the very beginning of the project. As a result a better system dynamics, controllability and optimum design was achieved.

The TWT design was also made following the main Wind Turbine International Standards: the IEC 61400 *(Wind Turbine Generator Systems),* the Germanisher Lloyd's *(Regulation for the Certification of Wind Energy Conversion Systems)* and the MEASNET European standards for measurements procedures.

The first TWT1650 prototype started working at Cabanillas Wind Farm (Spain) in August 2001. Since then more than 10 WTs were installed and a large amount of operational data was collected and used to improve the controllers of the machine. The main characteristics of the Torres Wind Turbine 1650 are described in Table 13.1.

Fig.13.19 TWT1650. (Courtesy of M.Torres)

Table 13.1 TWT1650 Characteristics.

Rotor	
Diameter	72 m
Swept Area	4072 m^2
Number of Blades	3
Position	Upwind
Nominal Rotor Speed	20 rpm
Range of Rotor Speed	Variable: 6 to 22 rpm
Control	Pitch controlled with independent electrical blade actuators
Tower	
Type	Tubular conical steel
Hub height	60 m
Weight	88000 Kg
Corrosion protection	Epoxy coating

Operation data	
Cut in wind speed	3 m/s
Nominal wind speed	12 m/s
Cut out wind speed (1)	25 m/s during 1 minute
Cut out wind speed (2)	30 m/s during 0.1 sec.
Survival speed	70 m/s
Generator	
Type	Direct Drive Multi-pole Synchronous Generator
Power to the grid	1650 kW
Voltage	650 V
Nacelle	
Construction	Monocoque in steel
Weight (rotor + hub)	73000 Kg
Yaw	Active system with electrical drives
Corrosion protection	Epoxy coating

The mechanical design of the TWT1650 was optimized using finite element calculations and test bed experimentation. The study focused on the analysis of fatigue, mechanical stress and structural resonances of the most critical elements of the system. Figure 13.20 shows a 3D drawing of the nacelle and Fig. 13.21 shows the finite elements grid of the generator.

The electrical diagram of the TWT1650 is shown in Fig. 13.22. It consists of a two three-phase reversible IGBT (*Insulated Gate Bipolar Transistor*) converters that connects the grid with the 2000 kW multi-pole synchronous generator, also designed by M.Torres (see Fig. 13.23). The design was optimized using finite elements calculations, advanced electrical simulators and a special test bed for full power experimentation (see Fig. 13.24).

The main advantages of the variable-speed direct-drive multi-pole system can be summarised as follows:

- *Power Quality Optimization* - It is able to control the power factor following the grid demand. The system also reduces the harmonics and flicker level, fulfilling the electrical standards.

- *Reliable Behavior Under Voltage Dips* - The machine is able to achieve E.ON standards, keeping the system connected under voltage dips. This fact improves grid stability and decreases the risk of blackouts which allows for a greater penetration of wind energy into the grid.

Fig.13.20 TWT1650 3D drawing (Courtesy of M.Torres)

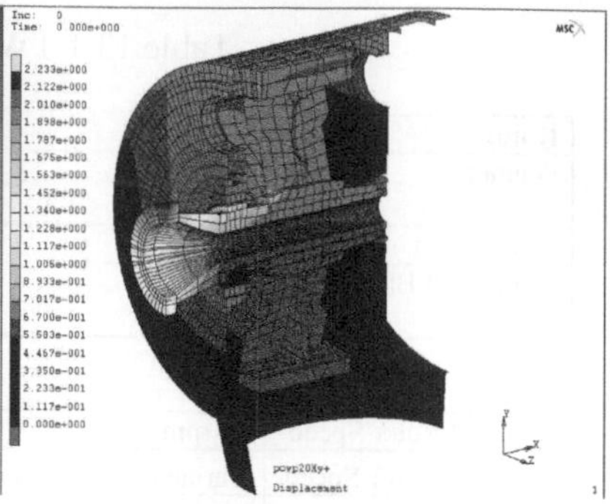

Fig.13.21 Finite elements analysis. of the generator.

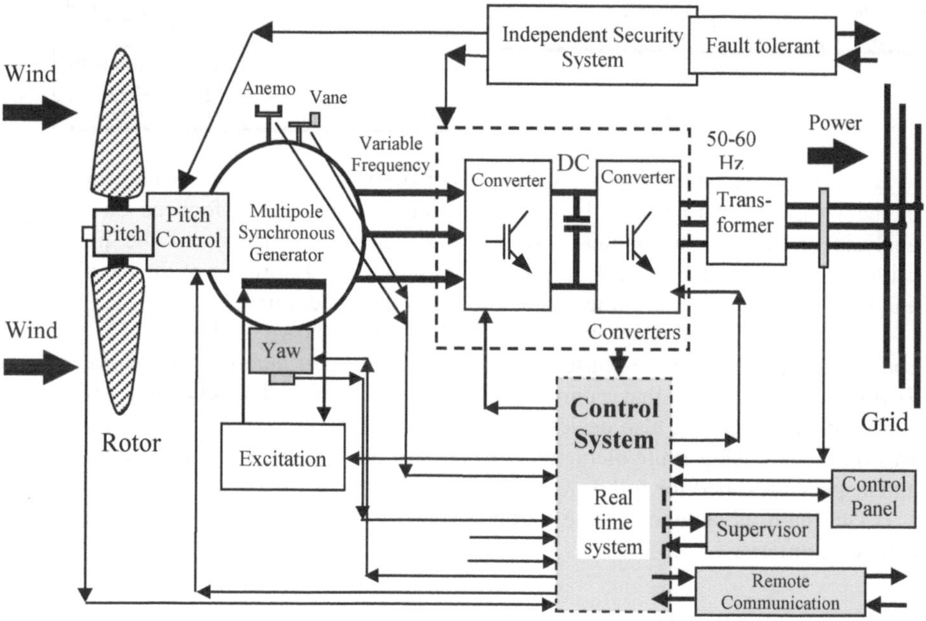

Fig. 13.22 Electrical and control diagram of the TWT1650.

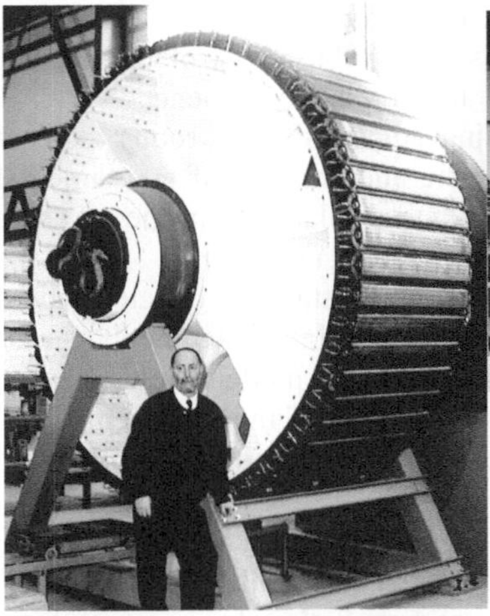

Fig. 13.23 Rotor of the multi-pole synchronous generator. (Courtesy of M.Torres).

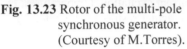

Fig. 13.24 Full Power Test Bed of WT designed by M.Torres. (Courtesy of M.Torres).

- *High Efficiency* - It produces more energy at every wind sector because the system is able to follow the maximum aerodynamic efficiency C_p at every wind speed (see Fig. 13.18, Zone 1).
- *Longer Machine Life* - The system minimises the mechanical fatigue by increasing the damping of the electrical torque with the controller.
- *Continuous Improvements* - It is easy to introduce new improvements and to parameterise the system for every Wind Farm by modifying the software.

13-3.3 CONTROL STRATEGIES

The main control objectives of the TWT are (1) the improvement of the power efficiency at every wind speed, (2) the attenuation of the transient mechanical loads and fatigue stresses, (3) the reduction of the electrical harmonics and flicker, and (4) the robustness against parameters variation.

Some critical problems arise in the design of the WT control system. These problems are (a) the difficulty to work safely with random and extreme gusts, (b) the complexity introduced by the strongly nonlinear multivariable and time variable mathematical model [Eqs. (13.14) and (13.15)], and (c) the impossibility to obtain a direct measurement of the wind speed experienced by the turbine because of the high uncertainty in the anemometer measurement and the strong influence of the blade movement (Leith and Leithead, 1997).[173]

The above set of motivations forced the control system designer to combine advanced control strategies such as QFT robust control techniques, adaptive schemes, multivariable methodology and predictive elements.

The controller design was made using advanced QFT robust control strategies, based on both, mathematical modeling and analysis of experimental data. The QFT templates of the system were first calculated from the physical model of the WT, tacking into account the parameter uncertainty. Once the machines were installed a large amount of operational data was collected (about 10000 data per second per WT, for over 3 years) to model the machine dynamics under very different situations. This large set of models (plants with uncertainty) was used to validate and increase the size of the QFT templates, which is the basis for continuously improving the controllers performance (Garcia-Sanz, 1998-2004).[174]

The variables that need to be controlled are the electrical power, torque, rotational rotor speed, pitch angle and pitch rotational speed (blades), yaw angle, power factor, generator current, dc voltage, current excitation, temperatures, etc. (Fig. 13.22). The next section contains some actual results of the rotor speed control at the nominal speed.

13-3.4 EXPERIMENTAL RESULTS

The control system of a pitch-regulated variable-speed wind turbine aims to maintain the rotor speed within a permitted range in above-rated wind speed (see Fig. 13.18, Zone 2). However, this requirement can be difficult to meet with conventional controllers during extreme gusts, particularly for large-scale machines where the pitch actuation capability may be comparatively quite limited.

This section investigates the experimental results of the QFT controllers with the 1.65 MW variable-speed turbine of M.Torres. Figure 13.25 shows the basic block diagram of the three cascade controllers used to control the rotor speed: rotor speed controller $C_1(s)$, pitch angle controller $C_2(s)$ and pitch speed controller $C_3(s)$.

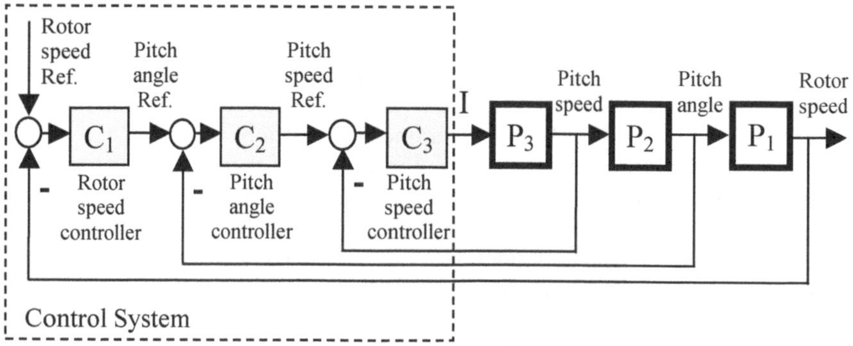

Fig. 13.25 Rotor speed control system block diagram.

The following two experimental cases, a medium wind speed and a very high wind speed, are presented:

Case 1: *Medium wind speed. 04.February.2003. Cabanillas Wind Farm (Spain)*

Figure 13.26 shows large wind gusts in a medium average wind speed context (15 m/s), with excursions from 13 m/s to 19 m/s. The control system shows a good performance, following the rotor speed set-point (20 rpm) correctly and rejecting the wind disturbances with a smooth movement of the blades.

Case 2: *Very high wind speed. 06.April.2003. Cabanillas Wind Farm (Spain)*

Figure 13.27 shows large wind gusts in a high average wind speed context (24 m/s), with excursions up to 30 m/s. The control system shows also a good performance, following the rotor speed set-point (20 rpm) correctly and rejecting the wind disturbances with a smooth movement of the blades.

Figures. 13.26 and 13.27 show, with experimental data of the TWT1650 at Cabanillas Wind Farm, how the QFT control strategies implemented in the WT are able to deal with very different operating points (from low to very high wind speed conditions), avoiding over-speed.

13-3.5 CONCLUSIONS PART II

This section introduced some general ideas about the control of variable-speed wind turbines and presented the design and experimental results of the TWT1650 wind turbine, designed by M.Torres and controlled with QFT strategies. The TWT1650 addressed a new technological concept: a gearless variable-speed pitch-controlled direct-drive multi-pole synchronous generator with a large rotor diameter (72 m), an output power of 1650 kW and an advanced robust control system. Part II showed the main advantages of the TWT and evaluated some of the most representative experimental results of the TWT1650 under medium and extreme wind conditions, using QFT robust control strategies.

13-4 SUMMARY

The purpose of this chapter is to present two additional real-world problems for which the QFT design technique was utilized to achieve the desired performance specifications. These two problems in addition to the real-world QFT control system designs of Chapters 10, 12 and 14 attempt to illustrate the concept of "*Bridging the Gap*" as stressed in Sec. 1-2 and in Chapters 9 and 11.

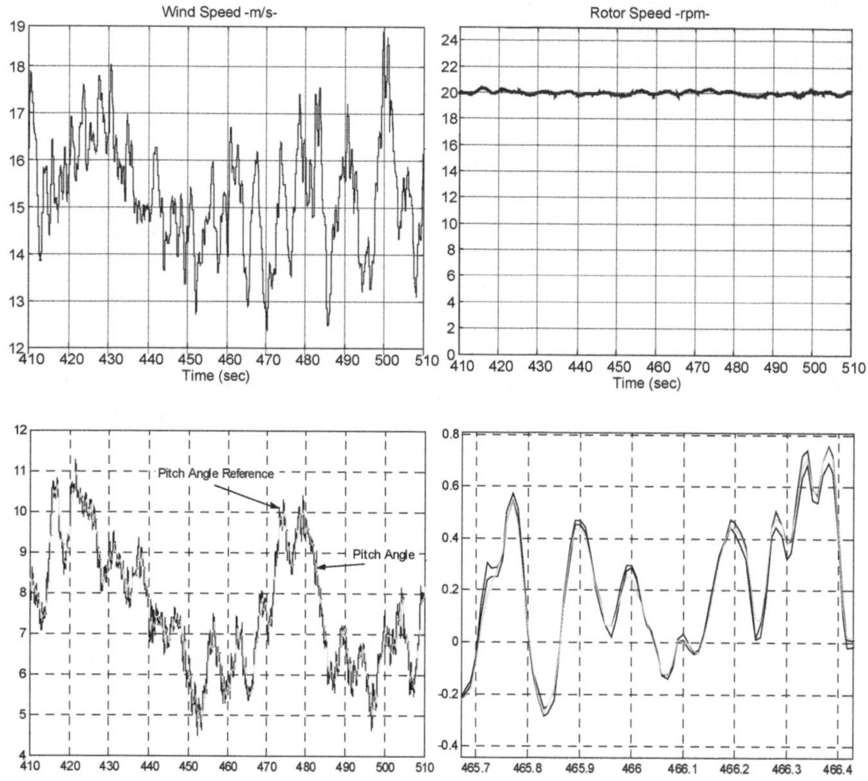

Fig. 13.26 Control under medium wind speed: Case 1 TWT experimental data.

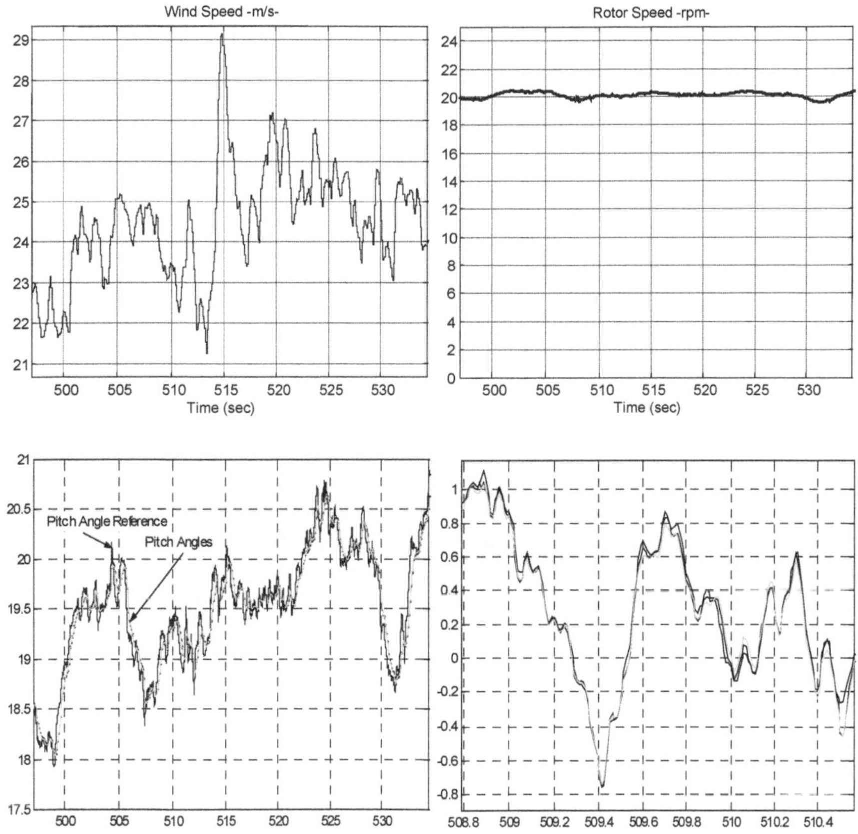

Fig. 13.27 Control under high wind speed: Case 2 TWT experimental data.

14

Weighting Matrices and Control Allocation

14-1 INTRODUCTION

When designing compensators for real systems accommodations must be made for attributes of systems that don't fit the paradigm of the control system design technique. It is often the case that the configuration of the real system is not compatible with the design technique. For instance, MIMO QFT design requires that the plant matrix, **P**(s) be square, that is, number of inputs must equal the number of outputs. Many real systems have more inputs than outputs,. For example, the inner loop of a flight control system (FCS) has three outputs [control roll rate (p), pitch rate (q), and yaw rate (r)] whereas many aircraft have more than three inputs, i.e. there are more than three control effectors.

In designing a QFT compensator one designs a fixed compensator that causes the closed-loop system to meet specifications throughout the envelope of operating conditions, the region of plant parameter uncertainty. There are many real systems, such as aircraft, where the plant model is nonlinear and is required to operate in a large envelope. While it is desirable, because of implementation and design concerns, to design a single fixed compensator for the full envelope, sometimes it is not possible and either multiple compensators must be designed or other methods must be employed, as is discussed later in this chapter.

One of the aims of MIMO QFT control design is to cause the closed-loop system to respond to commands to the system while minimizing cross-coupling between the different channels of the plant. There are many MIMO systems that are highly coupled and can present a challenge for QFT design. To perform a MIMO QFT design it is desirable to have a system where the determinate of the plant matrix, det **P**(s), is minimum phase (m.p.). This is

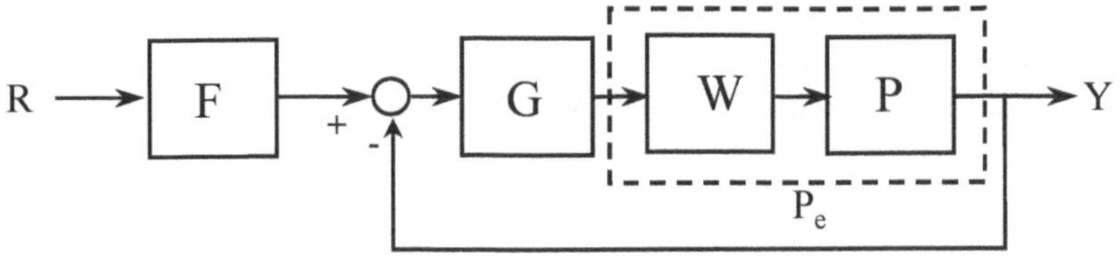

Fig. 14.1 Effective plant P_e.

desirable because non-minimum phase (n.m.p.) plants reduce the benefits of using feedback in the control system. For all of these examples, it is possible to use some form of a weighting matrix that is post-multiplied to **P**(s) to produce the effective plant, $\mathbf{P}_e(s) = \mathbf{P}(s)\mathbf{W}$, as shown in Fig. 14.1, during the QFT design, to correct undesirable attributes of the original systems.

A weighting matrix (**W**), as shown in Fig. 14.1, is useful because it can be used to precondition the plant to simplify the QFT design processes and, in some cases, is required to complete a QFT design. Fixed weighting matrices can be used for such purposes as achieving a square plant matrix, achieving a m.p. determinant of the effective plant ($\mathbf{P}_e$), and reducing cross coupling for the nominal plant. If the weighting matrix is variable and scheduled based on operating condition, then it can be used for such purposes as reducing cross-coupling for the full operating envelope, reducing the size of the templates, as well as other purposes discuss later in Sec. 14.3. Fixed weighting matrices can be constructed from constant elements or from dynamic elements. The variable weighting matrices can be generated from a set of table look-ups or be calculated based on the operating condition. An example of the latter form of variable weighting matrix is dynamic control allocation, which uses linear, or nonlinear, programming to compute control effector commands based on the operating condition and the inputs from the compensator. Chapter 7 contains a discussion of the techniques for designing **W** to remediate the problem of n.m.p. plants. This chapter discusses design of fixed and variable weighting matrices and how they can be used to achieve square $\mathbf{P}_e(s)$, how to reduce cross-coupling, how to achieve fixed compensator for the full operational envelope, and how to improve closed-loop system performance.

14-2 FIXED WEIGHTING MATRICES

Fixed weighting matrices have been used extensively for preconditioning plants for QFT designs.[179,180,181,182] They can be implemented with either constant or dynamic elements. Constant elements are more common because they are easier to design and implement. Discussed in the next few sections, is the design of fixed weighting matrices in a QFT design that must take into account all of the designed operating conditions. That is, a **W** is used to precondition $\mathcal{P}$ for all $\mathbf{P}\in\mathcal{P}$. This makes it likely that the designed **W** will not be the "optimal" implementation of **W** for many, or all, of the off nominal plant cases. Even though a fixed **W** is not optimal it is much simpler to implement than a variable **W** and can be very effective. Since the fixed weighting matrix is combined with each of the plants, $\mathbf{P}\in\mathcal{P}$, to form effective plants for the QFT design, the stability of the compensated system is guaranteed by meeting the stability bounds during the QFT design process. The following sections start with a discussion of the use of dynamic elements in fixed weighting matrices. Then three methods of a fixed weighting matrix design are presented, one based on engineering knowledge, one based on the frequency response of **P**, and one based on the pseudo-inverse of $\mathbf{P}\in\mathcal{P}$.

14-2.1 Fixed W with Dynamic Elements

Fixed weighting matrices can have constant or dynamic elements. Dynamic elements are elements of the weighting matrix that have poles and zeros, i.e. $w_{ij} = N(s)/D(s)$. When designing a fixed weighting matrix, dynamic elements can be used for purposes such as insertion of notch filters for bending modes and implementing low pass filters for limiting the bandwidth of command signals injected into selected plant inputs. Implementing dynamic elements in the weighing matrix can make the final compensator more effective, but also add to the complexity of the compensator. That is, dynamic elements in the weighting matrix are part of the control system. In many cases the desired effects of the dynamic weighting matrix elements can be designed into the QFT compensator and thus preventing possible over design by inserting dynamic elements into the control system before the start of the QFT design process. Any of the three methods described in the following sections can be used to design a fixed **W** with dynamic elements.

14-2.2 Fixed W – Engineering Knowledge Method

If the plant model, $\mathcal{P}$, is well known, the weighting matrix can be designed using engineering knowledge. Weighting matrices can be designed in this manner to achieve m.p. det $\mathbf{P}_e$, to reduce cross-coupling, and to produce a square $\mathbf{P}_e$. In the case where the plant is not square, a weighting matrix is needed to perform a QFT design. The weighting matrix in this case is used as a "control mixer" to map the system commands to the control effector commands, (see Fig. 14.2). The design of **W** can be accomplished by selecting the elements of **W**, i.e $w_{ij} \in \mathbf{W}$, based on knowledge of the plant.

For example, given a fighter aircraft with five control surfaces, two ailerons (δ_{al}, δ_{ar}), two elevators (δ_{el}, δ_{er}), and one rudder (δ_r), see Fig. 14.3, the designer can use knowledge of the reaction of the aircraft due to movement of

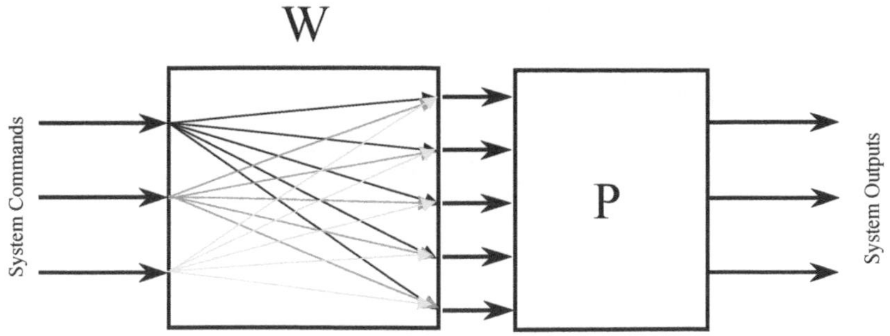

Fig. 14.2 Weighting matrix used as a control mixer.

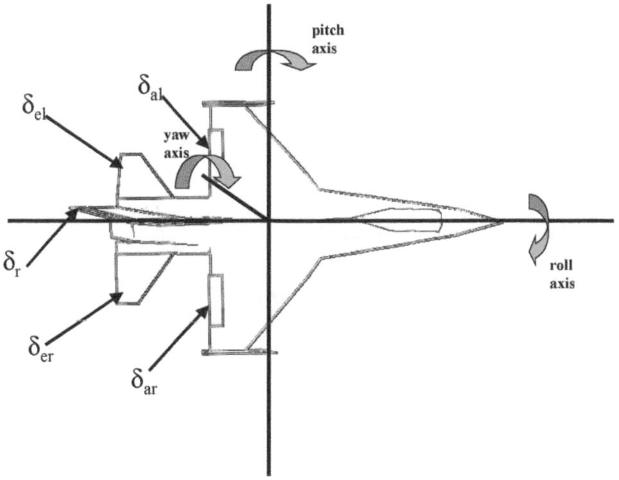

Fig. 14.3 Example aircraft.

the control surfaces to design **W**. For this design, the roll axis is defined to be positive through the nose. The pitch axis is defined as positive through the right wing and the yaw axis is defined as positive down through the center of gravity of the aircraft. In this case, the ailerons are primarily used to produce a moment about the roll axis, but can also produce a moment about the pitch axis. The horizontal elevators are primarily used to produce a moment about the pitch axis, but can also produce a moment about the roll axis. The rudder is used to produce a moment about the yaw axis. Note that there are other effects of deflecting these control surfaces, but they are neglected for this example. Assume that the desired commands, for the fighter, are roll rate (p), pitch rate (q), and yaw rate (r). As shown in Fig. 14.4, the inputs to **W** are:

$$\boldsymbol{u} = \begin{bmatrix} u_p & u_q & u_r \end{bmatrix}^T \tag{14.1}$$

The inputs to the plant are:

$$\boldsymbol{v} = \begin{bmatrix} \delta_{al} & \delta_{ar} & \delta_{el} & \delta_{er} & \delta_r \end{bmatrix}^T \tag{14.2}$$

The outputs from the plant are:

$$\boldsymbol{y} = \begin{bmatrix} p & q & r \end{bmatrix}^T \tag{14.3}$$

The equation relating inputs to the weighting matrix to outputs is:

$$\boldsymbol{v} = \boldsymbol{Wu} \tag{14.4}$$

and for the plant:

$$\boldsymbol{y} = \boldsymbol{Pv} \tag{14.5}$$

Thus. the effective plant, $\mathbf{P}_e$, is defined by:

$$\boldsymbol{y} = \boldsymbol{PWu} = \boldsymbol{P}_e\boldsymbol{u} \tag{14.6}$$

A candidate weighting matrix for this example is:

$$\boldsymbol{W} = \begin{bmatrix} 1.0 & -1.0 & 0.25 & -0.25 & 0.0 \\ 0.1 & 0.1 & 1.0 & 1.0 & 0.0 \\ 0.0 & 0.0 & 0.0 & 0.0 & 1.0 \end{bmatrix}^T \tag{14.7}$$

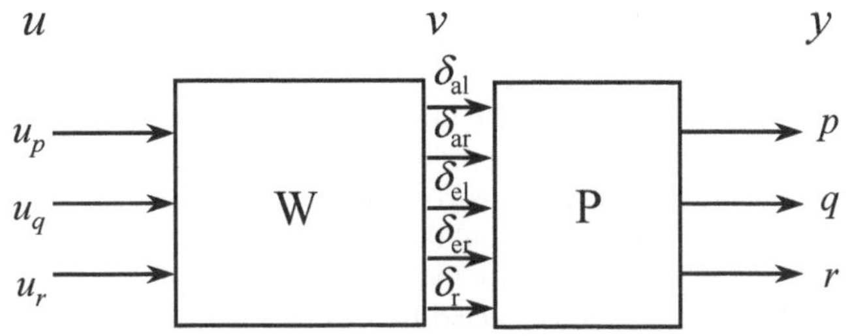

Fig. 14.4 Example aircraft effective plant.

When Eq. (14.7) is used as a weighting matrix, commands in the roll channel cause asymmetrical deflections of the left and right ailerons and elevators, which produces a moment about the roll axis. Because of effects neglected in this example, the use of elevators for producing roll moments is set to a fraction of that of the ailerons. Commands in the pitch channel result in symmetrical deflections of the left and right elevators and ailerons. Yaw commands result in the deflection of the rudder.

The signs of the elements of **W** are based on the mathematical model of the aircraft. For example, a positive deflection of the left aileron produces a positive moment about both the roll and pitch axes. A positive deflection of the right aileron produces a negative moment about the roll axis and a positive moment about the pitch axis. The magnitudes of the elements are based on engineering judgment. Since the ailerons are the most effective control surfaces in the roll axis, their gains are set to 1, i.e. $w_{1,1} = w_{1,2} = 1.0$. Deflection of the elevators produces rolling moment so they can be used to assist roll, but their primary use is to generate pitching moment so the gains are set accordingly, i.e. $w_{1,3} = w_{1,4} = 0.25$. The rudder produces a rolling moment, but also introduces other dynamics therefore, its gain is set to 0.0, i.e. $w_{1,5} = 0.0$.

Once the design is completed the resulting $\mathbf{P}_e$ for each $\mathbf{P} \in \mathcal{P}$ must be examined to insure that the design requirements are met. For example if **W** is used to achieve a m.p. det $\mathbf{P}_e$ (s) then each $\mathbf{P}_e \in \mathcal{P}_e$ must have a m.p. determinant. If the resulting $\mathbf{P}_e$ does not achieve the design requirements with the candidate **W**, then **W** must be modified, by trial and error, to find a **W** that meets the design requirements for all $\mathbf{P}_e \in \mathcal{P}_e$.

14-2.3 Fixed W – Frequency Response Method

If engineering knowledge is not sufficient to produce the required effective plants, the frequency responses of the individual plants can be used to aid in the design of the weighting matrix. The frequency responses are used to determine the effect of the movement of the control effectors on the plant outputs. This makes it possible to analyze the system for the required signs and gains for the weighting matrix elements, strategies for removing cross-coupling, and strategies for achieving a m.p. det $\mathbf{P}_e$.

By analyzing the magnitude and the phase of the frequency responses, the sign and gain of the elements of the weighting matrix, w_{ij}, can be determined. Individual frequency responses can be used to find areas where the system would benefit from the addition of dynamic w_{ij} elements. As with all fixed **W** methods, the candidate weighting matrix must be evaluated with all of the effective plants, $\mathbf{P}_e \in \mathcal{P}_e$ to insure that all of the effective plant meet the design requirements.

14-2.4 Fixed W – Pseudo-Inverse Method

The pseudo-inverse method, termed "Method of Specified Outputs" in Ref. 180, is based on manipulating the equations of the system to find a suitable weighting matrix. With this method, the nominal plant case, $\mathbf{P_o}$, is used to design a single $\mathbf{W}(s)$ and then the $\mathbf{W}(s)$ is evaluated with all of the plant cases, $\mathbf{P} \in \mathcal{P}$. From Fig. 14.4:

$$y = P(s)v \tag{14.8}$$

and:

$$v = W(s)u \tag{14.9}$$

Define the effective plant as:

$$P_e \equiv P(s)W(s) \tag{14.10}$$

Substituting Eq. (14.9) and Eq. (14.10) into Eq. (14.8) yields the input/output relationship of the effective plant:

$$y = P_e(s)u \tag{14.11}$$

Since the dynamics of the effective plant represent the dynamics of the system, a desired effective matrix, $\mathbf{P}_{e_{des}}(s)$, can be defined, where the elements of $\mathbf{P}_{e_{des}}(s)$, $p_{e_{des_{ij}}}(s)$, are transfer functions, that define the desired characteristics of the open loop system model. The desired characteristics for $\mathbf{P}_e(s)$ are: (1) a minimum phase determinant, (2) as diagonal as possible, and (3) to yield reasonable and stable responses to reasonable inputs. Therefore, to design $\mathbf{W}(s)$ select a $\mathbf{P}_{e_{des}}(s)$ and then set Eq. (14.10) equal to $\mathbf{P}_{e_{des}}(s)$ and solve for W(s):

$$P_{e_{des}}(s) \equiv P(s)W(s) \tag{14.12}$$

Since $\mathbf{P}(s)$ is non-square, define the right side pseudo-inverse of $\mathbf{P}(s)$ as:

$$P^{\#} = P^T\left(PP^T\right)^{-1} \tag{14.13}$$

where,

$$PP^{\#} = I \tag{14.14}$$

Post-multiplying both sides of Eq. (14.12) by Eq. (14.13) and rearranging yields:

$$W = P^{\#} P_{e_{des}} \tag{14.15}$$

Equation (14.15) defines a weighting matrix, **W**, which, when inserted into Eq. (14.12) results in the desired effective plant. Any of the plant cases from $\mathbf{P}_i \in \mathcal{P}$ can be used to design **W**. The resulting **W** produces $\mathbf{P}_{\mathbf{e_{des}}}(s)$ for the selected plant case. The resulting **W** must then be evaluated for all of the plant cases to make sure that the desired characteristics hold for each $\mathbf{P}_i \in \mathcal{P}$.

There are two drawbacks in using the pseudo-inverse method. They are: **W**(s) can be difficult to calculate and it can be difficult to find a single **W**(s) that results in an acceptable $\mathbf{P}_{\mathbf{e_i}}$ for all $\mathbf{P}_i \in \mathcal{P}$. Since calculating **W**(s) involves finding the inverse of a matrix of transfer functions, significant problems with numerical precision can arise. The problems can be alleviated by either approximating the results, as is done in Ref. 180, or using a program capable of symbolic math to accomplish the inverse. Finding an acceptable **W**(s) can be accomplished though an iterative process of combining a candidate **W**(s) with all of the plants, $\mathbf{P}_i(s) \in \mathcal{P}(s)$, to find the **W**(s) that best achieves the design goals. For an example of using the pseudo-inverse method, the reader is referred to Ref. 182.

14-3 VARIABLE WEIGHTING MATRICES

Variable weighting matrices are more complicated to implement than fixed weighting matrices, but they are able to tailor the effect of the weighting matrix to the operating condition, plant case. This means that variable weighting matrices have the potential of increasing the performance of the system. The use of variable weighting matrices introduces the possibility to do such things as, decrease cross-coupling and guarantee a minimum phase det $\mathbf{P}_e$, throughout the operational envelope. As discussed in Sec. 14-3.2, dynamic control allocation is a form of a variable weighting matrix. When dynamic control allocation is used it is possible to account for nonlinearities in the control effectors, identify operating conditions outside of the envelope, account for control effector failures, and optimize the use of effectors for different modes of operation. The stability of systems using variable weighting matrices is guaranteed by meeting the stability bounds during the QFT design process. Since the variable weighing matrices use the state of the system to select, or calculate the current **W**, checks must be in place to insure that the state of the system is identified correctly. The following sections concentrate on two implementation of variable weighting matrices, table look-ups and dynamic control allocation.

14-3.1 Table Look-up

The table look-up method of variable weighting matrix design and implementation is based on sensing the operating condition of the plant and then selecting a weighting matrix designed for that operating condition. It may be difficult, or impossible, to sense the operating condition directly. If direct sensing of the operating condition is not possible it may be possible that an estimate of the operating condition will be sufficient for weighting matrix selection during operation. If the operating condition is not available then a fixed weighting matrix must be used. Based on the operating condition the elements of the weighting matrix, $w_{ij}(s)$, can be obtained either by calculating them based on the stored equations or retrieved from storage in table look-ups. In either case the information can be in a structure such as simple "if...then" statements, linear or non-linear data tables, and neural networks. As with the determination of the operating condition, the method of weighting matrix data storage and retrieval depends on the capabilities of the hardware and software implementing the control system.

Design of a table look-up variable weighting matrix can be accomplished by using any of the methods describe in section 14.2. That is, each of the fixed weighting matrix design methods in section 14.2 is capable of producing a weighting matrix, $\mathbf{W}_i$, for each operating condition in the envelope of operation. For the QFT design, each of the weighting matrices is combined with the corresponding plant case, $\mathbf{P}_i$, to produce the set of effective plants for the design, i.e.:

$$\boldsymbol{P}_{e_i} = \boldsymbol{P}_i \boldsymbol{W}_i, \quad \forall\, i \in \{1,2,...,J\} \tag{14.16}$$

where J is the number of operating conditions.

This assumes that, during the operation of the closed loop system, the proper $\mathbf{W}_i$, for the operating condition, is present. It also assumes that the selected $\mathbf{W}_i$ is valid for operating conditions that are not included in the design of the QFT compensator. That is, the operating conditions selected for the design representing the full envelop of operation, but do not exhaustively cover the entire envelope. This means that the method used to retrieve the parameters for the weighting matrix has to perform some kind of interpolation to find the best fit of the weighting matrix to a given operating condition.

The table look-up form of variable weighting matrix design and implementation can be very effective in decoupling the effective plants and improving the performance of the closed-loop system. If some kind of control effector failure detection is implemented, then it is possible to accommodate control failures by implementing weighting matrices for each corresponding failure condition. But, these improvements come at the cost of more complexity in the design process and the overhead of implementing table-look ups in the final system.

14-3.2 Dynamic Control Allocation

While dynamic control allocation is discussed under the variable weighting matrices section, there is never an explicit weighting matrix calculated. Instead, dynamic control allocation calculates control effector commands based on the operating condition and the output from the compensator, **G**, at each time step. This has the effect of calculating an operating condition dependent on the weighting matrices at each time step. The use of dynamic control allocation makes it possible to accommodate control effector nonlinearities such as position and rate limits. In those systems where control effectors are the primary contributors to coupling between control channels, dynamic control allocation can have a profound effect in reducing or eliminating coupling. On systems where there are more control effectors than control channels, dynamic control allocation can exploit redundancies to optimize control effector responses to operate in selected modes. For example, control effector operation could be tuned to produce better fuel economy, while continuing to meet the requirements of the QFT designed closed loop system. This is discussed later in this section.

The form of dynamic control allocation discussed here relies on finding a pseudo-inverse of the control effectiveness matrix, **B**, to solve for the control effector commands required to produce the output commanded by the compensator, at each time step. Since the control effectiveness matrix is a function of the operating condition of the system, at a given time, then provisions must be made to sense the operating condition and generate the correct control effectiveness matrix. Given the state-space model of the system:

$$\dot{\boldsymbol{x}} = \boldsymbol{A}\boldsymbol{x} + \boldsymbol{B}\boldsymbol{u} \tag{14.17a}$$

$$\boldsymbol{y} = \boldsymbol{C}\boldsymbol{x} + \boldsymbol{D}\boldsymbol{u} \tag{14.17b}$$

solve Eq. (14.17a) for **x** and substitute into Eq. (14.17b) and then take the Laplace transform. Assume there is no direct feed through in the system so that $\boldsymbol{D} = [0]$. For many systems this is not an overly restrictive assumption. This results in:

$$\boldsymbol{y} = \boldsymbol{C}(s\boldsymbol{I} - \boldsymbol{A})^{-1}\boldsymbol{B}\boldsymbol{u} \tag{14.18}$$

The right pseudo-inverse of the control effectiveness matrix is now defined as:

$$\boldsymbol{B}^{\#} \equiv \boldsymbol{B}^{\mathrm{T}}\left(\boldsymbol{B}\boldsymbol{B}^{\mathrm{T}}\right)^{-1} \tag{14.19}$$

Where

$$BB^{\#} = I \tag{14.20}$$

and

$$u \equiv B^{\#} u' \tag{14.21}$$

Substituting Eq (14.21) into Eq. (14.18) results in:

$$y = C(sI - A)^{-1} BB^{\#} u' \tag{14.22}$$

A block diagram of Eq (14.22) is shown in Fig. 14.5. Substituting Eq. (14.20) into Eq. (14.22) and rearranging defines the effective plant to be:

$$P_{\mathrm{e}} \equiv C(sI - A)^{-1} \tag{14.23}$$

Thus, Eq. (14.23) can be used to calculate the effective plants, $\mathbf{P}_{\mathrm{e_i}}$, for every operating condition in the envelope. These effective plants can then be used in the QFT design process to design a closed loop control system. In order to implement this control system, operating condition dependent, the pseudo-inverse of the control effectiveness matrix, $B_i^{\#}$, must be applied to the inputs of the plant as shown in Fig 14.5.

Equation (14.22) is valid as long as the control effectors are not rate or position limited. That is, when the control effectors are effected by nonlinearities such as rate or position limits Eq. (14.22) is no longer valid. Since the pseudo-inverse is not unique it is possible to keep the control effectors operating in their linear region by introducing weights in the definition of the pseudo-inverse. For example, the inverse can be defined as:

$$B^{\#} \equiv D^{-1} B^{\mathrm{T}} \left(B D^{-1} B^{\mathrm{T}} \right)^{-1} \tag{14.24}$$

where **D** is a weighting matrix that can be used to emphasize the use of selected control effectors over others. This does not prevent the control effectors from becoming rate or position limited, but a penalty can be placed on large control deflections which can minimize the occurrence. To accommodate rate/position limits in control effectors a linear programming solution can be used.

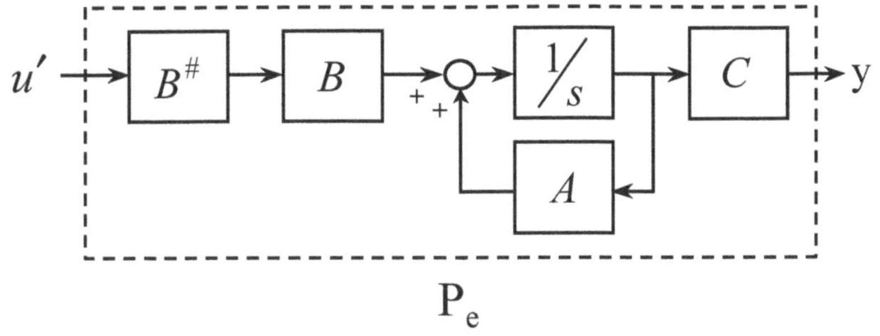

Fig. 14.5 Effective plant with pseudo-inverse of the B matrix.

In Ref. 180, Buffington developed a linear programming solution that is capable of dynamically solving for:

$$\boldsymbol{Bu} = \boldsymbol{u}' \tag{14.25}$$

while enforcing other constraints to ensure that the control effectors are not saturated, if possible. Substituting Eq, (14.25) into Eq. (14.18) and rearranging results in Eq. (14.26) which is defined as the effective plant derived using the pseudo-inverse of the control effectiveness matrix. In order to detect when there is not sufficient control power, $\boldsymbol{Bu}$, to meet the command from the compensator, $\boldsymbol{u}'$, Buffington used the following linear program:

$$\min_{u} \mathrm{J} = \begin{bmatrix} 0 \cdots 0 \; 1 \cdots 1 \end{bmatrix} \begin{bmatrix} \mathbf{u} \\ \mathbf{u}_s \end{bmatrix}$$

(14.26)

$$\text{subject to} \begin{bmatrix} \mathbf{u}_S \\ -\mathbf{u} \\ \mathbf{u} \\ -\mathbf{Bu} + \mathbf{u}_S \\ \mathbf{Bu} + \mathbf{u}_S \end{bmatrix} \geq \begin{bmatrix} \mathbf{0} \\ -\overline{\mathbf{u}} \\ \underline{\mathbf{u}} \\ -\mathbf{u}' \\ \mathbf{u}' \end{bmatrix}$$

where $\boldsymbol{u}_s$ is a slack variable, $\overline{\boldsymbol{u}}$ and $\underline{\boldsymbol{u}}$ represent the most restrictive actuator constraints, see Ref. 180. A *slack variable* is a variable that is added to a linear inequality to change it to an equality. For example,

$$x + y\_<\_15 \quad \text{becomes} \quad x + y + u = 15$$

where u takes up the slack between x + y and 15. If $J = 0$ then there is sufficient control power to meet the command, i.e. $\boldsymbol{Bu} = \boldsymbol{u}'$ can be satisfied. If there is not sufficient control power then a different compensator must be used or the commanded trajectory must be altered. If there is sufficient control power then the following linear program can be used to solve for $\boldsymbol{Bu} = \boldsymbol{u}'$:

$$\min_{u} \mathrm{J} = \boldsymbol{W}_{\mathrm{u}}\boldsymbol{u}_{\mathrm{s}}$$

$$\text{subject to} \begin{bmatrix} \boldsymbol{u}_{\mathrm{s}} \\ -\boldsymbol{u} \\ \boldsymbol{u} \\ -\boldsymbol{u} + \boldsymbol{u}_{\mathrm{s}} \\ \boldsymbol{u} + \boldsymbol{u}_{\mathrm{s}} \end{bmatrix} \geq \begin{bmatrix} \underline{\boldsymbol{0}} \\ -\overline{\boldsymbol{u}} \\ \underline{\boldsymbol{u}} \\ -\boldsymbol{u}_{\mathrm{pref}} \\ \boldsymbol{u}_{\mathrm{pref}} \end{bmatrix}$$

$$\boldsymbol{Bu} = \boldsymbol{u}' \tag{14.27}$$

$$\boldsymbol{W}_{\mathrm{u}}^{\mathrm{T}} \in \boldsymbol{IR}^{\mathrm{m}}$$

$$\boldsymbol{u}_{\mathrm{s}} \in \boldsymbol{IR}^{\mathrm{m}}$$

where $\boldsymbol{u}_{pref}$ contains the preferred control effector commands, and $\boldsymbol{W}_u$ contains weights used to select different modes of operation. By selecting appropriate combinations of $\boldsymbol{u}_{pref}$ and $\boldsymbol{W}_u$ the system can be configured to use redundant control authority to introduce different performance modes such as minimum drag or minimum radar signature modes.

The major drawback in using dynamic control allocation in a real system is the amount of processing time required to find solutions for the linear programming problems. The problems must be solved at each time step during operation of the closed-loop system. In Ref. 182 flight tests were performed with a control system that included linear programming in the real-time system. The linear programming algorithm that was implemented ran sufficiently fast to use at the desired update rate.

14-4 SUMMARY

Weighting matrices can be used to achieve a square effective plant matrix, reduce cross- coupling between control channels, achieve an m.p det $\mathbf{P}_e$, reduce the size of the templates, and increase the performance of the closed-loop system. Two methods of implementing weighting matrices, fixed and variable methods, are presented in this chapter. Three design methods for fixed weighting matrices are described, one using engineering knowledge of the plant, one that uses the frequency responses of the plants, and one that uses the pseudo-inverse of the plant. Fixed weighting matrices are desirable because they are less complicated to implement than variable controllers. The major drawback in using fixed weighting matrices is that they are not optimized for all of the plants over the operating envelope. If the plants are sufficiently different the system may benefit from the use of a variable weighting matrix.

Variable weighting matrices are desirable because they can be tuned for every plant case in the operating envelope. The drawbacks of using variable weighting matrices are that they can be complicated to implement, and they can require more computational capacity than is available to the control system. If linear programming solutions are used, then additional benefits such as accommodation for control effector nonlinearities, detection of conditions where there is insufficient control power, and the ability to take advantage of redundant control effectors to introduce performance modes.

APPENDIX A

TEMPLATES GENERATION[121]

A-1 TEMPLATE ANALYSIS

One of the main tasks when designing QFT robust controllers is the definition of the plant frequency response, taking into account the complete *n*-dimensional space of model plant uncertainty. Consider an uncertain plant ***P***, where $\boldsymbol{P} \in \wp$, and $\wp$ is the set of possible plants due to uncertainty, such that,

$$\wp = \left\{ \begin{array}{l} P(j\omega_i, a_0, \ldots a_l, b_0, \ldots b_m),\ a_0 \in [a_{0\_min}, a_{0\_max}],\ \ldots\ b_m \in [b_{m\_min}, b_{m\_max}]; \\ a_i, b_i \in \Re,\ \ \omega_i \in \Re^+,\ i, l, m \in N \end{array} \right\}$$

(A.1)

$\Im\boldsymbol{P}(j\omega_i)$ is the associated *template* for $\omega = \omega_i$, that is to say, the set of complex numbers representing the frequency response of the uncertain plant at a fixed frequency ω_i. In other words, it is the projection of the *n*-dimensional parameter space onto $\Re^2$, according to the plant transfer function ***P*** for a given frequency (see Fig. A.1).

In general, it is not true that the projection of the contour of the uncertain parameter space is the boundary of the template. In fact it could happen that inner points of the parameter space, after the projection, are points of the boundary of the template.

In this context, assuming the rank of the Jacobian matrix ***M*** [see Eq. (A.2)] of the projection of the uncertain parameter space onto the complex map $\Re^2$, according to the plant transfer function ***P***, is two, then the inner points of the initial space are also inner points of the projected space as illustrated in Fig. A.1.[121] To be more precise, research of the minimum dimension *k* that represent the template is reduced to the study of the rank of several sub-matrix of ***M***, so that,

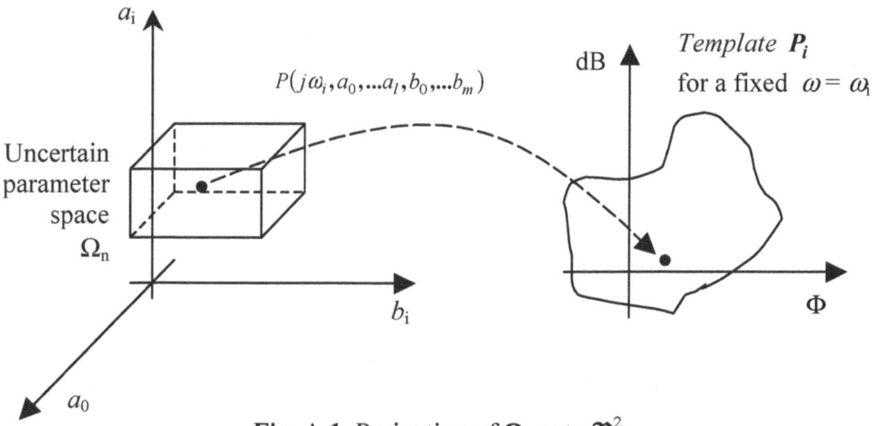

Fig. A.1 Projection of $\mathbf{\Omega}_n$ onto $\Re^2$.

$$M = \begin{bmatrix} \frac{\partial Re[P(j\omega_i, a_0, \ldots a_l, b_0, \ldots b_m)]}{\partial a_0} & \cdots & \frac{\partial Re[P(j\omega_i, a_0, \ldots a_l, b_0, \ldots b_m)]}{\partial b_m} \\ \frac{\partial Im[P(j\omega_i, a_0, \ldots a_l, b_0, \ldots b_m)]}{\partial a_0} & \cdots & \frac{\partial Im[P(j\omega_i, a_0, \ldots a_l, b_0, \ldots b_m)]}{\partial b_m} \end{bmatrix}_{2 \times n} \tag{A.2}$$

where $n = l + m + 2$.

To find such a minimum dimension k, an iterative procedure from $k = n$-1 to $k = 1$ is introduced.[121] It involves symbolic calculus. For each step it is checked whether the rank of the $\binom{n}{k+1}$ possible sub-matrices of k+1 columns of $\boldsymbol{M}$ is two, for every combination of the k+1 uncertain parameters that correspond with those k+1 columns of each sub-matrix. Such a checking must be satisfied for the whole 2^{n-k-1} possible combinations of both minimum and maximum of the n-k-1 non-considered parameters in each sub-matrix.

When the above condition is satisfied, then the projection onto $\Re^2$ of the inner points of the $\binom{n}{k+1}2^{n-k-1}$ sets $\boldsymbol{\Omega}_{k+1}$ will be within the template contour. Hence the $\binom{n}{k}2^{n-k}$ sets $\boldsymbol{\Omega}_k$ will be enough to define the template contour. The number of points that could remove from the template is

$\sum_{i=k+1}^{n}\left\{\binom{n}{i} 2^{n-i}(r-2)^{i}\right\}$, where i corresponds to an iteration that fulfills the rank condition, and where k goes from n-1 to 1.

A-2 EXAMPLE

In order to clarify the above ideas, let us consider an example that does not fulfill the rank condition [rank($\boldsymbol{M}$) = 2]. This case corresponds to the following second order system with time delay,

$$P(s)=\frac{1}{s^2+2\zeta\,\omega_n\,s+\omega_n^2}\;\exp(-\tau\,s) \tag{A.3}$$

where $\zeta = 0.02$, $\omega_n = [0.7, 1.2]$, $\tau = [0, 2]$. Figure A.2 shows the template that corresponds to the frequency of 1 rad/s. Cross points (+) are the projections of

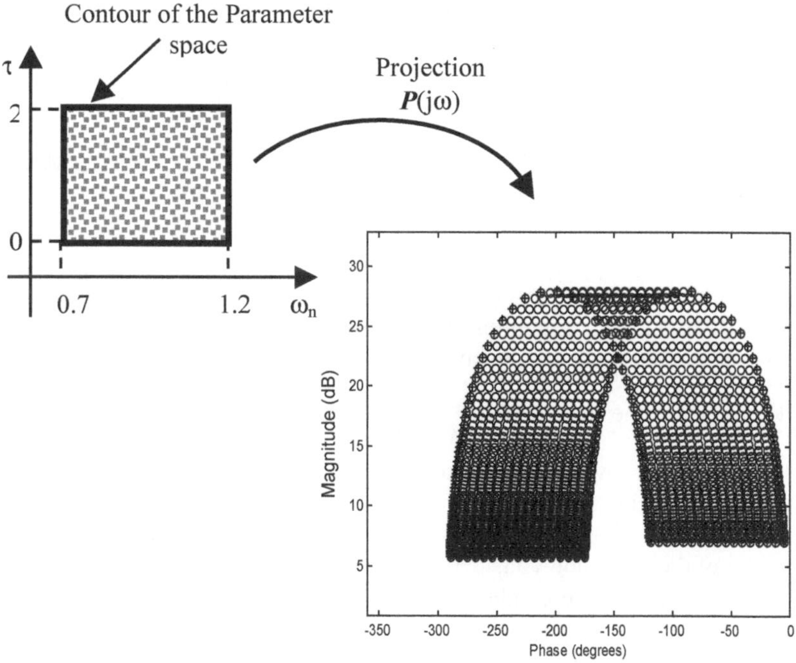

Fig. A.2 Projection onto $\Re^2$ of the whole 2-dimensional parameter space (o) and of the parameter space contour (+).

the parameter space contour and circle points (o) are the projections of the whole parameter space. It is observed that the projection of the contour (+) of the parameter space does not match the boundary at the top of the template. Using the previous method[121] it is possible to detect the problem. The top points of the template make rank($\boldsymbol{M}$) < 2.

APPENDIX B

INEQUALITY BOUND EXPRESSIONS[114, 115]

Consider the two-degrees-of-freedom feedback system shown in Fig. B.1. In a general case the transfer function ***P(s)*** represents an uncertain plant, where $\boldsymbol{P} \in \mathcal{P}$, and $\mathcal{P}$ is the set of possible plants due to uncertainty. The compensator ***G(s)*** and the pre-filter ***F(s)*** are designed to meet robust stability and robust performance specifications, following the desired reference ***R(s)***, rejecting the disturbances $\boldsymbol{D}_{1,2}(s)$ and the signal noise *N(s)*, using a limited control signal ***U(s)*** and minimizing the 'cost of feedback' (excessive bandwidth).[40, 128]

A fundamental step in the QFT methodology is the representation of the control objectives, modified with the model plant uncertainty, by some lines at every frequency on the Nichols Chart. These lines are called *Bounds* or *Horowitz-Sidi Bounds*. They synthesize the performance specifications and the model uncertainty and allow the designer to use only one plant, the nominal plant P_0(s), to design the compensator (controller).

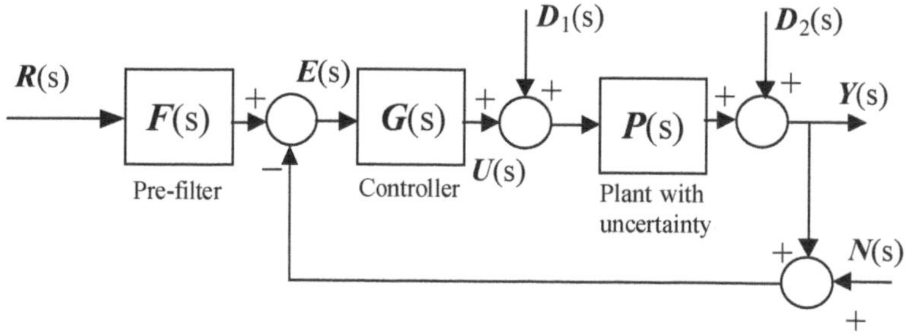

Fig. B.1 Control system structure.

Initially, designers used to calculated the bounds by using graphical manipulation of the specifications and the *templates* on the Nichols Chart, as it was introduced in Chapter 3. In 1993 Chait and Yaniv[114] developed an iterative algorithm to compute the bounds through quadratic inequalities. Afterwards some software pakages[89,90,91,121] implemented that algorithm for solving the bounds representation.

Closed loop specifications are usually described in terms of frequency functions $\delta_k(j\omega)$ that are imposed on the magnitude of the system transfer functions $|T_k(j\omega)|$. Table B.1 shows the stability and performance specifications in terms of transfer functions: $|T_k(j\omega)| \leq \delta_k(j\omega)$, $k = 1,...,5$. $\delta_1(j\omega)$ restricts the transfer function $|L/(1 + L)|$, $L = GP$, implying conditions on the robust stability, the control effort in the input disturbance rejection ($|U/D_1|$) and the sensor noise attenuation ($|Y/N|$) [Eq. (B.1)]. $\delta_2(j\omega)$ and $\delta_3(j\omega)$ constrain the transfer functions $|1/(1 + L)|$ and $|P/(1 + L)|$, respectively, for output disturbance rejection ($|Y/D_2|$) [Eq. (B.2)] and input disturbance rejection ($|Y/D_1|$) [Eq. (B.3)]. $\delta_4(j\omega)$ restricts the control signal $|G/(1 + L)|$ for the system output disturbance rejection ($|U/D_2|$), the noise attenuation ($|U/N|$) and the tracking of reference signals ($|U/RF|$) [Eq. (B.4)]. The upper $\delta_{5\text{sup}}(j\omega)$ and lower $\delta_{5\text{inf}}(j\omega)$ models constrain the signal tracking ($|Y/R|$) [Eq. (B.5)].

Table B.1 Feedback control specifications

Transfer functions and specification models	***Eq.***
$\lvert T_1(j\omega j\rvert = \left\lvert \frac{L(j\omega)}{1+L(j\omega)} \right\rvert = \left\lvert \frac{U(j\omega)}{D_1(j\omega)} \right\rvert = \left\lvert \frac{Y(j\omega)}{N(j\omega)} \right\rvert \leq \delta_1(j\omega),\ \omega \in \Omega_1$	**(B.1)**
$\lvert T_2(j\omega)\rvert = \left\lvert \frac{1}{1+L(j\omega)} \right\rvert = \left\lvert \frac{Y(j\omega)}{D_2(j\omega)} \right\rvert \leq \delta_2(j\omega),\ \omega \in \Omega_2$	**(B.2)**
$\lvert T_3(j\omega)\rvert = \left\lvert \frac{P(j\omega)}{1+L(j\omega)} \right\rvert = \left\lvert \frac{Y(j\omega)}{D_1(j\omega)} \right\rvert \leq \delta_3(j\omega),\ \omega \in \Omega_3$	**(B.3)**
$\lvert T_4(j\omega)\rvert = \left\lvert \frac{G(j\omega)}{1+L(j\omega)} \right\rvert = \left\lvert \frac{U(j\omega)}{D_2(j\omega)} \right\rvert = \left\lvert \frac{U(j\omega)}{N(j\omega)} \right\rvert = \left\lvert \frac{U(j\omega)}{R(j\omega)F(j\omega)} \right\rvert \leq \delta_4(j\omega),\ \omega \in$	**(B.4)**
$\delta_{5\text{inf}}(\omega) \leq \lvert T_5(j\omega)\rvert = \left\lvert F(j\omega)\frac{L(j\omega)}{1+L(j\omega)} \right\rvert = \left\lvert \frac{Y(j\omega)}{R(j\omega)} \right\rvert \leq \delta_{5\text{sup}}(j\omega),\ \omega \in \Omega_5$	**(B.5)**

Every plant in the ω_i-template can be expressed in its polar form as $P(j\omega_i) = p\ e^{j\theta} = p\angle\theta$. Likewise, the compensator polar form is $G(j\omega_i) = g\ e^{j\phi} = g\angle\phi$. By substituting them in Eqs. (B.1) through (B.5) and rearranging the inequalities, the quadratic inequalities of Eqs. (B.6) through (B.10) in Table B.2 are calculated.

For every frequency ω_i there is a constant $\delta_k = \delta_k(j\omega_i)$, and for a fixed plant $p\angle\theta$ in the ω_i-template and a fixed controller phase ϕ in [−360°,0°], the unknown parameter of the inequalities in Table B.2 is the controller magnitude g. Then, solving the equalities $a\,g^2 + b\,g + c = 0$ the set of ω_i-bounds for $\{\delta_{k=1,..,5}\}$ is computed. The two possible solutions, g_1 and g_2, of the quadratic inequalities of Table B.2, for every feedback problem in Table B.1, are shown in Table B.3.

Choosing the real and positive solutions of g_1 and g_2 in Table B.3 as effective compensator restrictions, the bounds can be classified in four cases: o, n, ō and u *typology bounds*[115] (see Fig. B.2 and Table B.4).

Table B.2 Bound quadratic inequalities

k	*Bound Quadratic Inequality*	*Eq.*
1	$p^2\left(1-\dfrac{1}{\delta_1^2}\right)g^2 + 2\,p\cos(\varphi+\theta)g + 1 \geq 0$	**(B.6)**
2	$p^2g^2 + 2\,p\cos(\phi+\theta)\,g + \left(1-\dfrac{1}{\delta_2^2}\right) \geq 0$	**(B.7)**
3	$p^2g^2 + 2\,p\cos(\phi+\theta)\,g + \left(1-\dfrac{p^2}{\delta_3^2}\right) \geq 0$	**(B.8)**
4	$\left(p^2-\dfrac{1}{\delta_4^2}\right)g^2 + 2\,p\cos(\phi+\theta)\,g + 1 \geq 0$	**(B.9)**
5	$p_e^2p_d^2\left(1-\dfrac{1}{\delta_5^2}\right)g^2 + 2p_ep_d\left(p_e\cos(\phi+\theta_d) - \dfrac{p_d}{\delta_5^2}\cos(\phi+\theta_e)\right)g + \left(p_e^2-\dfrac{p_d^2}{\delta_5^2}\right) \geq 0$	**(B.10)**

Table B.3 *G*-bound formulation

k	$g_{1,2}$ ***bound expressions***	***Eq.***
1	$g_{1,2} = \dfrac{1}{p\left(1-\dfrac{1}{\delta_1^2}\right)}\left(-\cos(\phi+\theta) \mp \sqrt{\cos^2(\phi+\theta)-\left(1-\dfrac{1}{\delta_1^2}\right)}\right)$	**(B.11)**
2	$g_{1,2} = \dfrac{1}{p}\left(-\cos(\phi+\theta) \mp \sqrt{\cos^2(\phi+\theta)-\left(1-\dfrac{1}{\delta_2^2}\right)}\right)$	**(B.12)**
3	$g_{1,2} = \dfrac{1}{p}\left(-\cos(\phi+\theta) \mp \sqrt{\cos^2(\phi+\theta)-\left(1-\dfrac{p^2}{\delta_3^2}\right)}\right)$	**(B.13)**
4	$g_{1,2} = \dfrac{1}{p\left(1-\dfrac{1}{p^2\delta_4^2}\right)}\left(-\cos(\phi+\theta) \mp \sqrt{\cos^2(\phi+\theta)-\left(1-\dfrac{1}{p^2\delta_4^2}\right)}\right)$	**(B.14)**
5	$g_{1,2} = \dfrac{-1}{p_e p_d\left(1-\dfrac{1}{\delta_5^2}\right)}\left(p_e\cos(\phi+\theta_d)-\dfrac{p_d}{\delta_5^2}\cos(\phi+\theta_e)\right)$ $\mp\sqrt{\left(p_e\cos(\phi+\theta_d)-\dfrac{p_d}{\delta_5^2}\cos(\phi+\theta_e)\right)-\left(1-\dfrac{1}{\delta_5^2}\right)\left(p_e^2-\dfrac{p_d^2}{\delta_5^2}\right)}$	**(B.15)**

Table B.4 Solutions to quadratic inequalities and bound typologies.

sign *a* = ***a*(*p*,*δ*)**	g_{min}= ***min***(g_1, g_2)	g_{max}= ***max***(g_1, g_2)	g	**Bound**	**Typology**
any	complex	complex	$g \geq 0$	No	--
any	real, <0	real, ≤0	$g \geq 0$	No	--
≥ 0	real, <0	real, ≥0	$g \geq g_{max}$	Upper	n
	real, >0	real, ≥0	$g \geq g_{max}$ and $g \leq g_{min}$	Outer	o
< 0	real, <0	real, ≥0	$g \leq g_{max}$	Lower	u
	real, >0	real, ≥0	$g \leq g_{max}$ and $g \geq g_{min}$	Inner	$\overline{o}$

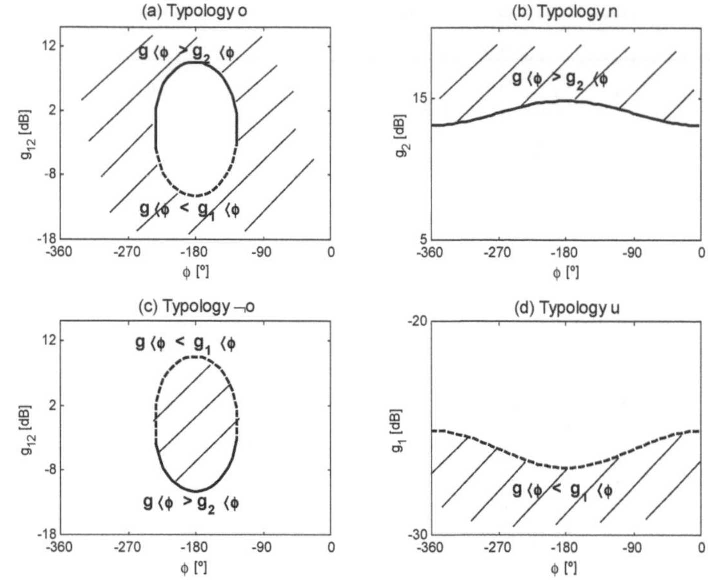

Fig. B.2 Bounds typologies.

A bound plotted with a *solid line* implies that $\boldsymbol{G}(\mathrm{j}\omega)$ (or $\boldsymbol{L}_\mathrm{o}(\mathrm{j}\omega)$) must lie above or on it to meet the specification, while a bound plotted with a *dashed line* means that $\boldsymbol{G}(\mathrm{j}\omega)$ (or $\boldsymbol{L}_\mathrm{o}(\mathrm{j}\omega)$) must lie below or on it.

The simultaneous meeting of general robust feedback control requirements (robust command tracking, disturbance rejection, robust stability and robust control effort minimization) can be analyzed with these bound typologies.[115]

For every control objective, two bound types are possible with dependence on the specification model ratio and the uncertainty size. All in all, three different bound typologies are found. They are:

1. Type *n* bound (single or multi-valued upper bounds). It tries to obtain the feedback benefits by raising the open-loop gain at low and medium frequencies.

2. Type *o* bound (single or multi-valued outer bounds). It encloses a forbidden area around the critical point (0dB, -180°) to ensure a robust stability. It allows reducing the open-loop high frequency gain to minimize the cost of feedback.

3. Type u bound (single or multi-valued lower bounds). It explicitly imposes the open-loop gain reduction needed to reduce the control effort due to different inputs at any frequency, and in particular the cost of feedback at high frequencies.

On this basis, some general hints can be established to quantify the trade-off of the control requirements simultaneously achievable.[115] For certain specification k, the least favorable intersection among the set of bounds for $\Im P(j\omega_i) = \{p\angle\theta\}$ is established in Table B.5 as $\max\{g_2\}$ for type n bounds, $\min\{g_1\}$ for type u bounds, or both for type o bounds; where g_1 and/or g_2 are the real and positive values calculated in Table 3.

Table B.5 Bound typologies for general feedback specifications.

k	$\delta_k(\omega_i)$	*Type*	*G-bound; g_1 and g_2*
1	$0<\delta_1<1$	u	$(g\angle\phi)\le(\min\{g_1\}\angle\phi)$; $\phi\in. =[-360^\circ,0^\circ]$
	$\delta_1>1$	o	$(g\angle\phi)\ge(\max\{g_2\}\angle\phi)$ $(g\angle\phi)\le(\min\{g_1\}\angle\phi)$ $\phi\in\Phi_{12}=[-180^\circ-\{\theta\}\mp\varepsilon]$
2	$0<\delta_2<1$	n	$(g\angle\phi)\ge(\max\{g_2\}\angle\phi)$; $\phi\in. =[-360^\circ,0^\circ]$
	$\delta_2>1$	o	$(g\angle\phi)\ge(\max\{g_2\}\angle\phi\ )$; $(g\angle\phi)\le(\min\{g_1\}\angle\phi)$ $\phi\in\Phi_{12}=[-180^\circ-\{\theta\}\mp\varepsilon]$
3	$p>\delta_3$	n	$(g\angle\phi)\ge(\max\{g_2\}\angle\phi)$; $\phi\in. =[-360^\circ,0^\circ]$
	$\{p\}<\delta_3$	o	$(g\angle\phi)\ge(\max\{g_2\}\angle\phi)$; $(g\angle\phi)\le(\min\{g_1\}\angle\phi)$ $\phi\in\Phi_{12}=[-180^\circ-\{\theta\}\mp\varepsilon]$
4	$p<1/\delta_4$	u	$(g\angle\phi)\le(\min\{g_1\}\angle\phi)$; $\phi\in. =[-360^\circ,0^\circ]$
	$\{p\}>1/\delta_4$	o	$(g\angle\phi)\ge(\max\{g_2\}\angle\phi)$; $(g\angle\phi)\le(g_1\angle\phi)$ $\phi\in\Phi_{12}=[-180^\circ-\{\theta\}\mp\varepsilon]$
5	$\dfrac{p_{max}}{p_{min}}>\delta_5$	n	$(g\angle\phi)\ge(\max\{g_2\}\angle\phi)$ $\phi\in\Phi=[-360^\circ,0^\circ]$
	$\dfrac{p_{max}}{p_{min}}<\delta_5$	o	$(g\angle\phi)\ge(\max\{g_2\}\angle\phi)$; $(g\angle\phi)\le(\min\{g_1\}\angle\phi)$ $\phi\in\Phi_{12}=[-180^\circ-\{\theta\}\mp\varepsilon]$

APPENDIX C

MIMO QFT CAD PACKAGE[15,47,48,69-71]

C-1 INTRODUCTION

This appendix presents the MIMO QFT CAD package. This CAD package implements algorithms for the design of robust multivariable control systems and includes provisions for rejecting external disturbance signals. Both analog- and discrete-time MIMO tracking control systems are considered. The CAD package , useable on a PC, is capable of carrying a robust control design from problem setup, through the design process, to a frequency domain analysis of the compensated MIMO system. For analog control problems, the design process is performed in the s-plane, while for discrete control problems the plants are discretized and the design process proceeds in the w'-plane. The following list is a sample of the major operations available in the MIMO QFT CAD Package:

1. Automation of weighting matrix selection
2. Discretization of the plants
3. Formation of the square effective plants
4. Inversion of the polynomial matrix required to obtain equivalent plants
5. Generation of templates
6. Selection of a nominal plant
7. Generation of the stability, tracking, cross-coupling disturbance, external disturbance rejection, gamma, and composite bounds
8. Selection of compensator transfer functions via loop shaping
9. Design of the pre-filter elements.

Once a control system design is completed, a frequency-domain analysis of the completed design is accomplished. If the results are satisfactory the design and plant models can be exported to the MATLAB SIMULINK® environment in order to validate the frequency domain design using time-domain simulations.

In addition to the above functionality, the bound generation routines and graphics have been enhanced to reduce over-design. During the process of generating tracking bounds, allocation of the amount of rejection of cross-coupling effects is automatically performed. The weighting matrix can be designed to include gain scheduling. The improved method (Method 2) may be applied for the general case of a *mxm* effective plant for both external disturbance rejection and tracking control problems. The CAD package can also be used to perform a QFT design for the special case of a MISO control system. The MIMO QFT CAD package is implemented using Mathematica® and runs on MS Windows and LINUX operating systems. Since both of the operating systems run on a PC they are both PC versions. This CAD package is now maintained by Steve Rasmussen and can be obtained by sending an email to Steve.Rasmussen@RasSimTech.com.

C-2 INTRODUCTION: OVERVIEW OF MULTIVARIABLE CONTROL

The CAD package, as outlined in Fig. C.1, is a design tool for applying the Quantitative Feedback Theory (QFT) technique to analog and digital multivariable tracking control and external disturbance rejection design problems involving MIMO plants having structured plant parameter uncertainty. For tracking control problems, a MIMO square effective plant $\boldsymbol{P}_e$ with m inputs and m outputs is to be controlled by use of a diagonal compensator $\boldsymbol{G}$ and a diagonal pre-filter $\boldsymbol{F}$ in the feedback structure shown in Fig. C.2. For external disturbance rejection problems, see Chapter 8 and Fig. C.3, a diagonal compensator $\boldsymbol{G}$ is designed such that the system rejects outside disturbances which are projected to the outputs of $\boldsymbol{P}_F$ through the disturbance plant model $\boldsymbol{P}_d$, as shown in Fig. C.4. The system structure of Fig. C.3 can be used for a control problem specifying both tracking and external disturbance rejection requirements. For both classes of control problems the closed-loop system is required to meet the appropriate stability and performance (tracking or external disturbance rejection) specifications.

C-3 CONTINUOUS-TIME VS. DISCRETE-TIME DESIGN (SEE BLOCK 9 IN FIG. C.1)

The design for a continuous-time system is done in the s-domain by a thoroughly defined analog QFT design process. These same procedures, as shown in Fig. C.1, are utilized for discrete-time systems that are described in the w'-domain (referred to in this section as the w-domain).

As indicated in Block 9, the user selects the analog or discrete design CAD package route. Once $\boldsymbol{F}(w)$ and $\boldsymbol{G}(w)$ are synthesized, they are transformed into the z-domain for implementation by a digital computer. Thus, the following sections, although they refer to the s-domain design, apply equally well to a w-domain design.

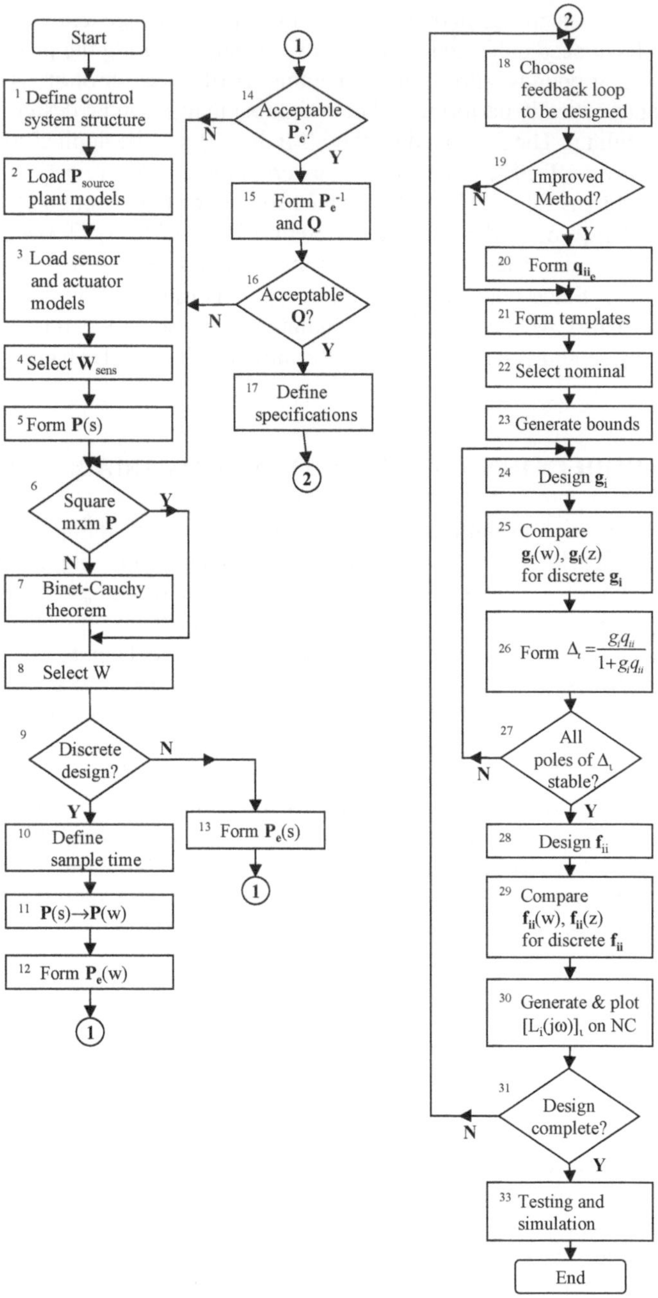

Fig. C.1 MIMO QFT flow chart for analog and discrete control systems.

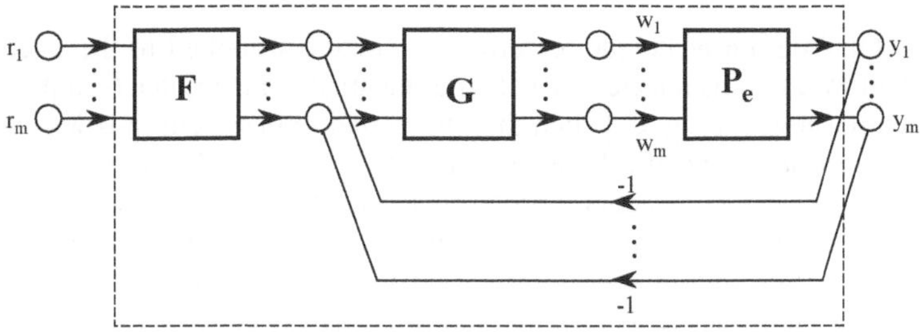

Fig. C.2 MIMO QFT control system with no external disturbance.

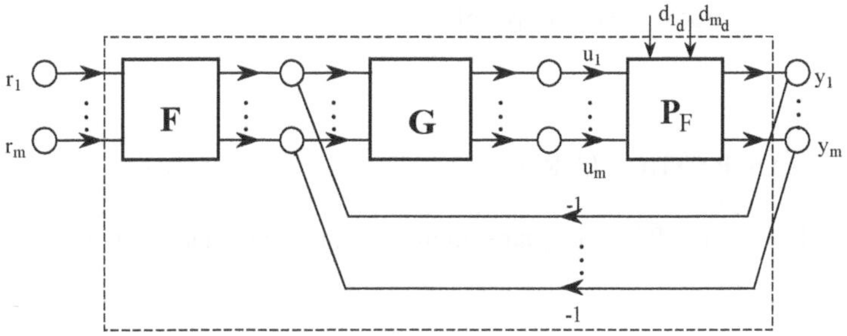

Fig. C.3 MIMO QFT control system with external disturbance [$\boldsymbol{r}(t) = \boldsymbol{0}$].

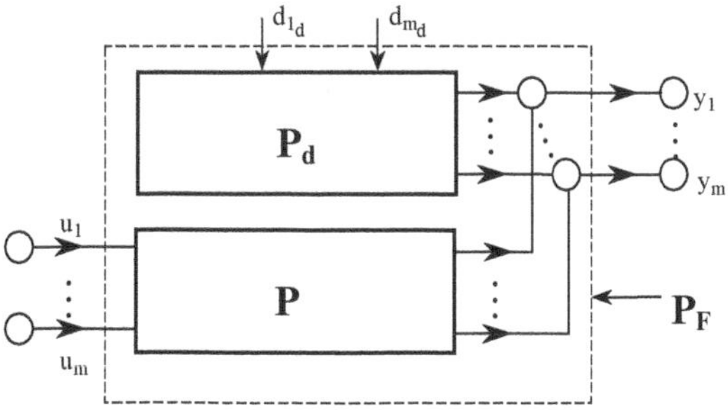

Fig. C.4 Partitioning of $\boldsymbol{P}_F$.

C-4 OVERVIEW OF THE MULTIVARIABLE EXTERNAL DISTURBANCE REJECTION PROBLEM

Using the QFT design technique, external disturbances applied to the uncertain MIMO plant $\boldsymbol{P}_F$ are to be rejected by use of a diagonal compensator $\boldsymbol{G}$ in the feedback structure shown in Fig. C.3 such that the closed loop system satisfies the stability and performance specifications. Obviously, a pre-filter $\boldsymbol{F}$ is not required for pure external disturbance rejection problems in which it is assumed that the tracking command input in Fig. C.3 is zero. Thus, the $m_d xm$ SISO equivalents, for the case in which $m_d = m = 3$, of the $m_d xm$ MIMO external disturbance rejection system of Fig. C.3 are shown in Fig. C.5. The plant model $\boldsymbol{P}_F$ is partitioned into two distinct plant models $\boldsymbol{P}$ and $\boldsymbol{P}_d$ as shown in Fig. C.4 for the QFT design process. The plant $\boldsymbol{P}_d$ models the transmission from the external disturbance inputs to the outputs of $\boldsymbol{P}_F$ and features only the external disturbance rejection problems. $\boldsymbol{P}_d$ does not affect the closed-loop stability of the m feedback loops in Fig. C.3. The plant $\boldsymbol{P}$ models the open-loop transmission of $\boldsymbol{P}_F$ in the feedback loop and takes the place of $\boldsymbol{P}_F$ in pure tracking control problems.

C-5 OPEN-LOOP STRUCTURE

When the plant matrix $\boldsymbol{P}$ is not square, then a square mxm plant $\boldsymbol{P}_e$ is formed from

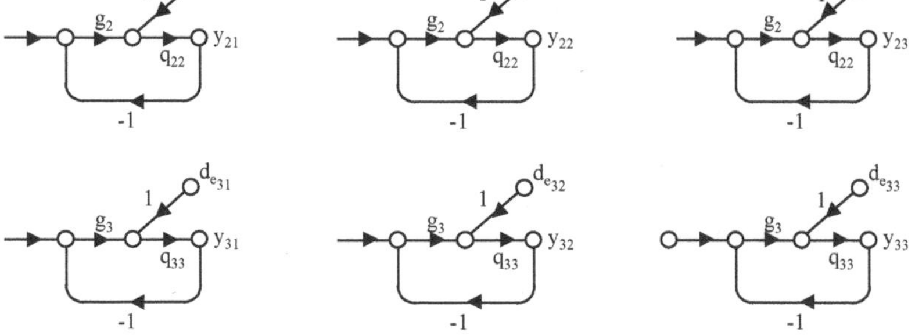

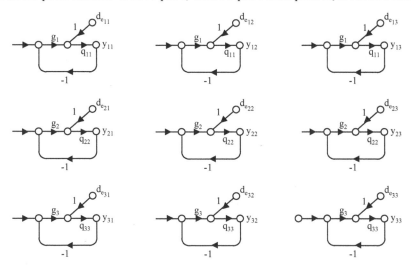

Fig. C.5 3x3 MISO equivalent loops for external disturbance inputs [$r(t) = 0$].

the *mxℓ* plant ***P*** by use of the *ℓxm* weighting matrix ***W*** as shown in the block diagram in Fig. C.6 for analog designs, and Fig. C.7 for discrete designs. Even if m = ℓ one may still use the weighting matrix ***W*** for the purpose of gain scheduling if needed. Thus:

$$\mathbf{P_e = PW} \tag{C.1}$$

The open-loop plant ***P*** is, in general, constituted in four component parts. A block diagram showing the placement of the loaded plant model $\boldsymbol{P}_L$, the actuator dynamics $\boldsymbol{T}_{ACT}$, the sensor dynamics $\boldsymbol{T}_{SEN}$, and the sensor gain matrix $\boldsymbol{W}_{SEN}$ is shown in Fig.C.8. The expression for the plant matrix ***P*** of dimension *mxℓ*, which in general is not square, is:

$$\boldsymbol{P} = \boldsymbol{T}_{SEN}\boldsymbol{W}_{SEN}\boldsymbol{P}_L\boldsymbol{T}_{ACT} \tag{C.2}$$

The disturbance plant model $\boldsymbol{P}_d$ is in general constituted in three component parts. A block diagram showing the placement of the loaded model $\boldsymbol{P}_{d_L}$, the sensor

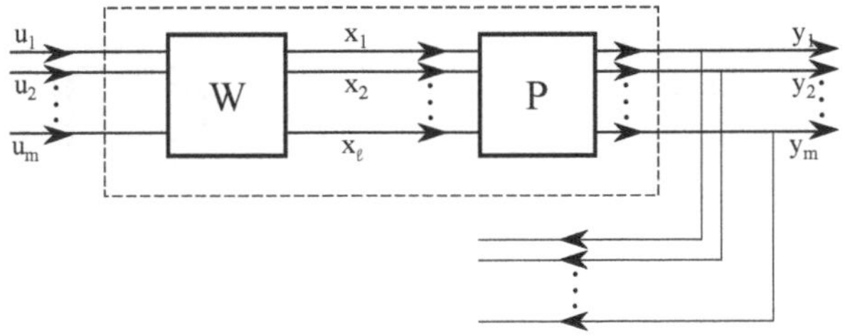

Fig. C.6 The formation of the analog plant $\boldsymbol{P}_e$.

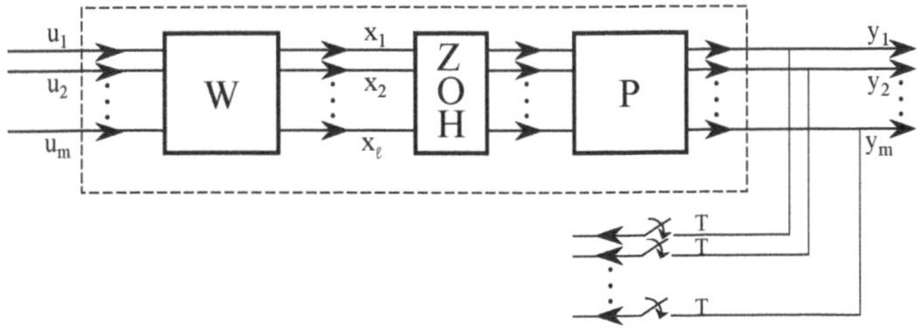

Fig. C.7 The formation of the discrete effective plant $\boldsymbol{P}_e$.

dynamics T_{SEN}, and the sensor gain matrix W_{SEN} is shown in Fig. C.9. The plant P_d of dimension mxm_d is, in general, not square. The expression for P_d is therefore:

$$P_d = T_{SEN} W_{SEN} P_{d_L} \quad \text{(C.3)}$$

According to the partitioning of the disturbance rejection system in Figs. C.3 through C.9, the bare plant is composed of two transfer function models P_L and P_{d_L} as shown in Fig. C.10. These matrices, which may not be square, are loaded in by the designer and placed into the control system structures used by the CAD package shown in Figs. C.8 and C.9, respectively

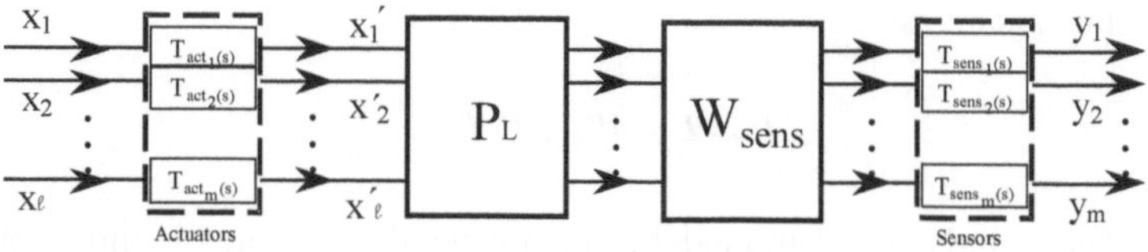

Fig. C.8 Components of the plant ***P***.

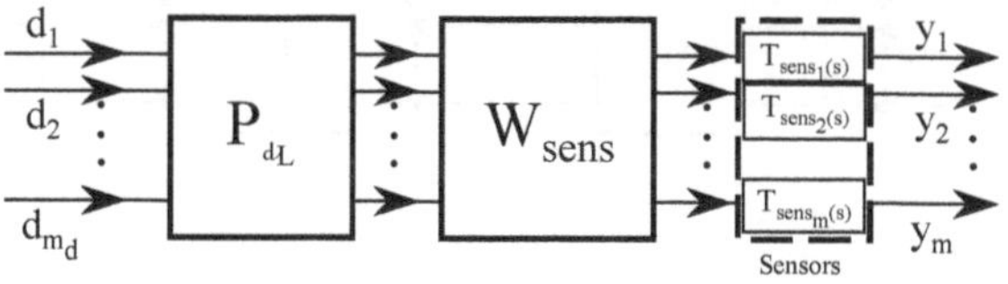

Fig. C.9 Components of the Plant $\boldsymbol{P}_d$.

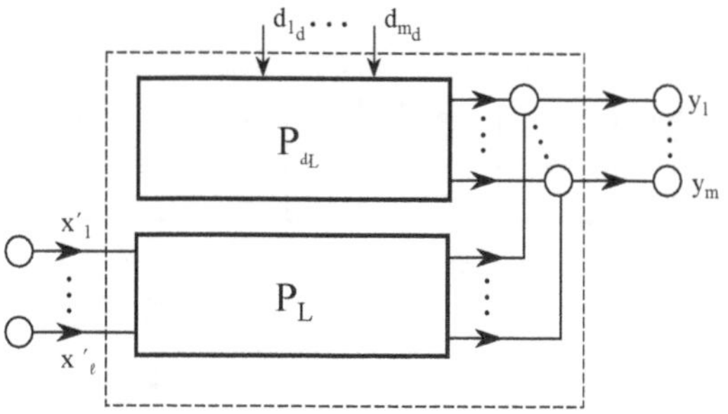

Fig. C.10 Partitioning of the bare plant model.

C-6 FORMATION OF PLANT MODELS FOR TRACKING CONTROL PROBLEMS (SEE BLOCKS 4-13 IN FIG. C.1)

For both analog and discrete control problems the plant model $\boldsymbol{P}$ to be controlled (see Fig. C.1) is in general constituted by the four analog component parts: the loaded plant model $\boldsymbol{P}_L$, actuator dynamics $\boldsymbol{T}_{ACT}(s)$, sensor dynamics $\boldsymbol{T}_{SENS}(s)$, and the sensor gain matrix $\boldsymbol{W}_{SEN}$. The plant $\boldsymbol{P}(s)$ for the J plant cases $\iota = 1,2,\ldots,J$ is therefore formed as follows:

$$\boldsymbol{P}_\iota(s) = \boldsymbol{T}_{SENS}(s)\boldsymbol{W}_{SENS_\iota}\boldsymbol{P}_{L_\iota}(s)\boldsymbol{T}_{ACT}(s) \tag{C.4}$$

For a discrete-time control problem, the analog plant $\boldsymbol{P}$ is embedded in the digital control system by placing a zero-order-hold before the inputs of the plant $\boldsymbol{P}$ and by sampling the feedback signal of the plant outputs.

The plant $\boldsymbol{P}$ of dimension $m x \ell$ is, in general, not square. Since QFT requires a square plant then the square mxm effective plant $\boldsymbol{P}_e$ is formed from the non-square $m x \ell$ plant $\boldsymbol{P}$ by use of the ℓxm weighting matrix $\boldsymbol{W}$. It is desirable to select elements of $\boldsymbol{W}$ such that the determinant of $\boldsymbol{P}_e(s)$ is m.p. For continuous-time designs, the Binet-Cauchy theorem (see Sec. 7-8) is applied to $\boldsymbol{P}_{e_\iota}(s)$ in order to determine whether a m.p. $det\ \boldsymbol{P}_{e_\iota}(s)$ is achievable by an appropriate $\boldsymbol{W}$. If so, it will result in all $(q_{ii})_\iota$ being m.p. For discrete-time designs one can apply the Binet-Cauchy theorem to $\boldsymbol{P}_\iota(s)$ in order to minimize the number of RHP zeros of $det\ \boldsymbol{P}_{e_\iota}$ in the w-plane by an appropriate $\boldsymbol{W}$. In some multivariable control problems, the degree of uncertainty in the plant $\boldsymbol{P}$ may render impossible a successful robust design. Thus, for these cases, (minimal) gain scheduling of $\boldsymbol{W}$ may be required to affect a QFT design by allowing a different weighting matrix $\boldsymbol{W}_\iota$ for each plant case ι. $\boldsymbol{P}_{e\iota}$ for plant case ι is formed from $\boldsymbol{P}_\iota$ and $\boldsymbol{W}_\iota$ as follows:

$$\boldsymbol{P}_{e_\iota} = [p_{ij}] = \boldsymbol{P}_\iota \boldsymbol{W}_\iota \tag{C.5}$$

For discrete-time control problems, each plant $\boldsymbol{P}_\iota(s)$ is discretized as each $\boldsymbol{P}_{e_\iota}$, is formed. To discretize $\boldsymbol{P}_\iota(s)$ an exact $\mathcal{Z}$-transform is performed, followed by the z- to w-transformation resulting in w-plane transfer functions, i.e.:

$$\boldsymbol{P}_\iota(z) \rightarrow \boldsymbol{P}_\iota(w) \tag{C.6}$$

The QFT design process then proceeds in the s-domain for a continuous design or in the w-domain for a discrete design using exactly the same design steps unless stated otherwise.

The effective plant matrix $\boldsymbol{P}_{e_\iota}$ must have full rank, viz., controllability and observability are assumed, and have diagonal elements that have the same sign for all plant cases as $\omega \to \infty$. These are conditions that any of the usual LTI design techniques must satisfy (see Sec. 7-8); they are not unique to QFT. The CAD package therefore allows the sign of the m diagonal plants to be examined for the J plant cases as $\omega \to \infty$ in table form. The CAD package also allows the designer to list the determinant of $\boldsymbol{P}_e$, one plant case at a time.

A nonzero determinant is indicative of full rank. The numerator factors of the determinant, which are zeros of the q_{ij}, are examined as well. Thus, these determinants determine the m.p. or n.m.p. character of the effective plants $(q_{ii})_\iota$. If any $\boldsymbol{P}_{e_\iota}$ is unacceptable based on the above criteria, the weighting matrix is revised, $\boldsymbol{P}_{e_\iota}$ is recomputed, and the tests are applied again.

C-7 INVERSE OF P_e (SEE BLOCK 15 IN FIG. C.1)

The polynomial matrix inverse is performed using the Mathematica inverse function:

$$\boldsymbol{P}_e^{-1} = \frac{adj\ \boldsymbol{P}_e}{det\ \boldsymbol{P}_e} = \{p_{ij}^*\} \tag{C.7}$$

The equivalent plants are then formed by inverting the elements p_{ij}^*, that is:

$$\boldsymbol{Q} = \frac{det\ \boldsymbol{P}_e}{adj\ \boldsymbol{P}_e} = \{q_{ij}\} = \left\{\frac{1}{p_{ij}^*}\right\} \tag{C.8}$$

The $\boldsymbol{Q}$ matrix elements become the equivalent plants of the MISO loops.

C-8 MISO LOOPS OF THE TRACKING CONTROL PROBLEM

By the principle of superposition, each MISO loop transmission (see Fig. C.11) consists of both a tracking and a cross-coupling component. When using a diagonal pre-filter, only the diagonal MISO loops have a transfer function component due to tracking:

$$t_{ii} = t_{r_{ii}} + t_{c_{ij}} \tag{C.9}$$

Off-diagonal loops, with $f_{ij} = 0$ and $i \neq j$, have a transfer function component due to cross-coupling only, i.e.:

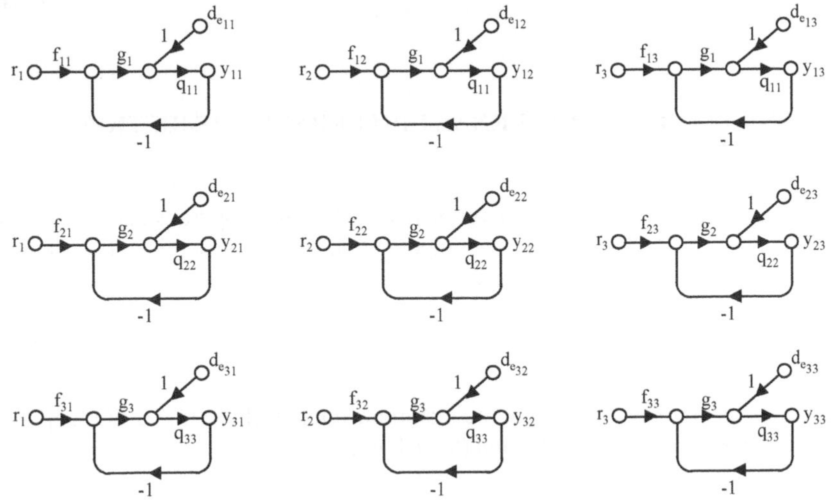

Fig. C.11 3x3 MISO equivalent loops containing both tracking and external disturbance inputs.

$$t_{ij} = t_{c_{ij}} \quad \text{where} \quad i \neq j \tag{C.10}$$

Expressions for tracking and cross-coupling transfer function components of the (*ij*) MISO loop are explicitly given by:

$$(t_{r_{ij}})_\iota = f_{ij}\left[\frac{g_i(q_{ii})_\iota}{1+g_i(q_{ii})_\iota}\right] = f_{ij}\left[\frac{(L_i)_\iota}{1+(L_i)_\iota}\right] \tag{C.11}$$

$$(t_{c_{ij}})_\iota = \frac{(c_{ij})_\iota (q_{ii})_\iota}{1+g_i(q_{ii})_\iota} = \frac{(c_{ii})_\iota (q_{ii})_\iota}{1+(L_i)_\iota} \tag{C.12}$$

where the index ι specifies one of the J LTI plants, i.e., $\iota = 1,2,\ldots,J$, and where $L_i = g_i q_{ii}$. The cross-coupling effect input, a function of all other controlled outputs, can be expressed by the equation:

$$c_{ij} = -\sum_{k \neq i}^{m} \left[\frac{t_{kj}}{q_{ik}} \right] \tag{C.13}$$

C-9 MISO LOOPS OF EXTERNAL DISTURBANCE REJECTION PROBLEM

For external disturbance rejection problems, each SISO loop transmission (see Fig. C.5) consists of only a disturbance component due to cross-coupling effects and the external disturbance forcing function:

$$t_{ij} = t_{d_{ij}} \tag{C.14}$$

The transfer function T_d relating the disturbance input of the (i,j) SISO loop to its output, [see Eqs. (8.14), (8.18), and (8.19)], is given by:

$$\left(t_{d_{ij}}\right)_\iota = \frac{(d_{e_{ij}})_\iota (q_{ii})_\iota}{1 + g_i (q_{ii})_\iota} = \frac{(d_{e_{ij}})_\iota (q_{ii})_\iota}{1 + (L_i)_\iota} \tag{C.15}$$

where the index ι specifies which LTI plant is being considered, i.e., $\iota = 1,2,...,J$, and where $L_i = g_i q_{ii}$. In comparing Eq. (C.15) to Eqs. (8.12) through (8.14) the following relationship holds:

$$\boldsymbol{P} = \boldsymbol{P_e} = \left(\boldsymbol{\Lambda} + \boldsymbol{B}\right)^{-1}$$

The SISO loop disturbance input, a function of the external disturbance input and all other controlled outputs (cross-coupling), is expressed [see Eqs. (8.18) and (8.19)] by the equation:

$$d_{e_{ij}} = (d_{ext})_{ij} - c_{ij} = \left[P_e^{-1} P_d\right]_{ij} - \sum_{k \neq j}^{m} \frac{t_{kj}}{q_{ik}} \tag{C.16}$$

where $\left[P_e^{-1} P_d\right]_{ij}$

is the (ij) element of the transfer function matrix product $\boldsymbol{P}_e^{-1}\boldsymbol{P}_d$.

C-10 *Q* MATRIX VALIDATION CHECKS (SEE BLOCK 16 IN FIG. C.1)

For analog and discrete tracking and external disturbance rejection problems, the ***Q*** matrix elements are tested to verify that the condition of diagonal dominance is satisfied. If diagonal dominance holds for all plant cases, then a QFT Method 1 design may be used. Otherwise, a QFT Method 2 (improved method) design must be used. If the results of this test are not satisfactory, then the weighting matrix ***W*** can be modified, and the ***Q*** matrix recomputed.

Additional tools for examining the equivalent plants q_{ii} of the **Q** matrix include a Bode plot function and a transfer function display subroutine. The Bode plot for a ***Q*** matrix element can be displayed for a specified set of plant cases or for all J plant cases together. The Bode plot for the set of J plant cases is useful for displaying variation in equivalent plant transmission as an aid in selecting representative template frequencies. Also, the CAD package allows the ***Q*** matrix transfer function elements to be displayed in factored form for any selected plant case.

To reduce the order of the ***Q*** matrix transfer functions the package performs automatic pole/zero cancellation. That is, it cancels nearly identical pole-zero pairs based on a user specified ratio of the distance between the pole-zero pair to the distance of the zero from the origin in both the right-half and left-half plane.

C-11 IMPROVED METHOD (SEE BLOCKS 19 AND 20 IN FIG. C.1)

The improved QFT Method 2, takes into account any correlation between the uncertainty in the designed MISO loops and the next row of MISO loops for which a design is to be performed. The standard approach of QFT Method 1 assumes worst case conditions and does not take this design information into account. The improved method requires the derivation of the *q-equivalent* plant transfer function for the next MISO loops yet to be designed. For both tracking and external disturbance rejection problems, the new set of transfer functions required by the improved method are generated using the equation:

$$q_{ije(n+1)} = q_{ije(n)} \left[\frac{1 + L_k}{1 + L_k - \gamma_{ij(n)}} \right] \qquad \textbf{(C.17)}$$

where the compensator g_k for row k of the MISO loops has been designed (L_k is known), and where:

$$L_k = g_k \, q_{kke(n)} \qquad \textbf{(C.18)}$$

$$\gamma_{ij(n)} = \frac{q_{kke(n)}\, q_{ije(n)}}{q_{kje(n)}\, q_{ike(n)}} \tag{C.19}$$

and $q_{kke(0)} = q_{kk}$ are generated during the matrix inversion of $\boldsymbol{P}_e$. The number in parentheses in the subscript of the improved method plants indicates the number of times this method has been applied to generate the plant transfer function. For example, the plant $q_{kke(1)}$ is obtained by applying the improved method once, and $q_{ije(n)}$ in Eq. (C.17) is obtained after applying the improved method n times. (i.e., n compensator elements of the matrix $\boldsymbol{G}$ were designed, and the improved method was applied after each design is completed.) Improved method plants are generated for all plant cases and for each row of MISO loops for which the compensator has not yet been designed. *The notation k on L_k is not to be confused with the standard notation i on L_i.* The notation k, as used here, denotes the index of the loop most previously designed, i.e., $k = i-1$ if loops are designed sequentially.

C-12 SPECIFICATIONS (SEE BLOCK 17 IN FIG. C.1)

C-12.1 STABILITY SPECIFICATIONS

A stability margin is specified for each row of MISO loops. The stability margin may be specified in terms of the gain margin gm, the phase margin angle γ or the corresponding M_L contour. The two remaining specifications are calculated from the one specification that has been entered. Only the M_L contour stability specification is stored in memory.

C-12.2 TRACKING PERFORMANCE SPECIFICATIONS

Frequency domain performance specifications are defined in the form of LTI transfer functions. For the diagonal MISO loops upper and lower bounds are specified as follows:

$$a_{ii} \le |\boldsymbol{t}_{ii}|_\iota \le b_{ii} \quad \text{for } \iota = 1,2,\ldots, J \tag{C.20}$$

For the off-diagonal MISO loops the following upper bound is specified:

$$|\boldsymbol{t}_{ij}|_\iota \le b_{ij} \quad \text{for } \iota = 1,2,\ldots, J \tag{C.21}$$

C-12.3 EXTERNAL DISTURBANCE REJECTION PERFORMANCE SPECIFICATIONS

For all MISO loops the following upper bound is specified:

$$|\, t_{ij} \,|_{\iota} \leq b_{ij} \quad \text{for } i \neq j, \;\; \iota = 1,2,\ldots,J \tag{C.22}$$

This upper bound is determined based upon Eqs. (8.33) and (8.34) in Chapter. 8. (Note that upper case letters for the bounds are used therein.)

For analog design problems, all performance specifications are defined as s-domain transfer functions. In the case of a digital control problem, the performance specifications are approximated in the bandwidth of interest by making the substitution $s \to w$. If the sampling rate is not sufficiently high for the above assumption to hold, the s-domain transfer functions yielding the desired performance specifications, are transformed to the z-domain and then transformed into the w-domain.

C-12.4 GAMMA BOUND SPECIFICATIONS

The improved method requires the derivation of the effective q plant transfer function, i.e. Eq. (C.17). By proper design of each compensator g_k, new RHP poles will not be introduced in q_{iie}. By requiring the magnitude of the denominator of Eq. (C.17) be larger than a small value ε_k, sign changes in the denominator are prevented and new RHP poles are not introduced. This constraint is given by:

$$\varepsilon_k \leq |\, 1 + \boldsymbol{L}_k - \gamma_{ij} \,| \tag{C.23}$$

A unique minimum value ε_k is specified by the designer for each of the $k = 1,2,\ldots,m$ channels.

C-13 BOUNDS ON THE NC (SEE BLOCK 23 IN FIG. C.1)

For a given row i of MISO loops, and for a template frequency $\omega = \omega_i$, several bounds may be included in the set plotted on the NC. These bounds include stability bound, an allocated tracking bounds, cross-coupling bounds, external disturbance bounds, and gamma bounds when using the improved method. The allocated tracking and cross-coupling bounds are generated such that the proper reduction in over design is achieved when using the improved method. This set of bounds can be replaced by a single composite bound before beginning a design.

C-13.1 STABILITY BOUNDS

A stability bound is generated for each template. The stability bounds constrain the maximum closed-loop transmission of the MISO loop with unity gain pre-filter to have a bounded magnitude of:

$$\left| \frac{g_i (q_{ii})_\iota}{1 + g_i (q_{ii})_\iota} \right| \leq M_L \tag{C.24}$$

The bound is plotted for a given frequency by plotting the path of the nominal point while traversing the M_L contour with the template generated for that frequency.

C-13.2 CROSS-COUPLING BOUNDS

For tracking problems, cross-coupling bounds are generated for each template, one for each off-diagonal MISO loop in the row of MISO loops for which the compensator is to be designed. Each bound is generated based on the constraint:

$$| t_{ij} | = \left| c_{ij} \left[\frac{q_{ii}}{1 + L_i} \right] \right| \leq b_{ij} \quad \text{for } i \neq j \tag{C.25}$$

which is a function of all other cross-coupling controlled outputs. The specifications dictate that c_{ij} is less than an upper bound for each plant case ι. Thus, for Method 1:

$$\left(| c_{ij} |_{max} \right)_\iota = \sum_{k \neq j}^{m} \left[\frac{| b_{kj} |}{| q_{ik} |_\iota} \right] \tag{C.26}$$

When the improved method has been applied, over design is substantially reduced by modifying Eq. (C.26) as follows:

$$\left(| \mathbf{c}_{ij} |_{\max} \right)_\iota = \sum_{\substack{k \neq i \text{ for} \\ L_k,\, f_{kj} \text{ unknown}}} \left[\frac{| \mathbf{b}_{kj} |}{| \mathbf{q}_{ik(n)} |_\iota} \right] + \sum_{\substack{k \neq i \text{ for} \\ L_k,\, f_{kj} \text{ unknown}}} \left[\frac{| \mathbf{f}_{kj} |}{| \mathbf{q}_{ik(n)} |_\iota} \right] \tag{C.27}$$

where $f_{kj} = 0$ when $k \neq j$ for the case of a diagonal pre-filter $\boldsymbol{F}$ required by the MIMO QFT CAD package.

Based on Eqs. (C.25) - (C.27), a lower bound is placed on $|1 + L_i|$ as follows:

$$|1+L_i| \geq \frac{|c_{ij}|_{max}|q_{ii}|}{|b_{ij}|} \tag{C.28}$$

By substituting $L_i = 1/m$ into Eq. (C.28), the latter is transformed such that the bound is plotted on the inverse NC, i.e.,

$$\left|\frac{m}{1+m}\right| \leq \frac{|b_{kj}|}{\left(|c_{ij}|_{max}\right)_l |q_{ii}|_l} \tag{C.29}$$

Equation (C.29) is the basis upon which the cross-coupling bounds on L_{io} are generated. The bounds are generated such that the correlation between $\boldsymbol{m}$ and $\boldsymbol{q}_{ii}$ in Eq. (C.29) is properly taken into account over the range of the plant parameter uncertainty which is outlined by the template when the nominal loop transmission does not violate these bounds.

C-13.3 GAMMA BOUNDS

Gamma bounds are generated for each template, where the compensator for row j is to be designed after the compensator for row i of the MISO loops is designed. It is desired that the magnitude of the denominator of the effective plant q_{22e} calculated using Eq. (C.17) for an improved method design, not be smaller than a specified minimum value despite plant uncertainty. Thus, the gamma bound is generated based upon satisfying:

$$|1+L_k - \gamma_{ij}| \geq \varepsilon_k \tag{C.30}$$

where $\boldsymbol{L}_k$ and γ_{ij} are given, respectively, by Eqs. (C.18) and (C.19).

Satisfying the constraint in Eq. (C.30) prevents a sign change (preventing the introduction of RHP poles) in the characteristic equation of the improved method plants' of Eq. (C.17). This enhances the ability to design a stabilizing compensator for each successive feedback loop. Given the range of plant uncertainty defined by the templates and variation of γ_{ij} among the plant cases, each gamma bound is generated such that Eq. (C.30) is satisfied when the nominal loop transmission $\boldsymbol{L}_{io}$ does not violate the gamma bound.

C-13.4 ALLOCATED TRACKING BOUNDS

For tracking problems, allocated tracking bounds are used to insure that the variation in closed loop frequency domain transmission t_{ii} of the diagonal MISO loop does not exceed the variation δ_R permitted by the performance specifications. Variation in the closed loop transmission of the diagonal MISO loop results from uncertainty in the response due to tracking and from the presence of the cross-coupling effects:

$$t_{ii} = t_{r_{ii}} + t_{c_{ij}} \tag{C.31}$$

where $t_{r_{ii}}$ and $t_{d_{ij}}$ are given, respectively, by Eqs. (C.11) and (C.12). The constraint on L_i used to determine a point on the tracking bound is:

$$Lm[T_{R_{max}} + (t_{c_{ii}})_{max}] - Lm[T_{R_{min}} - (t_{c_{ii}})_{max}] \leq \delta_R \tag{C.32}$$

where the transmission T_R, with unity gain pre-filter, is:

$$T_R = \frac{L_i}{1 + L_i} \tag{C.33}$$

and the most extreme transmission due to cross-coupling effects is:

$$(t_{c_{ii}})_{max} = \frac{|(\boldsymbol{c}_{ii})_{max}|\,|\boldsymbol{T}_{R_{max}}|}{|\boldsymbol{g}_i|} \tag{C.34}$$

Because points on the cross-coupling bound (if generated) are identical to those on the allocated tracking bound for the value of $(c_{ii})_{max}$ in Eq. (C.34), only an allocated tracking bound is generated for diagonal MISO loops. By constraining L_i to be above the bound, the actual variation in t_{ii} is less than $(t_{c_{ii}})_{max}$.

C-13.5 EXTERNAL DISTURBANCE REJECTION BOUNDS

External disturbance bounds are plotted for each template, one for each MISO loop in the row of MISO loops for which the compensator is to be designed [see Eqs. (C.14) and (C.15)]. The disturbance entering the (ij) MISO loop resulting from the external disturbance entering through $\boldsymbol{P}_d$ and from the cross-coupling transmissions is given by Eq. (C.16). The specifications dictate that $\boldsymbol{d}_{e_{ij}}$ is less than an upper bound for each plant case ι in the set of J plants; i.e.:

$$|\mathbf{d}_{e_{ij}}| \leq \left|[\mathbf{P}_e^{-1}\mathbf{P}_d]\right|_\iota + \sum_{k \neq i}^{m} \left[\frac{|\mathbf{t}_{kj}|}{|\mathbf{q}_{ik}|_\iota} \right] \tag{C.35}$$

For a Method 2 design in which the improved method has already been applied, (say n times so far) the calculation of

$$\left(|\mathbf{d}_{eij}|_{\max}\right)_{\iota}$$

is modified as follows: by replacing the term $|t_{kj}|/|q_{ik}|_{\iota}$ in the summation above (a) with $|b_{kj}|/|q_{ik(n)}|_{\iota}$ which is utilized when the improved method has not yet been applied, and (b) with

$$\frac{|\,\mathbf{g}_k^{-1}[\mathbf{P}_e^{-1}]_{kj}\,|}{|\,q_{ik(n)}|_{\iota}} \qquad \textbf{(C.36)}$$

when the improved method has been applied to take into account the designed open loop transmission L_k.

C-13.6 COMPOSITE BOUNDS

A set of composite bounds is formed based on any or all of the tracking, stability, cross-coupling effects, external disturbance rejection, and gamma bounds. The composite bound is formed, for a given frequency, by retaining the most restrictive portion of all the bounds.

C-14 COMPENSATOR DESIGN (SEE BLOCKS 24-25 IN FIG. C.1)

The compensator for an analog system (a controller for a discrete system) is designed to satisfy design specifications for the entire row of MISO loops in which the compensator is used. Since $L_{io} = g_i q_{iio}$ is the same for all MISO loops in a given row, bounds for all MISO loops are plotted together on the NC. The compensator design may thus be performed for an entire row of MISO loops using a single design iteration based on the composite bounds plotted on the NC.

The open-loop transmission is shaped by adding, deleting, or modifying the poles and zeros of the compensator and by allowing adjustment of the gain until an acceptable loop shape is obtained. Stability is checked during loop shaping by examining the nominal closed-loop MISO transmission in factored form. All closed-loop s- or w-domain poles should be in the left-half-plane. For a discrete design, $g_i(w)$ is transformed by a bilinear transformation to $g_i(z)$. As a validation check, the Bode plots of $g_i(w)$ and $g_i(z)$ are compared for $0 < \omega < 2/3(\omega_s/2)$. If very close, then one can proceed, since robustness will be maintained in the z-

domain. Next, proceed with the formation of $\Delta_i(z) = L_i(z)/[1 + L_i(z)]$ in order to ensure that all the poles of $\Delta_i(z)$ are inside the unit circle (see Block 26 in Fig. C.1). If not, g_i needs to be modified in order to achieve a stable system for all cases.

C-15 PRE-FILTER DESIGN (SEE BLOCKS 28-29 IN FIG. C.1)

The proper design of the compensator guarantees that the variation in closed loop transmission due to uncertainty for t_{ii} is acceptable, but does not guarantee that the transmission is within the upper and lower performance tolerances a_{ii} and b_{ii}. The pre-filter is therefore required to translate the closed loop transmission t_{ii} such that it satisfies the upper and lower performance tolerances.

The pre-filter design begins with the determination of $T_{R_{max}}$ and $T_{R_{min}}$, the maximum and minimum closed loop transmission due to tracking T_R with unity gain pre-filter, respectively, at each template frequency ω_i using Eq. (C.33). As is the case for tracking bounds on the NC, a portion of the permitted range of variation of t_{ii} is allocated to the cross-coupling effects. Thus, restricted tolerances are placed on t_{r_i}

$$b'_{ii} = b_{ii} - |(t_{c_{ii}})_{max}| \tag{C.37}$$

$$a'_{ii} = a_{ii} + |(t_{c_{ii}})_{max}| \tag{C.38}$$

The filter bounds on the nominal t_{r_i} are as follows:

$$Lm\,(b'_{ii}) - Lm\,(T_{R_{max}}) \tag{C.39}$$

$$\mathrm{Lm}\,(a'_{ii}) - \mathrm{Lm}\,(T_{R_{min}}) \tag{C.40}$$

Once the filter bounds are generated, a pre-filter is synthesized such that the Bode plot of the nominal t_{r_i} lies between the two filter bounds and satisfies $t_{ri}(s) = 1$ in the limit as $s \to 0$. For a discrete design, $F(w)$ is transformed by a bilinear transformation to $F(z)$. As a validation check, the Bode plots of $F(w)$ and $F(z)$ are compared for $0 < \omega < 2/3(\omega_s/2)$ (see Block 29 in Fig. C.1). If very close, i.e. if the plots are essentially on top of one another then one can proceed, since robustness will be maintained in the z-domain.

C-16 DESIGN VALIDATION (SEE BLOCKS 25-33 IN FIG. C.1)

The CAD package provides a number of tests to validate that the completed MIMO design meets the stability and performance specifications for the J plant cases. The first (Block 27): poles of Δ_i for each feedback loop i are checked to validate that all poles of the characteristic equation are stable. If some plants are n.m.p., one cannot rely on the loop shapes on the NC.

For the second test (Block 30), an array of the J open-loop SISO loop transmissions $(L_i)_\iota = g_i(q_{ii})_\iota$ for all plant cases $\iota = 1,2,...,J$ are plotted on the NC along with the M_L contour to validate that the stability requirements are satisfied for each feedback loop i. If no open loop transmission violates the M_L contour, then the stability specifications are satisfied.

For tracking control problems an mxm array of Bode magnitude plots is generated for the mxm matrix of elements t_{ij} of the closed-loop transfer function matrix $\boldsymbol{T}$, where:

$$\mathbf{T} = [\mathbf{I} + \mathbf{P_e G}]^{-1}\mathbf{P_e GF} \tag{C.41}$$

where $\boldsymbol{I}$ is the identity matrix, and $\boldsymbol{P_e}$, $\boldsymbol{F}$, and $\boldsymbol{G}$ are the mxm plant matrix, the mxm diagonal pre-filter matrix, and the mxm diagonal compensator matrix, respectively.

Each Bode plot illustrates the frequency domain transmission t_{ij}, for the set of J plant cases along with the tracking performance specifications, to allow the designer to validate that performance specifications placed on the closed-loop system have been met over the frequency range of interest. For each diagonal t_{ii}, J Bode magnitude plots are plotted along with the performance bounds a_{ii} and b_{ii}. For each off-diagonal t_{ij}, J Bode magnitude plots are plotted along with the performance bound b_{ij}.

For external disturbance rejection problems, a set of J Bode magnitude plots are plotted for each t_{ij}, based upon the disturbance rejection specifications b_{ij}. The mxm closed-loop transfer function matrix $\boldsymbol{T}$, whose elements t_{ij} are the transmissions plotted on the Bode plots, is formed for the J plant cases based on the equation:

$$\mathbf{T} = [\mathbf{I} + \mathbf{P_e G}]^{-1}\mathbf{P_d} \tag{C.42}$$

where $\boldsymbol{I}$ is the identity matrix, and $\boldsymbol{P_d}$, $\boldsymbol{P_e}$, and $\boldsymbol{G}$ are the mxm_d external disturbance plant matrix, the mxm effective plant matrix, and the mxm diagonal compensator matrix, respectively.

For the final validation, MATLAB SIMULINK models are generated based on the completed design, one for each of the J plant cases. The designer can then insert nonlinear elements such as saturation or rate limits, and add anti-windup protection. The model is then simulated to verify that the time domain figures of merit specifications are satisfied. For a continuous-time control problem, an analog simulation is performed. For a discrete-time control problem, a hybrid simulation is performed based on the s-domain effective plants $\boldsymbol{P}_{e\iota}(s)$, and the z-domain compensator $\boldsymbol{G}(z)$, and pre-filter $\boldsymbol{F}(z)$.

C-17 SUMMARY

The MIMO QFT CAD package has been developed for both analog- and discrete-time control system design based on Mathematica. The design procedure is automated. This includes problem setup, equivalent plant formation, compensator and pre-filter design and design validation for *mxm* MIMO systems. Bound generation routines have been optimized to reduce over design. The package has been extended to handle in a unified way external disturbance rejection problems as well as tracking control problems.

APPENDIX D

TOTAL-PC CAD PACKAGE

D-1 INTRODUCTION

TOTAL-PC is ideally suited as an educational tool while others are more suited for the practicing engineer. In general, most packages are command-driven interactive programs through the keyboard or a file designation. Some software products include extensive system building capabilities, very high-order plant modeling, nonlinear system construction, and a multitude of data displays as the *MIMO QFT* CAD package of Appendix C. Some of these packages require extensive knowledge to use their full capabilities, whereas others are not comprehensive in their capabilities.

The computer program called TOTAL-PC is discussed in this appendix as a specific example of a CAD package. This CAD package along with the MATLAB® and the MIMO QFT CAD packages are the ones employed in the text examples. The disk attached to this text contains the TOTAL-PC CAD package and its associated USERS Manual. This CAD package can also be used by students for their basic control theory courses. Questions in the use of this package should be addressed to Dr. Robert Ewing whose e-mail address is: ewingrl@sensors.wpafb.af.mil Users of this CAD package should periodically check with Dr. Ewing for updated versions.

D-2 OVERVIEW OF TOTAL-PC

TOTAL[D.1,D..2] is an option-number package that reflects a hand calculator or line-terminal environment for interface speed and agility. This *FORTRAN* package was originally developed at the Air Force Institute of Technology (AFIT) in the late 1970s, and provides an extensive set of control system analysis and design CAD capabilities. During the early 1990s the improved version of TOTAL-PC has enhanced the analysis and design of control systems. This CAD package contains over 150 commands (see Table D.1), which are divided according to general functional categories. It divides each of its options (commands) into various groups. It contains the conventional analog and discrete control system analysis and design options and the QFT design options.

D-3 QFT CAD PACKAGE

Up to about 1986 there were essentially no Quantitative Feedback Theory (QFT) CAD packages designed specifically to assist in doing a complete QFT design for a control system. The TOTAL QFT CAD package,[D..3] for MISO systems was designed as an educational tool as illustrated in Fig. D.1. In 1992 the MIMO QFT CAD package[D.4] of Appendix C, developed at AFIT, has accelerated the utilization of the QFT technique for the design of robust multivariable control systems.

REFERENCES

D.1 Larimer, S.J.: "An Interactive Computer-Aided Design Program for Digital and Continuous System Analysis and Synthesis (TOTAL)," M.S. thesis, GE/GGC/EE/78-2, School of Engineering, Air Force Institute of Technology, Wright-Patterson AFB, Ohio, 1978.

D.2 Ewing, R.: "TOTAL/PC," School of Engineering, Air Force Institute of Technology, P Street, Wright-Patterson AFB, OH 45433-7765, 1992.

D.3..Lamont, G.B.: "ICECAP/QFT," School of Engineering, Air Force Institute of Technology, P Street, Wright-Patterson AFB, OH 45433-7765, 1986.

D.4..Sating, R.R.: "Development of an Analog MIMO Quantitative Feedback Theory P(QFT) CAD Package," M.S. Thesis, AFIT/GE/ENG/92J-04, Graduate School of Engineering, Air Force Institute of Technology, Wright-Patterson AFB, OH, June 1992.

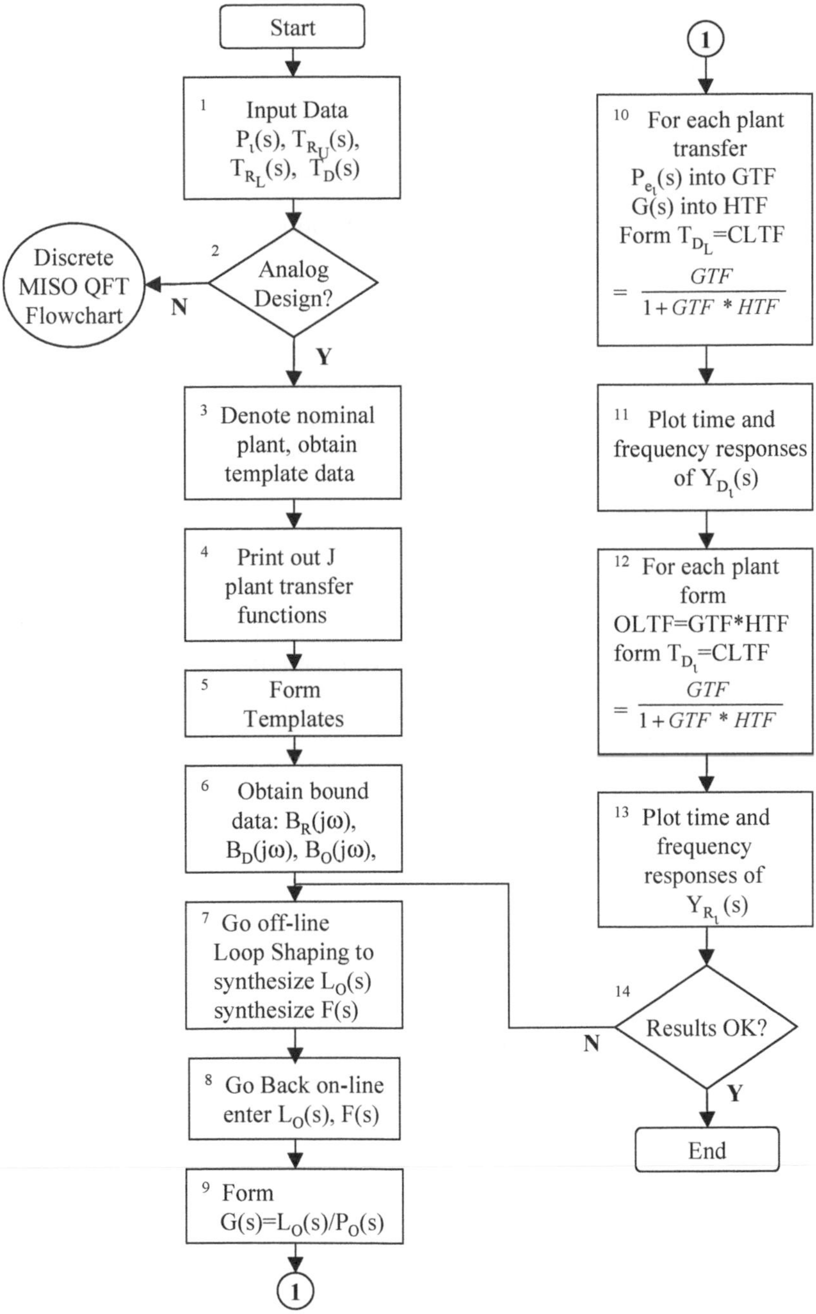

Fig. D.1 CAD flowchart for MISO analog QFT design.

Table D.1 Some TOTAL-PC options

No.	Option
	Transfer-function input options
0	List options
1	Recover all data from file memory
2	Polynomial form-GTF (forward transfer function)
3	Polynomial form-HTF (feedback transfer function)
4	Polynomial form-OLTF (open-loop transfer function)
5	Polynomial form-CLTF (closed-loop transfer function)
6	Factored form-GTF
7	Factored form-HTF
8	Factored form-OLTF
9	Factored form-CLTF
	Matrix input options for state equations
10	List options
11	AMAT-Continuous plant matrix
12	BMAT-Continuous input matrix
13	CMAT-Output matrix
14	DMAT-Direct transmission matrix
15	KMAT-State variable feedback matrix
16	FMAT-Discrete plant matrix
17	GMAT-Discrete input matrix
18	Set up state space model of system
19	Explain use of above matrices
	Block diagram manipulation and state-space options
20	List options
21	Form OLTF=GTF*HTF (in cascade)
22	Form CLTF=(GAIN*GTF)/(1+GAIN*GTF*HTF)
23	Form CLTF=(GAIN*OLTF)/(1+GAIN*OLTF)
24	Form CLTF=GTF+HTF (in parallel)
25	GTF(s) and HTF(s) from continuous state space model

26	GTF(z) and HTF(z) from discrete state space model
27	Write adjoint (*s**I**-AMAT) to file answer
28	Find HTF from CLTF & GTF for CLTF=GTF*HTF/(1+GTF*HTF)
29	Find HTF from CLTF & GTF for CLTF=GTF/(1+GTF*HTF)

Time-response options, continuous $F(t)$ and discrete $F(kT)$

30	List options
31	Tabular listing of $F(t)$ or $F(kT$
32	Plot F(t) or $F(kT)$ at user's terminal
33	Printer plot (written to file answer)
34	Calcomp plot (written to file plot)
35	Print time or difference equation [$F(t)$ or $F(kT)$]
36	Partial fraction expansion of CLTF (or OLTF)
37	LIST T-PEAK, T-RISE, T-SETTLING, T-DUP, M-PEAK, final value
38	Quick sketch at user's terminal
39	Select input: step, ramp, pulse, impulse, sin ωT

Root-locus options

40	List options
41	General root locus
42	Root locus with a gain of interest
43	Root locus with a zeta (damping ratio) of interest
44	List n points on a branch of interest
45	List all points on a branch of interest
46	List locus roots at a gain of interest
47	List locus roots at a zeta of interest
48	Plot root locus at user's terminal
49	List current values of all root locus variables

Frequency-response options

50	List options
51	Tabular listing
52	Two cycle scan of magnitude (or dB)
53	Two cycle scan of phase (degrees or radians)
54	Plot $F(j\omega)$ at user's terminal
55	Create GNUPlot-Frequency plot
56	Create GNUPlot-Root locus plot

57	Tabulate points of interest: peaks, breaks, etc.
58	Create GNUPLOT-Nichols Log magnitude/angle plot
59	Chalk Pitch Axis HQ Criterion Analysis

Polynomial operations

60	List options
61	Factor polynomial (POLYA)
62	Add polynomials (POLYC=POLYA+POLYB)
63	Subtract polynomials (POLYC=POLYA-POLYB)
64	Multiply polys (POLYC=POLYA*POLYB)
65	Divide polys (POLYC+REM=POLYA/POLYB)
66	Store Polynomial (POLY_) into POLYD
67	Expand roots into a polynomial
68	$(s+a)^n$ Expansion into a polynomial
69	Activate polynomial calculator

Matrix operations

70	List options
71	ROOTA=eigenvalues OF AMAT
72	CMAT=AMAT+BMAT
73	CMAT=AMAT-BMAT
74	CMAT=AMAT*BMAT
75	CMAT=AMAT inverse
76	CMAT=AMAT transposed
77	CMAT=identity matrix ***I***
78	DMAT=zero matrix ***0***
79	Copy one matrix to another

Digitization options

80	List options
82	CLTF(s) to CLTF(z) by first difference approximation
83	CLTF(s) to CLTF(z) by Tustin transformation
84	CLTF(z) TO CLTF(s) by impulse invariance
85	CLTF(z) TO CLTF(s) by inverse first difference
86	CLTF(z) TO CLTF(s) by inverse Tustin
87	Find FMAT and GMAT from AMAT and BMAT
89	CLTF(X) to CLTF(Y) by X=ALPHA* (Y+A)/(Y+B)

Miscellaneous options

90	LIST OPTIONS
91	Rewind and update Memory file with current data
93	List current switch settings (ECHO, ANSWER, etc.)
96	List special commands allowed in option mode
97	List variable name directory
98	List main options of total
99	Print new features bulletin
119	Augment AMAT:[AMAT]=[AMAT]_GAIN*[BMAT]*[KMAT]
121	Form OLTF=CLK*CLNPOLY/[CLDPOLY-CLK*CLNPOLY](G-EQUIVALENT)
129	(Integral of (CLTF SQUARED))/2PI=

Double-precision discrete transform options†

140	LIST OPTIONS
141	CLTF(*s*) TO CLTF(*z*) (IMPULSE VARIANCE)
142	CLTF(*s*) TO CLTF(*w*) *W*=(*Z*-*1*)/(*Z*+*1*)
143	CLTF(*s*) TO CLTF(*w'*) *W'* =(*2*/*T*)(*Z*-*1*)/(*Z*+*1*)
144	HI-RATE CLTF(*z*) TO LO-RATE CLTF(*z*)
145	HI-RATE CLTF(*w*) TO LO-RATE CLTF(*w*)
146	HI-RATE CLTF(*w'*) TO LO-RATE CLTF(*w'*)
147	OPTION 144 (AVOIDING INTERNAL FACTORING)
148	OPTION 144 (ALL CALCULATIONS IN *Z* PLANE)
	†Option 37 is not available for the discrete domain.

Quantitative Feedback Technique (QFT)

150	Define-Plant
151	List options
152	Define-TLTF: (Lower Tracking Ratio)
153	Define-TUTF: (Upper Tracking Ratio)
154	Define-TDTF: (Disturbance)
155	Select-Nominal plant: (1-10)
156	Define-FTF: (Pre-filter function)
157	Define-LOTF: (Nominal loop function)
158	Display Plant, TLTF, TUTF, TDTF, FTF, LOTF
159	QFT design/report macro

APPENDIX E

DISCRETE QFT DESIGN PROCESS

E-1 DISCRETE QFT DESIGN (w'- AND z-DOMAIN)

Figures E.1 and E.2 present a QFT CAD flow chart for the s-, w-, and z-domain designs. This appendix, which complements Sec. 4-6, provides additional information for each of the blocks in the flow chart as follows:

Block 1: The input s-domain data consists of:

1. Variable plants: $P_\iota(s)$.
2. Boundary transfer functions: $T_{R_U}(s)$, $T_{R_L}(s)$, and $T_D(s)$.

Block 2: "Is this an s- or w-domain design?"

Blocks 2 and 27: For the analog s-domain design see Appendix D.

The following descriptions deal with the w'-domain design procedure of Chapter 4.

Block 3: Form $P_{e_\iota}(s) = P_\iota(s)/s$ [see Eq. (4.42) where $\iota = 1,2,\ldots J$]

Block 4: Obtain $P_{e_\iota}(z)$ [Eq. (4.41)]

Block 5: Form $T_{R_U}(w)$, $T_{R_L}(w)$, and $T_D(w)$ by replacing s by w in the boundary transfer functions of Block 1.

Block 6: For this block the following steps are involved:

1. Transform $P_{e_\iota}(z)$, via $z = (Tw + 2)/(-Tw + 2)$, to $P_{e_\iota}(w)$.

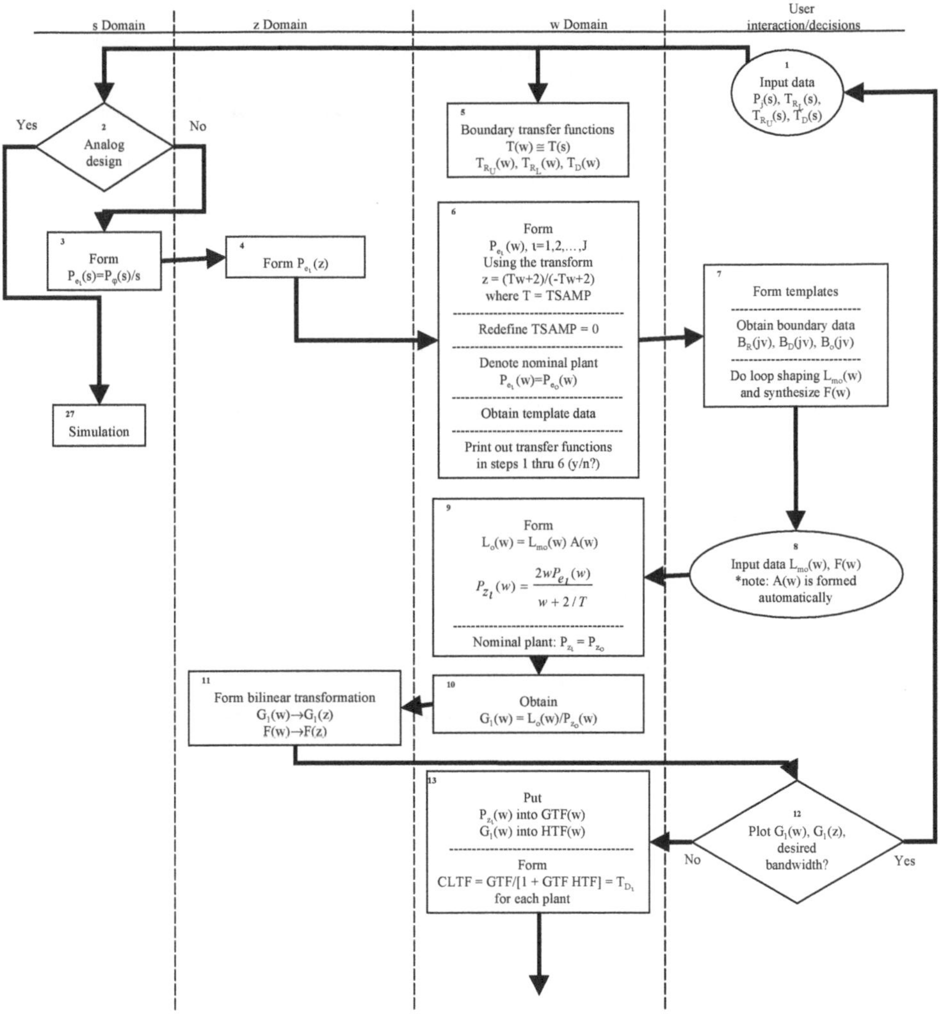

Fig E.1 QFT simulation flow chart for the *s*-, *w*-, and *z*-domains: Part 1.

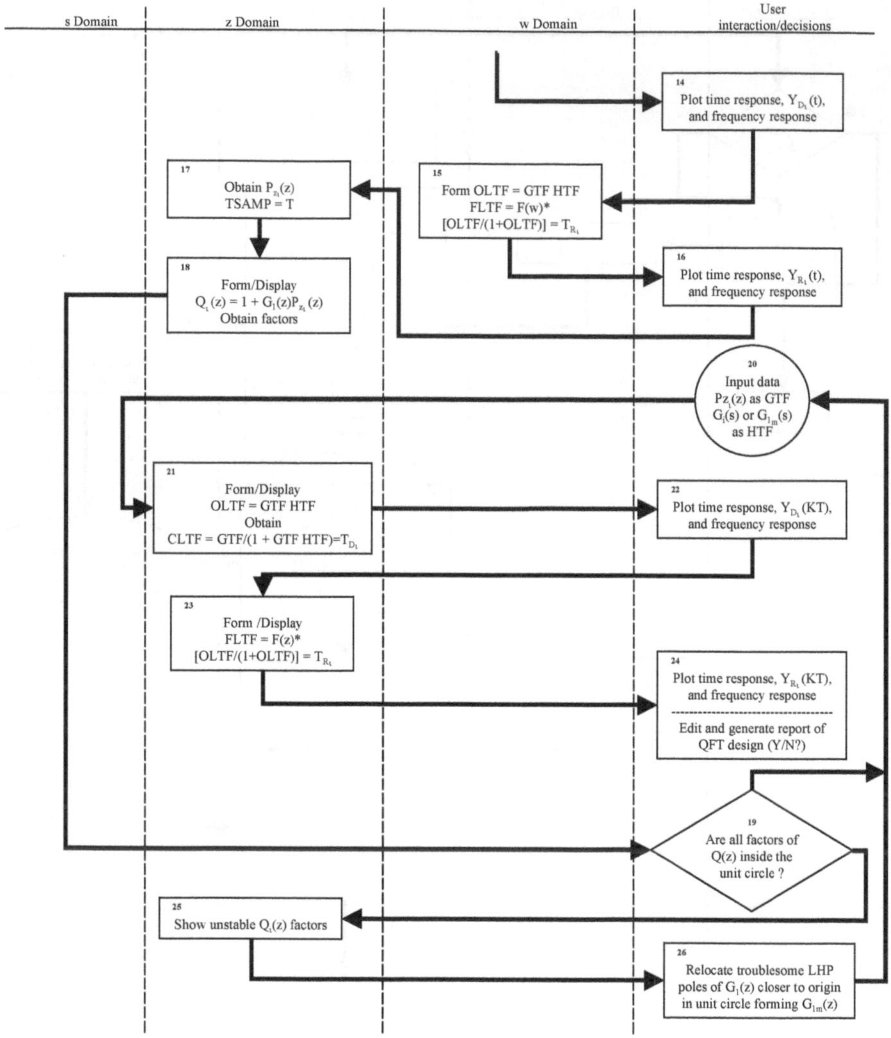

Fig. E.2 QFT simulation flow chart for the *s*-, *w*-, and *z*-domains: Part 2.

2. Denote the nominal plant $P_{e_o}(w)$.
3. Obtain data to construct the templates.
4. Based on the utilization of the **QFT options** redefine **TSAMP=0**.

Block 7: Exit From TOTAL-PC

At this point of the design process the designer exits the CAD package. The templates are constructed and plots for $\boldsymbol{B}_R(jv_i)$, $\boldsymbol{B}_D(jv_i)$ and $\boldsymbol{B}_{mo}(jv_i)$ bounds, and the $\mathbf{B}'_h$, $\boldsymbol{B}_h$, and the $\boldsymbol{B}_S$ contours are obtained. Next the optimal loop transmission transfer function $L_{mo}(w)$ and the pre-filter $F(w)$ are synthesized.

Block 8: Reenter TOTAL-PC

1. Enter $L_{mo}(w)$ and $F(w)$.
2. $A(w)$ is formed automatically.

Block 9:

1. Form $L_o = -L_{mo}(w)A(w)$ [see Eq. (4.74)].
2. Form $P_{z_\iota}(w) = P_{e_\iota}(w)[2w/(w + 2/T)]$ [see Eqs. (4.70) and (4.110)].
3. Ensure the nominal plant $P_{z_\iota} = P_{z_o}$.

Block 10:

1. With **CANCEL,ON** obtain $G_1(w) = L_o(w)/P_{z_o}(w)$ [see Eq. (4.90].

Block 11:

1. Via bilinear transformation obtain $G_l(z)$ and $F(z)$.

Block 12:

1. Obtain Bode plots and data for $G_l(w)$, $F(w)$, $G_l(z)$, and $F(z)$ for the frequency range $0 < v_i < v_{24}$ corresponding to

$$0 < \omega < \omega_{24}$$

where ω_{24} is the value of the frequency at the -24 dB point on $B_L(v)$ plot (see Fig. 4.9).

Block 13: [See Eq. (4.118)]

1. Enter $P_{z_\iota}(w)$ as **GTF.**
2. Enter $G_\iota(w)$ as **HTF.**
3. With **CANCEL,ON** obtain $T_{D_\iota}(w) =$ **GTF** /(1 + **GTF**x**HTF**) = **CLTF** for each plant ι.

Block 14:

1. Set forcing functions $r(t)$ and $d(t)$ to a unit step forcing function.
2. Obtain the time response data to determine the figures of merit (F.O.M.) and the plots for $y_{D_\iota}(t)$.
3. Obtain the frequency response data and the plots for $|\mathbf{T}_{D_\iota}(j\omega)|$.

Block 15: With CANCEL,ON

1. Form **OLTF** = **GTF**x**HTF** [= $L_\iota(w)$].
2. Form **FLTF** = $F(w)$[**OLTF**/(1 + **OLTF**)] [= $T_{R_\iota}(w)$] [see Eq.(4.106)].

Block 16:

1. Obtain the time response data to determine the FOM and the plots for $y_{R_\iota}(t)$.
2. Obtain the frequency-response data for the plots of $|\mathbf{T}_{R_\iota}(j\omega)|$

Block 17:

1. Set **TSAMP = T** and obtain $P_{z_\iota}(z)$ [see Eqs. (4.40) and (4.102)].

Block 18:

1. Set **CANCEL,ON** and form $Q_\iota(z) = 1 + G_\iota(z)P_{z_\iota}(z)$.
2. Obtain the factors for each $Q_\iota(z)$.

Block 19:

1. The determination if all J control ratios (tracking and disturbance) are stable. If all J plants yield stable responses proceed to Block 20, if not proceed to Block 25.

Block 20:

1 Enter $P_{z_\iota}(z)$ as **GTF**.
2 Enter $G_\iota(z)$ as **HTF**.

Block 21:

1. Form **OLTF** = **GTFxHTF**
2. With **CANCEL,ON** obtain **CLTF** [= **GTF**/(1 + **GTFxHTF**)] [$= T_{D_\iota}(z)$] [see Eq.(4.113)].

Block 22 through 24: With **CANCEL,ON**

1. Plot time and frequency responses for $Y_{D_\iota}(kT)$.
2. A printout of **OLTF** is given in order to determine if "proper cancellation" has occurred. If not, the user is allowed to delete zero(s) and pole(s) that justify cancellation.
3. Form **FLTF** = *F(z)*[**OLTF**/(1 + **OLTF**)] [$= T_{R_\iota}(z)$] [see Eq. (4.112)].
4. Set forcing functions *r*(t) and *d*(*t*) to a unit step forcing function.
5. Obtain discrete domain time response data in order to determine the FOM and the plots for $y_{R_\iota}(kT)$.
6. Obtain the frequency response data for T_{D_ι} and T_{R_ι}.
7. Query: Do you want a printout of input and calculated data of Blocks 1 to 24 or 1-19, 25, 26, or 20-24?
8. Exit from TOTAL-PC.

Block 25:

From Block 18 show all unstable factors of $Q_\iota(z)$.

Block 26:

Relocate the pole(s) on the negative real axis of $G_\iota(z)$, that result in the closed-loop unstable pole(s), closer to the origin within the unit circle as discussed in Sec.4-5.6. Repeat Blocks 19 through 24 and exit from TOTAL-PC. Note that if a printout is desired, the user is allowed to edit data before printing.

APPENDIX F

MISO DESIGN EXAMPLE

F-1 MIS0 DESIGN PROBLEM

Given the plant transfer function:

$$P(s) = \frac{k}{s+a}$$

The specifications on $|\boldsymbol{T}_R(j\omega)|$ is given in the following table:

ω	0	0.5	1	2	5	10	20
Lm $\mathbf{B}_U(j\omega)$	0	1	1	0	-4	-10	-20
Lm $\mathbf{B}_L(j\omega)$	0	-2	4	8	20	(-40) ∞	(-80) ∞

where $M_L < 3$ dB, $\lambda > 3$ and there is no specifications on T_D. The nominal plant is:

$$P_o(s) = 1/s \quad (k = 1, a = 0)$$

F-2 PROBLEM

Design $L_o(s)$ and $F(s)$ to meet the specifications.

F-3 SOLUTION

F-3.1 TEMPLATE GENERATION

The set of J plant transfer functions $P_\iota(a_\iota, k_\iota)$, where $P_\iota \in \mathcal{P}$ and $\iota = 1, 2, \ldots, J$, used to obtain the templates are:

$$P_1(3,1),\ P_2(0,1),\ P_3(-3,1),\ P_4(-3,100),\ P_5(0,100),\ \text{and}\ P_6(3,100)$$

The templates all have the shape shown in Fig. F.1 where $V = 40$ dB and as ω increases the templates become narrower. It is left to the reader to obtain the templates in order to verify the solution to this problem.

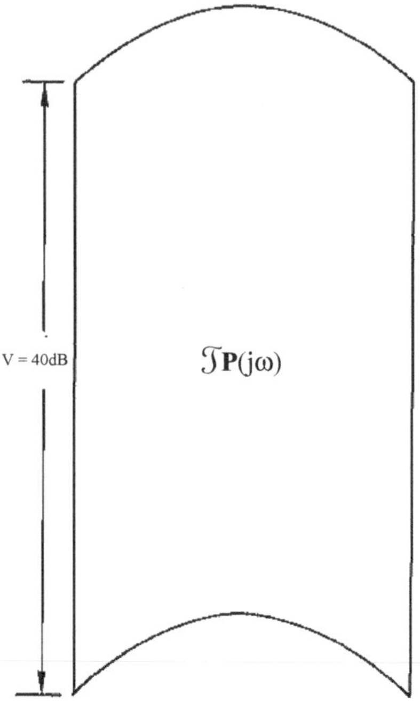

Fig. F.1 Template $\mathfrak{T}P(j\omega_i)$

F-3.2 DETERMINATION OF BOUNDARIES $B_R(j\omega_i)$

The templates are used to determine the boundaries $B_R(j\omega_i)$ and, along with the M_L-contour and the value of V, to determine the U-contour as shown in Fig. F.2.

F-3.3 DETERMINATION OF $L_o(s)$

The design of an optimum $L_o(s)$ is not unique. A possible design of an $L_o(s)$ is shown in Fig. F.2. The designs at the onset should be based upon a qualitative analysis of

$$L(s) = G(s)P(s) = \frac{k}{s+a}G(s) \tag{F.1}$$

and

$$L_o(s) = P_o(s)G(s) = \frac{kG(s)}{s} \tag{F.2}$$

Trial 1 -- A Type 0 $L_o(s)$ function requires the compensator to be of the format of:

$$G(s) = \frac{Ks(\ )\ldots(\ )}{(\ )\ldots(\ )} \tag{F.3}$$

which from Eqs. (F.l) and (F.2) results in, respectively:

$$L(s) = \frac{kK(\ )\ldots(\ )}{(s+a)(\ )\ldots(\ )} \tag{F.4}$$

and

$$L_{o1}(s) = \frac{kK(\ )\ldots(\ )}{(\ )\ldots(\ )} \tag{F.5}$$

When $a \neq O$, Eq. (F.4) results in

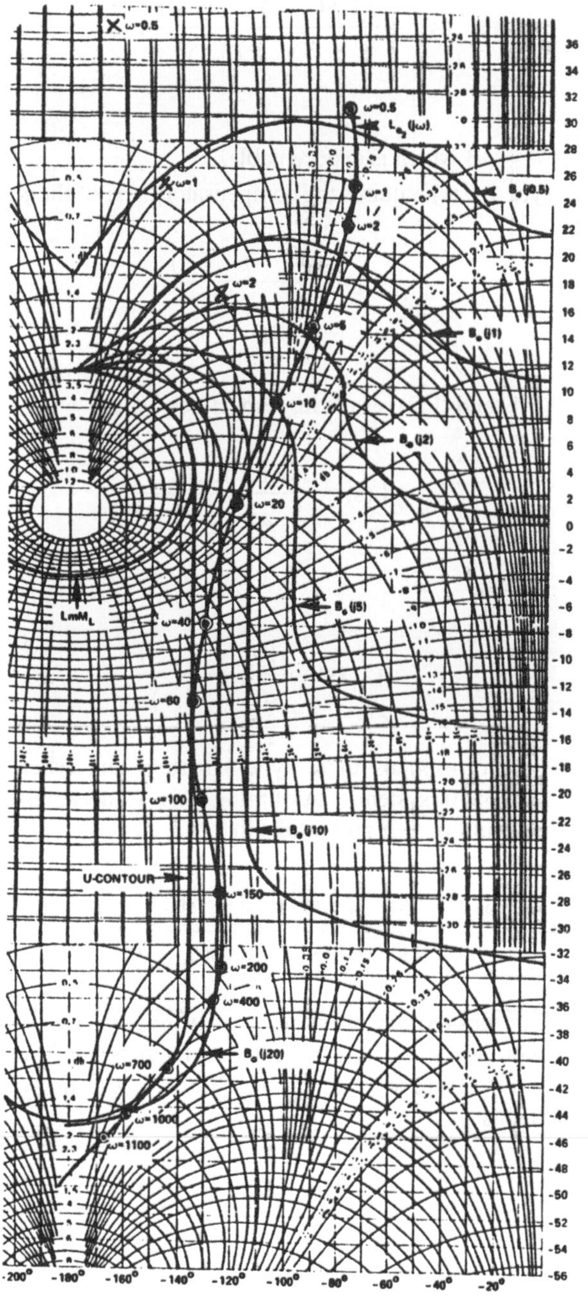

Fig. F.2 Plots of $B_R(j\omega_i)$ and $L_o(j\omega)$.

$$T_{R_1}(s) = \frac{kKs(\quad)\ldots(\quad)}{(\quad)\ldots(\quad)} F(s) \qquad \textbf{(F.6)}$$

and $Lm\ \boldsymbol{T}_{R_1}(j\omega) = -\infty$ dB which is unacceptable.

Trial 2 A Type 1 $L_o(s)$ requires the compensator to be of the format:

$$G(s) = \frac{K(\quad)\ldots(\quad)}{(\quad)\ldots(\quad)} \qquad \textbf{(F.7)}$$

which from Eqs. (F.1) and (F.2) results in, respectively:

$$L(s) = \frac{kK(\quad)\ldots(\quad)}{(s+a)(\quad)\ldots(\quad)} \qquad \textbf{(F.8)}$$

and

$$L_{o_2}(s) = \frac{kK(\quad)\ldots(\quad)}{s(\quad)\ldots(\quad)} \qquad \textbf{(F.9)}$$

When $a = 0$, $T_R(s)$ corresponds to a Type 1 system and when $a \neq O$ $T_R(s)$ corresponds to a Type 0 system. Assuming this is acceptable an $L_o(s)$ is synthesized, as shown in Fig. F.2, and results in

$$G(s) = \frac{8.504\text{x } 10^6 (s+1)(s+12)(s+120)}{(s+2)(s+10)(s+26.5)(s+720 \pm j960)} \qquad \textbf{(F.10)}$$

In order to achieve the desired specification $e(\infty) = r(\infty) - c(\infty) = 0$, for $r(t) = u_{-1}(t)$, then the compensator $G(s)$ must be a Type 1 transfer function in order to yield a Type 1 system for $a \neq O$ and a Type 2 system for $a = 0$. The synthesized loop transmission function is

$$L_{o_3} = \frac{kK(\quad)\ldots(\quad)}{s^2(\quad)\ldots(\quad)} \qquad \textbf{(F.11)}$$

for which three points (denoted by X) are shown in Fig. F.2. It is left to the reader to complete the synthesis of a $L_o(s)$.

F-3.4 DESIGN OF *F*(*s*)

Based upon the design procedure in Chapter 3, the following non-unique filter is obtained:

$$L(s) = F(s) = \frac{1.5}{s+1.5} \frac{kK(\ )\ldots(\ )}{(s+a)(\ )\ldots(\ )} \qquad \textbf{(F.12)}$$

F-3.5 SIMULATION OF DESIGN

A simulation of the system for $L_o(s)$ is made for each $P_\iota(a_\iota k_\iota)$, where $\iota = 1,\ldots,6$, based upon Eqs. (F.10) and (F.12). The simulations indicate that all the specifications on $|T_R(j\omega)|$ are met except, as expected, when $a \neq O$ and $\omega = 0$. The reader can simulate the system utilizing his $L_{o3}(s)$.

APPENDIX G

DIAGONAL MIMO DESIGN EXAMPLE

G-1 DESIGN PROBLEM

Given: A *2x2* plant where

$$Q(s) = \frac{1}{s}\begin{bmatrix} k_{11} & k_{12} \\ k_{21} & k_{22} \end{bmatrix} \tag{G.1}$$

has the following independent structured plant parameter uncertainties:

$$1 \le k_{11} \le 2 \qquad 0.5 \le k_{12} \le 2$$

$$5 \le k_{21} \le 10 \qquad 0.5 \le k_{22} \le 1$$

The diagonal dominance condition (Sec.5-7) requires, as $\omega \to \infty$, that

$$|\mathbf{q}_{11}\mathbf{q}_{22}| < |\mathbf{q}_{12}\mathbf{q}_{21}| \Rightarrow 2 < 2.5$$

Specifications:

(a) There is only one command input: $r_2(t)$

(b) $Lm\, \boldsymbol{t}_{12} < -20$ dB for all ω

(c) For $Lm\, \boldsymbol{t}_{22}$:

The values of $a_{22}(j\omega)$ are considered to be zero for $\omega > 5$, and thus

$$\underset{\omega>5}{\text{Lm}}[a_{22}(j\omega)] \approx -\infty$$

(d) For $\mathbf{L}_1 = g_1 q_{11}$ and $\mathbf{L}_2 = g_2 q_{22}$, respectively:

(e)

$$\text{Lm } M_{L_1} \leq \text{Lm}\left[\frac{L_1}{1+L_1}\right] \leq 3 \text{ dB} \tag{G.2}$$

$$\text{Lm } M_{L_2} \leq \text{Lm}\left[\frac{L_2}{1+L_2}\right] \leq 3 \quad \text{dB} \tag{G.3}$$

G-2 PROBLEM

Design $L_{1_o}(s)$, $L_{2_o}(s)$ and $f_{22}(s)$.

G-3 SOLUTION

G-3.1 L_{1_o}(s) DESIGN

For loop one design, see Fig. G.1, let $f_{12} = 0$ (i.e., decouple output 1 from input 2). With $f_{12} = 0$ then the loop 1 design becomes strictly a cross-coupling rejection problem (see Fig 5.12). Thus, from Eqs. (5.36) thru (5.46) and Eqs. (5.55) thru (5.58), where for an impulse input $t_{ij} = y_{ij}$:

$$t_{12} = \frac{q_{11}(\overset{0}{\cancel{g_1 f_{12}}} + c_{12})}{1 + g_1 q_{11}} = \frac{q_{11} c_{12}}{1 + g_1 q_{11}} \tag{G.4}$$

Table G.1

ω	0	0.5	1	2	5	10	20
Lm $b_{22}(j\omega)$	0	1	1	0	-4	-10	-20
Lm $a_{22}(j\omega)$	0	-2	-4	8	20	(-40) ∞	(-80) ∞

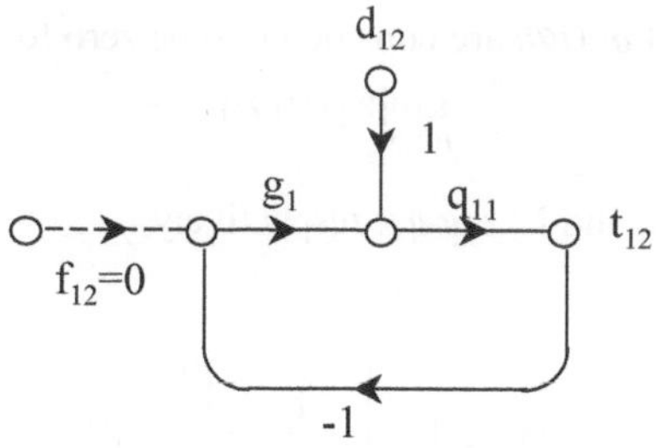

Fig. G.1 Equivalent MISO loop one.

The given disturbance rejection specification

$$|t_{12}| = \left|\frac{q_{11}c_{12}}{1+g_1q_{11}}\right| \leq 0.1 \tag{G.5}$$

is rearranged to

$$|1+g_1q_{11}| \geq \left|\frac{q_{11}c_{12}}{0.1}\right| \tag{G.6}$$

Since

$$c_{12} = -t_{22}/q_{22} \tag{G.7}$$

Equation (G.7) is substituted into Eq. (G.6) to yield

$$|1+g_1\,q_{11}| \geq \left|\frac{10q_{11}t_{22}}{q_{12}}\right| \tag{G.8}$$

Applying the maximum tracking specification value for $|t_{22}|$ yields:

$$|1+g_1\,q_{11}| \geq \left|\frac{10b_{22}q_{11}}{q_{12}}\right| \tag{G.9}$$

Thus, this constraint imbeds the cross-coupling rejections bounds specifications on $y_{12}(t)$. As a consequence the cross-coupling bounds $B_{c_l}(j\omega_i)$ become the composite bounds $\mathbf{B}_{o_l}(j\omega_i)$. For the plant uncertainty a template can be generated for q_{11} which is good (for this particular design problem) for all ω. The template is simply a 6 dB vertical line. Also, from the specification $\text{Lm}\, M_{L_1} \le 3$ dB and the uncertainty in q_{11} at $\omega = \infty$, the universal high frequency boundary can be drawn as shown in Fig. G.4. The nominal plant q_{11} is chosen to be the ιth plant which is at the bottom of the *6 dB* template. Thus,

$$q_{11}(s) = \frac{1}{s} \tag{G.10}$$

Generally, for lower values of ω, since $|g_1 q_{11}| >> 1$ then Eq. (G.9) can be simplified to:

$$|\mathbf{g}_1| \ge \left|\frac{10b_{22}}{q_{12}}\right| \tag{G.11}$$

Multiplying both sides of Eq. (G.11) by q_{11_o} yields:

$$|\mathbf{L}_{1_o}| = |\mathbf{g}_1 \mathbf{q}_{11_o}| \ge \left|\frac{10b_{22}\mathbf{q}_{11_o}}{q_{12}}\right| \tag{G.12}$$

where $L_{1_o} = g_1 q_{11_o}$. The worst case scenario must be chosen from the specifications in order to apply Eq. (G.12). Therefore, for

$$q_{12_{min}} = \frac{0.5}{s} \text{ and } q_{11_o} = \frac{1}{s}$$

Eq. (G.12) becomes

$$|\mathbf{L}_{1_o}| \ge |20b_{22}| \tag{G.13}$$

or

$$\text{Lm}\,\mathbf{L}_{1_o} \ge 26 + \text{Lm}\,\mathbf{b}_{22} = \mathbf{B}_{o_1}(j\omega_i) \tag{G.14}$$

Thus Eq. (G.14) reveals that for this example the composite bounds $\mathbf{B}_{o1}(j\omega j$ are straight lines whose magnitudes are a function of frequency. Based upon Eq. (G.14) and the given specifications on t_{22} in Table G.1 the bounds are:

TABLE G.2

ω_i	1	0.5	1	2	5	10	20
Lm $\mathbf{B}_o(j\omega_i)$	26	27	27	26	22	N/A	N/A

Since the value of $\mathbf{b}_{22}(j\omega)$ is very small for $\omega > 5$, the tracking frequency range is effectively $0 < \omega < 5$. The value $Lm\ \boldsymbol{a}_{22}(j\omega)$ yields $\delta_R(j\omega) = Lm\ \boldsymbol{b}_{22}(j\omega) - Lm\ \boldsymbol{a}_{22}(j\omega) \to \infty$ dB for $\omega > 5$ and the tracking problem will always be satisfied in this frequency range. It is therefore necessary to shape the loop transmission $L_{1_o}(j\omega)$ only for $0 \leq \omega \leq 5$. Only the cross-coupling rejection problem remains for $\omega > 5$.

Since the p_{ij} elements of **P**, and in turn all the elements of **Q**, and the nominal q_{11_o} are Type 1 functions then $L_{1_o}(s)$ should be chosen as a Type 1 or higher function. A Type 1 function is chosen for $L_i(s)$. An initial simple $L_{1_o}(s)$ design containing only one pole at the origin, one complex-pole pair, and one zero resulted in ω_ϕ = 65 rps. In order to reduce the value of the phase margin frequency, ω_ϕ, the following $L_{1_o}(s)$ is synthesized:

$$L_{1_o}(s) = \frac{54038(s+25)}{s(s+10)(s+84 \pm j112)} \qquad \textbf{(G.15)}$$

where ω_n = 140 rps and is shown in Fig. G.2. Note that the approximation made in Eq. (G.12) is justified because all the values in Table G.2 are at least a magnitude above one. Since $L_{1_o} = g_1 q_{11_o}$ then

$$g_1(s) = \frac{L_{1_o}(s)}{q_{11_o}(s)} = \frac{540838(s+25)}{(s+10)(s+84 \pm j112)} \qquad \textbf{(G.16)}$$

G-3.2 DESIGN (IMPROVED METHOD) FOR $L_{2_e}(s)$

The equations for loop 2, see Fig. G.3 and Eqs. (5.36) thru (5.39) and Eq. (5.58), are

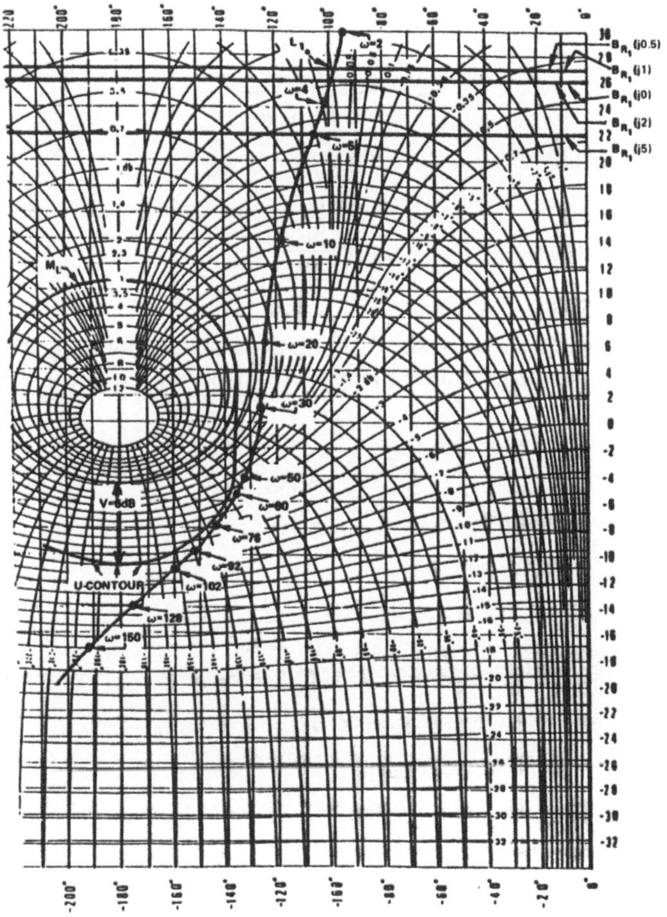

Fig. G.2 The $L_{1_o}(j\omega)$ design.

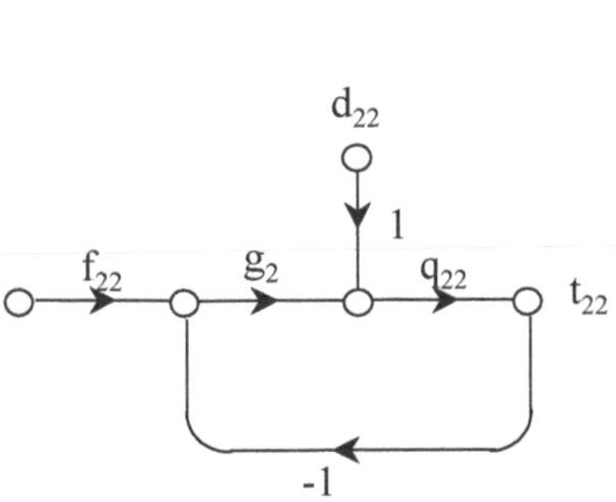

Fig. G.3 Loop 2 MISO equivalent.

$$t_{22} = \frac{q_{22}(g_2 f_{22} + c_{22})}{1 + g_2 q_{22}} \quad \textbf{(G.17)}$$

where $c_{22} = -t_{12}/q_{21}$ and

$$t_{12} = \frac{q_{11} c_{12}}{1 + g_1 q_{11}} \quad \textbf{(G.18)}$$

Instead of making the necessary substitutions and performing the associated algebra the following generalized loop 2 equations, see Eq. (7.4), are utilized in the design.

$$\gamma_{12} = \frac{q_{11} q_{22}}{q_{21} q_{12}} \quad \textbf{(G.19)}$$

$$q_{22_e} = \frac{q_{22}(1 + L_1)}{1 + L_1 - \gamma_{12}} \quad \textbf{(G.20)}$$

$$L_{2_e} = g_2 q_{22_e} \quad \textbf{(G.21)}$$

and

$$c_{22_e} = \frac{g_1 f_{12} p_{21}(1 - \gamma_{12})}{1 - \gamma_{12} + L_1} = 0 \quad \textbf{(G.22)}$$

since $f_{12} = 0$. Thus loop 2 is only a tracking problem (see Fig. 5.11).

Before generating the templates for q_{22_e} it is necessary to ascertain whether all the poles and zeros of q_{22_e} lie in the LHP s-plane over the region of uncertainty. The proper design of $L_1(s)$ guarantees that the zeros of $1 + L_1(s)$ are in the LHP. Since an m.p. plant is assumed, then $q_{22_e}(s)$ will also be m.p. However, it is possible for $1 + L_1(s) - \gamma_{12}(s)$ of Eq. (G.20) to have RHP zeros which become RHP poles of $q_{22_e}(s)$ and therefore of $L_{2_e}(s)$. This possibility must be verified before proceeding with the design. The knowledge of the presence of RHP poles is necessary in order to interpret correctly the data for $|\boldsymbol{q}_{22_e}(j\omega_i)|$ and $\angle \boldsymbol{q}_{22_e}(j\omega_i)$ over the frequency range of interest (see Sec. 18.22 of Ref. 1).

Plant uncertainty case 10 of Table G.3 represents an unstable plant. At low frequencies the templates are rectangles *6 dB* in height and essentially 360° wide. That is, the template for $\omega = 0.5$ is essentially 6 *dB* in height and 360° wide, and for $\omega = 20$ *rad/sec* it is 6 dB in height, 345° wide at the top and 352° wide at the bottom. The templates for $\omega = 130$ *rad/sec* and $\omega = 150$ *rad/sec* are approximately *180°* and *138°* wide respectively. In order to assure a negative phase angle at all frequencies for $\mathbf{L}_{2_e}(j\omega)$, it is necessary to generate the templates at high frequencies such that they shrink sufficiently in width in order to allow a gap on the right side of the *U*-contour for $\mathbf{L}_{2_e}(j\omega)$. This permits the loop transmission plot to "squeeze" through and achieve the desired value of $\gamma = 45°$ (for this example). This analysis of the plant template data reveals:

1. That the width (change in angle) of the templates is the crucial factor in the shaping of L_{2_e} of the template.
2. That the template shapes go from being "very wide" (essentially rectangular), as shown in Fig. G.4, at "low" frequencies to the shape shown by the shaded area in Fig. G.4 at "high" frequencies.

As shown in Fig. G.4, that in going from $\omega = 0.5$ to $\omega = 300$ *rad/sec* the templates start to "droop" down to the right. This droop becomes more exaggerated if more template points are plotted from other unstable plants within $\mathcal{Q}$. These unstable plants occur for this example, when the condition

$$\gamma_{12} = \frac{k_{11}k_{22}}{k_{21}k_{12}} > 0.5 \tag{G.23}$$

is satisfied. This can be seen by partitioning the characteristic equation of Eq. (G.20) to yield $L_l/(1 - \gamma_{12}) = -1$ for which a root-locus analysis can be made as the value of γ_{12} is varied. The "very wide" characteristics of some of these templates can best be appreciated by the reader by plotting a positive and negative real pole (*a* and *−a*) in the *s*-plane. The reader is then urged to analyze the angular contribution of $\angle[1 + j(\omega/a)]$ and $\angle[1 + j(\omega/-a))]$ as the frequency is varied from zero to infinity.

In order to simplify the process of generating the bounds rectangular (solid curve) templates are used as shown in Fig. G.4. Although making this assumption yields an *over design*.

Table G.3

Case	k_{11}	k_{12}	k_{21}	k_{22}	Comments
Combinations using plant maximums and minimums					
1	1	0.5	5	0.5	
2	1	0.5	5	1	
3	1	0.5	10	0.5	
4	1	0.5	10	1	
5	1	2	5	0.5	
6	1	2	5	1	
7	1	2	10	0.5	
8	1	2	10	1	
9	2	0.5	5	0.5	
10	2	0.5	5	1	← unstable plant #10
11	2	0.5	10	0.5	
12	2	0.5	10	1	
13	2	2	5	0.5	
14	2	2	5	1	
15	2	2	10	0.5	
16	2	2	10	1	← nominal plant #15
Random points					
17	1.5	1	7	1	
18	2	1.5	9	0.75	
19	1.75	0.75	6	0.75	
20	1.25	1.25	8	0.5	

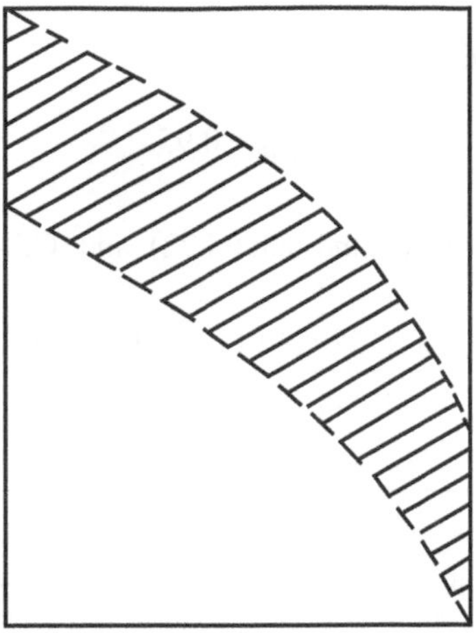

Fig. G.4 Template shapes.

The data for the templates (see Fig. G.5) that play the key role in achieving the desired value of γ is given in Table G.4. Note that at ω = 500 the template is 20 dB in height which is the value of *V* (UHFB) used to obtain the *U*-contour.

Table G.4

	Δ dB	**Δθ°**
150	14.5	138
170	15.8	81
180	15	61
200	16.6	37
300	18	10
500	20	-0

The data in Table G.4 is used to obtain the templates which are used to obtain the $\mathbf{B}_{R_e}(j\omega_i) = \mathbf{B}_{O_e}(j\omega_i)$ bounds of Fig. G.6 for loop 2 design. Note that for < 150 rad/sec the bounds are considered to be essentially straight lines. That is, these bounds are represented by the straight line tangent to the top of the Lm M_L contour since all the templates for ω < 150 rad/sec are greater than 135° in width. A stable plant is chosen to be the nominal plant see Table G.3, and is located at the lower right-hand corner of the templates. Thus, the plant gain values

k_{ij}, for the nominal plant, are $k_{11} = 2$, $k_{12} = 2$, $k_{21} = 10$, and $k_{22} = 0.5$. For these nominal values, where $\gamma_{12} = 0.05$, let

$$\alpha = 1 - \gamma_{12} = 0.95 \tag{G.24}$$

Thus, from Eqs. (G.15), (G.19) through (G.22), and (G.24) obtain

$$q_{22_e} = \frac{k22\,[s^4 + 178\,s^3 + 21280\,s^2 + (196000 + 540838\,k_{11})s + 13520950\,k_{11}]}{s[\alpha s^4 + 178\alpha s^3 + 21280\alpha s^2 + (196000\alpha + 540838\,k_{11})s + 13520.95\,k_{11}]}$$
$$= \frac{0.5263158(s^4 + 178\,s^3 + 21280\,s^2 + 1277676s + 27041900}{s^5 + 178s^4 + 21280s^3 + 667303s^2 + 14232578.955s}$$
$$= \frac{0.5263158(s + 49.2349 \pm j6.0882)(s + 39.7651 \pm j96.9859)}{s(s + 17.3915 \pm j24.9674)(s + 71.6085 \pm j101.2169)} \tag{G.25}$$

In order to make $g_2(s)$ "fairly simple" the nominal plant is used in Eq. (G.21) which results in a non-unique g_2 being designed rather than L_{2_e} where

$$L_{2_e}(s) = g_2(s)\,q_{22_e}(s) \tag{G.26}$$

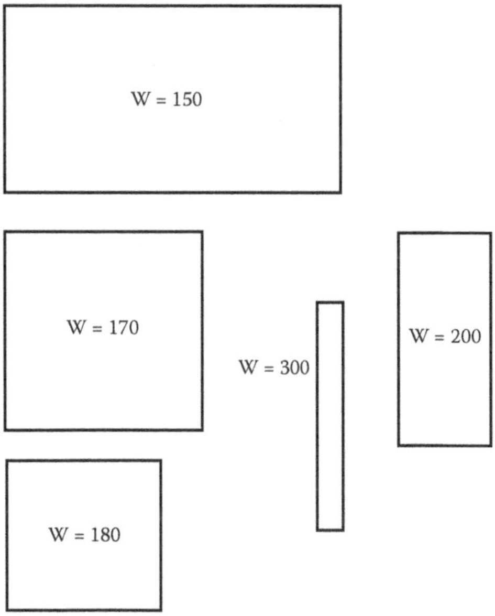

Fig. G.5 Tracking templates for loop 2 (using a rectangular approximation).

A synthesized (non-unique) $L_{2_e}(s)$ is shown in Fig. G.6. The hump in L_{2_e} in the vicinity of $\omega = 150$ *rad/sec* is caused by the complex zeros ($\omega_n = 104.8$ *rad/sec*) of the nominal plant. These zeros have a low damping ratio ($\zeta \approx 0.38$) which causes the "rapid" change in the phase angle. Usually a complicated $g_2(s)$ results when adjusting $L_{2_e}(j\omega)$ so that it crosses the $B_{O_e}(j150)$ bound near the right side of the NC with $L_{2_e}(j\omega)$ being on the corresponding $B_{O_e}(j\omega_i)$ bound. However, in order to obtain a simpler $g_2(s)$, the approach in this example is to design $g_2(s)$ so that $L_{2_e}(s)$ crosses the first few bounds at a phase angle of approximately $-90°$. As it turns out, for this example, $g_2(s)$ does not contribute very much to the phase angle at the $B_{O_e}(j\omega_i)$ crossing for $\omega \leq 150$ *rad/sec*; i.e., the phase angle is due essentially to the nominal plant. The resultant $g_2(s)$ is:

$$g_2(s) = \frac{3.3155\text{x}\,10^{10}(s+1000)}{(s+500)(s+3500 \pm j6062.1778)} \qquad \textbf{(G.27)}$$

where $\omega_n = 7000$ and $\zeta = 0.5$. A second design for $g_2(s)$ can be made by choosing smaller values for the real pole and zero of Eq. (G.27) in order to have $L_{2_e}(s)$ track down the right side of the *U*-contour more closely. In general, a more complicated $g_2(s)$ may be synthesized that approaches the optimal loop transmission function

$$L_2(j\omega_i)_{opt} = O(j\omega) = |O(j\omega)| \angle O(j\omega) \qquad \textbf{(G.28)}$$

whose m,agnitude lies on $B_{O_e}(j\omega_i)$, for each value of ω_i, and whose phase angle $\angle O(j\omega_i)$ lies on the right side of the *U*-contour.

G-3.3 PRE-FILTER $F_{22}(s)$ DESIGN

The pre-filter design procedure of Chapter 3 requires the use of the templates to obtain the values of $Lm\ t_{2max}$ and $Lm\ t_{2min}$ where $t_2 = L_2/(1 + L_2)$ over the frequency range $0 < \omega < 20$ *rad/sec*. In this frequency range, for this example, the templates are between 345° and 360° wide. As a consequence, the location on the NC where these templates are placed result in small positive values for $Lm\ t_2$ (≤ 0.1 *dB*). Thus, small variations exist for $Lm\ t_2$, i.e., $\Delta t_2 = Lm\ t_{2max} - Lm\ t_{2min}$ will be a very small number in the range $0 < \omega < 20$ *rad/sec*. For example, the bottom of the template, for $\omega = 20$, would be placed on the $Lm\,\mathbf{L}_{2_o}$ curve at the $Lm\,L_{2e}$ point to yield $Lm\,t_{2max} \approx 0.1\,dB$ and a value for $Lm\,t_{2min}$ something less than 0.1 dB. Thus, for $\omega < 20$ Δt_2 is a very small number and $|t_2| \approx 1$ for $\mathcal{P} = \{\mathbf{P}\}$. Therefore, a figure corresponding to Fig. 3.23, for this example, are obtained by plotting the data from Table G.l, i.e.,

$$Lm\,\mathbf{T}_{R_U} - Lm\,\mathbf{T}_{\max} = Lm\;\mathbf{b}_{22} \text{ and } Lm\,\mathbf{T}_{R_L} - Lm\,\mathbf{T}_{\min} = Lm\,\mathbf{a}_{22}$$

The filter chosen is:

$$f_{22}(s) = \frac{1}{s+1} \tag{G.29}$$

Note: care must be taken in determining the location of the pole of Eq. (G.29) since it may be necessary to lower the value of $|f_{22}|$ at large values of ω (≥ 10), where $a_{22} = 0$, since this affects the value of $|t_{12}|$, etc. [(see Eqs. (G.7) through (G.9)].

G-4 SIMULATION

The computer data for y_{22} for a number of plants from $\mathcal{P}$ resulted in $|t_2| \approx 1$ for $0 \leq \omega \leq 20$. Table G.5 presents the time response characteristics of $y_{22}(t)$, for a unit step forcing function, for three cases from Table G.3. These cases are chosen on the following basis:

Table G.5

Case	M	t_p, s	t_s, s	Final value	Figure
10 (Unstable plant)	1.000	3.916	2.19709	1.000	G.8
15 (Nominal)	1.000	3.913	2.19726	1.000	G.7
16 (High gain)	1.000	3.913	2.19722	1.000	G.9

(1) *Nominal plant* -- The loop transmission is obtained based upon the nominal plant.
(2) *Unstable plant* -- A "worst" case situation representative of the unstable plant region of $\mathcal{P}$.
(3) *High gain plant* -- A high gain plant is chosen because it is representative of the top of the templates on the "nominal plant side."

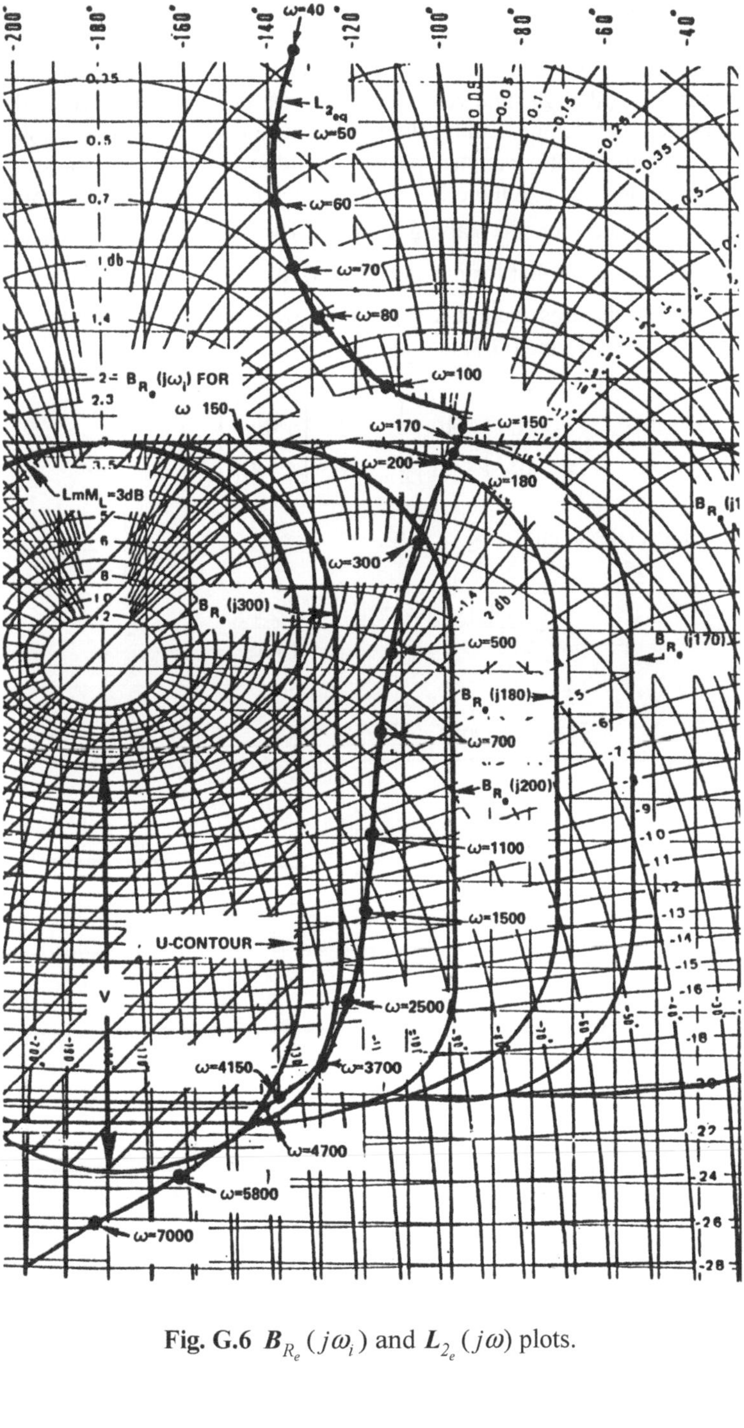

Fig. G.6 $B_{R_e}(j\omega_i)$ and $L_{2_e}(j\omega)$ plots.

Since all the other cases of Table G.3 are extremely similar to one of the three cases of Table G.5 they are not simulated.

The frequency domain specifications, as demonstrated by the simulation, are met for loop 2. Although time response specifications are not prescribed, $y_{22}(t)$ exhibits respectable rise and settling times. Because of the gross over design any other plant from the plant parameter space $\mathcal{P}$ should meet the specifications.

In designing loop one the cross-coupling rejections bounds are dropped to an arbitrary low value for the shaping of L_{1_o}. The synthesized L_{1_o} for $\omega = 20$ lies on $\boldsymbol{B}_{o_1}(j20) = 6\ dB$ but for $\omega = 10$ $Lm\ \boldsymbol{L}_1(j10) \approx 14.5\ dB < \boldsymbol{B}_{o_1}(j10) = 16\ dB$ Thus, a check on the cross-coupling rejection $\boldsymbol{t}_{12}(j\omega)$ performance for loop one is made. From Eqs. (G.4) and (G.7) obtain:

$$t_{12} = \frac{-t_{22}}{q_{12}}\left[\frac{q_{11}}{1+L_1}\right] = -\frac{st_{22}}{k_{12}}\left[\frac{q_{11}}{1+L_1}\right] \tag{G.30}$$

where $q_{12} = k_{12}/s$, $c_{22} = 0$, and

$$t_{22} = \frac{q_{22}(g_2 f_{22} - c_{22})}{1+L_{2_e}} = \frac{f_{22} L_{2_e}}{1+L_{2_e}} \tag{G.31}$$

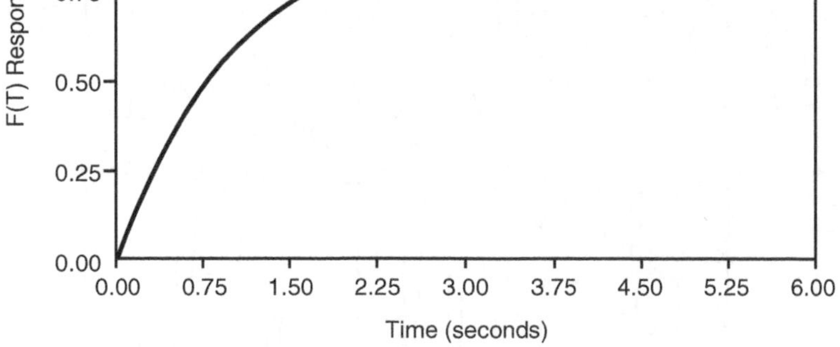

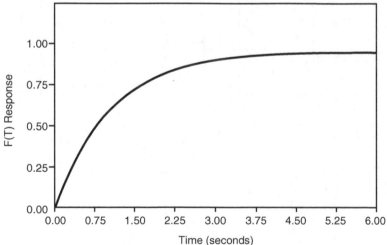

Fig. G.7 Time responses $y_{22}(t)$ for $r_2(t) = u_{-1}(t)$: nominal plant case.

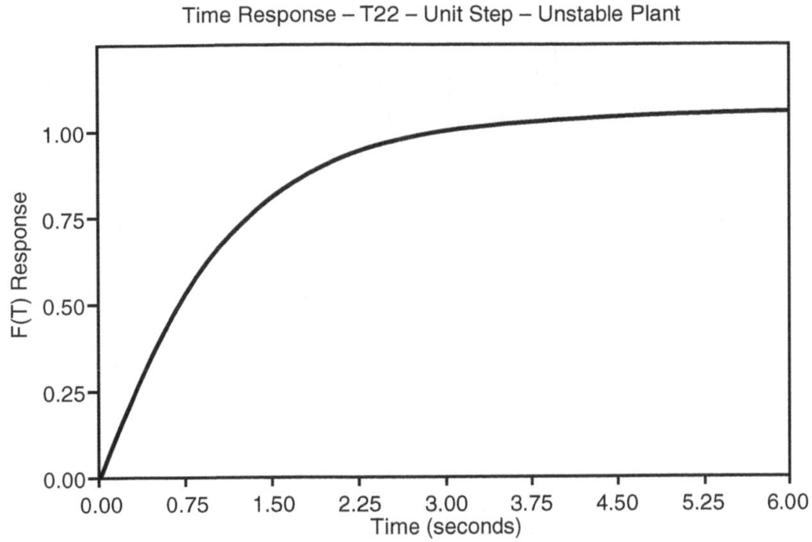

Fig. G.8 Time response $y_{22}(t)$ for $r_2(t) = u_{-1}(t)$: unstable plant case.

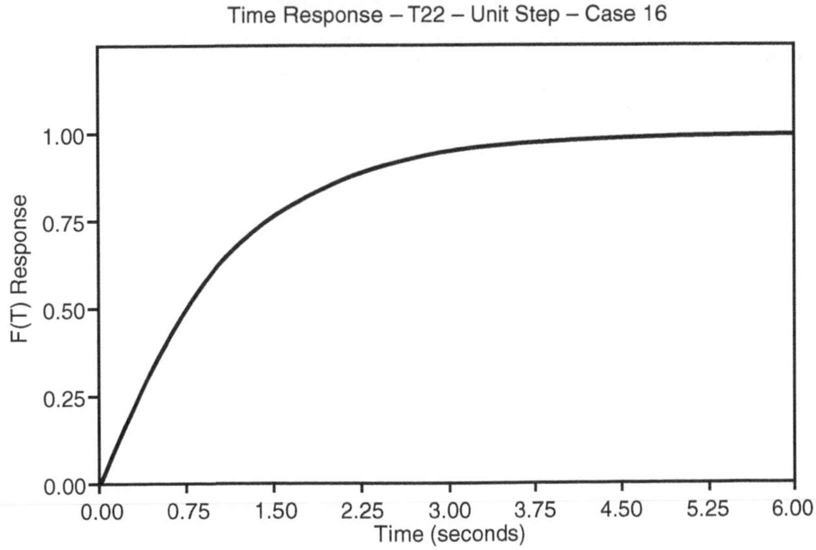

Fig. G.9 Time response $y_{22}(t)$ for $r_2(t) = u_{-1}(t)$: high gain case.

Substituting Eq. (G.31) into Eq. (G.30) yields

$$t_{12} = -\left[\frac{f_{22} L_{2e}}{1+L_{2e}}\right]\left[\frac{s}{k_{12}}\right]\left[\frac{q_{11}}{1+L_1}\right] \qquad \textbf{(G.32)}$$

Only the plant parameters associated with the unstable q_{22_e} plant (case 10) are used in Eq. (G.31) for a simulation. No other plants are simulated to determine if the cross-coupling rejection specification is met. The frequency response plot for Eq. (G.32) using the parameters for the unstable plant is shown in Fig. G.10. As seen from this figure the $-20\ dB$ specification is met. The simulation yields $t_{12}(t_p) = -0.0579$ at $t_p \approx 0.026\ s$, $|t_{12}(t_s)| = 0.002$ (2% of the specified maximum magnitude of 0.1) at $t_s = 2.61\ s$, and $t_{12}(\infty) = 0$.

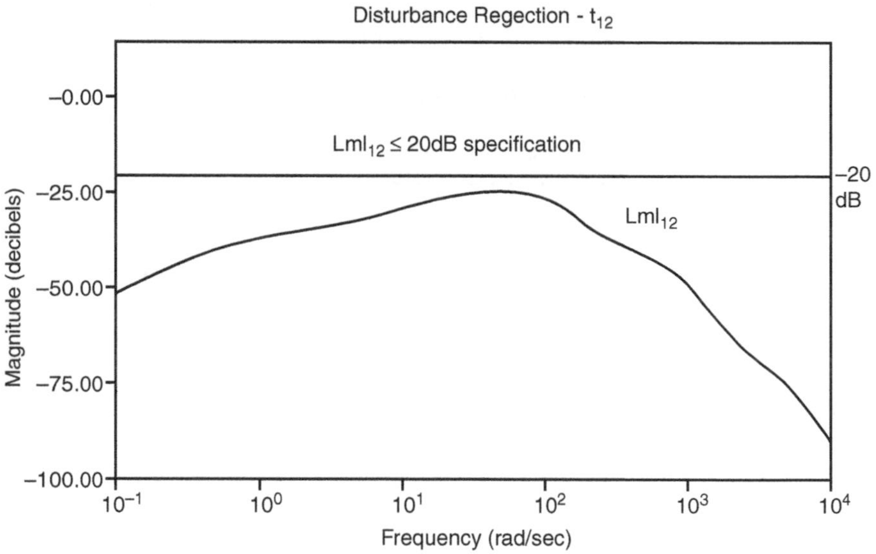

Fig. G.10 Frequency response $t_{12}(j\omega)$: unstable case

G-5 SUMMARY

In some problems $\gamma_{12} << 1$ and therefore $q_{22_e} \approx q_{22}$ [see Eq. (G.20)] which simplifies the design. Note that for the example of this appendix $\gamma_{12max} = 0.8$. Thus, for $0.5 < \gamma_{12} < 0.8$ there exists an unstable q_{22_e}, i.e., q_{22_e} has RHP poles. An unstable plant with a pole $p_1 > 0$ requires a loop transmission with a crossover frequency[77] $\omega_\phi \cong > 2p$. Therefore, L_2 has to have a large crossover frequency. In order for L_{2_e} to have a lower value for its ω_ϕ it is necessary to design L_1 so that $1 + L_1 - \gamma_{12}$ does not have RHP zeros. This makes L_1 costlier, the benefit going to L_{2_e} (or g_2). This freedom can be used as a trade-off feature to be used according to the relative sensor noises, etc.

APPENDIX H

NON-DIAGONAL MIMO QFT TRACKING DESIGN EXAMPLE

H-1 HEAT EXCHANGER TEMPERATURE CONTROL SYSTEM [97, 99, 112]

A heat exchanger temperature control system of a pasteurization process is analyzed in this Appendix as an example to applied the non-diagonal MIMO QFT tracking design methodology introduced in chapter 10. The plant PCT23, manufactured by Armfield (UK),[127] is a miniature version of a real High Temperature Short Time (HTST) pasteurization process (Fig. H.1). It consists of a bench-mounted process unit connected to a control console to provide access to the measured and control signals.

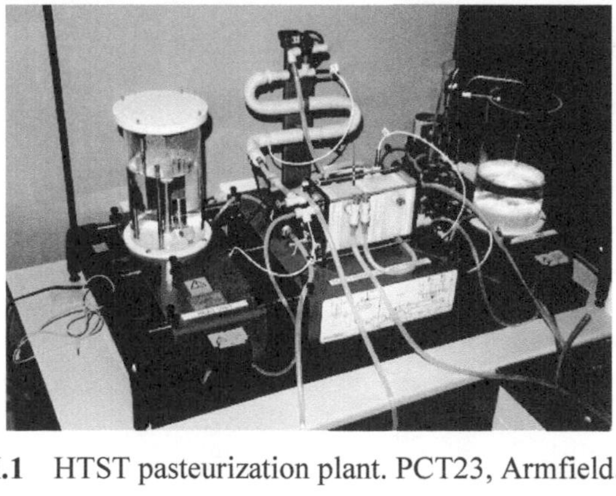

Fig. H.1 HTST pasteurization plant. PCT23, Armfield.[127]

The considered problem consists of a raw product circuit with a constant flow (N_2) that is heated in a countercurrent heat exchanger using a hot water flow (N_1) coming from a closed circuit with a water heater tank, as shown in Fig. H.2. The case is a 2x2 plant. The variables to be controlled are the temperatures T_1 [°C] and T_2 [°C]. The manipulated variables are the hot water flow N_1 [ml/min] and the power Q [kW] spent in the electrical resistance of the water heater tank. The raw product flow N_2 [ml/min] is constant.

H-2 MATHEMATICAL MODEL

The plant model is non-linear with variations of the rate N_1/N_2.[112] However, an uncertain linear system may be used to design the control system.[97, 99] The plants p_{12}, p_{21} and p_{22} are modeled by a set of uncertain parametric plants with the structure of Eq. (H.1) and the parameters of Table H.1. The plant p_{11} is modeled by a non-parametric uncertain set of plants, with the frequency response of Fig. H.3 and the nominal plant described by Eq. (H.1) and Table H.2.

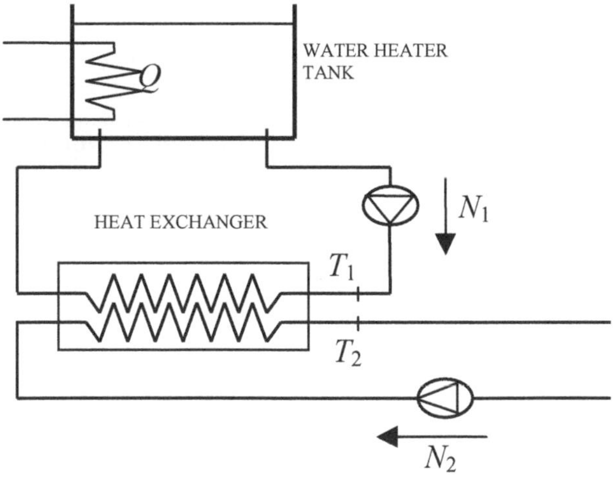

Fig. H.2 HTST pasteurization plant diagram.

$$\begin{bmatrix} T_1[^\circ\mathrm{C}] \\ T_2[^\circ\mathrm{C}] \end{bmatrix} = \begin{bmatrix} p_{11}(s) & p_{12}(s) \\ p_{21}(s) & p_{22}(s) \end{bmatrix} \begin{bmatrix} N_1[\mathrm{ml/sec}] \\ Q[\mathrm{kW}] \end{bmatrix} = \tag{H.1}$$

$$\begin{bmatrix} \dfrac{n_{11,b}s + n_{11,a}}{s^2 + d_{11,b}s + d_{11,a}} e^{-\tau_{11}s} & \dfrac{n_{12,a}}{s + d_{12,a}} e^{-\tau_{12}s} \\ \dfrac{n_{21,b}s + n_{21,a}}{s^2 + d_{21,b}s + d_{21,a}} e^{-\tau_{21}s} & \dfrac{n_{22,a}}{s + d_{22,a}} e^{-\tau_{22}s} \end{bmatrix} \begin{bmatrix} N_1 \\ Q \end{bmatrix}$$

It is important to emphasize that this system includes the non-minimum phase element $p_{11}(s)$ and time delays, which are unfavorable for the inclusion of non-diagonal compensators as mentioned in Sec. 10.3.3. However this example comes in useful in order to reinforce that methodology solving a challenging problem.[97]

Table H.1 Coefficients of parametric uncertainties

Parameter	Minimum	Maximum	Nominal
$n_{21,a}$	$0.0387\cdot10^{-6}$	$0.494\cdot10^{-6}$	$0.236\cdot10^{-6}$
$n_{21,b}$	$0.739\cdot10^{-3}$	$0.853\cdot10^{-3}$	$0.806\cdot10^{-3}$
$d_{21,a}$	$17.2\cdot10^{-6}$	$26.1\cdot10^{-6}$	$22.3\cdot10^{-6}$
$d_{21,b}$	$20.8\cdot10^{-3}$	$24.3\cdot10^{-3}$	$22.6\cdot10^{-3}$
τ_{21}	8	12	10
$n_{12,a}$	$40.7\cdot10^{-3}$	$48.2\cdot10^{-3}$	$43.6\cdot10^{-3}$
$d_{12,a}$	$0.770\cdot10^{-3}$	$0.825\cdot10^{-3}$	$0.797\cdot10^{-3}$
τ_{12}	20	24	22
$n_{22,a}$	$23.9\cdot10^{-3}$	$29.0\cdot10^{-3}$	$26.0\cdot10^{-3}$
$\mathrm{d}_{22,a}$	$0.663\cdot10^{-3}$	$0.841\cdot10^{-3}$	$0.699\cdot10^{-3}$
τ_{22}	60	84	72

Table H.2 Coefficients of plant p_{11}

Parameter	Case A	Case B	Case C	Nominal
$n_{11,a}$	$-1.01 \cdot 10^{-6}$	$-7.56 \cdot 10^{-6}$	$-1.01 \cdot 10^{-6}$	$-6.89 \cdot 10^{-6}$
$n_{11,b}$	$0.0295 \cdot 10^{-3}$	$2.30 \cdot 10^{-3}$	$0.22 \cdot 10^{-3}$	$1.05 \cdot 10^{-3}$
$d_{11,a}$	$0.0225 \cdot 10^{-3}$	$0.358 \cdot 10^{-3}$	$0.0151 \cdot 10^{-3}$	$0.157 \cdot 10^{-3}$
$d_{11,b}$	0.036	0.273	0.0282	0.225
τ_{11}	16	9	4	16

H-3 PERFORMANCE SPECIFICATIONS

The non-diagonal MIMO QFT methodology introduced in Chapter 10 is applied to design a robust compensator for controlling the two temperatures of the heat exchanger described in the last section. The desired performance specifications for the two control loops are the following:

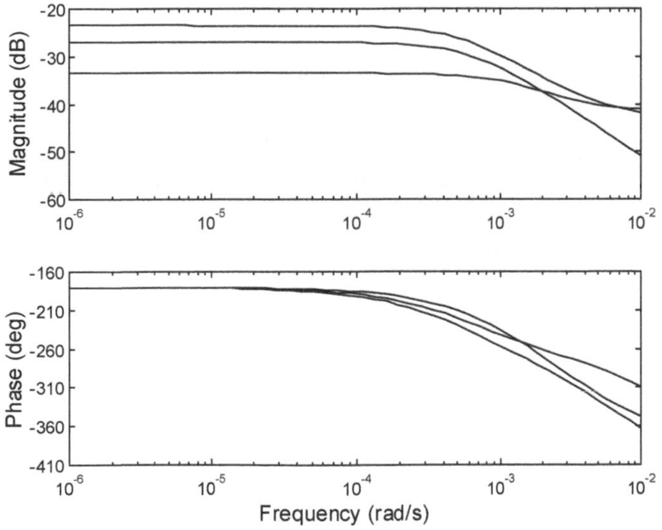

Fig. H.3 Frequency response of plant $p_{11}=T_1/N_1$.

- *Robust stability*: $\left|t_{ii}(j\,\omega)\right| \leq 1.2$, i = 1,2 $\forall\omega$, that involves at least 50° lower phase margin and at least 1.833 (5.26 dB) lower gain margin.

- *Control effort constraints*: control signals have to be lower than 150 ml/min and 1 kW for N_1 and Q respectively for maximum disturbances.

- *Coupling*: reduction of coupling effect as much as possible.

- *Tracking specifications*: tracking model for temperature T_1 with the following tolerances:
  - overshoot smaller than 25%.
  - settling time lower than 6000 seconds for a 5% of error.

That is to say:

$$a(\omega) \leq \left|t_{11}^{\mathrm{y/r}}(j\,\omega)\right| \leq b(\omega) \tag{H.2}$$

where,

$$b(\omega) = \left|\frac{(j\omega+1)}{10000(j\omega)^2 + 91.2\,j\omega + 1}\right| \tag{H.3}$$

$$a(\omega) = \left|\frac{0.0065}{[10000(j\omega)^2 + 36.8\,j\omega + 0.0064]\;[j\omega/(0.0008 + 1)]}\right| \tag{H.4}$$

H-4 DESIGN PROCEDURE

- Step A *Input/Output pairing and loop ordering*

The first step of the method is the Relative Gain Analysis (RGA). It allows to pair inputs and outputs signals. Due to RGA properties for a 2x2 system, only the λ_{11} element is needed to be calculated.[113] Computing it from the identified multivariable models within the parameter uncertainty, the results for the element λ_{11} yields the following:

$$\lambda_{11} \in [0.58\;;\;0.95], \quad \lambda_{11}^{\mathrm{nominal}} = 0.76 \tag{H.5}$$

This analysis involves that the best choice is to control temperature T_1 with the hot water flow N_1, and to control temperature T_2 with the electrical power Q.

Once the pairing has been established, the second question to decide is which loop has to be closed first. To do so, it is very advisable to take a look into the system dynamics. Note that the non-diagonal controller g_{21} should reduce the effect of control effort N_1 on the temperature T_2 through plant p_{21}. However, the plant p_{22} presents a very slow dynamic behavior and then it is not possible to get any benefit from the non-diagonal element g_{21}. Thus, the chosen element g_{21} is null (see Fig. H.4) and the first loop to be closed is temperature T_2.[113]

- Step B.1 *Design of the second diagonal compensator element, g_{22}(s).*

Figure H.5 shows the templates that represent the p_{22}(s) model, T_2(s)/Q(s), with the identified parameter uncertainty and for the set of frequencies of interest. Figures H.6 and H.7 present the stability and control effort bounds, respectively, that take into account the model uncertainty and the desired specifications for Loop 2. Figure H.8 shows the intersection of all bounds (the worst case for every frequency and phase). Through the application of the standard loop-shaping technique, the compensator of Eq. (H.6) is found. It satisfies all the performance specifications (see Fig. H.9). An integral element is also included in the compensator to remove steady-state errors.

$$g_{22}(s) = \frac{0.05269s + 20.9\times10^{-6}}{s} \quad \textbf{(H.6)}$$

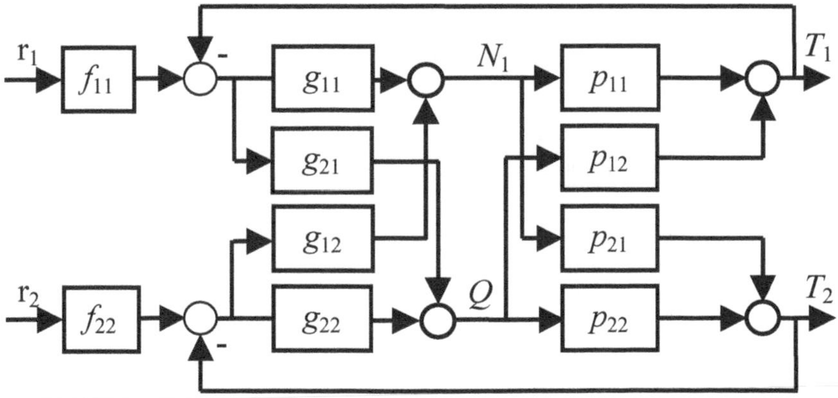

Fig. H.4 System block diagram.

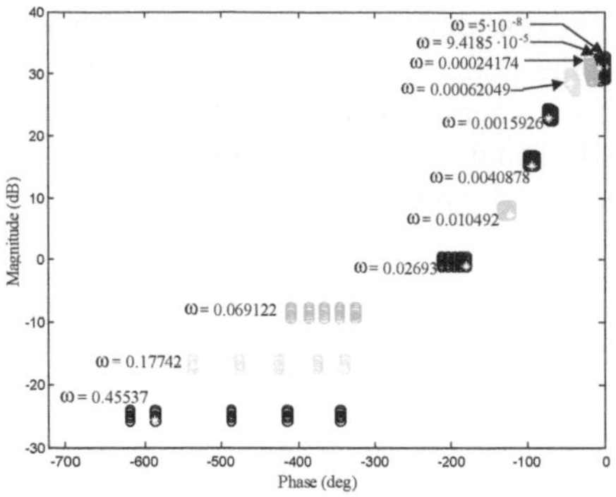

Fig. H.5 Templates of p_{22}.

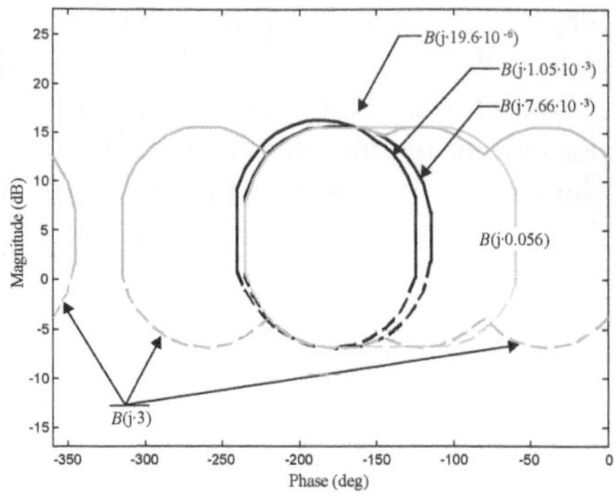

Fig. H.6 Robust stability bounds for loop 2.

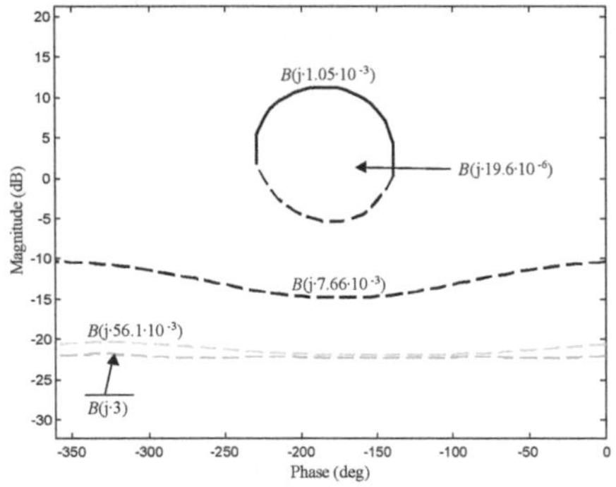

Fig. H.7 Control effort constraint bounds: Loop 2.

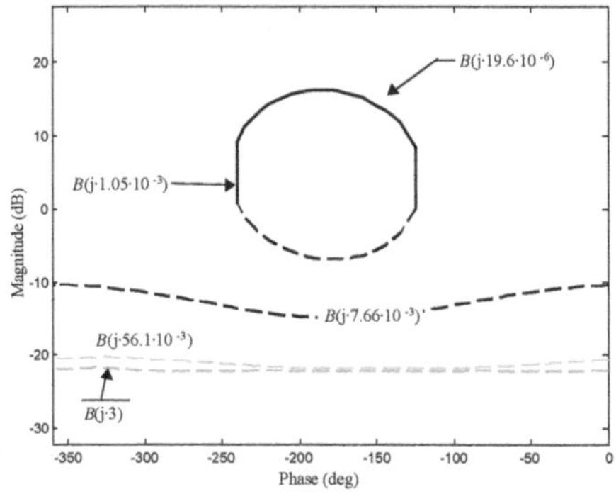

Fig. H.8 Worst case (intersection) of all bounds: Loop 2.

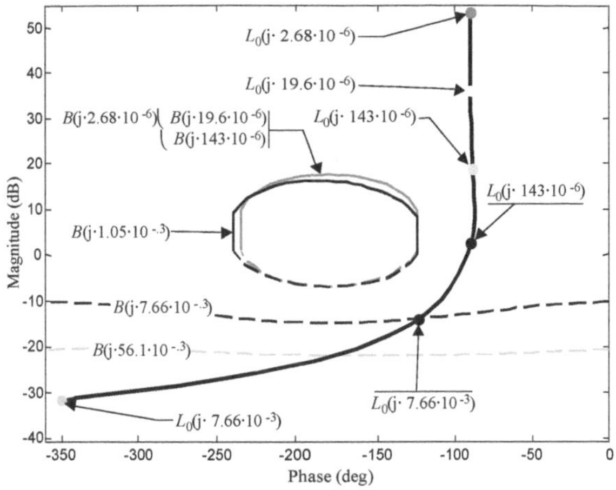

Fig. H.9 Loop-shaping [$L_0 = g_{22}(s)\,p_{22}(s)$]: Loop 2.

- Step C.1 *Design of the non-diagonal compensator element,* $g_{12}(s)$.

Taking into account the optimum compensator of Eq. (10.16), the compensator g_{12} of Eq. (H.7) is designed, minimizing the coupling effect c_{12} that can be obtained from Eq. (10.14) and maximizing the quality function of Eq. (10.20). Figure H.10 shows the frequency plot of the resulting coupling reduction.

$$g_{12}(s) = \frac{68.44s + 27.2\times10^{-3}}{s} \qquad \textbf{(H.7)}$$

- Step B.2 *Design of the first diagonal compensator element,* $g_{11}(s)$.

This step involves the design of the compensator of the first loop. However, note that the dynamic behavior that now can be observed acting on input N_1 and obtaining measures from T_1 is not from the stand-alone plant p_{11}, but from the following equivalent plant p_{11}^{e} [see Eq. (10.21)],

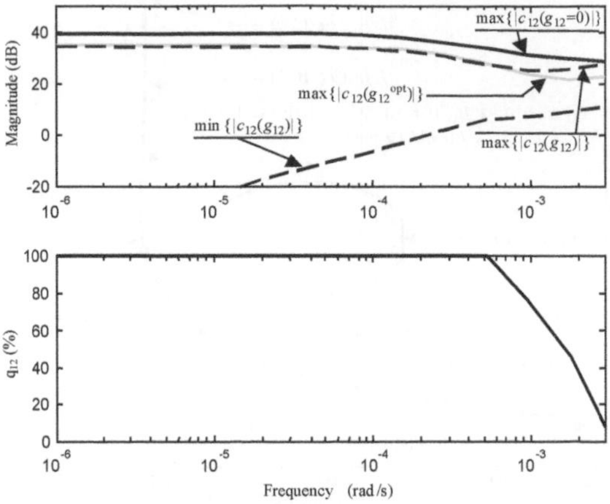

Fig. H.10 q_{12}(%) quality function for the coupling reduction.

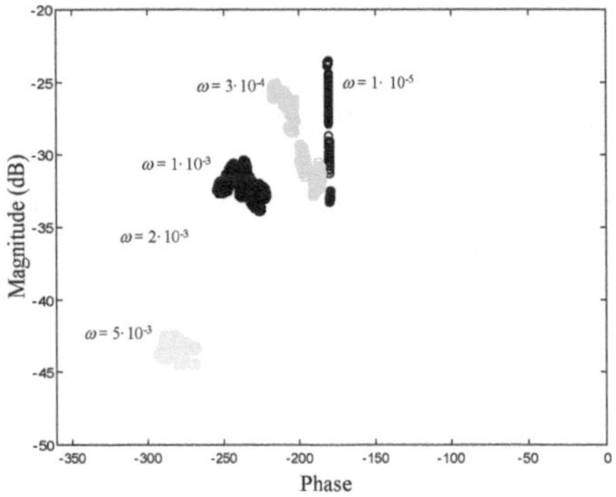

Fig. H.11 Templates of p_{11}^{e}.

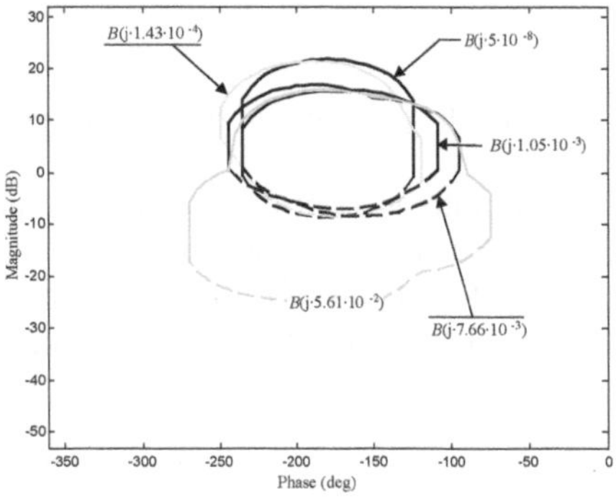

Fig. H.12 Robust stability bounds for loop 1.

$$p_{11}^{e} = p_{11} - \frac{(p_{11}g_{12} + p_{12}g_{22})p_{21}}{1 + p_{21}g_{12} + p_{22}g_{22}} \tag{H.8}$$

Figure H.11 shows the templates that represent the p_{11}^{e} equivalent model, $[T_1(s)/N_1(s)]$, with the identified parameter uncertainty and for the set of frequencies of interest. Figures H.12 to H.14 present the stability, control effort,

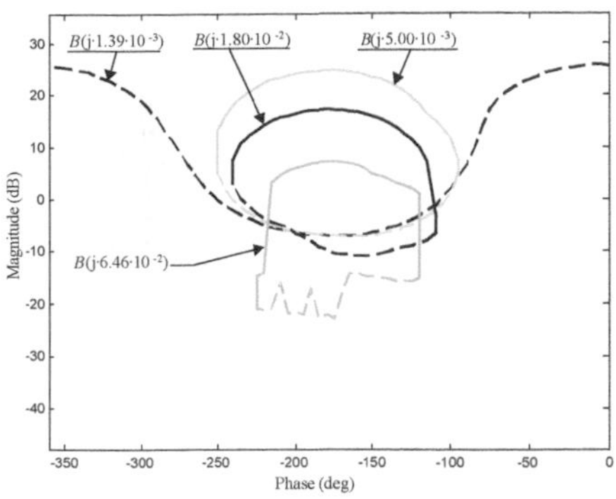

Fig. H.13 Control effort constraint bounds for loop 1.

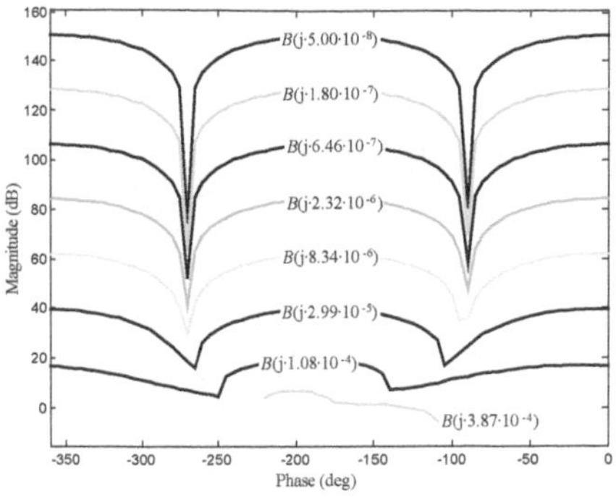

Fig. H.14 Robust tracking bounds for loop 1.

and tracking bounds, respectively, that take into account the model uncertainty and the desired specifications for Loop 1.
Figure H.15 shows the intersection of all bounds: the worst case for every frequency and phase. Through standard loop-shaping procedures, the compensator of Eq. (H.9) is determined, satisfying all the performance specifications (see Fig. H.16). The designed compensator includes an integral element in order to remove steady-state errors as indicated in Eq. (H.9).

$$g_{11}(s) = \frac{-1.886s - 18.9\times 10^{-3}}{s} \qquad \textbf{(H.9)}$$

The last step is the design of the open loop pre-filter $\boldsymbol{F}$ to achieve tracking specifications. The designed pre-filter is,

$$f_{11}^{non\,diag}(s) = \frac{1.4875\times 10^{6}\,s^{2} + 780.5\,s + 1}{0.694\times 10^{6}\,s^{2} + 1583\,s + 1} \qquad \textbf{(H.10)}$$

- <u>Pre-filters</u>.

On the other hand, to compare the practical results, a classical diagonal compensator is also designed using the multivariable technique introduced in chapter 7. It is composed of the feedback compensator g_{22} of Eq. (H.6) to regulate the temperature T_2, and two new elements: the feedback controller g_{11} of Eq. (H.11) to control temperature T_1, and the pre-filter of Eq. (H.12).

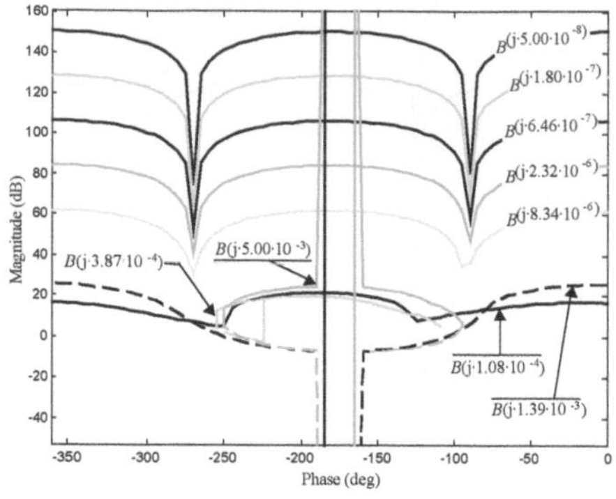

Fig. H.15 Worst case (intersection) of all bounds for loop 1.

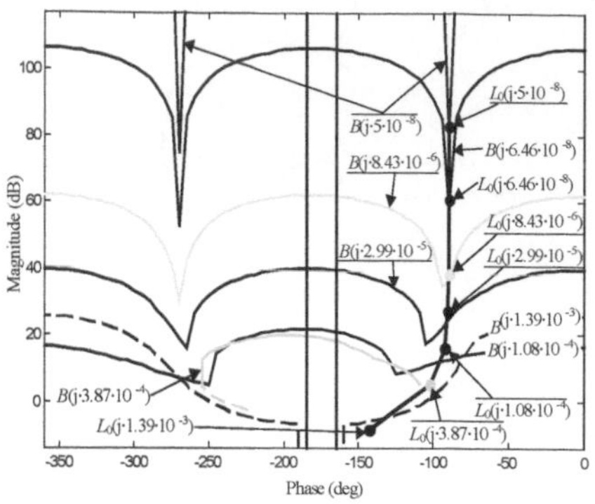

Fig. H.16 $[L_0 = g_{11}(s)\ p^{e}_{11}(s)]$ Loop-shaping for Loop 1.

$$g_{11}(s) = \frac{-12.17s - 18.9\times10^{-3}}{s} \tag{H.11}$$

$$f_{11}^{diag}(s) = \frac{1.161s^2 + 2.134\times10^3 s + 6.25\times10^{-6}}{s^2 + 4.7\times10^3 s + 6.25\times10^{-6}} \tag{H.12}$$

Note, that the bandwidth of the compensator g_{11} [see Eq. (H.11)] is larger than that achieved by the use of the non-diagonal technique, shown in Eq. (H.9). This is because both of them must achieve the same specifications, but for different equivalent plants: Eq. (H.8) for the non-diagonal technique and Eq. (H.8), with $g_{12} = 0$, for the diagonal technique.

H-5 EXPERIMENTAL RESULTS

Some real experiments[97] have been carried out implementing the previously described compensators into digital form and using a sampling time of 4 seconds. The tests taken with the prototype[127] show the profit from using the non-diagonal element, despite of the non-fulfillment of the sufficient conditions required to apply the proposed methodology (Sec. 10.3.3).

Figure H.17 presents the step responses of a real experiment for a control system design, utilizing a diagonal compensator, described by Eqs. (H.6), (H.11), and (H.12). Figure H.18 presents the step responses for the same control system but utilizing the design of a non-diagonal compensator, described by Eqs. (H.6), (H.7), (H.9), and (H.10). The latter one definitely reduces the disturbance introduced by the coupling effect on T_2.

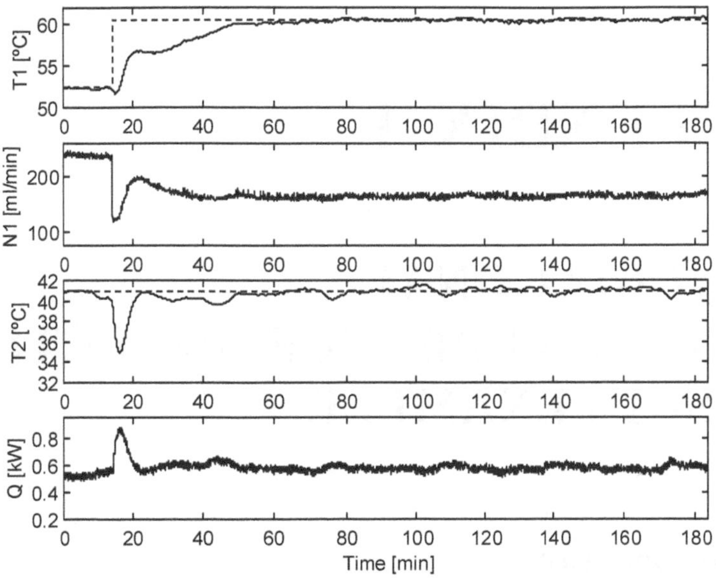

Fig. H.17 Response with diagonal compensator.

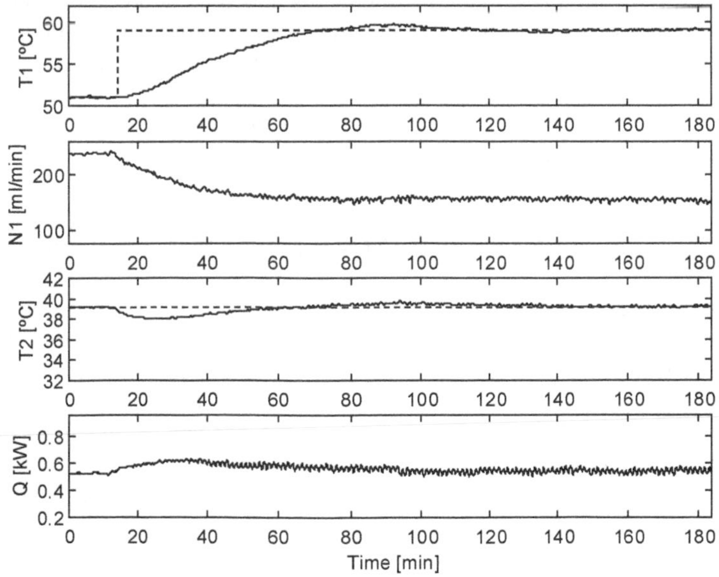

Fig. H.18 Response with non-diagonal compensator.

APPENDIX I

NON-DIAGONAL QFT DISTURBANCE REJECTION DESIGN EXAMPLE

I-1 INTRODUCTION

Heat exchangers play an essential role in industries that use renewable sources for energy generation and water heating. This appendix addresses the control of a heat exchanger of a solar water heating system. One of the main control objectives of such systems is to achieve simultaneous and accurate control of some temperatures. A 2x2 model of the heat exchanger of a solar plant and a non-diagonal MIMO QFT robust compensator able to govern the system with external disturbances at plant output are developed in the appendix.

I-2 MODEL OF A HEAT EXCHANGER FOR A SOLAR PLANT

Solar water heaters can adopt several configurations. Figure I.1 presents one of the most typical schemes, where a heat exchanger is placed between the solar energy collector and the water storage tank. The collector heats an antifreeze solution (glycol), which is sent to a heat exchanger through a primary closed circuit. A secondary circuit starts in the heat exchanger and heats a water storage tank that will be used for domestic purposes. The two manipulated variables are the collector q_c and the storage q_t pump volumetric flux rates. The two controlled variables are the temperature of the storage tank T_t and the temperature at the exit of the collector T_0. This section introduces a simplified

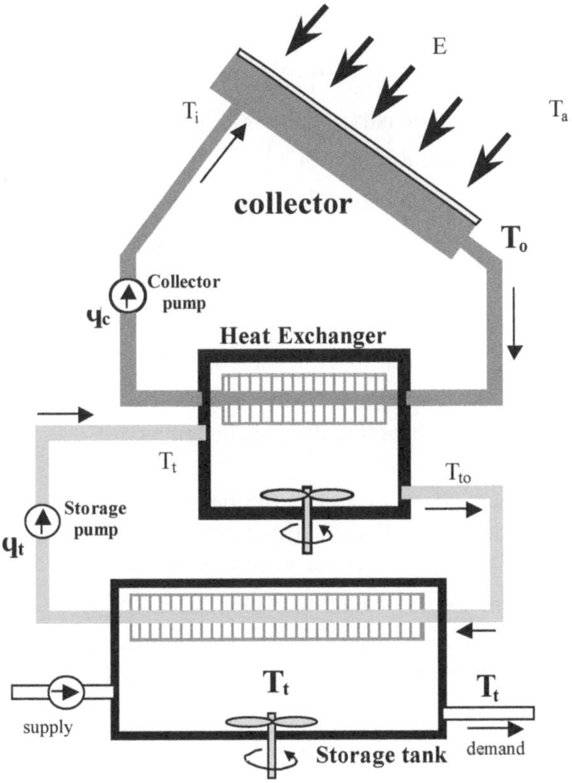

Fig. I.1. Diagram of a solar water heating system

mathematical model of the solar water heating system, focused on control purposes.

The system is modeled by the following expressions,

$$Q_u(t) = \dot{m}_c\, c_{pc}\left[T_o(t) - T_i(t)\right] = A_c F_r\left\{E(t) - U_L\left[T_i(t) - T_a(t)\right]\right\} \tag{I.1}$$

$$Q_{ex}(t) = \varepsilon\, \dot{m}_t c_{pt}\left[T_o(t) - T_t(t)\right] = L_t(t) \tag{I.2}$$

$$T_o(t) - T_i(t) = \varepsilon\left[T_o(t) - T_t(t)\right] \tag{I.3}$$

$$\dot{m}_t\, c_{pt}\frac{dT(t)}{dt} = \dot{m}_c c_{pc}\left[T_o(t) - T_i(t)\right] - U_t\left[T_t(t) - T_a(t)\right] - L_t(t) \tag{I.4}$$

where the variables and coefficients are:

$\dot{m}_c = \rho_c q_c =$ glycol flow rate $[\text{Kg s}^{-1}]$; $\dot{m}_t = \rho_t q_t =$ Water flow rate $[\text{Kg s}^{-1}]$
$\rho_c = 1094$ Kg m^{-3} = glycol density; $\rho_t = 1000$ Kg m^{-3} = water density
$q_c = 0.00115$ m^{3}s^{-1} ; $q_t = 0.864\ 10^{-3}$ m^{3}s^{-1}
$c_{pc} = 3850$ J Kg^{-1} ºC^{-1} = glycol specific heat ; $V_t =$ 0.1 m^3 = tank volume of water
$c_{pt} = 4190$ J Kg^{-1} ºC^{-1} = water specific heat ; E = solar radiation
$\varepsilon \in [0.4\ ,\ 0.6]$ = heat exchanger effectiveness ; $F_R = 0.8$
$U_c \in [6\ ,\ 8]$ w m^{-2} ºC^{-1} = overall loss coefficient ; $A_c \in [4\ ,\ 10]$ m^2 = collector area
$U_t = 2500$ w ºC^{-1} = overall heat transfer coefficient area

This set of equations describes the behavior of the solar water heating system. Eq.(I.1) models the collector. Equations (I.2) and (I.3) are related to the effectiveness of the heat exchanger. The storage unit is defined by Eq.(I.4).

If the system works around an operating point $(Q_t^o, Q_c^o, T_t^o, T_c^o)$, then the transfer functions matrix, which relates the controlled variables (T_t and T_o) with the manipulated ones (Q_c and Q_t), is,

$$\begin{bmatrix} \Delta T_t(s) \\ \Delta T_o(s) \end{bmatrix} = \begin{bmatrix} p_{11}(s) & p_{12}(s) \\ p_{21}(s) & p_{22}(s) \end{bmatrix} \begin{bmatrix} \Delta Q_t(s) \\ \Delta Q_c(s) \end{bmatrix} \tag{I.5}$$

where,

$$p_{11}(s) = \frac{\Delta T_t(s)}{\Delta Q_t(s)} = \frac{k_{11}}{s+\tau} \tag{I.6}$$

$$p_{12}(s) = \frac{\Delta T_t(s)}{\Delta Q_c(s)} = \frac{k_{12}}{s+\tau} \tag{I.7}$$

$$p_{21}(s) = \frac{\Delta T_o(s)}{\Delta Q_t(s)} = \frac{k_{21}}{s+\tau} \tag{I.8}$$

$$p_{22}(s) = \frac{\Delta T_o(s)}{\Delta Q_c(s)} = \frac{k_{22}\ (s+\gamma_{22})}{s+\tau} \tag{I.9}$$

and,

$$\tau = \frac{M\,A + Z\,C}{A\,a} \;\; ; \;\; k_{11} = \frac{Y}{A} \;\; ; \;\; k_{12} = \frac{-Z\,B}{A\,a} \;\; ; \;\; k_{21} = \frac{-Y\,C}{A\,a} \;\; ; \;\; k_{22} = \frac{-B}{A}$$

with,

$$A = b\varepsilon Q_c^o + fU_c(1-\varepsilon) \;\; ; \;\; B = b\varepsilon(T_o^o - T_t^o) \;\; ; \;\; C = fU_c\varepsilon - b\varepsilon Q_c^o$$

$$Y = \varepsilon d(T_t^o - T_o^o) \;\; ; \;\; Z = -fU_c(1-\varepsilon) - \varepsilon d Q_c^o \;\; ; \;\; M = U_t + fU_c\varepsilon - \varepsilon d Q_t^o$$

$$a = \rho_t c_{pt} V_t \;\; ; \;\; b = \rho_c c_{pc} \;\; ; \;\; d = \rho_t c_{pt} \;\; ; \;\; f = A_c F_r$$

In this example the operating point is:

$$Q_t^o = 0.864\ 10^{-3}\ \text{m}^3\ \text{s}^{-1} \;\; ; \;\; Q_c^o = 0.00115\ \text{m}^3\ \text{s}^{-1} \;\; ; \;\; T_t^o = 35\ ^\circ\text{C} \;\; ; \;\; T_o^o = 53^\circ\text{C};$$

The $p_{ij}(s)$ transfer functions (Eqs. I.6 through I.8) present the parameter uncertainty indicated in Table I.1.

Table I.1 Parameter uncertainty

Parameter	Min	Max
ε	0.4	0.6
U_c [w °C^{-1} m^{-2}]	6	8
A_c [m^2]	4	10

On the other hand, the system can suffer several disturbances like a change in feed temperature or rate, a change in pressure or a variation in product demand. These disturbances cause the process output (temperatures T_t and T_o) to move from the desired operating value.

I-3 NON-DIAGONAL MIMO QFT COMPENSATOR DESIGN

The non-diagonal MIMO QFT methodology introduced in Chapter 10 is applied to design a robust compensator for controlling, against external disturbances at plant output, the solar water heating system described in the last section.

Performance specifications:

The desired closed-loop performance specifications for the solar water heating system are the following:

- Robust stability in every channel:

$$\left|\frac{L_i(s)}{1+L_i(s)}\right| \le 1.2 \quad i=1,2 \tag{I.10}$$

where $L_i(s) = p_{ii}(s)\, g_{ii}(s)$. This means at least 50° phase margin and at least 1.833 (5.26 dB) gain margin.

- Reduction of coupling effects as much as possible.

- Robust disturbance rejection at plant output so that,

$$\left|\frac{y_i(j\omega)}{do_i(j\omega)}\right| \le 0.1, \quad \omega < 0.05\ rad/s, \quad i=1,2 \tag{I.11}$$

Design Procedure:

- Step A *Input/Output pairing and loop ordering*

The first step is the Relative Gain Analysis (RGA). Computing it for more than 600 plants generated within the parameter uncertainty, the results show that the best possible pairing are: $\left[T_t^o - Q_t^o\right], \left[T_c^o - Q_c^o\right]$.

- Step B.1 *Design of the first diagonal compensator element,* $g_{11}(s)$.

By using the $\left(p_{11}^*\right)^{-1}$ model with uncertainty and the desired specifications for the first Loop, the compensator of Eq. (I.12) is found, satisfying all the performance objectives.

$$g_{11}(s) = \frac{-0.0002s - 1.2\times10^{-5}}{s^2 + 0.6s} \tag{I.12}$$

- Step C.1 *Design of the non-diagonal compensator element,* $g_{21}(s)$.

Taking into account the optimum compensator for the disturbance rejection at plant output problems of Eq. (10.37), the compensator g_{21} of Eq. (I.13) is designed minimizing the coupling effect c_{21}.

$$g_{21}(s) = \frac{0.001}{s+5} \qquad \textbf{(I.13)}$$

- Step B.2 *Design of the second diagonal compensator element*, $g_{22}(s)$.

The following equivalent plant $\left(p_{22}^{*e}\right)^{-1}$ derived from Eq. (10.46) is calculated,

$$\left[p_{22}^{*e}\right]_2 = \left[p_{22}^{*}\right]_1 - \frac{\left(\left[p_{21}^{*}\right]_1 + \left[g_{21}\right]_1\right)\left(\left[p_{12}^{*}\right]_1 + \left[g_{12}\right]_1\right)}{\left[p_{11}^{*}\right]_1 + \left[g_{11}\right]_1} \qquad \textbf{(I.14)}$$

The compensator of Eq. (I.15) is determined that satisfies all the performance specifications for the above equivalent plant with uncertainty.

$$g_{22}(s) = \frac{-0.06s - 0.006}{s^2 + 7s} \qquad \textbf{(I.15)}$$

- Step C.2 *Design of the non-diagonal compensator element*, $g_{12}(s)$.

The compensator g_{12} of Eq. (I.16) is designed minimizing the coupling effect c_{12}.

$$g_{12}(s) = \frac{-0.001(s+0.005)}{s+0.2} \qquad \textbf{(I.16)}$$

I-4 RESULTS

Transient responses of the controlled system with external disturbances at plant output in the first loop are shown in Figs. I.2 and I.3. In case (a), the fully populated matrix compensator designed with the above methodology is implemented, whereas in case (b) a classical diagonal compensator is applied (see Eqs. I.17 and I.18). At time t = 400 sec., a unit step disturbance do_1 is added at plant output y_1. Figs. I.2 and I.3 show how the response to the disturbance is more satisfactory in case (a): the non-diagonal compensator.

Figures I.4 and I.5 show the transient responses of the closed-loop system with external disturbances at the plant output for the second loop. Case (a) shows the result of the fully populated matrix compensator and case (b) of a classical diagonal compensator (see Eqs. I.17 and I.18). At time t = 400 sec., a unit step disturbance do_2 is applied at plant output y_2. The results yield that the response to the disturbance input is better, once again, with the fully populated compensator.

The diagonal compensator used to compared with are two classical structures (PI+filters) so that,

$$g_{11}(s) = \frac{-0.000533s - 8\times 10^{-5}}{s^2 + 8s} \qquad \textbf{(I.17)}$$

$$g_{22}(s) = \frac{-0.0057s + 0.00057}{s^2 + 10s} \qquad \textbf{(I.18)}$$

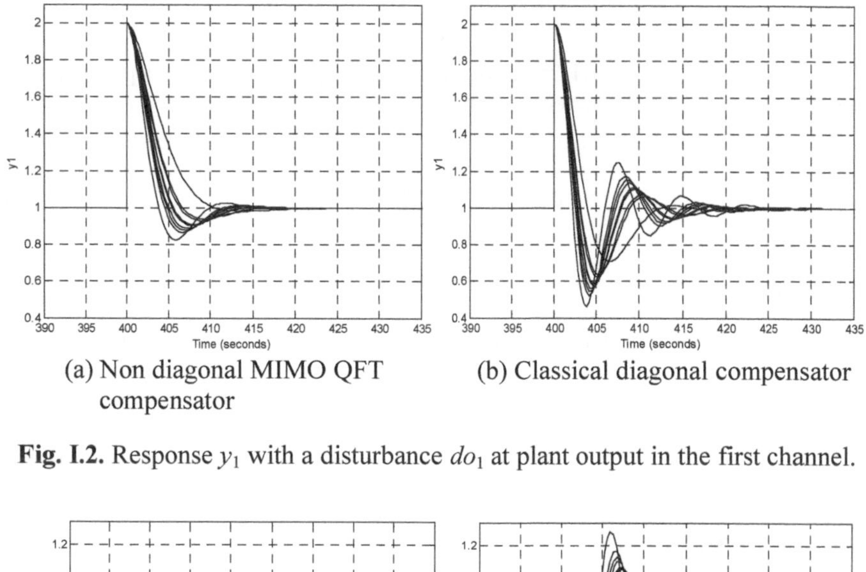

(a) Non diagonal MIMO QFT compensator (b) Classical diagonal compensator

Fig. I.2. Response y_1 with a disturbance do_1 at plant output in the first channel.

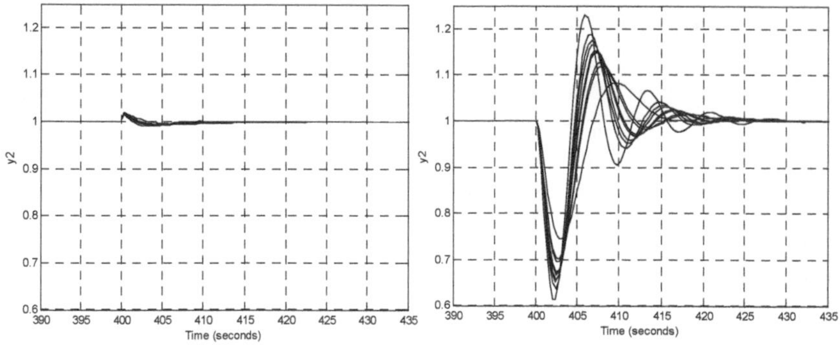

(a) Non diagonal MIMO QFT compensator (b) Classical diagonal compensator

Fig. I.3. Response y_2 with a disturbance do_1 at plant output in the first channel.

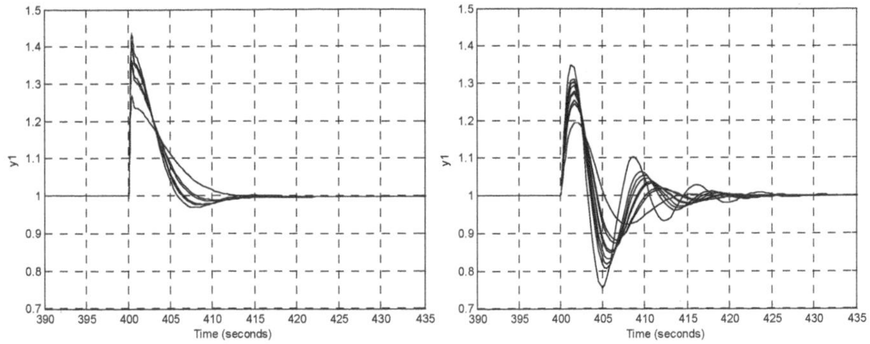

(a) Non diagonal MIMO QFT compensator (b) Classical diagonal compensator

Fig. I.4. Response y_1 with a disturbance do_2 at plant output in the second channel.

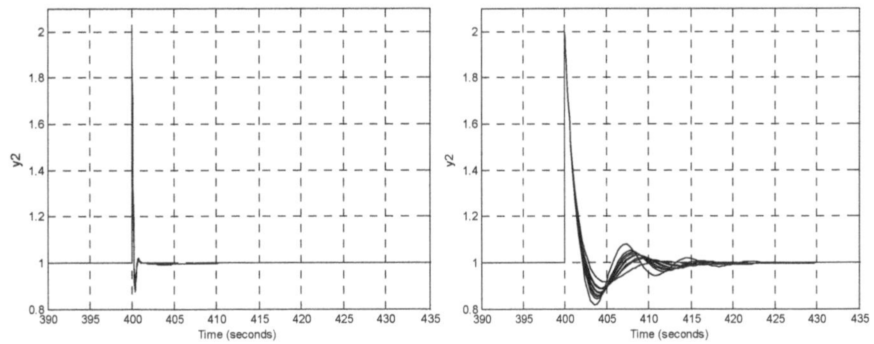

(a) Non diagonal MIMO QFT compensator (b) Classical diagonal compensator

Fig. I.5. Response y_2 with a disturbance do_2 at plant output in the second channel.

I-5 CONCLUSIONS

This appendix discussed the control of a heat exchanger placed in a solar water heating system and influenced by external disturbances at plant output. Due to the multivariable condition of the heat exchangers (several temperatures must be controlled with several manipulated variables) the control strategy selected was the non-diagonal MIMO QFT methodology presented in Chapter 10. The fully populated matrix compensator designed was found effective in achieving the given specifications. It not only coped with plant uncertainties but also enhanced the rejection of external disturbances, attenuating the coupling between control loops.

APPENDIX J

ELEMENTS FOR LOOP SHAPING

J-1 INTRODUCTION

After calculating the stability and performance QFT bounds, the following step consists of designing (loop shaping) the compensator $\boldsymbol{G}(\mathrm{s})$ that allows the nominal open loop $\boldsymbol{L}_o(\mathrm{s}) = \boldsymbol{G}(\mathrm{s})\ \boldsymbol{P}_o(\mathrm{s})$ to satisfy the worst case of all bounds. This is probably one of the most difficult steps of the methodology for the beginner. For this reason this appendix introduces a fundamental study of the most common elements that can be added to the compensator $\boldsymbol{G}(\mathrm{s})$. The appendix shows how these elements modify the position of $\boldsymbol{L}_o(\mathrm{s})$ in the Nichols Chart (NC),[1,131] looking at the magnitude and phase values at the most representative frequencies of the element.

J-2 SIMPLE GAIN: K

The effect of a simple gain K is to shift $\boldsymbol{L}(j\omega)$ up or down if $K > 1$ or $K < 1$ respectively. The distance that $\boldsymbol{L}(j\omega)$ is vertically shifted in the Nichols Chart is $[20 \log_{10}(K)]$ in dB. To make the calculations in dB much easier, bearing in mind that multiplying $\boldsymbol{L}(j\omega)$ by a gain $K = 2$ (or 1/2) corresponds to shifting it by a distance of $20 \log_{10}(K) = 6\ dB$ (or $-6\ dB$) in the NC (see Table J.1). Figure J.1 shows the effect of a simple gain $K = 2$, where $L_1 = K\ L_o$.

Table J.1 Magnitude: Module versus *dB*

Module	0.125	0.25	0.5	1	2	4	8	16	32	multiply by 2
dB	-18.1	-12.0	-6.0	0.0	6.0	12.0	18.1	24.1	30.1	= add + 6 *dB*

Module	0.001	0.01	0.1	1	10	100	1000	10000	multiply by 10
dB	-60.0	-40.0	-20.0	0.0	20.0	40.0	60.0	80.0	= add + 20 *dB*

J-3 REAL POLE: $1/[(s/p)+1]$

The effect of a real pole, located at $[-p]$ *rad/sec*, is to shift $\boldsymbol{L}(j\omega)$ down by $-10\log_{10}(1+\omega^2/p^2)$ *dB* and left by $-\tan^{-1}(\omega/p)$ *deg*. That is to say, at the frequency $\omega = p$ the pole shifts $\boldsymbol{L}(j\omega)$ down by -3 *dB* and left by -45 *deg*. Figure J.2 shows the effect of a real pole $p = 1$, where $L_1(s) = \{1/[(s/p)+1]\}\, L_0(s)$.

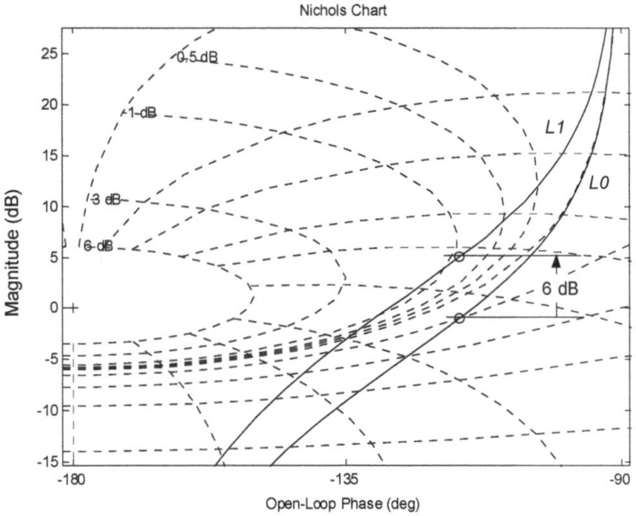

Fig. J.1 Effect of a simple gain $K = 2$, where $L_1 = K\, L_0$. At every frequency the gain shifts $\boldsymbol{L}(j\omega)$ up by +6 *dB*.

J-4 REAL ZERO: [(s/z)+1]

The effect of a real zero, located at [-z] *rad/sec*, is to shift $L(j\omega)$ up by 10 $\log_{10}(1+\omega^2/z^2)$ *dB* and right by $\tan^{-1}(\omega/z)$ *deg*. That is, at the frequency $\omega = z$ the zero shifts $L(j\omega)$ up by +3 *dB* and right by +45 *deg*. Figure J.3 shows the effect of a real zero at z = 1, where $L_1(s) = [(s/z) + 1]L_0(s)$.

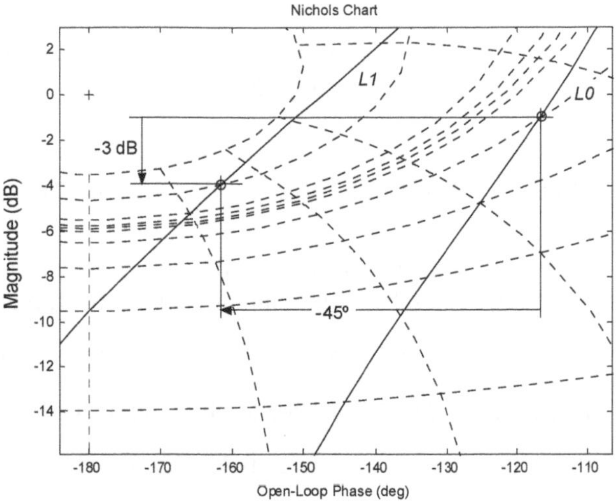

Fig. J.2 Effect of a real pole $p = 1$, where $L_1(s) = \{1/[(s/p)+1]\}\ L_0(s)$.

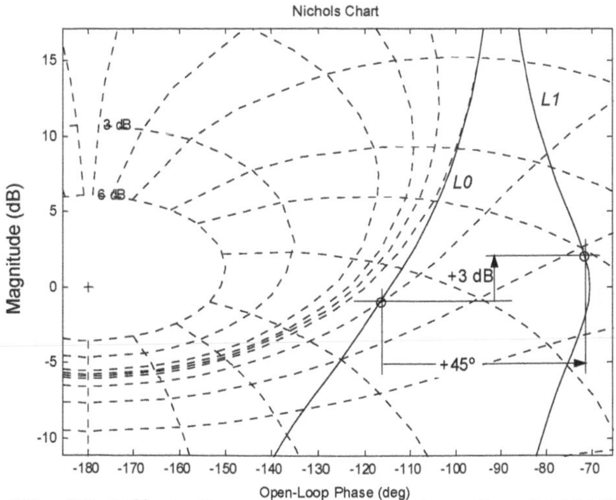

Fig. J.3 Effect of a real zero $z = 1$, where $L_1 = [(s/z)+1]\ L_0$.

J-5 INTEGRATOR: $1/s^n$; $n = 1,2,...$

The effect of n integrators, is to shift $L(j\omega)$ left by –90n *deg* at every frequency. Notice that $L(j\omega)$ is not vertically shifted.

J-6 DIFFERENTIATOR: s^n ; $n = 1,2,...$

The effect of n differentiators, is to shift $L(j\omega)$ right by +90n *deg* at every frequency. Notice that $L(j\omega)$ is not vertically shifted.

J-7 COMPLEX POLE: $\left[\omega_n^2/(s^2+2\zeta\,\omega_n\,s+\omega_n^2)\right]$, $(\zeta < 0.707)$

The effect of a complex pole, with a natural frequency ω_n and a damping factor ζ, is to shift $L(j\omega)$ up by a maximum magnitude (resonance) of $-20\log_{10}\left[2\zeta\sqrt{1-\zeta}\right]$ *dB* at the frequency $\omega = \omega_n\sqrt{1-2\zeta^2}$ *rad/sec* (ζ< 0.707). Futhermore, at that frequency, $L(j\omega)$ is shifted left by $-90^\circ + \sin^{-1}\left(\zeta/\sqrt{1-\zeta^2}\right)$ *deg*. Figure J.4 shows the effect of a complex pole with ω_n = 1 *rad/sec* and ζ = 0.4, where,

$$L_1(s) = \left[\frac{\omega_n^2}{s^2}\omega_n^2/(s^2+2\zeta\,\omega_n\,s+\omega_n^2)\right] L_o(s) \tag{J.1}$$

At the resonance frequency ω = 0.82 *rad/sec* the complex pole shifts $L(j\omega)$ up by 2.7 *dB,* and left by –89.5 *deg* (see Table J.2).

Table J.2 Magnitude and phase at the resonance frequency

ζ	0.001	0.01	0.1	0.2	0.3	0.4	0.5	0.6	0.7
M dB (resonance)	53.98	33.98	14.02	8.14	4.85	2.70	1.25	0.35	0.00
ϕ *deg* (resonance)	-90.00	-89.99	-89.90	-89.79	-89.68	-89.55	-89.38	-89.15	-88.63

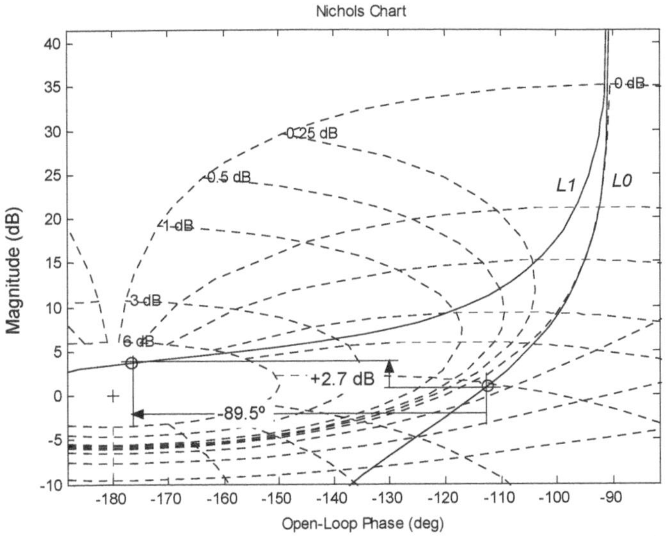

Fig. J.4 Effect of a complex pole with $\omega_n = 1$ *rad/sec* and $\zeta = 0.4$, where $L_1(s) = \left[\omega_n^2 / (s^2 + 2\zeta\,\omega_n\, s + \omega_n^2)\right] L_0(s)$

J-8 COMPLEX ZERO: $(s^2 + 2\zeta\,\omega_n\, s + \omega_n^2)/(\omega_n^2)$

The effect of a complex zero, with a natural frequency ω_n and a damping factor ζ, is to shift $L(j\omega)$ down by a maximum magnitude (resonance) of $20\log_{10}\left[2\,\zeta\,\sqrt{1-\zeta^2}\right]$ *dB* at the frequency $\omega = \omega_n\sqrt{1-2\zeta^2}$ *rad/sec* ($\zeta <$ 0.707). Futhermore, at that frequency, $L(j\omega)$ is shifted right by $90^\circ - \sin^{-1}\left[\zeta/\sqrt{1-\zeta^2}\right]$ *deg*. Figure J.5 shows the effect of a complex zero with $\omega_n = 1$ *rad/sec* and $\zeta = 0.4$, where,

$$L_1(s) = \left[(s^2 + 2\zeta\,\omega_n\, s + \omega_n^2)/\omega_n^2\right] L_o(s) \qquad \textbf{(J.2)}$$

At the resonance frequency $\omega = 0.82$ *rad/sec* the complex zero shifts $L(j\omega)$ down by -2.7 *dB*, and right by +89.5 *deg* (see Table J.3).

Table J.3 Magnitude and phase at the resonance frequency

ζ	0.001	0.01	0.1	0.2	0.3	0.4	0.5	0.6	0.7
M dB (resonance)	-53.98	-33.98	-14.02	-8.14	-4.85	-2.70	-1.25	-0.35	-0.00
ϕ *deg* (resonance)	90.00	89.99	89.90	89.79	89.68	89.55	89.38	89.15	88.63

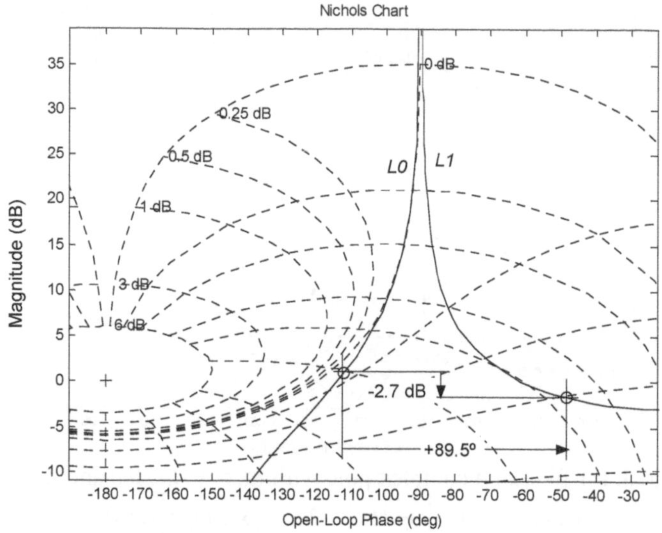

Fig. J.5 Effect of a complex zero with $\omega_n = 1$ *rad/sec* and $\zeta = 0.4$, where $L_1(s) = \left[(s^2 + 2\zeta\,\omega_n\,s + \omega_n^2)/\,\omega_n^2\right]L_0(s)$

J-9 LEAD NETWORK: $[(s/z)+1]/[(s/p)+1]$, $(z < p)$

The effect of a lead network, with a zero located at $[-z]$ *rad/sec* and a pole located at $[-p]$ *rad/sec* ($z < p$), is to shift $\boldsymbol{L}(j\omega)$ right by a maximum phase of $\phi_{\max} = 90 - 2\tan^{-1}\sqrt{z/p}$ *deg* at the frequency $\omega = \sqrt{z\,p}$ *rad/sec.* Futhermore, at that frequency, $\boldsymbol{L}(j\omega)$ is shifted up by $10\log_{10}(p/z)$. Figure J.6 shows the effect of a simple lead network with $p = 10$ and $z = 1$, where $L_1 = \{[(s/z)+1]/[(s/p)+1]\}L_0$. At the frequency $\omega = 3.16$ *rad/s* the lead network shifts $\boldsymbol{L}(j\omega)$ up by +10 *dB* and right by +54.9 *deg* (see Table J.4).

Table J.4 Magnitude and phase at $\omega = \sqrt{z\,p}$ *rad/sec*

z/p	0.001	0.01	0.1	0.2	0.3	0.4	0.5	0.6	0.7	0.8	0.9	1
$\phi_{\max}$ (*deg*)	86.4	78.6	54.9	41.8	32.6	25.4	19.5	14.5	10.2	6.4	3.0	0.0
M dB	30.00	20.00	10.00	6.99	5.23	3.98	3.01	2.22	1.55	0.97	0.46	0.00

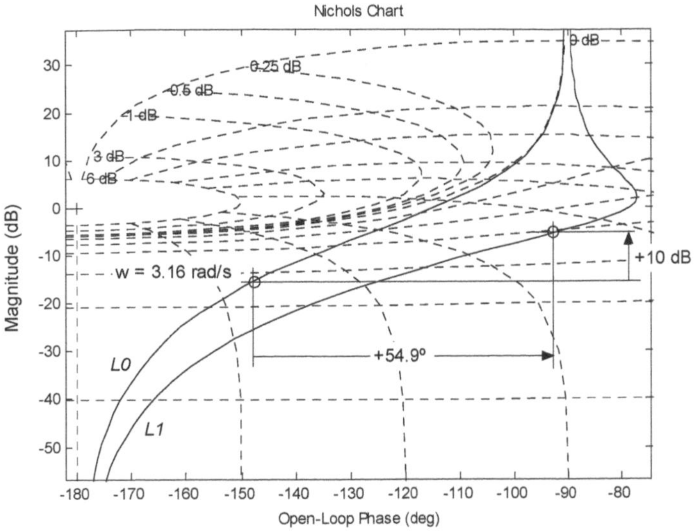

Fig. J.6 Effect of a lead network with $p = 10$ and $z = 1$, where $L_1(s) = \{[(s/z)+1]/[(s/p)+1]\}\ L_0(s)$.

J-10 LAG NETWORK: $[(s/z) + 1]/[(s/p) + 1]$, $(z > p)$

The effect of a lag network, with a zero located at $[-z]$ *rad/sec* and a pole located at $[-p]$ rad/sec $(z > p)$, is to shift $\boldsymbol{L}(j\omega)$ left by a maximum phase of $\phi_{\max} = -90 + 2\tan^{-1}\sqrt{z/p}$ *deg* at the frequency $\omega = \sqrt{z\,p}$ *rad/sec* (see Table J.5). Futhermore, at that frequency, $\boldsymbol{L}(j\omega)$ is shifted down by $-10\ \log_{10}(p/z)$.

Table J.5 Magnitude and phase at $\omega = \sqrt{z\,p}$ *rad/sec*

p/z	0.001	0.01	0.1	0.2	0.3	0.4	0.5	0.6	0.7	0.8	0.9	1
$\phi_{\max}$ *(deg)*	-86.4	-78.6	-54.9	-41.8	-32.6	-25.4	-19.5	-14.5	-10.2	-6.4	-3.0	0.0
M dB	-30.00	-20.00	-10.00	-6.99	-5.23	-3.98	-3.01	-2.22	-1.55	-0.97	-0.46	0.00

Figure J.7 shows the effect of a simple lag network with $p = 0.2$ and $z = 2$, where $L_1(s) = \{[(s/z)+1]/[(s/p)+1]\}L_0(s)$. At the frequency $\omega = 0.632$ *rad/s* the lead network shifts $\boldsymbol{L}(j\omega)$ down by -10 *dB* and left by -54.9 *deg*.

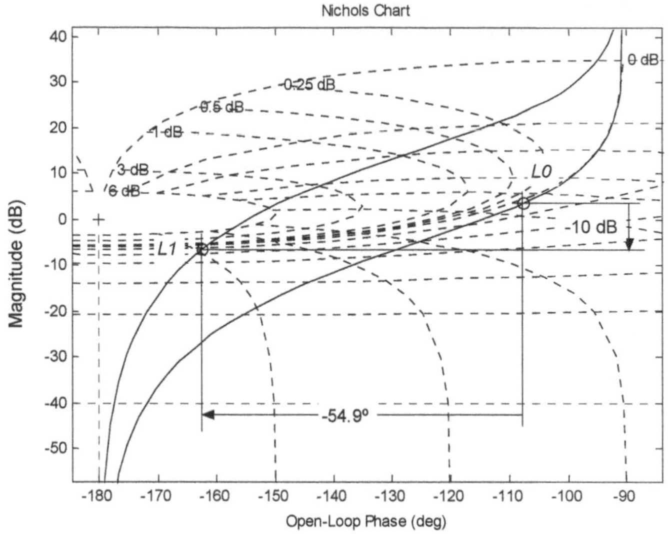

Fig. J.7 Effect of a lag network with $p = 0.2$ and $z = 2$, where $L_1(s) = \{[(s/z)+1]/[(s/p)+1]\}L_0(s)$.

J-11 NOTCH FILTER: $[s^2 + 2\zeta_1\omega_n s + \omega_n^2]/[s^2 + 2\zeta_2\omega_n s + \omega_n^2]$, $\zeta_1 < \zeta_2$

The effect of a notch filter, with a natural frequency ω_n and two damping factors ζ_1 and ζ_2, is to shift $\boldsymbol{L}(j\omega)$ down by a maximum magnitude of $20\log_{10}(\zeta_1/\zeta_2)$ *dB* at the frequency $\omega = \omega_n$ *rad/sec* (see Table J.6). At that frequency, $\boldsymbol{L}(j\omega)$ is not horizontally shifted. Figure J.8 shows the effect of a notch filter with $\zeta_1 = 0.1$, $\zeta_2 = 1$ and $\omega_n = 2$ *rad/sec*, where the notch filter shifts $\boldsymbol{L}(j\omega)$ down by -20 *dB* and,

$$L_1(s) = \left[(s^2 + 2\zeta_1\omega_n s + \omega_n^2)/(s^2 + 2\zeta_2\omega_n s + \omega_n^2)\right]L_0(s) \qquad \textbf{(J.3)}$$

Table J.6 Magnitude and phase at $\omega = \omega_n$ *rad/sec*

ζ_1/ζ_2	0.001	0.01	0.1	0.2	0.3	0.4	0.5	0.6	0.7	0.8	0.9	1
M dB	-60.0	-40.0	-20.0	-13.98	-10.46	-7.96	-6.02	-4.44	-3.1	-1.94	-0.92	0.0
ϕ *deg*	0	0	0	0	0	0	0	0	0	0	0	0

The transfer function $L_0(s)$ used in the above examples is:

$$L_0(s) = 1/[s(0.5s+1)] \tag{J.4}$$

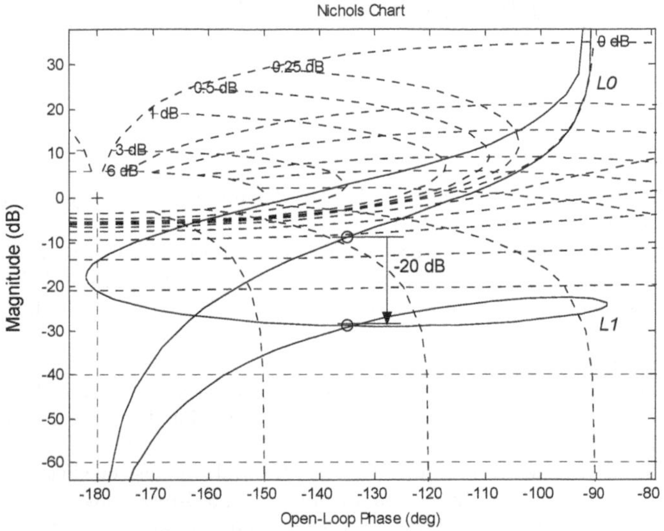

Fig. J.8 Effect of a notch filter with $\zeta_1 = 0.1$, $\zeta_2 = 1$ and $\omega_n = 2$ *rad/sec*, where $L_1(s) = \left[(s^2 + 2\zeta_1\,\omega_n\,s + \omega_n^2)/(s^2 + 2\zeta_2\,\omega_n\,s + \omega_n^2)\right]L_0(s)$

REFERENCES

1. D'Azzo, J.J., and C. H. Houpis, "Linear Control System Analysis and Design," McGraw-Hill, NY, 4th Ed. 1995.
2. Houpis, C. H., "Quantitative Feedback Theory (QFT) for the Engineer: A Paradigm for the Design of Control Systems for Uncertain Systems," WL-TR-95-3061, AF Wright Laboratory, Wright-Patterson AFB, OH, 1995 (Available from National Technical Information Service, 5285 Port Royal Road, Springfield, VA 22151, document number AD-A297571.)
3. Horowitz I. M, and M. Sidi, "Synthesis of Feedback Systems with Large Plant Ignorance for Prescribed Time Domain Tolerances," Int. J. of Control, vol. 16, pp. 287-309, 1973.
4. Horowitz, I. M. and C. Loecher, "Design *3x3* Multivariable Feedback System with Large Plant Uncertainty," Int. J. Control. vol. 33, pp. 677-699, 1981.
5. Horowitz, I. M., "Optimum Loop Transfer Function in Single-Loop Minimum Phase Feedback Systems," Int. J. Control, vol. 22, pp. 97-113, 1973.
6. Ibid, "Synthesis of Feedback Systems with Non-Linear Time Uncertain Plants to Satisfy Quantitative Performance Specifications ," IEEE Proc., vol. 64, pp. 123-130, 1976.
7. Ibid, "Quantitative Synthesis of Uncertain Multiple-Input Multiple-Output Feedback Systems," Int. J. Control, vol. 30, pp. 81-106, 1979.
8. Thompson, D. F., and O. D. I. Nwokah, "Optimal Loop Synthesis in Quantitative Feed back Theory," Proceedings, of the American Control Conference, San Diego, CA, pp. 626-631, 1990.
9. Houpis, C.H. and P.R. Chandler, Editors, "Quantitative Feedback Theory Symposium Proceedings ," WL-TR-92-3063, Wright Laboratories, Wright-Patterson AFB, OH, 1992.
10. Keating, M.S., M. Pachter, and C. H. Houpis, "Damaged Aircraft Control System Design Using QFT, "Proceedings of the National Aerospace and Electronics Conference (NAECON), Dayton, OH, vol. 1, pp. 621-628, May, 1994.
11. Reynolds, O. R., M Pachter, and C.H. Houpis, "Design of a Subsonic Flight Control System for the Vista F-16 Using Quantitative Feedback Theory," Proceedings of the American Control Conference, pp. 350-354, 1994.
12. Trosen, D.W., M. Pachter, and C. H. Houpis, "Formation Flight Control Automation," Proceedings of the American Institute of Aeronautics and Astronautics (AIAA) Conference, pp. 1379-1404, Scottsdale, AZ, 1994.

13. Houpis, C.H. and G. Lamont, "Digital Control Systems: Theory, Hardware, Software," McGraw-Hill, NY, 2nd Ed., 1992.
14. Horowitz, I.M. and Y.K. Liao, "Quantitative Feedback Design for Sampled-Data System," Int. J. Control, vol. 44, pp. 665-675, 1986.
15. Sating, R.R., I.M. Horowitz, and C.H. Houpis, "Development of a MIMO QFT CAD Package (Version 2)," Graduate School of Engineering, Air Force Institute of Technology, Wright-Patterson AFB, OH 45433, USA, Presentation at the 1993 American Control Conference.
16. Boyum, K.E. M. Pachter, and C.H. Houpis, "High Angle of Attack Velocity Rolls," 13th IFAC Symposium on Automatic Control in Aerospace, pp. 51-57, Palo Alto, CA, Sept. 1994.
17. Schneider, D.L., "QFT Digital Flight Control Design as Applied to the AFIT/F-16," M.S. Thesis, AFIT/GE/ENG/66D-4, Graduate School of Engineering, Air Force Institute of Technology, 1986.
18. Horowitz, I.M., "Advanced Control Theory and Applications," The Weizmann Institute of Science, Rehovot, Israel, unpublished notes, 1982.
19. Ibid, "Synthesis of Feedback Systems," Academic Press, 1963.
20. Ibid, "Improved Design Technique for Uncertain Multiple Input-Output Feedback Systems," Int. J. of Control, vol. 6, pp. 977-988, 1982.
21. Ibid, "A Synthesis Theory for Linear Time-Variable Time-Varying Feedback Systems with Plant Uncertainty," IEEE Trans, AC-20, pp. 454-463, 1975.
22. Iibid, "A Quantitative Inherent Reconfiguration Theory for a Class of Systems ," Int. J. of Sys. Sci., vol. 16, pp. 1377-1390, 1985.
23. Ibid, "The Singular –G Method in Unstable Non-minimum-phase Feedback Systems," Int. J. of Control, vol. 44, pp. 533-541, 1986.
24. Horowitz, I. M. and T. Kopelman, "Multivariable Flight Control Design with Uncertain Parameters, "The Weizmann Institute of Science, Rehovot, Israel, Final Report, Oct. 1981.
25. Blakelock, John H., "Automatic Control of Aircraft and Missles," 2nd Ed. Wiley-Interscience, 1991.
26. Horowitz, I.M. and Y.K. Liao, "Limitations of Non-minimum Phase Feedback Systems, Int. J. of Control, vol. 40, pp. 1003-1013, 1984.
27. Horowitz, I. M. and M. Sidi, "Optimum Synthesis of Non-minimum-Phase Feedback Systems with Parameter Uncertainty," Int. J. of Control, vol. 27, pp. 361-386, 1978.
28. Ibid, "Synthesis of Cascaded Multiple-Loop Feedback Systems with Large Plan Parameter Ignorance," Automatica, vol. 9, pp. 589-600, 1973.
29. Horowitz, I.M., S Oldak, and O. Yaniv, "Important Property of Non-minimum Phase MIMO Feedback Systems," Int, J. of Control, vol. 44, pp. 677-688, 1986.
30. Horowitz, I.M. et al, "Multivariable Flight Control Design with Uncertain Parameters (YF16CCV)," AFWAL-TR-83-3036, Air Force Wright Aeronautical Laboratories, Wright-Patterson AFB, OH, 1982.
31. Yaniv, O., and I.M. Horowitz, "A Quantitative Design Method for MIMO Linear Feedback Systems Having Uncertain Plants," Int. J. of Control, vol. 43, pp. 401-421, 1986.
32. Horowitz, I.M., "Design of Feedback Systems with Non-minimum-Phase Unstable Plants," Int. J. Sys. Sciences, vol. 10, pp. 1025-1040, 1979.
33. Reynolds, O.R., M. Pachter and C.H. Houpis, "Full Envelope Flight Control System Design Using QFT," Proceedings of the American Control Conference, Baltimore, MD, pp. 350-354, June 1994.

34. Rasmussen, S.J. and C.H. Houpis, "Development Implementation and Flight of a MIMO Digital Flight Control System for an Unmanned Research Vehicle Using Quantitative Feedback Theory," Proceedings of the ASME Dynamic Systems and Control, Winter Annual Meeting of ASME, Chicago, IL, Nov. 1994.
35. Bossert, D.E., "Design of Pseudo-Continuous-Time Quantitative Feedback Theory Robot Controllers," M.S. Thesis, AFIT/GE/ENG/89D-2, Graduate School of Engineering, Air Force Institute of Technology, Wright-Patterson AFB, OH, Dec. 1989.
36. Sating, R.R., "Development of an Analog MIMO Quantitative Feedback Theory (QFT) CAD Package," M.S. Thesis, AFIT/GE/ENG/92J-04, Graduate School of Engineering, Air Force Institute of Technology, Wright Patterson AFB, OH, 1992.
37. Houpis, C.H. and M. Pachter, "Application of QFT to Control System Design – An Outline for Engineers," Int. J. of Robust and Nonlinear Control, vol. 7, pp. 515-531, June 1997.
38. Horowitz, I. M., "Quantitative Feedback Theory," Proceedings of IEE, vol. 129D, No. 6, Nov. 1982.
39. Bailey, F.N. J.J. W. Helton and O. Merino, "Alternative Process in Frequency Domain Design of Single Loop Feedback Systems with Plant Uncertainty," Proceedings of the American Control Conference, Baltimore, MD, June 1994.
40. Bode, H.W., "Network Analysis and Feedback Amplifier Design," Van Nostrand, NY, 1945.
41. Houpis, C.H., "Quantitative Feedback Theory (QFT): Technique for Designing Multivariable Control Systems," AFWAL-TR-86-3107, Air Force Wright Laboratories, Wright-Patterson AFB, OH, Jan. 1987. (Available from Defense Technical Information Center, Cameron Station, Alexandria, VA 22314, document number AD-A176883.)
42. Symposium On Quantitative Feedback Theory and Other Frequency Domain Methods and Applications Proceedings, University of Strathclyde, Glasgow, Scotland, Aug. 1997.
43. Trosen, D.W., "Development of an Prototype Refueling Automatic Flight Control System Using Quantitative Feedback Theory," M.S. Thesis, AFIT/GE/ENG/93-J-03, Graduate School of Engineering, Air Force Institute of Technology, Wright-Patterson AFB, OH, Jan. 1993.
44. Franchek, M. and G.K. Hamilton, "Robust Controller Design and Experimental Verification of I.C. Engine Speed Control," Int. J. of Robust and Nonlinear Control, vol. 7, pp. 609-628, 1997.
45. Pachter, M., C.H. Houpis, and K. Kang, "Modelling and Control of an Electro-hydrostatic Actuator," Int. J. of Robust and Nonlinear Control, vol. 7, pp. 591-608, 1997.
46. Phillips, S., M. Pachter, and C.H. Houpis, "A QFT Subsonic Envelope Flight Control System Design," National Aerospace Electronics Conference (NAECON), Dayton, OH, May 1995.
47. Sheldon, S.N., and C. Osmon, "Piloted Simulation of An F-16 Flight Control System Designed Using Quantitative Feedback Theory," Symposium on Quantitative Feedback Theory and Other Frequency Domain Methods and Applications Proceedings, University of Strathclyde, Glasgow, Scotland, Aug. 1997.
48. Osmon, C., M. Pachter, and C.H. Houpis, "Active Flexible Wing Control Using QFT," IFAC 13th Triennial World Congress, vol. H, pp. 315-320, San Francisco, CA, July 1996.

49. Ewing, R.L., et al, "Object-Oriented Design and Programming of QFT CAD Environment," Quantitative Feedback Theory Symposium Proceedings, WL-TR-92-3063, Wright Laboratory, Wright-Patterson AFB, OH, Aug. 1992.
50. Ostolaza, J.X. and M. Garcia-Sanz, "Control of an Activated Sludge Wastewater Treatment Plant with Nitrification-Denitrification Configuration Using QFT Technique," Symposium on Quantitative Feedback Theory and Other Frequency Domain Methods and Applications Proceedings, University of Strathclyde, Glasgow, Scotland, Aug. 1997.
51. Horowitz, I. M. AFIT Lecture notes
52. Ibid, "Survey of Quantitative Feedback Theory (QFT)," Int. J. of Control, vol. 53, No. 2, pp. 255-291, 1991.
53. Fontenrose, P.L. and C.E. Hall, "Development and Flight Testing of a QFT Pitch Rate Stability Augmentation System," J. of Guidance, Control and Dynamics, vol. 19, No. 5, pp. 1109-1115, Sept-Oct., 1996.
54. Chait, Y. and C.V. Hollot, "A Comparison of QFT and H_∞ Techniques," Proceedings of the ASME Winter Meeting, Dallas, TX, 1990.
55. Betzold, R.W., "Multiple Input-Multiple Output Flight Control Design with Highly Uncertain Parameters, Application to the C-135 Aircraft," M.S. Thesis, AFIT/GE/EE (83D-11), Graduate School of Engineering, Air Force Institute of Engineering, Wright-Patterson AFB, OH, Dec. 1983.
56. Franklin, G.F. and J.D. Powell, "Digital Control of Dynamic Systems," Addison-Wesley, Reading, MA, 2nd Ed, 1988.
57. Tustin, A., "A Method of Analyzing the Behavior of Linear Systems in Terms of Time Series," JIEE (London), vol. 94, Pt IIA, 1947.
58. Breiner, M., and I.M. Horowitz, "Quantitative Synthesis of Feedback Systems with Uncertain Nonlinear Multivariable Plants," Int. J. Syst. Sci., vol. 12, pp. 539-563, 1981.
59. Gera, A. and I.M. Horowitz, "Optimization of the Loop Transfer Function," Int. J. of Control, vol. 31, p. 389, 1980.
60. Powell, J.D., E. Parsons and G. Tashka, "A Comparison of Flight Control Design Methods," Guidance and Control Conference, San Diego, CA, Aug. 1976.
61. Rosenbrock, H.H., "Computer-Aided Control System Design," Academic Press, 1974.
62. Horowitz, I.M., "Synthesis of Feedback Systems with Non-linear Time Uncertain Plants to Satisfy Quantitative Performance Specifications," IEEE Proc., vol. 64, pp. 81-106, 1979.
63. Zames, G. and D. Bensousan, "Multivariable Feedback Sensitivity and Optimal Robustness," IEEE Trans., AC-28, pp. 1030-1035, 1983.
64. Lancaster, P., "Theory of Matrices," Academic Press, 1969.
65. Lacey, D.J., Jr., "A Robust Digital Flight Control System for an Unmanned Research Vehicle Using Discrete Quantitative Feedback Theory," M.S. Thesis, AFIT/GE/ENG/91D, Graduate School of Engineering, Air Force Institute of Technology, Wright-Patterson AFB, OH, Dec. 1991.
66. Jayasuriya, S. and Y. Zhao, "Stability of Quantitative Feedback Designs and the Existence of Robust QFT Controllers," In Proceedings of Quantitative Feedback Theory Symposium, C. H. Houpis and P. R. Chandler, editors, pp. 503-541, Wright-Patterson AFB, OH, Aug. 1992. USAF.
67. Ibid, "Stability of Quantitative Feedback Designs and the Existence of Robust QFT Controllers," Int. J. of Nonlinear Control, 1993.

68. Nwokah, O. D. I., "Synthesis of Controllers for Uncertain Multivariable Plants for Described Time Domain Tolerances," Int. J. of Control, vol. 40, pp. 1189-1206, 1984.
69. Ibid, "Strong Robustness in Uncertain Multivariable Systems," IEEE Conference on Decision and Control, Austin, TX, Dec. 1988.
70. Nwokah, O.D.I., S. Jayasuriya and Y. Chait, "Parametric Robust Control by Quantitative Feedback Theory," AIAA J. of Guidance, Control and Dynamics, 15(1):207-214, 1992.
71. Nwokah, O.D.I., R.E. Nordgren and G.S. Grewel, "Optimal Loop Transmission Functions in SISO Quantitative Feedback Theory," Proceedings of the American Control Conference, Baltimore, MD, June 1994.
72. Nwokah O.D.I. and D.F. Thompson, "Algebraic and Topological Aspects of Quantitative Feedback Theory," Int. J. of Control, 50:1057-1069, 1989.
73. Cacciatore, V.J., "A Quantitative Feedback Theory FCG Design for the Subsonic Envelope of the VISTA F-16 Including Configuration Variation and Aerodynamic Control Effector Failures," M.S. Thesis, AFIT/GE/ENG/95D-04, Graduate School of Engineering, Air Force Institute of Technology, Wright-Patterson AFB, OH, Dec. 1995.
74. Horowitz, I.M. and U. Shaked, "Superiority of Transfer Function over State-Variable Methods in Linear Time Invariant Feedback System Design," IEEE Trans. Autom. Control, vol. AC-20, pp. 84-97, 1975.
75. Horowitz, I.M., et al, "Research in Advanced Flight Control Design," Rep. AFFDL-TR-79-3120, Air Force Wright Aeronautical Laboratories, Wright-Patterson AFB, OH, 1979.
76. Bernstein, D.S., "A Student's Guide to Classical Control," IEEE Control Systems, vol. 17, no. 4, pp. 96-98, Aug. 1997.
77. Clough, B.T., "Reconfigurable Flight Control System for a STOL Aircraft using Quantitative Feedback Theory," M.S. Thesis, AFIT/GE/ENG/85D-8, Graduate School of Engineering, Air Force Institute of Technology, Wright-Patterson AFB, OH, Dec. 1985.
78. Swift, Gerald A., "Model Identification and Control System Design for the Lambda Unmanned Research Vehicle," M.S. Thesis, Air Force Institute of Technology, Wright-Patterson AFB, OH, Sept. 1991.
79. Wheaton, David G., "Automatic Flight Control System for an Unmanned Research Vehicle Using Discrete Quantitative Feedback Theory," M.S. Thesis, Air Force Institute of Technology, Wright-Patterson AFB, OH, Dec. 1990.
80. Robertson, Scott D., "A Real-Time Hardware-in-the-Loop Simulation of an Unmanned Aerial Research Vehicle,"echnical Report WL-TR-93-9005, Wright-Patterson AFB, OH: Wright Laboratory, (Aug 1992).
81. Boje E and O.D.I. Nwokah; "Quantitative Multivariable Feedback Design for a turbofan engine with Forward Path Decoupling," Symposium on Quantitative Feedback Theory and Other Frequency Domain Methods, pp. 185-191, University of Strathclyde, Glasgow, Scotland, pp. 192-207, August 1997.
82. Garcia-Sanz, M., "Quantitative Non-Diagonal MIMO Controller Design for Uncertain Systems," 1999 International Symposium on Quantitative Feedback Theory and Robust Frequency Domain Methods, University of Natal, Durban, South Africa, 26-27 August 1999.
83. Horowitz, I.M., "Quantitative Feedback Design Theory (QFT)," QFT Pub., 660 South Monaco Parkway, Denver, Colorado 80224-1229, 1992.

84. Yaniv, O., "Quantitative Feedback Design of Linear and Non-linear Control Systems," Kluver Academic Pub., ISBN: 0-7923-8529-2, 1999.
85. Sidi, M., "Design of Robust Control Systems: From classical to modern practical Approaches," Krieger Publishing, 2002.
86. Houpis, C.H. (Guest Editor), "Quantitative Feedback Theory Special Issue," Int. J. of Robust and Nonlinear Control. Vol. 7, N. 6. Wiley, June 1997.
87. Eitelberg, E. (Guest Editor), "Isaac Horowitz Special Issue," Int. J. of Robust and Nonlinear Control. Part 1, vol. 11, no. 10, August 2001 and Part 2, Vol. 12, N. 4. Wiley, April 2002.
88. Garcia-Sanz, M. (Guest Editor), "Robust Frequency Domain Special Issue," Int. J. of Robust and Nonlinear Control. vol. 13, no. 3. Wiley, March 2003.
89. Borghesani, C., Y. Chait, and Yaniv, O., "Quantitative Feedback Theory Toolbox – For use with MATLAB". 2nd Edition, Terasoft, 2002.
90. Gutman P.O., "Qsyn - the Toolbox for Robust Control Systems Design for use with Matlab, User's Guide," El-Op Electro-Optics Industries Ltd, Rehovot, 1996.
91. Barreras, M., P. Vital, and M. Garcia-Sanz, "Interactive Tool for Easy Robust Control Design," IFAC International Workshop on Internet Based Control Education, IBCE'01, pp. 83-88, UNED, Madrid, Spain, December 2001.
92. Bristol E.H., "On a new measure of interactions for multivariable process control," IEEE Trans. on Automatic Control, vol. 11, pp. 133-134, 1966.
93. Franchek M.A, P. Herman and O.D.I. Nwokah, "Robust non-diagonal controller design for uncertain multivariable regulating systems," ASME Journal of Dynamic Systems, Measurement and Control, vol. 119, pp. 80-85, 1997.
94. Horowitz I, and M. Sidi, "Practical design of feedback systems with uncertain multivariable plants," Int. J. of Systems Science, Vol. 11, N. 7, pp. 851-875, 1980.
95. Horowitz I., "Quantitative synthesis of uncertain multiple input-output feedback systems," Int. J. of Control, vol. 30, pp. 81-106, 1979.
96. Skogestad S. And K. Havre, "The use of RGA and condition number as robustness measures," Proceedings of the European Symposium of Computer-Aided Process Engineering, Rodhes, Greece, 1996.
97. Garcia-Sanz, M. and I. Egana, "Quantitative Non-Diagonal Controller Design for Multivariable Systems with Uncertainty," Int. J. of Robust and Nonlinear Control; Isaac Horowitz Special Issue Part 2, 12: 321-333, 2002.
98. Egana I., J. Villanueva, and M. Garcia-Sanz, "Quantitative Multivariable Feedback Design for a SCARA Robot Arm," 5th International Symposium on Quantitative Feedback Theory and Robust Frequency Domain Methods, pp. 67-72, Pamplona Spain, August 2001.
99. Egana I., "Diseno de controladores multivariables QFT de matriz completa," PhD, Universidad Publica de Navarra, July 2002.
100. Garcia-Sanz, M., I. Egana, and J. Villanueva, "Interval Modelling of a SCARA robot for Robust Control," 10th Mediterranean Conference on Control and Automation, MED2002 Conference, IEEE, Lisboa, Portugal, July 2002.
101. Egana, I. and M. Garcia-Sanz, "Controller Design Sequence for Multivariable Feedback,". 10th Mediterranean Conference on Control and Automation, MED2002 Conference, IEEE, Lisboa, Portugal, July 2002.
102. Barreras, M. and M. Garcia-Sanz, "Model Identification of a multivariable industrial furnace," 13th IFAC Symposium on System Identification, SYSID'03, Rotterdam, The Netherlands, August 2003.

103. Garcia-Sanz, M. and J.X.Ostolaza, "QFT-Control of a Biological Reactor for Simultaneous Ammonia and Nitrates Removal," International Journal on Systems, Analysis, Modelling, Simulation, SAMS, No. 36, pp. 353-370. OPA, Gordon & Breach Science Publishers, ISSN: 0232-9298, Malaysia, 2000.
104. Suescun, J., X. Ostolaza, M. Garcia-Sanz, and E. Ayesa, "Control Strategies for DN Activated Sludge Plants. Part II: Biodegradation Control," Water Science and Technology, vol. 43, no. 1, pp. 209-216, 2001.
105. Ostolaza, X and M. Garcia-Sanz, "Control of an Activated Sludge Wastewater Treatment Plant with Nitrification-Denitrification Configuration using QFT Technique," 3rd Int. Symp. on Quantitative Feedback Theory QFT and other Frequency Domain Methods and Applications, pp. 163-170, Glasgow, United Kingdom, August 1997.
106. Ostolaza, X and M. Garcia-Sanz, "QFT-Robust Control of a Wastewater Treatment Plant," IEEE Conference on Control Applications, pp. 21-25, Trieste, Italy. ISBN: 0-7803-4104-X. IEEE Catalog number: 98CH36104, September 1998.
107. Garcia-Sanz, M., "A Robust Model of Central Heating Systems as a Realistic Test-bed for Analyzing Control Strategies," Applied Mathematical Modelling, Elsevier Sc., USA, vol. 21, no. 9, pp. 535-545, 1997.
108. M.Torres, Diseños Industriales, Inc. Crta. Pamplona-Huesca Km 9. 31119 Torres de Elorz, Spain, (http://www.mtorres.es).
109. Franchek, M.A. and O.D.I. Nwokah, "Robust multivariable control of distillation columns using non-diagonal controller matrix," DSC-Vol. 57-1, IMECE, ASME Dynamics systems and control division, pp. 257-264, 1995.
110. Yaniv, O., "MIMO QFT using non-diagonal controllers," Int. J. of Control, vol. 61, no. 1, pp. 245-253, 1995.
111. Boje E. and O.D.I. Nwokah, "Quantitative multivariable feedback design for a turbofan engine with forward path decoupling," Int. J. of Robust and Non-linear Control, vol. 9, no. 12, pp. 857-882, 1999.
112. Ibarrola, J.J., J.C. Guillen, J.M. Sandoval and M. Garcia-Sanz, "Modelling of a High Temperature Short Time Pasteurization Process," Food Control, vol 9, no. 5, pp. 267-277, Elsevier Science Ltd., United Kingdom, 1998.
113. Marlin, E.T., "Process Control: Designing Processes and Control Systems for Dynamic Performance," McGraw-Hill: New York, pp. 682-717, 1995.
114. Chait Y. and O Yaniv, "Multi-input/single-output computer-aided control design using the Quantitative Feedback Theory," Int. J. Robust and Non-linear Control, vol. 3, pp. 47-54, 1993.
115. Gil-Martinez, M. and M. Garcia-Sanz, "Simultaneous Meeting of Robust Control Specifications in QFT," Int. J. of Robust and Non-linear Control. Special Issue: "Robust Frequency Domain," Wiley, USA. vol. 13, no. 3. March 2003.
116. Houpis, C.H., Chandler, P. (Editors), "1st International Symposium on Quantitative Feedback Theory," Wright Patterson Air Force Base, Dayton, Ohio, USA, August 1992.
117. Nwokah, O.D.I., Chandler, P. (Editors), "2nd International Symposium on Quantitative Feedback Theory," .Purdue University, West Lafayette, Indiana, USA, August 1995.
118. Petropoulakis, L., Leithead, W.E. (Editors), "3rd International Symposium on Quantitative Feedback Theory and other Robust Frequency Domain Methods," University of Strathclyde, Glasgow, UK, August 1997.

119. Boje, E., and Eitelberg, E. (Editors), "4th International Symposium on Quantitative Feedback Theory and Robust Frequency Domain Methods," University of Natal, Durban, South Africa, August 1999.
120. Garcia-Sanz, M. (Editor), "5th International Symposium on Quantitative Feedback Theory and Robust Frequency Domain Methods," Public University of Navarra, Pamplona, Spain. ISBN: 84-95075-56-3, August 2001.
121. Garcia-Sanz, M. and P. Vital, "Efficient Computation of the Frequency Representation of Uncertain Systems," 4th International Symposium on Quantitative Feedback Theory QFT and Robust Frequency Domain Methods, pp. 117-126, ISBN: 1-86840-330-0, University of Natal, Durban, South Africa, August 1999.
122. Garcia-Sanz, M. and J.C. Guillen, "Automatic loop-shaping of QFT robust controllers via genetic algorithms," 3rd IFAC Symposium on Robust Control Design, ROCOND'00, F3A/107, Praha, Czech Republic, June 2000.
123. Chait, Y., Q. Chen and C.V. Hollot, "Automatic loop-shaping of QFT controllers via linear programming," 3rd International Symposium on Quantitative Feedback Theory and other Robust Frequency Domain Methods. University of Strathclyde, Glasgow, UK, pp. 13-28, August 1997.
124. Garcia-Sanz, M., J.C. Guillen, and J.J. Ibarrola, "Robust controller design for time delay systems with application to a pasteurisation process," Control Engineering Practice, IFAC, Elsevier Science Ltd, N° 9, pp. 961-972, USA, 2001.
125. Garcia-Sanz, M. and J.C. Guillen, "Smith Predictor For Uncertain Systems In The QFT Framework," Lecture Notes in Control and Information Sciences, Ed. Springer Verlag. vol 243. Progress in System and Robot Analysis and Control Design. Chapter 20, pp. 243-250. ISBN: 1-85233-123-2, 1999.
126. Adept Technology, Inc., 3011 Triad Drive, Livermore, California 94551. http://www.adept.com
127. Armfield Limited Bridge House, West Street, Ringwood, BH24 1DY, England. http://www.armfield.co.uk
128. Horowitz I.M, "Optimum loop transfer function in single-loop minimum-phase feedback systems," Int. J. of Control, Vol. 18, N. 1, pp. 97-113, 1973.
129. Ballance, D.J., and W. Chen, "Symbolic computation in value sets of plants with uncertain parameters," UKACC International conference on control '98, pp. 1322-1327, 1998.
130. Skogestad, S. and I. Postlethwaite, "Multivariable Feedback Control. Analysis and Design," Wiley, ISBN 0-471-94330-4, 1996.
131. D'Azzo, J.J., C.H. Houpis, and S. N. Sheldon, "Linear Control System Analysis and Design With Matlab," Marcel Dekker, NY, 5th Ed., 2003.
132. Garcia-Sanz, M.J. Brugarolas, and I. Eguinoa, "Quantitative Analysis of Controller Fragility in the Frequency Domain," IASTED Int. Symp. on Modelling, Identification and Control, Grindelwald, Switzerland, 2004.
133. Garcia-Sanz, M. and E. Torres, "Control y experimentación del aerogenerador síncrono multipolar de velocidad variable TWT1650" (in Spanish) accepted for publication, RIAI 2004.
134. Garcia-Sanz, M., M. Barreras, I. Egana, and C.H. Houpis, "External Disturbance Rejection in Uncertain MIMO Systems with QFT Non-diagonal Controllers," Int. Symposium on Quantitative Feedback Theory QFT and Robust Frequency Domain Methods, Cape Town, South Africa, December, 2003.

135. De Bedout, J.M. and M.A. Franchek. "Stability conditions for the sequential design of non-diagonal mu1tivariable feedback controllers," Int. J. of Control, vol. 75, no. 12, pp. 910-922, 2002.
136. Kerr, M.L., S. Jayasuriya and S.F, Asokanthan, "On stability in non-sequential MIMO QFT designs," ASME Journal of Dynamic Systems, Measured and Control, 2005.
137. Garcia-Sanz, M., I Egana, and M. Barreras, "Design of Quantitative Feedback Theory Non-Diagonal Controllers for Use in Uncertain Multiple-input Multiple-output Systems," IEE Proc. Control Theory and Applications, vol. 152, pps. 177-178, 2005.
138. Karpenko, M. and N. Sepehri, "QFT Design of a PI Controller with Dynamic Pressure Feedback for Positioning a Pneumatic Actuator," Proc. of the American Control Conference, Boston, MA, pp. 5084-5089, June 2004.
139. Kerr, M., "Robust Control of an Articulating Flexible Structure Using MIMO QFT," PhD. Dissertation, The University of Queensland, Australia, January, 2004.
140. Chen-yang Lan, M.L. Kerr, and S. Jayasuriya, "Synthesis of Controllers for Non-minimum Phase and Unstable Systems Using Non-sequential MIMO Quantitative Feedback Theory," Proc. of the American Control Conference, Boston, MA, pp. 4139-4144, June 2004.
141. Smith, O.J.M., "Closer control of loops with dead time," Chem. Eng. Progr., 53, 217-219, 1957.
142. Astrom K.J., C.C. Hang, and B.C. Lim, "A new Smith predictor for controlling a process with an integrator and long dead-time," IEEE Trans. Automat. Contr., vol. 39, pp. 343–345, Feb. 1994.
143. Matausek M.R. and A.D. Micic, "A modified Smith predictor for controlling a process with an integrator and long dead-time," IEEE Trans. Automat. Contr., vol. 41, pp. 1199–1203, Aug. 1996.
144. Normey-Rico, J.E. and E.F.Camacho, "Smith predictor and modifications: A comparative study," in Proc. Euro. Control Conf. ECC'99 Karlsruhe, Germany Aug. 31–Sept. 3, 1999.
145. Ioannides, A.C., G.J. Rogers and V. Latham, "Stability limits of a Smith controller in simple systems containing a dead-time," Int. .J. Control, 29, 557-563, 1979.
146. Palmor, Z., "Stability properties of Smith dead-time compensator controllers," Int. J. Control, 32, 937-949, 1980.
147. Horowitz, I., "Some properties of delayed controls (Smith regulator)," Int. J. Control, 38, 977-990, 1983.
148. Yamanaka, K. and E. Shimemura, "Effects of mismatched Smith controller on stability in systems with time-delay," Automatica, 23, 787-791, 1987.
149. Santacesaria, C. and R. Scattolini, "Easy tuning of Smith Predictor in Presence of Delay Uncertainty," Automatica, 29 (6), 1595-1597, 1993.
150. de Paor, A.M., & M.J. O'Malley, "The zero-order hold equivalent transfer function for a time-delayed process," Int. J. Control 61(3), 657-665, 1995.
151. Henze M., C.P.L. Grady, W. Gujer, G.V.R. Marais and T. Matsuo. "A General Model for Single-sludge Wastewater Treatment Systems," Wat. Res., 20, 505-515, 1987.
152. Lindberg, C.F., "Control and estimation strategies applied to the activated sludge process," PhD dissertation. Uppsala University, 1997.
153. Urrutikoetxea A. and J.L García de las Heras, "Secondary Settling in Activated Sludge. A Lab-scale Dynamic Model of Thickening," Med. Fac. Landbow, 59 (4ª), Univ. Gent, 1994.
154. Bierman, G.J., "Factorization Methods for Least Square Estimation," Academic Press, New York, 1977.

155. Garcia-Sanz, M. and F.Y. Hadaegh, "Coordinated Load Sharing QFT Control of Formation Flying Spacecrafts. 3D Deep Space and Low Earth Keplerian Orbit problems with model uncertainty," NASA-JPL, Jet Propulsion Laboratory Document, D-30052, 66 pages, August 2004, Pasadena, California, USA.
156. Eitelberg, E., "Load Sharing Control," NOY Publ. Durban, South Africa, 1999.
157. European Wind Energy Association, EWEA, "Wind Force 12, a blue print to achieve 12% of the world's electricity from wind power by 2020," Forum of Energy & Development, Greenpeace and BTM Consult ApS, Oct. 2004.
158. Leithead, W.E., and M.C.M. Rogers, , "Drive-train characteristics of constant speed HAWT's: Part I – Representation by simple dynamic models," Wind Engineering, vol. 20, no. 3, pp. 149-174, 1996.
159. Leithead, W.E., and M.C.M. Rogers, "Drive-train characteristics of constant speed HAWT's: Part II – Simple characterization of dynamics." Wind Engineering, Vol. 20, No. 3, pp. 175-201, 1996.
160. Bongers, P.M.M., and T.G. Van Engelen, "A theoretical model and simulation of a wind turbine," Wind Engineering, vol. 11, no. 6, pp. 344-350, 1987.
161. Sheinman, Y. and A. Rosen, "A dynamic model for performance calculations of grid-connected horizontal axis wind turbines. Part II: Validation," Wind Engineering, vol. 15, no. 4, pp 229-239, 1991.
162. Sheinman, Y. and A. Rosen, "A dynamic model for performance calculations of grid-connected horizontal axis wind turbines. Part I: Description model," Wind Engineering, vol. 15, no. 4, pp 211-228, 1991.
163. De la Salle, S.A., Reardon, D., Leithead, W.E. and M. J. Grimble, "Review of wind turbine control," Int. J. of Control, vol. 52, no. 6, pp. 1295-1310, 1990.
164. Leith, D.J., and W. E. Leithead, "Appropriate realization of gain-scheduled controllers with application to wind turbine regulation," Int. J. of Control, vol. 65, no. 2, pp. 223-248, 1996.
165. Torres, E. and M. Garcia-Sanz, "Experimental Results of the Variable Speed, Direct Drive Multipole Synchronous Wind Turbine: TWT1650," Wind Energy, vol. 7, no. 2, pp. 109-118, 2004.
166. Garrad Hassan, "Bladed for windows," Bristol, England, 1997, 2003.
167. Germanisher Lloyd., "Rules and regulations." Hamburg, Germany, 1994, 2004.
168. International Electro-technical Commission. IEC 61400 "Wind Turbine Generator Systems."
169. MEASNET, Network of measuring institutes in Europe. Leuven, 1996.
170. Freris, L.L., "Wind energy conversion systems." Prentice Hall, 1990.
171. Burton, T., Sharpe, D., Jenkins, N. and E. Bossanyi, "Handbook of wind energy." John Wiley, 2001.
172. Manwell, J.F., McGowan, J.G. and A.L. Rogers, "Wind energy explained: theory, design and application." John Wiley, 2002.
173. Leith, D.J., and W.E. Leithead, "Implementation of Wind Turbine Controllers." International Journal of Control, Vol. 66, pp. 349-380, 1997.
174. Garcia-Sanz, M. "Control systems of the TWT1500 and TWT1650," Internal reports of M.Torres, 1998-2004.
175. Garcia-Sanz, M. and I. Eguinoa, "Improved Non-diagonal MIMO QFT Design Technique Considering Non-minimum Phase Aspects," 7th Int. Symp.on QFT and Robust Frequency Domain Methods, University of Kansas, Lawrence, KS, USA, August 2005.

176. Garcia-Sanz, M., A. Huarte, and A. Asenjo, "One-point Feedback Robust Control for Distributed Parameter Systems," 16th IFAC World Congress, Prha, Czech Republic, 2005.
177. Garcia-Sanz, M., A. Huarte, and A. Asenjo, "QFT Approach to Control One-point feedback Distributed Parameter Systems," 7th Int. Symp. on QFT and Robust Frequency Domain Methods, University of Kansas, Lawrence, KS, USA, August 2005.
178. M. Barreras, and M. Garcia-Sanz, "Multivariable QFT Controller Design for Heat Exchangers of Power Plants," 2nd Int. Conference on Renewable Energy and Power Quality ICREPQ'04, Barcelona, Spain, 2004.
179. Hamilton, S. W., "QFT Digital Controller for an Unmanned Research Vehicle with an Improved Method for Choosing the Control Weightings," M.S. Thesis, Air Force Institute of Technology, Wright-Patterson AFB, OH, Dec. 1987.
180. Buffington, J. M., "Modular Control Law Design for the Innovative Control Effectors (ICE) Tailless Fighter Aircraft Configuration 101-3," Technical Report AFRL-VA-WP-TR-1999-3057, Wright-Patterson AFB, OH, May 1999.
181. Schierman, J. D., et al, "Integrated Adaptive Guidance and Control for Re-Entry Vehicles with Flight-Test Results" J. of Guidance Control and dynamics, Vol. 27, No. 6, pp. 975-988, Nov-Dec., 2004
182. Phillips, W. D., "Selection of a Frequency Sensitive QFT Weighting Matrix Using the Method of Specified Outputs," M.S. Thesis, Air Force Institute of Technology, Wright-Patterson AFB, OH, Dec. 1988.

PROBLEMS

CHAPTER 2

2.1. The plant transfer function of Fig. 2.9 is:

$$P(s) = \frac{k}{s\,(s+a)(s+b)}$$

where the parameter variations are given by:

$$610 \le k \le 1050 \;\;;\;\; 1 \le a \le 15 \;\;;\;\; 150 \le b \le 170$$

Plot *Lm* ***P***$(j\omega)$ and $\angle$***P***$(j\omega)$ vs ω on semilog graph paper for several plants for the frequency range of $0.05 < \omega < 500$ rad/s.

2.2. The plant transfer function of Fig. 2.9 is:

$$P(s) = \frac{k\,(s+b)}{s\,(s+a)}$$

where the parameter variations are given by:

$$0.5 \le k \le 4 \;\;;\;\; 0.5 \le a \le 1 \;\;;\;\; 1.1 \le b \le 1.6$$

Plot *Lm* ***P***$(j\omega)$ and $\angle$***P***$(j\omega)$ vs ω on semilog graph paper for several plants for the frequency range of $0.05 < \omega < 500$ rad/s.

2.3. The plant transfer function of Fig. 2.9 is:

$$P(s)=\frac{k\ (s+b)}{s^2\ (s+a)}$$

where the parameter variations are given by:

$$1 \le k \le 5 \quad ; \quad 0.5 \le a \le 1 \quad ; \quad 2 \le b \le 5$$

Plot *Lm* ***P***$(j\omega)$ and $\angle$***P***$(j\omega)$ vs ω on semilog graph paper for several plants for the frequency range of $0.05 < \omega < 100$ rad/s.

CHAPTER 3

3.1. Obtain the tracking models $T_{R_U}(s)$ and $T_{R_L}(s)$ to satisfy the following specifications:

a) *For the upper bound* $M_p = 1.11$ *and* $t_s = 1.21$ *s*
For the lower bound $t_s = 1.71$ *s*

b) For the upper bound $M_p = 1.13$ and $t_s = 1.17$ s
For the lower bound $t_s = 2$ s

c) For the upper bound $M_p = 1.2$ and $t_s = 1.2$ s
For the lower bound $t_s = 1.48$ s

d) For the upper bound $M_p = 1.15$ and $t_s = 1.34$ s
For the lower bound $t_s = 1.69$ s

3.2. Determine the disturbance bound model $T_D(s)$, for Case 2 disturbance (disturbance at the plant input), that satisfies the following specifications:

a) $|y_D(t_p)| = 0.1$ for $t_p = 65$ ms
b) $|y_D(t_p)| = 0.08$ for $t_p = 50$ ms

Hint: Recommend that the following transfer function be synthesized:

$$T_D(s)=\frac{K_x s}{s^2+2\varsigma\omega_n s+\omega_n^2}=\frac{K_x s}{(s+a)^2+b^2}$$

where

$$t_p = \frac{\cos^{-1}(\xi)}{\omega_n \sqrt{1-\xi^2}}$$

and where a value of ζ be selected in the range of $0.5 \le \zeta \le 0.7$ (see Ref. 1).

3.3. The plant transfer function of Fig. 2.9 is:

$$P(s) = \frac{k(s+b)}{s^2(s+a)}$$

where the parameter variations are given by:

$$1 \le k \le 5 \quad ; \quad 0.5 \le a \le 1 \quad ; \quad 2 \le b \le 5$$

(a) Obtain templates $\Im P(j\omega_i)$ for the following frequencies: 0.05, 0.3, 1, 5, 10 and 100 rad/s

(b) Determine the value of V [see Eq. (3.20)].

(c) Determine the values of δ_p for the frequencies of 5, 10, and 100 rad/s. Compare these values with the value of V determined in (b).

3.4. Figure 3.1 is modified by inserting a unity-feedback loop around the plant $\mathcal{P}$ as shown in the figure below.

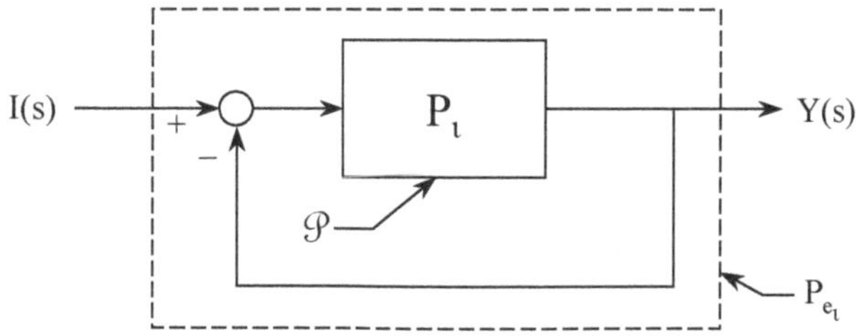

The effective plant that is to be utilized for a QFT design is given by:

$$P_e(s) = \frac{Y(s)}{I(s)} = \frac{P_\iota}{1+P_\iota}$$

(a) Obtain the transfer functions of $P_e(s)$ using the model of Prob. 3.3.

(b) Obtain the templates $\Im P_e(j\omega_i)$ for the same values of frequencies given in Prob. 3.3(a).

(c) Compare the size of the templates of this problem with those of Prob. 3.3. By putting a unity-feedback loop around the plant $\mathcal{P}$ what effect did it have on the size of the templates for the plant $\mathcal{P}_e$?

The reader is highly urged to do a few of the following QFT design problems with a minimum use of a QFT CAD package. By doing Steps 4 through 11 of Sec. 3-4 by use of TOTAL-PC (Appendix F) or a similar software, as done in Secs. 3-6 through 3-20 except for Fig. 3.26, will enhance the readers' understanding of the basic QFT concepts of Chapter 3. In all these design problems show all open-loop Nichols Chart plots and closed-loop Nichols Chart and Bode plots as shown in Design Example 1. Also, show the time response plots for all J plants.

3.5. For Fig. 3.1 synthesize $G(s)$ and $F(s)$, where the plant is given in Prob. 3.3, to satisfy the following specifications:

Robust Stability. $\left|\dfrac{\mathbf{P}(j\omega)\mathbf{G}(j\omega)}{1+\mathbf{P}(j\omega)\mathbf{G}(j\omega)}\right| \le 1.25, \quad \forall\omega \in [0,\infty].$

Tracking specifications:

- For the upper bound $M_p = 1.07$ and $t_s = 0.825$ s
- For the lower bound $t_s = 0.824$ s

Disturbance rejection at plant input (Case 2):

$$T_D(s) = 1.0562 / [(s + 0.6322)_2 + 0.843^2], \quad \omega < 100 \text{ rad/s}$$

3.6. Repeat Prob. 3.5 but where the plant P_e and the feedback structure are given in Prob. 3.4. Compare the order of the compensator and pre-filter of this problem with those of Prob. 3.5.

3.7. Repeat Design Example 1 (Case 1 disturbance) of Sec. 3-18 for the tracking specifications of (a) Prob. 3.1(a), (b) Prob. 3.1(b), (c) Prob. 3.1(d), and (e) Prob. 3.1(e).

3.8. Repeat Design Example 2 (Case 2 disturbance) of Sec. 3-18 for the tracking specifications of (a) Prob. 3.1(a), (b) Prob. 3.1(b), (c) Prob. 3.1(d), and (e) Prob. 3.1(e).

3.9. In Design Example 1 (Case 1 disturbance) of Sec. 3-17 $\boldsymbol{B}_o = \boldsymbol{B}_D$. Repeat this example for which the disturbance rejection specification is ignored which results in $\boldsymbol{B}_o = \boldsymbol{B}_R$. How close does the resulting QFT designed system satisfy the disturbance response specification for both Design Examples 1 and 2?

3.10. For the analog control system of Fig. 3.1 the uncertain plant is:

$$P(s) = \frac{k}{s(s+a)(s+b)}$$

where the parameter variations are given by:

$$610 \le k \le 1050 \;\; ; \;\; 1 \le a \le 15 \;\; ; \;\; 150 \le b \le 170$$

(1) The tracking specifications are given by Prob. 3.1(d):

- For the upper bound $M_p = 1.15$ and $t_s = 1.34$ s
- For the lower bound $t_s = 1.69$ s

(2) The design specifications for Case 2 disturbance rejection (plant input disturbance rejection) are those given by Prob. 3.2(a):

$$\alpha_p < |y_D(t_p)| = 0.1 \text{ for } t_p = 65 \text{ ms}, \;\; y_D(\infty) = 0, \;\; \omega < 10 \text{ rad/s}$$

(3) The desired phase margin angle (γ), which sets the value of M_L, is 40°.

(a) Synthesize the compensator G(s) and the pre-filter F(s) for the uncertain plant of Fig. 3.1.
(b) Obtain y_R(t) and y_{DI}(t) and compare their respective FOM with the desired FOM.

3.11. Repeat Prob. 3.10 for a Case 1 disturbance (disturbance at the plant output).

3.12. For the analog control system of Fig. 3.1 the uncertain plant is:

$$P(s) = \frac{k}{s(s+a)(s+b)}$$

where the parameter variations are given by:

$$300 \le k \le 480 \quad ; \quad 1 \le a \le 4 \quad ; \quad 120 \le b \le 150$$

(1) The design specifications for Case 2 disturbance rejection (plant input disturbance rejection) are those given by:

$\alpha_p < |y_D(t_p)| = 0.1$, $y_D(\infty) \neq 0$, $\omega < 10$ rad/s

(2) The desired phase margin angle (γ), which sets the value of M_L, is 40°.

(a) Synthesize the compensator G(s) for the uncertain plant of Fig. 3.1.
(b) Obtain y_{DI}(t) and compare its respective FOM with the desired FOM.

3.13. For the analog control system of Fig. 3.1 the uncertain plant is:

$$P(s) = \frac{k}{s(s+a)(s+b)}$$

where the parameter variations are given by:

$$300 \le k \le 480 \quad ; \quad 1 \le a \le 4 \quad ; \quad 120 \le b \le 150$$

(1) The design specifications for Case 2 disturbance rejection (plant input disturbance rejection) are those given by:

$\alpha_p < |y_D(t_p)| = 0.1$, $y_D(\infty) = 0$, $\omega < 10$ rad/s

(2) The desired phase margin angle (γ), which sets the value of M_L, is 45°.

(a) Synthesize the compensator G(s) for the uncertain plant of Fig. 3.1.
(b) Obtain y_{DI}(t) and compare its respective FOM with the desired FOM.

3.14. For the analog control system of Fig. 3.1 the uncertain plant is:

$$P(s) = \frac{k}{s\,(s+a)(s+b)}$$

where the parameter variations are given by:

$$900 \le k \le 1200 \quad ; \quad 2 \le a \le 10 \quad ; \quad 120 \le b \le 150$$

(1) The tracking specifications are given by Prob. 3.1(c):

- For the upper bound $M_p = 1.2$ and $t_s = 1.2$ s
- For the lower bound $t_s = 1.48$ s

(2) The design specifications for Case 2 disturbance rejection (plant input disturbance rejection) are those given by Prob. 3.2(a):

$$\alpha_p < |y_D(t_p)| = 0.1\ ,\quad y_D(\infty) = 0,\quad \omega < 10 \text{ rad/s}$$

(3) The desired gain margin, which sets the value of M_L, is 4 dB.

(a) Synthesize the compensator G(s) and the pre-filter F(s) for the uncertain plant of Fig. 3.1.
(b) Obtain y_R(t) and y_{DI}(t) and compare their respective FOM with the desired FOM.

3.15. For the analog control system of Fig. 3.1 the uncertain plant is:

$$P(s) = \frac{k(s+b)}{s(s+a)}$$

where the parameter variations are given by:

$$0.5 \le k \le 4 \quad ; \quad 0.5 \le a \le 1 \quad ; \quad 1.1 \le b \le 1.6$$

(1) The tracking specifications are given by Prob. 3.1(d):

- For the upper bound $M_p = 1.15$ and $t_s = 1.34$ s
- For the lower bound $t_s = 1.69$ s

(2) The design specifications for Case 2 disturbance rejection (plant input disturbance rejection):

$$\alpha_p < |y_D(t_p)| = 0.1\ ,\ \ y_D(\infty) = 0\ ,\ \ \omega < 10 \text{ rad/s}$$

(3) The desired phase margin angle (γ), which sets the value of M_L, is 40°.

(a) Synthesize the compensator G(s) and the pre-filter F(s) for the uncertain plant of Fig. 3.1.

(b) Obtain y_R(t) and y_{DI}(t) and compare their respective FOM with the desired FOM.

3.16. Figure 3.1 is modified by inserting a unity-feedback loop around the plant P as shown in the figure below.

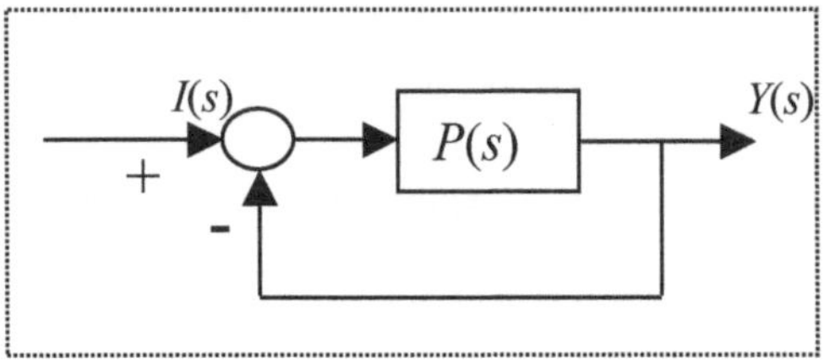

The plant $\mathcal{P}$ is defined as:

$$P(s) = \frac{k(s+b)}{s(s+a)}$$

where the parameter variations are given by:

$$0.5 \le k \le 4\ \ ;\ \ 0.5 \le a \le 1\ \ ;\ \ 1.1 \le b \le 1.6$$

(1) The tracking specifications are given by Prob. 3.1(d):

- For the upper bound $M_p = 1.15$ and $t_s = 1.34$ s
- For the lower bound $t_s = 1.69$ s

(2) The design specifications for Case 2 disturbance rejection (plant input disturbance rejection):

$$\alpha_p < |y_D(t_p)| = 0.1\ ,\ \ y_D(\infty) = 0\ ,\ \ \omega < 10 \text{ rad/s}$$

(3) The desired phase margin angle (γ), which sets the value of M_L, is 40°.

(a) Synthesize the compensator $G(s)$ and the pre-filter $F(s)$ for the uncertain plant of Fig. 3.1.

(b) Obtain $y_R(t)$ and $y_{DI}(t)$ and compare their respective FOM with the desired FOM.

3.17. A design has been made for the following figure:

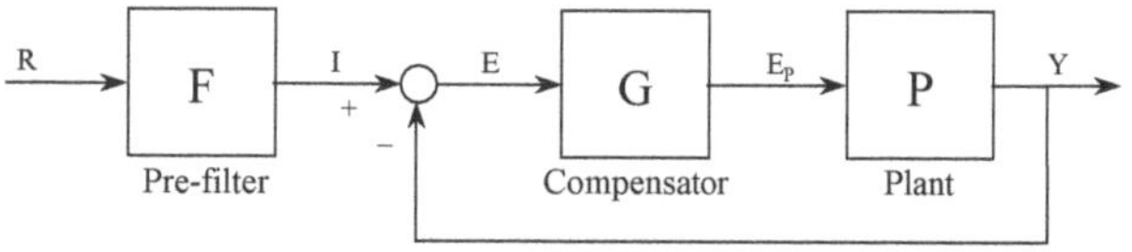

Where the transfer function $P(s)$, $F(s)$, and $G(s)$ are known. It turns out to be more convenient to implement the system by the one shown in the following figure:

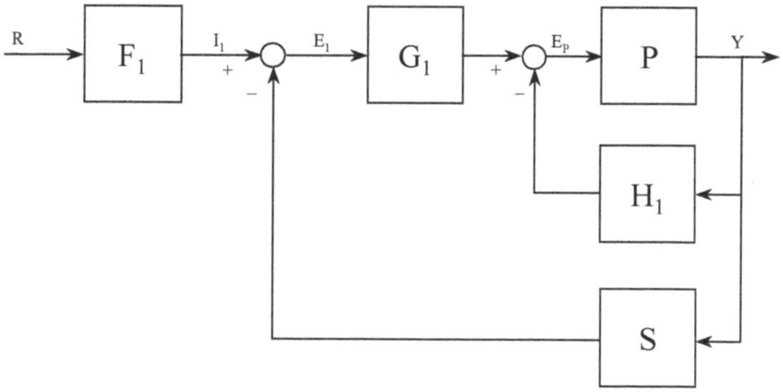

$H_1(s)$ is a tachometer and $S(s)$ is a sensor, and these transfer functions are known. Find $F_1(s)$ and $G_1(s)$ as functions of the known transfer functions $F(s)$, $G(s)$, $H_1(s)$, and $S(s)$ so that the two figures are equivalent. *Note*: the loop transmissions for both control systems must be identical.

3.18. In the following Fig. A $P(s) = 1/s$ and $G_1(s) = 1/(s + 2)$. In Fig. B below $P(s) = 1/s$ and $H(s) = s/(s+ 1)$.

(a) Find $G_2(s)$ so that $Y(s)/R(s)$ is the same for both configurations.

(b) Find $S_P^T(s)$, the system sensitivity to variations in P, for each configuration. If these are different, which one yields smaller values of $|\,S_P^T(j\omega)\,|$ over a range of finite values of ω?

(c) Which configuration is the best for minimizing the parameter variation effects on $Y(s)$?

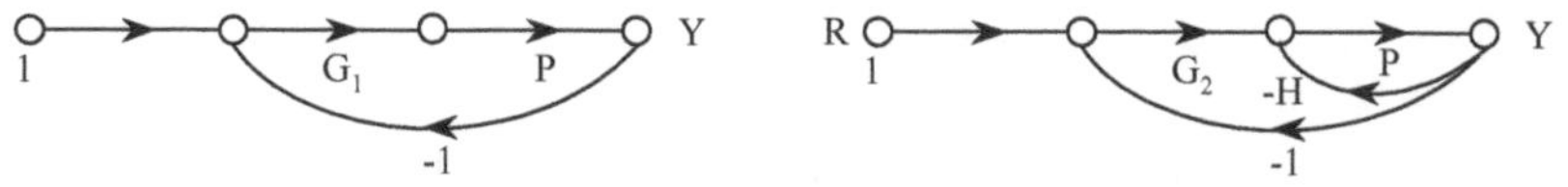

A B

3.19. For Prob. 3.17 the plant transfer function is $P(s) = K / [s(s + a)]$ where $1 \le K \le 5$ and $1 \le a \le 10$. [*Note*: see Fig.2.6.]

(a) Obtain templates for $\boldsymbol{P}(j\omega_i)$ where

$$\boldsymbol{P}_e(j\omega_i) = \frac{\boldsymbol{P}(j\omega_i)}{1 + \boldsymbol{P}(j\omega_i)\boldsymbol{H}(j\omega_i)}$$

is defined as the effective plant and where $H_i(s) = s$, for values of ω_i = 0.5, 1, and 2. Obtain template data for points A, B, C, and D of Fig. 3.6.

(b) Explain why the sizes of the templates for $\Im\boldsymbol{P}(j\omega_i)$ and $\Im\boldsymbol{P}_e(j\omega_i)$ are different. See the discussion on sensitivity in Sec. 6-2.2.

3.20. For the analog control system of Fig. 3.1 the uncertain plant is:

$$P(s) = \frac{k}{s(s+a)(s+b)}$$

where the parameter variations are given by:

$$3060 \le k \le 3360 \quad ; \quad 3 \le a \le 9 \quad ; \quad 140 \le b \le 170$$

(1) The tracking specifications are given by Prob. 3.1(b):

- For the upper bound $M_p = 1.13$ and $t_s = 1.17$ s
- For the lower bound $t_s = 2$ s

(2) The design specifications for Case 1 disturbance rejection (plant output disturbance rejection) are:

$$\alpha_p < |y_D(t_p)| = 0.1\,,\quad y_D(\infty) = 0,\quad \omega < 20 \text{ rad/s}$$

(3) The desired gain margin, which sets the value of M_L, is 5 dB.

(a) Synthesize the compensator G(s) and the pre-filter F(s) for the uncertain plant.
(b) Obtain y_R(t) and y_{D2}(t) and compare their respective FOM with the desired FOM.

3.21. For the analog control system of Fig. 3.1 the uncertain plant is:

$$P(s) = \frac{k}{s(s+a)(s+b)}$$

where the parameter variations are given by:

$$3060 \le k \le 3360 \;\;;\;\; 3 \le a \le 9 \;\;;\;\; 140 \le b \le 170$$

(1) The tracking specifications are given by Prob. 3.1(a):

- For the upper bound $M_p = 1.11$ and $t_s = 1.21$ s
- For the lower bound $t_s = 1.71$ s

(2) The desired phase margin angle (γ), which sets the value of M_L, is 45°.

(3) The design specifications for Case 1 disturbance rejection (plant output disturbance rejection) are:

$$\alpha_p < |y_D(t_p)| = 0.1\,,\quad y_D(\infty) = 0,\quad \omega < 20 \text{ rad/s}$$

(1) Synthesize the compensator G(s) and the pre-filter F(s) for the uncertain plant.
(2) Obtain y_R(t) and y_{D2}(t) and compare their respective FOM with the desired FOM.

3.22. For the analog control system of Fig. 3.1 the uncertain plant is:

$$P(s) = \frac{k}{s\,(s+a)(s+b)}$$

where the parameter variations are given by:

$$3060 \le k \le 3360 \;\; ; \;\; 3 \le a \le 9 \;\; ; \;\; 140 \le b \le 170$$

(1) The tracking specifications are given by Prob. 3.1(a):

- For the upper bound $M_p = 1.11$ and $t_s = 1.21$ s
- For the lower bound $t_s = 1.71$ s

(2) The desired phase margin angle (γ), which sets the value of M_L, is 45°.

(3) The design specifications for Case 1 disturbance rejection (plant output disturbance rejection) are:

$$T_D(s)=(s^2+3.0540\,s)/(s^2+6.1090\,s+9.5790),\;\; \omega < 20 \text{ rad/s}$$

(a) Synthesize the compensator G(s) and the pre-filter F(s) for the uncertain plant.
(b) Obtain y_R(t) and y_{D2}(t) and compare their respective FOM with the desired FOM.

3.23. Consider the feedback structure of Fig. 3.1 and the plant transfer function P(s) according to the equation:

$$P(s) = \frac{ck(as^2 + ds + b)}{fs\left[cas^2 + d(c+a)s + b(c+a)\right]}$$

where the parameter variations are given by:

$$k \in [100, 800]\; ; \; a \in [1400, 11000]\; ;$$
$$b \in [58000, 115000]\; ; \; c = 65.6\; ; \; d = 377\; ; \; f = 3.07$$

(a) Obtain the complete template $\Im P(j\omega)$ for the frequency $\omega = 10$ rad/s.
(b) Compare the result with the template generated by using only the contour of the parameter uncertainty space.

3.24. Repeat Prob. 3.23 for the next transfer function:

$$P(s) = \frac{k}{s^2 + 2\zeta\omega_n s + \omega_n^2} e^{-\tau s}$$

where the parameter variations are given by:

$$k = 6\ ;\ \zeta = 0.1\ ;\ \omega_n \in [0.6, 1.4]\ ;\ \tau \in [0.1, 2].$$

and for the frequency $\omega = 0.9$ rad/s.

3.25. Repeat Prob. 3.23 for the next transfer function:

$$P(s) = \frac{b_1 s^3 + b_2^{\ 2} s^2 + 12s + (12 + 5\,b_3)}{12\,s^4 + (b_4 + 5)\,s^3 + (b_5 b_1^{\ 2})s^2 + (b_3 + 8b_6)\,s + (3b_7)} e^{-\tau s}$$

where the parameter variations are given by:

$$b_1 \in [1, 2]\ ;\ b_2 \in [2, 3]\ ;\ b_3 \in [21, 25]\ ;\ b_4 \in [1, 9]\ ;$$
$$b_5 \in [1, 2]\ ;\ b_6 \in [1, 2]\ ;\ b_7 \in [1, 2]\ ;\ \tau \in [\pi/6, \pi/3],$$

and for the frequency $\omega = 10$ rad/s.

3.26. Repeat Prob. 3.23 for the next transfer function:

$$P(s) = \frac{k}{as + b} e^{-\tau s}$$

where the parameter variations are given by:

$$k \in [1, 2]\ ;\ a \in [1, 5]\ ;\ b \in [10, 12]\ ;\ \tau \in [0.1, 0.2]$$

and for the frequency $\omega = 3$ rad/s.

3.27. Explore if the transfer function $P_1(s)$ with three uncertain parameters can be reduced to the transfer function $P_2(s)$ with only two related uncertain parameters, in order to reduce the number of operations needed to compute the template.

$$P_1 = \left\{ \begin{array}{ll} P_1(s) = \dfrac{c}{as+b} \;, & a \in [1,\ 5] \\ & b \in [10,\ 12],\ c \in [1,\ 2] \end{array} \right\}$$

$$P_2 = \left\{ \begin{array}{ll} P_2(s) = \dfrac{c/a}{s+b/a} \;, & (c/a) \in \left[\dfrac{\min c}{\max a} = 0.2,\ \dfrac{\max c}{\min a} = 2 \right] \\ & (b/a) \in \left[\dfrac{\min b}{\max a} = 2,\ \dfrac{\max b}{\min a} = 12 \right] \end{array} \right\}$$

Obtain, on the Nichols Chart, the templates at $\omega = 5$ rad/s of both transfer functions. Explain the effect of the reduction (selection of P_2 instead of P_1), if any, in the controller design.

3.28. Consider the bench-top helicopter shown in the figure. It is a laboratory scale plant with 3 Degrees of Freedom (3 DOF), roll angle ϕ, pitch angle θ, and yaw angle ψ, each one measured by an absolute encoder. Two electrical DC motors are attached to the helicopter body, making two propellers turn. The total force F caused by aerodynamics makes the total system turn around an angle measured by an encoder. A counterweight of mass M helps the propellers lift the body weight due to mass m.

The dynamics of the pitch angle is obtained by applying Lagrange's equations to the mechanical scheme, so that,

$$Fl_1 - mg\left[(h+d)\sin\theta + l_1\cos\theta\right] + Mg(l_2 + l_3\cos\alpha)\cos\theta + Mg(l_3\sin\theta - h)\sin\theta - b_e\frac{d\theta}{dt} = J_e\frac{d^2\theta}{dt^2}$$

where h, d, l_1, l_2 and l_3 are lengths; m, the sum of both motors' mass and M the counterweight mass; b_e is the dynamic friction coefficient; g the gravity acceleration; J_e the inertial moment of the whole system around the pitch angle θ, and α a fixed construction angle. The total non-linear model obtained from the previous equation can be simplified by linearising around the operational point $\theta_0 = 0$. It yields a second order transfer function between the pitch angle θ and the motor control signal U:

$$P(s) = \frac{\theta(s)}{U(s)} = \frac{k\omega_n^2}{s^2 + 2\zeta\omega_n s + \omega_n^2}$$

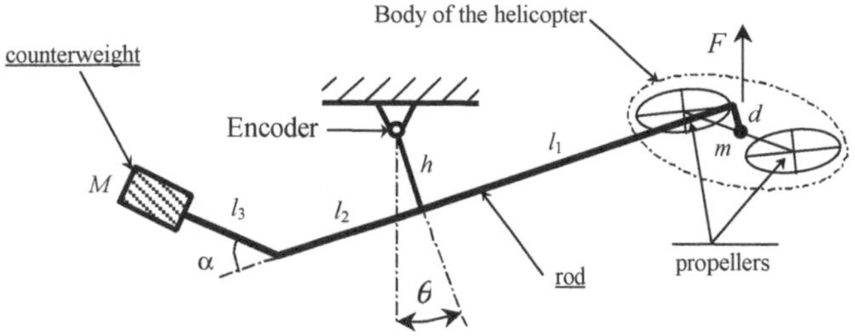

Using system identification techniques from experimental data the parameter variations obtained are: $k \in [0.01, 0.099]$; $\zeta \in [0.1, 0.16]$; $\omega_n \in [0.55, 0.58]$. The motor control signal U presents a saturation limit of ±10 V.

Design a compensator to fulfill the following control objectives: (a) to minimize the reference tracking error, (b) to increase the damping and increase the system stability, (c) to reduce the overshoot, (d) to reject the high frequency noise at feedback sensors, and (e) to deal with the actuator constraints.

3.29. From Sec. 10.9, consider the second angle $\delta_2(s)$ of the 5 Degree-Of-Freedom SCARA robot (see figure). The transfer function that describes $\delta_2(s)$ in terms of the motor signal $u_2(s)$ is:

$$\frac{\delta_2(s)}{u_2(s)} = p_{22}(s) = \frac{(\alpha_1 + 2\alpha_3 h)s + v_1}{s\Delta(s)} \frac{1}{k}$$

where: $\Delta = \chi_2 s^2 + \chi_1 s + \chi_0$

and:

$$\chi_2 = \alpha_2(\alpha_1 + 2\alpha_3 h) - (\alpha_2 + \alpha_3 h)^2$$
$$\chi_1 = \alpha_2 v_1 + v_2(\alpha_1 + 2\alpha_3 h)$$
$$\chi_0 = v_1 v_2$$

The parameter variations are given by:

$k = 75$ ct/N·m ; $h \in [-1, 1]$; $\alpha_1 k \in [719, 813]$ ct s^2/rad;
$\alpha_2 k \in [186, 200]$ ct s^2/rad; $\alpha_3 k \in [134, 230]$ ct s^2/rad;
$v_1 k \in [67, 381]$ ct s/rad; $v_2 k \in [11.6, 91.9]$ ct s/rad;

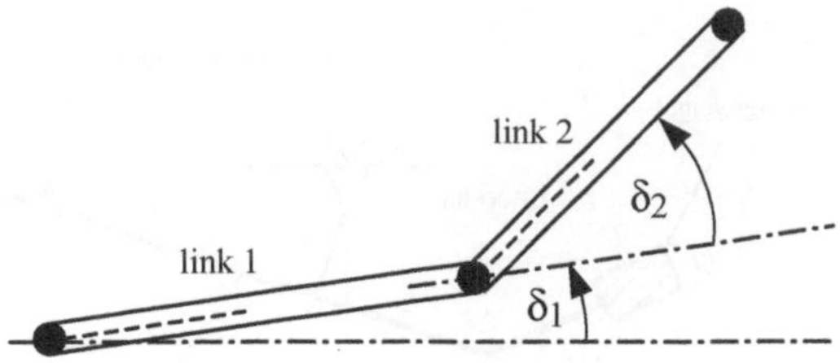

Fixing the other four Degree-Of-Freedom, design a compensator to control the second angle $\delta_2(s)$, fulfilling the following performance specifications,

- *Robust Stability*. $\left| \dfrac{\boldsymbol{P}(j\omega)\boldsymbol{G}(j\omega)}{1+\boldsymbol{P}(j\omega)\boldsymbol{G}(j\omega)} \right| \le 1.2 \;,\; \forall \omega \in [0,\infty]$.

- *Control effort constraint*. Control signals have to be lower than 32767 [ct] for a disturbance rejection at the plant output of about 20°.

- *Disturbance rejection at plant input*. The maximum allowed error has to be 30° for torque disturbances of 1000 [ct].

- *Tracking specifications*. $\left|\boldsymbol{T}(j\omega)\right| = \left| \dfrac{\boldsymbol{P}(j\omega)\boldsymbol{G}(j\omega)}{1+\boldsymbol{P}(j\omega)\boldsymbol{G}(j\omega)} \boldsymbol{F}(j\omega) \right|$ has to achieve tracking tolerances defined by, $a(\omega) \le \left|\boldsymbol{T}(j\omega)\right| \le b(\omega)$

where,
$$b(\omega) = \left| \frac{12.25\left[(j\omega)/30+1\right]}{(j\omega)^2 + 5.25(j\omega) + 12.25} \right|$$

$$a(\omega) = \left| \frac{2.25}{\left[(j\omega)^2 + 4.5(j\omega) + 2.25\right]\left[(j\omega)/10+1\right]} \right|$$

3.30. Consider an inverted pendulum mounted on a cart, as shown in the figure. The cart must be moved so that the mass M_2 is always in the upright position. The transfer function that describes the angular rotation $\theta(s)$ in terms of the motor signal $u(s)$ is linearised around the equilibrium point $\theta(t) = 0$ and $d\theta(t)/dt = 0$, so that:

$$\frac{\theta(s)}{u(s)} = P(s) = \frac{kb_1 s}{a_3 s^3 + a_2 s^2 + a_1 s + a_0} h_\theta$$

where,

$b_1 = l_1 m_e;$
$a_3 = l_1^2 m_e^2 - I_e M_e;$
$a_2 = -(C_0 I_e + C_1 M_e);$
$a_1 = g\, l_1 m_e M_e - C_0 C_1;$
$a_0 = C_0\, g\, l_1 m_e;$
$m_e = m_1/2 + M_2;$
$M_e = M_1 + m_1 + M_2;$
$I_e = m_1 l_1^2 / 3 + m_1^3 l_1^4 / 36 + M_2 m_1^2 l_1^4 / 9$

and where l_1 is the stick length, M_1 the cart mass, M_2 the end mass, m_1 the stick mass, and C_0 and C_1 the friction coefficients between the cart and the rail and between the cart and the stick respectively.

The parameters of the inverted pendulum model used here have been estimated by experimental testings and by taking into account all the possibilities of the real system: three different rod lenghts and masses (l_1, m_1), two different cylindrical weight masses (M_2), and a wide range of friction coefficients (C_0, C_1). Accordingly, with those possibilities, the parameter variations are:

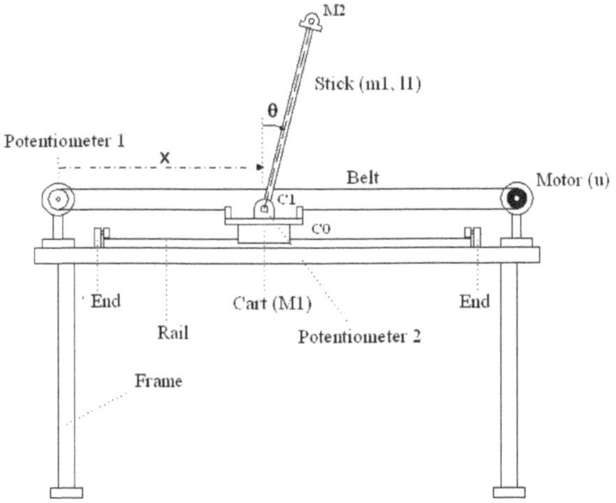

$l_1 \in [0.05, 0.65]$ m; $m_1 \in [0.01, 0.2]$ kg;
$M_2 \in [0.02, 0.5]$ kg; $k = 2.6$ N/V
$C_0 \in [1, 6.2]$ kg/s; $C_1 \in [0.015, 0.035]$ kgm^2/s; $M_1 = 3.2$ kg
$h_\theta = 3.7$ V/rad; $g = 9.8$ m/s^2

Design a compensator $G(s)$ to keep the pendulum in the upright position for the following stability and performance specifications:

Stability: $$\left|\frac{P(j\omega)\,G(j\omega)}{1+P(j\omega)\,G(j\omega)}\right| \le 1.4\,,\ \forall \omega \in [0,\infty]$$

Disturbance rejection : $$\left|\frac{P(j\omega)}{1+P(j\omega)\,G(j\omega)}\right| \le 0.19,\ \forall \omega \in [0,10] \text{ rad/s}$$

Control effort limitation:

$$\left|\frac{\boldsymbol{G}(j\omega)}{1+\boldsymbol{P}(j\omega)\boldsymbol{G}(j\omega)}\right| \le W_5(j\omega j\omega\ \forall \omega \in [0,10] \text{ rad/s} \quad \text{(see figure).}$$

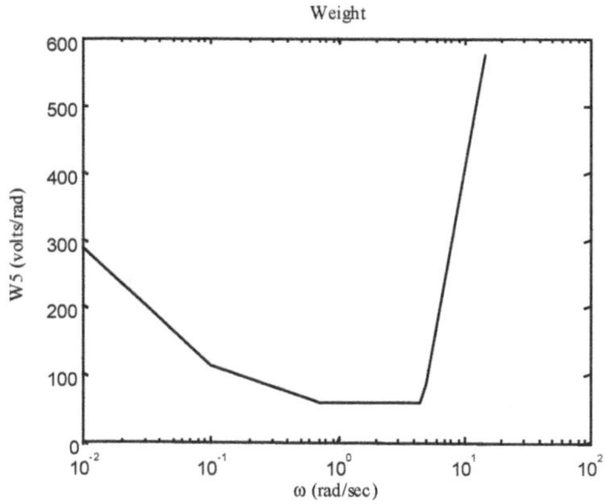

3.31. Consider a dc field-controlled motor (see figure) whose nominal characteristics are showed in the following table. Assuming that the angular velocity Ω is the plant output to be controlled, and the plant input the field voltage V_f, the dc motor in Laplace terms obeys:

$$P(s) = \frac{\Omega}{V_f} = \frac{K_m / R_f}{Js + b}$$

where the electrical time constant L_f/R_f has been neglected compared to the field time constant J/b.

Usually the motor parameters differ from the nominal ones due to temperature, aging and some other special effects. Because of that, it is supposed a 20% variation in b and a 5% variation in K_m/R_f from their nominal values.

In addition, the larger divergence observed in the proposed real application takes place at the rotor inertia J. Its nominal value of $J = 0.01$ refers to unload driving. However, considering the load, J is expected to be in the range [0.01 0.1], between unload and full load.

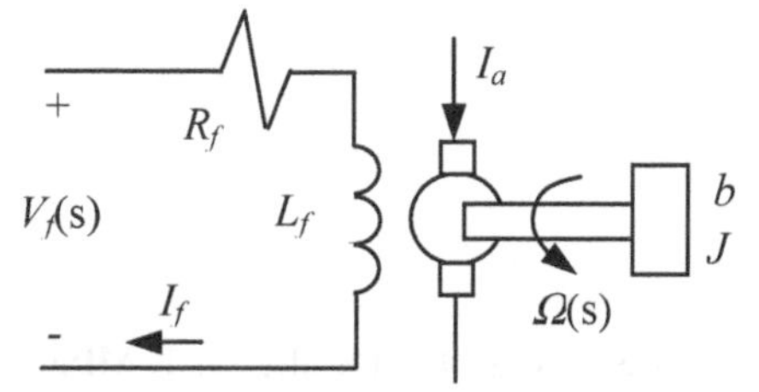

NOMINAL CHARACTERISTICS
OF A DC FIELD-CONTROLLED MOTOR

Parameter	Nominal Values
Unload Rotor Inertia, J	0.01 kg m^2
Friction, b	0.1 N m s
Motor constant, K_m	0.05 N m /A
Field Resistance, R_f	1 Ω
Field Inductance, L_f	<<0.1 H

Synthesize the compensator G(s) and the pre-filter F(s) to fulfill the required specifications, so that:

i.- Robust stability

$$\left| \frac{\boldsymbol{P}(\mathrm{j}\omega)\boldsymbol{G}(\mathrm{j}\omega)}{1+\boldsymbol{P}(\mathrm{j}\omega)\boldsymbol{G}(\mathrm{j}\omega)} \right| \leq 1.3 \ , \ \ \forall \omega \in [0, \infty]$$

ii.- Robust reference tracking. $\left|T(j\omega)\right| = \left|\dfrac{P(j\omega)G(j\omega)}{1+P(j\omega)G(j\omega)} F(j\omega j\right|$ has to achieve tracking tolerances defined by, $a(\omega) \le \left|T(j\omega)\right| \le b(\omega)$

where, $$b(\omega) = \left|\frac{0.66\left[(j\omega)+30\right]}{(j\omega)^2+4(j\omega)+19.75}\right|$$

$$a(\omega) = \left|\frac{8400}{\left[(j\omega)+3\right]\left[(j\omega)+4\right]\left[(j\omega)+10\right]}\right|$$

iii.- Gain at high frequency: Reduce the 'cost of feedback':

$$|L(j\omega_{hf})| < -20\text{dB}, \ \omega_{hf} \ge 100.$$

CHAPTER 4

Reference should be made to Fig. 4.8 for the QFT MISO sampled-data control system design problems of this chapter.

4.1. Consider the classical benchmark problem shown in the next figure. It is a mechanical frictionless system composed of two carts of mass M_1 and M_2, coupled by a link of stiffness γ. The problem is to control the position $x_2(t)$ of the second cart by applying a force $u(t)$ to the first cart.

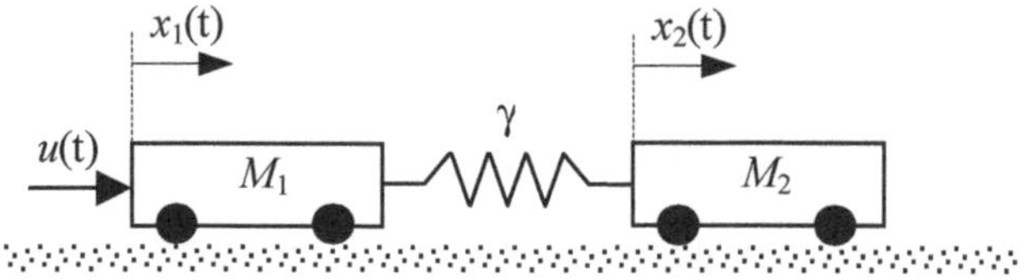

The parameter uncertainty is given by:

$M_1 \in [0.9, 1.1]$; $M_2 \in [0.9, 1.1]$; $\gamma \in [0.4, 0.6]$

Synthesize the compensator G, where the sampling time is $T = 0.01$ s, to control the position $x_2(t)$, fulfilling the following robust stability specification:

$$\left|\frac{\boldsymbol{P}(\mathrm{j}\omega)\boldsymbol{G}(\mathrm{j}\omega)}{1+\boldsymbol{P}(\mathrm{j}\omega)\boldsymbol{G}(\mathrm{j}\omega)}\right| \leq 1.2\ , \ \ \forall\omega \in [0,\infty] .$$

4.2. Synthesize the compensator G and the pre-filter F for the uncertain plant of Prob. 3.5 where the sampling time is $T = 0.002$ s.

4.3. Synthesize the compensator G and the pre-filter F for the uncertain plant of Prob. 3.6 where the sampling time is $T = 0.002$ s.

4.4. Synthesize the compensator G and the pre-filter F for the uncertain plant of Prob. 3.10 where the sampling time is $T = 0.002$ s.

4.5. Synthesize the compensator G and the pre-filter F for the uncertain plant of Prob. 3.11 where the sampling time is $T = 0.001$ s.

4.6. Synthesize the compensator G for the uncertain plant of Prob. 3.12 where the sampling time is $T = 0.01$ s.

4.7. Synthesize the compensator G for the uncertain plant of Prob. 3.13 where the sampling time is $T = 0.001$ s.

4.8. Synthesize the compensator G and the pre-filter F for the uncertain plant of Prob. 3.14 where the sampling time is $T = 0.002$ s.

4.9. Synthesize the compensator G and the pre-filter F for the uncertain plant of Prob. 3.15 where the sampling time is $T = 0.002$ s.

4.10. Synthesize the compensator G and the pre-filter F for the uncertain plant of Prob. 3.16 where the sampling time is $T = 0.01$ s.

4.11. Synthesize the compensator G and the pre-ilter F for the uncertain plant of Prob. 3.20 where the sampling time is $T = 0.002$ s.

4.12. Synthesize the compensator G and the pre-filter F for the uncertain plant of Prob. 3.21 where the sampling time is $T = 0.002$ s.

4.13. Synthesize the compensator G and the pre-filter F for the uncertain plant of Prob. 3.22 where the sampling time is $T = 0.01$ s.

CHAPTER 5

5.1. The plant matrix and the desired control ratio matrix for a 2x2 system are:

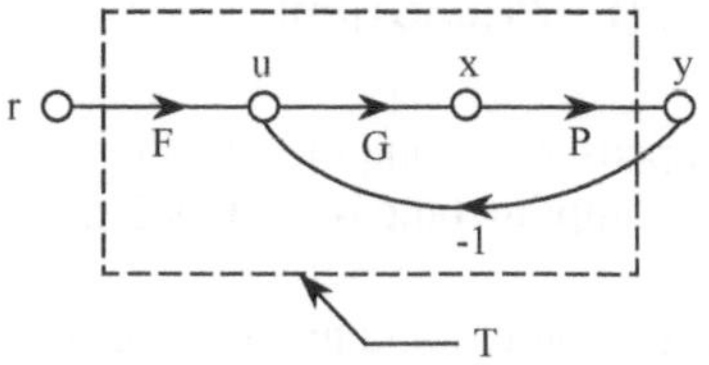

$$\mathbf{P} = \begin{bmatrix} \dfrac{1}{s+1} & \dfrac{0.2}{(s+1)(s+2)} \\ \dfrac{0.5}{s+1} & \dfrac{0.5s}{(s+1)(s+2)} \end{bmatrix} \qquad \mathbf{T} = \begin{bmatrix} \dfrac{2}{s+2} & 0 \\ 0 & \dfrac{2}{s+2} \end{bmatrix}$$

The one degree of freedom is shown in the figure where $\boldsymbol{I}$ is the identity matrix and $\boldsymbol{I} = \boldsymbol{F}$.

(a) Determine $\boldsymbol{P}^{-1}$.
(b) Determine $\boldsymbol{G}$ that results in achieving the desired $\boldsymbol{T}$.
(c) Suppose the actual compensator transfer function is:

$$\mathbf{G} = \begin{bmatrix} \dfrac{(0.5+\varepsilon_1)s}{s+2} & g_{12} \\ \dfrac{-(0.5+\varepsilon_2)}{s-0.2} & g_{22} \end{bmatrix} \quad \text{where } \varepsilon_1 > 0,\ \varepsilon_2 > 0$$

Determine $t_{11}(s)$ and $t_{21}(s)$. Do these control ratios yield stable responses? Hint: Analyze $\boldsymbol{T} = [\boldsymbol{I} + \boldsymbol{P}\boldsymbol{G}]^{-1}\boldsymbol{P}\boldsymbol{G}$, especially $\boldsymbol{PG}$.

(d) For this part assume $t_{11}(s) = t_{22}(s) = (s - 0.2)/[(s + 2)(\tau s + 1)]$ and $t_{12}(s) = t_{21}(s) = 0$ where $\tau > 0$. Determine $\boldsymbol{G}(s)$ and compare it with $\boldsymbol{G}$ of part (b).

5.2. In *2x2* MIMO system, the system poles due to feedback, are the zeros of $\Delta = \det[\boldsymbol{I} + \boldsymbol{PG}]$. Assume that $\boldsymbol{G}$ is a diagonal matrix. Prove that the zeros of interest are those of

$$\Delta' = = (1 + g_1 q_{11})(1 + g_2 q_{22}) - \frac{q_{11} q_{22}}{q_{12} q_{21}}$$

Hint: show that $(\Delta/p_{11}) = (\Delta'/q_{11})$.

5.3. Let Fig. 5.1 represent a *2x2* MIMO plant where the plant is described by:

$$\ddot{y}_1 + A\,\dot{y}_1 + By_1 + C\,y_2 = E_1\,\dot{u}_1 + E_2\,u_1 + E_3\,u_2$$

$$Jy_2 + K\,\dot{y}_1 + Hy_1 = Qu_2$$

(a) Determine ***P***.

(b) Determine $\boldsymbol{P}^{-1}$ directly from ***P***. Suppose ***P*** has the form

$$\mathbf{P} = \frac{\begin{vmatrix} p_{11} & p_{12} \\ p_{21} & p_{22} \end{vmatrix}}{d} \quad \text{thus} \quad \mathbf{P}^{-1} = \frac{\begin{vmatrix} p_{22} & -p_{12} \\ -p_{21} & p_{11} \end{vmatrix} d}{\Delta\Delta}$$

where $\Delta = p_{11}p_{22} - p_{12}p_{21}$. By dividing d into Δ you should obtain the quotient $E_1Qs + E_2Q$ and no remainder. This reveals that for most practical MIMO plants of the above form if $\boldsymbol{P}^{-1}$ is derived from ***P*** then the common poles of the elements of ***P*** should not appear as common zeros of the elements of $\boldsymbol{P}^{-1}$.

(c) Utilizing Eq. (5.6) prove that the remainder is zero. Hint: $\boldsymbol{P} = \boldsymbol{D}^{-1}\boldsymbol{N}$ and $\boldsymbol{U} = \boldsymbol{P}^{-1}\boldsymbol{Y}$.

5.4. A MISO control system involving the uncertain plant $\mathcal{P}$ for which the compensator G_1 is to be designed to satisfy the desired performance specifications is shown in Fig. A. It is suggested that before the compensator is designed, that the system structure be modified by inserting a feedback unit around the plant as shown in Fig. B. It has been stated that by so doing will enhance the degree of system robustness to plant parameter variations.

Given :

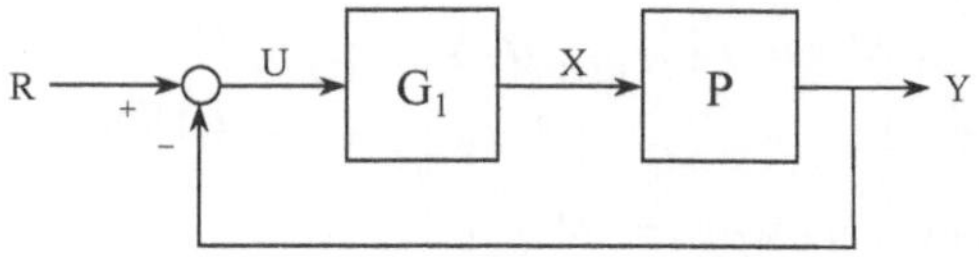

A

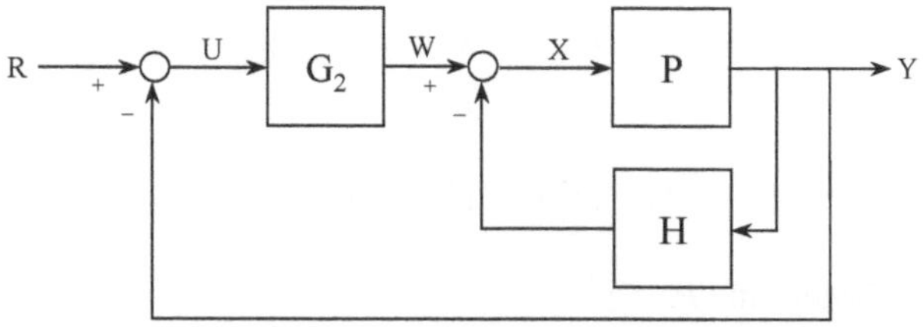

B

$$P(s) = \frac{1}{s}, \quad G_1(s) = \frac{1}{s+2}, \quad \text{and} \quad H(s) = \frac{s}{s+1}$$

Find $G_2(s)$ so $T(s) = Y(s)/R(s)$ is the same for both system structures.

5.5. Given

$$\boldsymbol{P}(s) = \begin{bmatrix} \dfrac{9}{s+1} & \dfrac{-10}{s+1} \\ \dfrac{-8}{s+2} & \dfrac{9}{s+2} \end{bmatrix}$$

The following weighting function $\boldsymbol{W}$ is selected to achieve an effective decoupled plant $\boldsymbol{P}_e = \boldsymbol{PW}$:

$$\boldsymbol{W} = \begin{bmatrix} 9 & 10 \\ 8 & 9 \end{bmatrix}$$

With no parameter uncertainty in $\boldsymbol{P}$ and/or $\boldsymbol{W}$ the det $[\boldsymbol{I} + \boldsymbol{PW}]$has excellent stability margins. Show that with uncertainty in the parameters in $\boldsymbol{W}$ may lead to an unstable system. Hint: let

$$W = \begin{bmatrix} 9+a & 10+b \\ 8+c & 9+d \end{bmatrix}$$

and suppose a = $d = -0.1$ and $b = c = 0.1$.

CHAPTER 6

NOTE: For all MIMO problems for Chapters 6 through 8 only diagonal **G**(s) and **F**(s) matrices are to be used.

6.1. For Prob. 5.4 find the system sensitivity function S_P^T for each structure. If these sensitivity functions are different, which structure is least sensitive to plant parameter variations and why?

6.2. In a $2x2$ MIMO system $r_1 \neq 0$ and $r_2 = 0$, thus only t_{11} ant t_{21} are involved in the design. The specifications on t_{11} are:

ω	0.1	0.2	0.5	1	2	5	10
b_{11}	1.02	1.07	1.09	0.97	0.65	0.25	0.05
a_{11}	0.95	0.92	0.80	0.55	0.20	0	0

The specification on t_{21} is that $b_{21} = 0.01$ for all values of ω and the specification on each L_i satisfies

$$Lm\left|\frac{L_i}{1+L_i}\right| \leq 2.3 \text{ dB}$$

The plant is described as follows:

$$q_{11} = k_{11},\ q_{12} = k_{12},\ q_{21} = k_{21},\ q_{22} = k_{22}$$

$$0.2 \leq k_{11} \leq 1,\ 0.5 \leq k_{12} \leq 2,\ 2 \leq k_{21} \leq 5,\ 0.1 \leq k_{22} \leq 0.5$$

Notes:

(1) The q's are assumed as real numbers (gain uncertainty only) thus $\gamma_{max} = [(1)(0.5)]/[(0.5)(2)] = 0.5$.
(2) It is intended that this problem be done without the use of a CAD package. See Sec. 6-5.3

Choose the following nominal values:

$$q_{11_o} = 0.2 \qquad q_{22_o} = 0.1$$

(a) Find the bounds on L_{1_o} at $\omega = 0.1$, 0.2, 0.5, 1 and 2.
(b) Use

$$\left|\frac{1}{1+\mathbf{L}_2}\right| \le \left|\left(\frac{b_{21}}{b_{11}}\right)\left(\frac{q_{21}}{q_{22}}\right)\right|$$

to find the bounds on L_{2_o} for the same values of ω as in part (a). Hint: Use the "inverse NC" for this purpose, i.e, let $\ell_2 = 1/\mathbf{L}_2$ to obtain the form $\ell_2/(1 + \ell_2)$.

6.3. Given a 2x2 plant whose **Q** matrix is:

$$Q = \begin{bmatrix} \dfrac{10a}{cs+b} & \dfrac{4b}{cs+b} \\ \dfrac{-3c}{bs+c} & \dfrac{3ac}{0.25s+c} \end{bmatrix}$$

where:

$$a \in [0.8, 1.2],\ b \in [8, 10],\ c \in [5, 6]$$

$$|q_{11}q_{22}| \le |q_{12}q_{21}| \text{ as } s \to \infty,\ r_1(t) = 0$$

The specifications based upon input $r_1(t) = u_{-1}(t)$ are:

$$|a_{11}(s)| \le |t_{11}^R(s)| \le |b_{11}(s)|$$

with:

$$a_{11}(s) = \frac{16}{(s^2 + 7.6s + 16)(s/10 + 1)}$$

$$b_{11}(s)=\frac{16\ (s/5+1)}{(s^2+5.2s+16)}$$

The specifications based upon input $r_2(t) = u_{-1}(t)$ are:

$$|a_{22}(s)| \le |t_{22}^R(s)| \le |b_{22}(s)|$$

with:

$$a_{22}(s)=\frac{100}{(s^2+16s+100)(s/20+1)}$$

$$b_{22}(s)=\frac{100\ (s/25+1)}{(s^2+14s+100)}$$

Additional specifications on L_1 and L_2 are, $Lm\left|\frac{L_i}{1+L_i}\right| \le 5\,\text{dB}$, for $i = 1,2$

(a) Find $L_{1_o}(s)$, $L_{2_o}(s)$ and $f_{22}(s)$ using *Method* 1 when $f_{12}(s) = 0$.

(b) Design $g_1(s)$ and $g_2(s)$.

6.4. Repeat Prob.6.3 by redesigning $L_2(s)$ where $b \in [3, 5]$ and $L_1(s)$ of Prob.6.3 is not affected by the change in k_{22}.

6.5. Given the *2x2* plant

$$P=\frac{1}{s}\begin{bmatrix} k_{11} & k_{12} \\ k_{21} & k_{22} \end{bmatrix}$$

with the uncertainties:

$$1 < k_{11} < 2 \qquad 0.5 < k_{12} < 1 \qquad 0.25 < k_{21} < 0.5 \qquad 2 < k_{22} < 4$$

and with one command input $r_2(t) = u_{-1}(t)$. Design $L_{1_o}(s)$, $L_{2_o}(s)$, and $f_{22}(s)$, by use of *Method 1*, to satisfy the following specifications:

$$b_{22} = T_{R_U}(s) = \frac{1.4432(s + 7.78)}{s + 1.95 \pm j2.725}$$

$$a_{22} = T_{R_L}(s) = \frac{21.50525}{(s + 2.036)(s + 1.8 \pm j2.70601)}$$

$$\text{Lm}\,\text{M}_{\text{L}_1} \le \text{Lm}\left[\frac{\mathbf{L}_1}{1+\mathbf{L}_1}\right] \le 3\text{ dB} \qquad \text{Lm}\,\text{M}_{\text{L}_2} \le \text{Lm}\left[\frac{\mathbf{L}_2}{1+\mathbf{L}_2}\right] \le 3\text{ dB}$$

and where $|\boldsymbol{t}_{12}| < -20\ dB$ for all ω. Use the MIMO QFT CAD package of Appendix C to do the design and obtain the Nichol plots for both loops, the open-loop and close-loop Bode plots for this *2x2* system, and simulate your design in order to obtain the time domain performance results. Note that t_{21} need not be specified since $r_1(t) = 0$. Note the phase margin frequency ω_ϕ for each loop.

6.6. Repeat Prob. 6.5 where

$$\boldsymbol{P} = \begin{bmatrix} \dfrac{k_{11}}{s + 0.01} & \dfrac{k_{12}}{s} \\ \dfrac{k_{21}}{s} & \dfrac{k_{22}}{s + 0.01} \end{bmatrix}$$

6.7. Repeat Prob. 6.5 where

$$0.5 \le k_{11} \le 1, \quad 0.5 \le k_{12} \le 1, \quad 0.1 \le k_{21} \le 0.2, \quad \text{and} \quad 1 \le k_{22} \le 2$$

6.8 Given the *2x2* MIMO control system of Fig. 5.5 whose $\iota = 1, 2, 3$, $J = 4$ plants are:

$$P_1(s) = \begin{bmatrix} \dfrac{1}{s+1} & \dfrac{0.095}{s+2} \\ \dfrac{0.11}{s+2} & \dfrac{1.5}{s+1.5} \end{bmatrix} \qquad P_2(s) = \begin{bmatrix} \dfrac{1.6}{s+1.4} & \dfrac{0.09}{s+1.9} \\ \dfrac{0.11}{s+2.1} & \dfrac{2.1}{s+1.8} \end{bmatrix}$$

$$P_3(s) = \begin{bmatrix} \dfrac{1.4}{s+1.8} & \dfrac{0.085}{s+1.8} \\ \dfrac{0.12}{s+2.2} & \dfrac{1.9}{s+2.2} \end{bmatrix} \qquad P_4(s) = \begin{bmatrix} \dfrac{1.2}{s+2.2} & \dfrac{0.08}{s+1.7} \\ \dfrac{0.13}{s+2.3} & \dfrac{1.7}{s+2.6} \end{bmatrix}$$

where

$$b_{11} = \frac{11.70}{s + 1.95 \pm j2.81} \qquad a_{11} = \frac{39.76}{(s + 3.398)(s + 2.7 \pm j2.1)}$$

$$b_{22} = \frac{3.082s + 21.574}{s^2 + 4.88s + 21.574} \qquad a_{22} = \frac{691.5}{s^3 + 57.75s^2 + 387.5s + 691.5}$$

$b_{ij} = 0.1$ where $i \neq j$, $\omega_{\phi_1} \leq 10\ rad/sec$, and $\omega_{\phi_2} \leq 8 rad/sec$.

(a) Verify diagonal dominance for this plant. (b) Using *Method 1* synthesize the compensators and pre-filters, assuming diagonal $\boldsymbol{G}(s)$ and $\boldsymbol{F}(s)$ matrices, where the command inputs are: $r_1(t) = r_2(t) = u_{-1}(t)$. Note: the stability specicification for each loop is to be based upon the peak overshoot value of b_{11} and b_{22}, respectively. (c) Obtain the $\boldsymbol{y}(t)$ responses. Note the BW frequency ω_ϕ for each loop.

6.9. In order illustrate the conservatism of the QFT design technique repeat Prob. 6. 8 by ignoring the specification b_{ij}; that is, the optimal bounds will be based only upon the stability and tracking specifications. Compare the results with those of Prob. 6.8.

6.10. Repeat Prob. 10.4, designing a diagonal compensator $\boldsymbol{G}$(s) according to the MIMO QFT *Method 1*.

6.11. Repeat Prob. 10.5, designing a diagonal compensator $\boldsymbol{G}$(s) according to the MIMO QFT *Method 1*.

6.12. Repeat Prob. 10.7, designing a diagonal compensator $\boldsymbol{G}$(s) according to the MIMO QFT *Method 1*.

CHAPTER 7

7.1. Repeat Prob. 6.3 by *Method 2* and compare the compensators and the pre-filters for each loop.

7.2. Repeat Prob. 6.4 by *Method 2* and compare the compensators and the pre-filters.

7.3. Repeat Prob. 6.5 by *Method 2* and compare the results.

7.4. Repeat Prob. 6.6 using *Method 2* and compare the compensators and the pre-filters.

7.5. Repeat Prob. 7.2 for Prob. 6.8 and using L_{1o} of Prob. 6.8. Compare the results with those of Prob. 7.2.

7.6. Repeat Prob. 6.3 to redesign L_{2_o} by use of *Method 2* and using L_{1_o} of Prob. 6.3 where $0.5 \leq k_{22} \leq 1.2$. Compare the results with those of Prob. 6.3.

7.7. Repeat Prob. 6.8 using Method 2 and compare the results with those of Prob. 6.8.

7.8. In order to illustrate the conservatism of the QFT design technique repeat Prob. 7.7 by ignoring the specification b_{ij}; that is, the optimal bounds are to be based only upon the stability and tracking specifications. Compare the results with those of Prob. 7.7.

7.9. In order to illustrate the conservatism of the QFT design technique repeat Prob. 7.8 by ignoring the specification b_{ij}; that is, the optimal bounds are to be based only upon the stability and tracking specifications. Compare the results with those of Prob. 7.8.

7.10. Repeat Prob. 10.4, designing a diagonal compensator $\boldsymbol{G}$(s) according to the MIMO QFT *Method* 2.

7.11. Repeat Prob. 10.5, designing a diagonal compensator $\boldsymbol{G}(s)$ according to the MIMO QFT *Method* 2.

7.12. Repeat Prob. 10.7, designing a diagonal compensator $\boldsymbol{G}(s)$ according to the MIMO QFT *Method* 2.

CHAPTER 8

External disturbance rejection problems where[$r(t) = 0$].

8.1. The $\boldsymbol{P}$ matrices of Fig. 8.1 are those given in Prob. 6.8. The $\boldsymbol{P}_D$ matrices are:

$$\boldsymbol{P}_{D_1} = \begin{bmatrix} \frac{1}{s+2} & \frac{0.095}{s+1.1} \\ \frac{0.11}{s+1.6} & \frac{2}{s+3} \end{bmatrix} \qquad \boldsymbol{P}_{D_2} = \begin{bmatrix} \frac{2}{s+4} & \frac{0.0855}{s+1.9} \\ \frac{0.121}{s+1.1} & \frac{4}{s+5} \end{bmatrix}$$

$$\boldsymbol{P}_{D_3} = \begin{bmatrix} \frac{2}{s+3} & \frac{0.081}{s+1.8} \\ \frac{0.132}{s+1.2} & \frac{3}{s+4} \end{bmatrix} \qquad \boldsymbol{P}_{D_4} = \begin{bmatrix} \frac{3}{s+5} & \frac{0.068}{s+1.7} \\ \frac{0.169}{s+1.3} & \frac{2.5}{s+6} \end{bmatrix}$$

The specifications are as follows: (a) use the M_L specifications of Prob. 6.9, (b) $b_{ij} \le 0.1$ where $i \ne j$, $(B_{d_e})_{ij} = 0.1 \ge |t_{d_{ij}}|$ [see Eq. (8.33)], $\omega_{\phi_1} \le 30$ rad/sec, and $\omega_{\phi_2} \le 40$ rad/sec. Use Method 1 to synthesize $\boldsymbol{G}(s)$. For $\boldsymbol{d}_{ext}(t) = u_{-1}(t)$ simulate your system to obtain the $\boldsymbol{y}(t)$ responses.

8.2. Repeat Prob.8.1 using *Method 2* and compare the results with those of Prob. 8.1.

Note: For the Tracker/External disturbance rejection problems unit step tracking and external disturbance forcing functions are to be utilized.

8.3. Use the tracking specificationss of Prob. 6.8 and the external. disturburbance specifications of Prob. 8.1. Use *Method 1*.

8.4. Repeat Prob. 8.3 using *Method 2* and compare the phase margin frequencies with those of Prob. 8.3.

8.5. Consider a solar thermal energy application for domestic hot water supplies as is showed in Fig. I.1.[178] The basic elements are the collector, the heat exchanger, the storage tank and two closed circuits with antifreeze solutions. The manipulated variables are q_c and q_t, which are the collector and the storage pump volumetric flux rates. The variables to control are T_t, and T_0 which represent the temperatures at the storage tank and at the exit of the collector, respectively.

A detailed analysis of a solar process system is a complicated problem. Nevertheless a simplified analysis yields very useful results when focusing on control purposes. The linearized MIMO transfer function matrix that represents the controlled variables (T_t and T_o) in terms of the manipulated ones (Q_c and Q_t), around the operating point $Q_t^o, Q_c^o, T_t^o, T_o^o$, is,

$$\begin{bmatrix} \Delta T_t(s) \\ \Delta T_o(s) \end{bmatrix} = \begin{bmatrix} p_{11}(s) & p_{12}(s) \\ p_{21}(s) & p_{22}(s) \end{bmatrix} \begin{bmatrix} \Delta Q_t(s) \\ \Delta Q_c(s) \end{bmatrix}$$

where,

$$p_{11}(s) = \frac{\Delta T_t(s)}{\Delta Q_t(s)} = \frac{k_{11}}{s+\tau} \quad ; \quad p_{12}(s) = \frac{\Delta T_t(s)}{\Delta Q_c(s)} = \frac{k_{12}}{s+\tau}$$

$$p_{21}(s) = \frac{\Delta T_o(s)}{\Delta Q_t(s)} = \frac{k_{21}}{s+\tau} \quad ; \quad p_{22}(s) = \frac{\Delta T_o(s)}{\Delta Q_c(s)} = \frac{k_{22}(s+\gamma_{22})}{s+\tau}$$

and,

$$\tau = \frac{MA + ZC}{Aa} \; ; \; k_{11} = \frac{Y}{A} \; ; \; k_{12} = \frac{-ZB}{Aa} \; ;$$

$$k_{21} = \frac{-YC}{Aa} \; ; \; k_{22} = \frac{-B}{A}$$

with,

$$A = b\varepsilon Q_c^o + fU_c(1-\varepsilon) \; ; \; B = b\varepsilon\,(T_o^o - T_t^o) \; ; C = fU_c\varepsilon - b\varepsilon Q_c^o$$

$$Y = \varepsilon d(T_t^o - T_o^o) \; ; \; Z = -fU_c(1-\varepsilon) - \varepsilon dQ_c^o$$

$$M = U_t + f\,U_c\,\varepsilon - \varepsilon dQ_t^o \; ; \; a = \rho_t c_{pt} V_t$$

$$b = \rho_c c_{pc} \; ; \; d = \rho_t c_{pt} \; ; \; f = A_c F_r$$

The equilibrium point is: $Q_t^o = 0.864\ 10^{-3}$ m^3 s^{-1}; $Q_c^o = 0.00115$ m^3 s^{-1}; $T_t^o = 35$ °C; $T_o^o = 53$°C. The parameter variations are: $\varepsilon \in [0.41, 0.63]$; $U_c \in [6, 7]$ w °C^{-1} m^{-2}; $A_c \in [5, 10]$ m^2.

Variation in solar radiation, in feed temperature, pressure or in product demand can move the process outputs (temperatures T_t and T_o) from the desired operating values. In Appendix I, is designed a non-diagonal MIMO compensator to reject external disturbances at the plant output. For this problem design a diagonal MIMO compensator to reject external disturbances at the plant input. The closed-loop performance specifications are:

- Robust stability in each channel: $\left|\frac{L_i(s)}{1+L_i(s)}\right| \le 1.3 \quad i = 1, 2$,

  where $L_i(s) = p_{ii}(s)\, g_{ii}(s)$.

- Robust disturbance rejection at plant input so that:

$$\left|\frac{y_i(s)}{di_i(s)}\right| \le 0.13, \quad \omega < 0.05\ rad/s, \quad i = 1, 2$$

8.6. Repeat Prob. 10.6, designing a diagonal compensator $\boldsymbol{G}(s)$ according to the MIMO QFT *Method 2*.

8.7. Repeat Prob. 10.8, designing a diagonal compensator ***G***(s) according to the MIMO QFT *Method 2.*

8.8. Repeat Prob. 10.9, designing a diagonal compensator ***G***(s) according to the MIMO QFT *Method 2.*

CHAPTER 10

10.1. Consider a 3x3 MIMO system whose transfer function matrix is:

$$\begin{bmatrix} y_1(s) \\ y_2(s) \\ y_3(s) \end{bmatrix} = \begin{bmatrix} \dfrac{0.1e^{-0.4s}}{0.92s+1} & \dfrac{2(3s+1)}{4s+1} & \dfrac{-1}{2s+1} \\ \dfrac{1e^{-0.1s}}{7s+1} & \dfrac{1}{3s+1} & \dfrac{-0.1e^{-0.2s}}{0.87s+1} \\ \dfrac{-2(s+1)}{0.92s+1} & \dfrac{-3e^{-0.4s}}{0.54s+1} & \dfrac{1e^{-0.3s}}{6s+1} \end{bmatrix} \begin{bmatrix} m_1(s) \\ m_2(s) \\ m_3(s) \end{bmatrix}$$

- Using the original Relative Gain Analysis introduced by Bristol,[92] find the best input/output pairing.
- Analyse the existing coupling between loops.
- Suppose that the three control loops are designed and working. In that case study the effect of an instantaneous opening of the loop $y_2(s)$.
- Design a non-diagonal MIMO QFT compensator so that the system reaches a good level of stability, disturbance rejection and de-coupling (no uncertainty).

10.2. Consider two gaseous elements A and B. In the figure below the tank 1 has a 80% of A (X_1 = 80%) and a 20% of B whereas tank 2 has a 20% of A (X_2 = 20%) and a 80% of B. The hydraulic system represented in the figure mixes two fluxes, F_1 and F_2, which come from the tanks 1 and 2 respectively. The product has a composition of X% of A and a flux F, so that,

$$F = F_1 + F_2$$

$$F\,X = F_1\,X_1 + F_2\,X_2$$

The inputs are the fluxes F_1 and F_2, and the controlled outputs are the flux F and the concentration X of element A.

In order to control around $X_{ref} = 60\%$ and $F_{ref} = 200$ mol/h:

- Find the best input/output pairing by using the original Relative Gain Analysis introduced by Bristol.[92]
- Analyze the existing coupling between loops.
- Find the coupling worst case according to the set point X_{ref}.

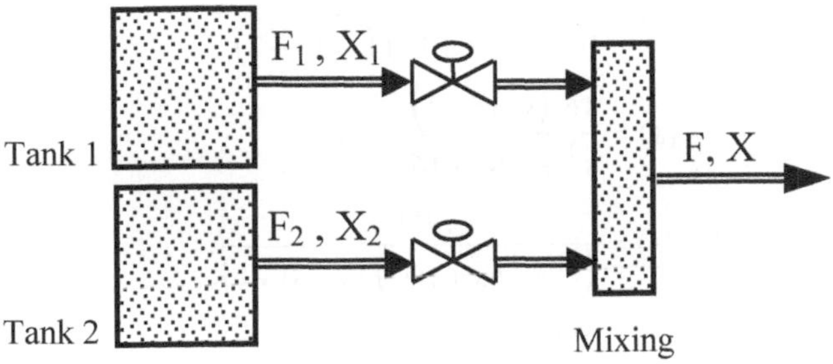

10.3. Repeat Prob. 8.5 by using a non-diagonal MIMO QFT compensator to fulfill the previous performance specifications and also to reduce the coupling effect as much as possible.

10.4. Consider a 2x2 linear multivariable system with uncertainty for the following transfer function matrix:

$$\boldsymbol{P}(s) = \begin{bmatrix} p_{11}(s) & p_{12}(s) \\ p_{21}(s) & p_{22}(s) \end{bmatrix} = \begin{bmatrix} \dfrac{10\alpha\beta}{2s+\beta} & \dfrac{-4\beta}{s+\beta} \\ \dfrac{2\alpha\gamma}{s+\gamma} & \dfrac{3\alpha\gamma}{0.5s+\gamma} \end{bmatrix}$$

$\alpha \in [0.8, 1.2]$; $\beta \in [8, 10]$; $\gamma \in [5, 6]$

Calculate the RGA matrix for the whole set of parameter uncertainty and for every ω, and design a non-diagonal MIMO compensator to fulfill the required specifications, so that:

i.- *Robust stability*:

$$\left| \frac{\left(p_{ii}^{*e}\right)^{-1} g_{ii}}{1+\left(p_{ii}^{*e}\right)^{-1} g_{ii}} \right| \leq 1.4, \quad i = 1,\ 2$$

ii.- *Robust reference tracking:*

$$\left|B_{ii}^{L}(j\omega)\right| \leq \left|t_{ii}^{Y/R}(j\omega)\right| \leq \left|B_{ii}^{U}(j\omega)\right|,\ i = 1,2$$

where,

$$B_{11}^{L}(s) = \frac{16}{\left(s^2 + 7.6s + 16\right)\left(s/10 + 1\right)}$$

$$B_{11}^{U}(s) = \frac{16\left(s/15 + 1\right)}{s^2 + 3.6s + 16}$$

$$B_{22}^{L}(s) = \frac{100}{\left(s^2 + 16s + 100\right)\left(s/20 + 1\right)}$$

$$B_{22}^{U}(s) = \frac{100\left(s/25 + 1\right)}{s^2 + 14s + 100}$$

iii.- *Minimizing the coupling effects*:

Reduce the interaction, $t_{12}^{Y/R}$ and $t_{21}^{Y/R}$ as much as possible.

10.5. The mathematical model of a 3x3 industrial furnace (see figure) was derived from modeling techniques that apply for both first principles equations and input-output operating records collected during furnace operation.[102] For the compensator design, the overall dynamics of the model are captured by a symmetric, tri-diagonal, transfer function matrix, so that,

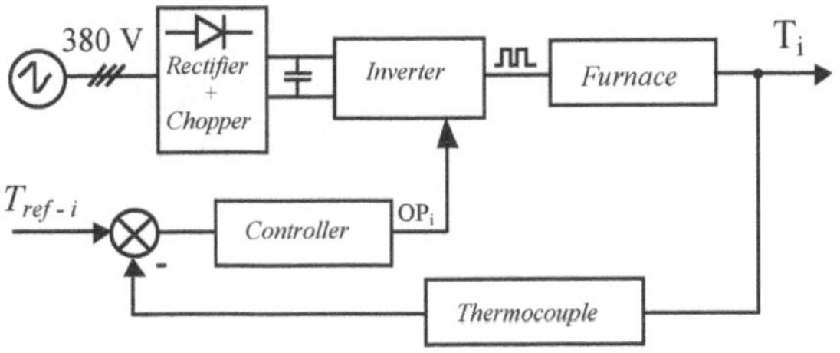

$$\begin{bmatrix} T_1(t) \\ T_2(t) \\ T_3(t) \end{bmatrix} = \begin{bmatrix} p_{11}(t) & p_{12}(t) & 0 \\ p_{21}(t) & p_{22}(t) & p_{23}(t) \\ 0 & p_{32}(t) & p_{33}(t) \end{bmatrix} \begin{bmatrix} OP_1(t) \\ OP_2(t) \\ OP_3(t) \end{bmatrix}$$

where $p_{ij}(t)$ are the elements of the matrix of the process $\boldsymbol{P}(t)$, which relates system outputs, temperatures $T_i(t)$ measured in the three zones, with the inputs, power signals $OP_i(t)$ supplied to the resistances (see figure). The furnace is a multivariable system, stable and minimum phase. It presents a high closed-loop level of interaction and parametric uncertainty due to the variety of pieces to be cured.

The multivariable QFT theory, as applied in this book, deals with the inverse plant of the process. Thus, in the interest of further calculations, it is advisable to work with the corresponding inverse matrix,

$$\begin{bmatrix} OP_1(\mathrm{t}) \\ OP_2(\mathrm{t}) \\ OP_3(\mathrm{t}) \end{bmatrix} = \begin{bmatrix} p^*_{11}(\mathrm{t}) & p^*_{12}(\mathrm{t}) & 0 \\ p^*_{12}(\mathrm{t}) & p^*_{22}(\mathrm{t}) & p^*_{23}(\mathrm{t}) \\ 0 & p^*_{23}(\mathrm{t}) & p^*_{33}(\mathrm{t}) \end{bmatrix} \begin{bmatrix} T_1(\mathrm{t}) \\ T_2(\mathrm{t}) \\ T_3(\mathrm{t}) \end{bmatrix}$$

where $p^*_{\mathrm{ij}}(\mathrm{t})$ are the elements of the inverse matrix of the process $\mathbf{P}^{-1}(\mathrm{s}) = \mathbf{P}^*\mathrm{s})$. Using the classical differential equations of heat transmission and thermodynamics and its electrical equivalency, the elements of the $\mathbf{P}^{-1}(\mathrm{t})$ matrix are,

$$p^*_{11}(s) = \frac{d_{11}(s)n_{12}(s) + d_{12}(s)n_{11}(s)}{n_{11}(s)n_{12}(s)}$$

$$p^*_{12}(s) = p^*_{21}(s) = \frac{-d_{12}(s)}{n_{12}(s)}$$

$$p^*_{22}(s) = \frac{d_{22}(s)n_{12}(s)n_{23}(s) + d_{12}(s)n_{22}(s)n_{23}(s) + d_{23}(s)n_{22}(s)n_{12}(s)}{n_{22}(s)n_{12}(s)n_{23}(s)}$$

$$p^*_{23}(s) = p^*_{23}(s) = \frac{-d_{23}(s)}{n_{23}(s)}$$

$$p^*_{33}(s) = \frac{d_{33}(s)n_{23}(s) + d_{23}(s)n_{33}(s)}{n_{33}(s)n_{23}(s)}$$

where,

$$d_{\mathrm{ij}}(s) = s\left(b_{\mathrm{ij}}s + c_{\mathrm{ij}}\right), \quad \mathrm{i,j} = 1,2,3$$
$$n_{\mathrm{ij}}(s) = \left(a_{\mathrm{ij}}s + 1\right), \quad \mathrm{i,j} = 1,2,3$$

b_{ij}, c_{ij} and a_{ij} are the design parameters of the furnace, whose values change with each piece to be manufactured. They depend on the load material, size and distribution and also on the initial conditions of the experiments carried out. So, the parameter variations are:

$a_{11} \in [1, 2\ 10^4]$; $a_{12} \in [1, 1000]$; $a_{22} \in [1, 15000]$;
$a_{23} \in [1, 500]$; $a_{33} \in [1, 2\ 10^4]$.
$b_{11} \in [1, 5\ 10^6]$; $b_{12} \in [1, 9\ 10^6]$; $b_{22} \in [1, 7\ 10^6]$;
$b_{23} \in [1, 2\ 10^6]$; $b_{33} \in [1, 5\ 10^6]$.
$c_{11} \in [4000, 10^4]$; $c_{12} \in [7000, 2\ 10^4]$; $c_{22} \in [5000, 7000]$;

$c_{23} \in [1000, 5000]$; $c_{33} \in [2000, 10^4]$.

The array of frequencies of interest for the system is:

$$\omega = [5\ 10^{-6}, 10^{-5}, 2\ 10^{-5}, 5\ 10^{-5}, 10^{-4}, 5\ 10^{-4}, 10^{-3}, 10^{-2}]\ \text{rad/sec}$$

The overall objective is to control the furnace temperature with the power supplied by the heat sources, following as accurately as possible and simultaneously in every zone a pre-determined curing cycle which varies depending on the piece to be manufactured. The multivariable phenomena of the presence of model uncertainties and the exothermal processes generated by composite materials to be manufactured add a significant complexity to the problem.

Due to the characteristics of the system, only four off-diagonal terms and three diagonal elements need to be designed.

$$\mathbf{G}(s) = \begin{bmatrix} g_{11}(s) & g_{12}(s) & 0 \\ g_{21}(s) & g_{22}(s) & g_{23}(s) \\ 0 & g_{32}(s) & g_{33}(s) \end{bmatrix}$$

The compensator $\boldsymbol{G}$(s) and the pre-filter $\boldsymbol{F}$(s) are designed to satisfy the next performance specifications:

i.- Robust stability

$|\, t_{ii}(j\omega) \,| \leq 1.2$ for i = 1, 2, 3, $\forall\omega$, where the terms $t_{ii}(j\omega)$ are the diagonal elements of the matrix $T_{y/r}$. This condition implies at least 50° lower phase margin and at least 1.833 (5.26 dB) lower gain margin.

ii.- Robust reference tracking

The desired system output, i.e. the temperature $y(t) = \boldsymbol{T}(t)$, is required to lie between the specified upper and lower bounds $B(t)_U$ and $B(t)_L$ respectively. The corresponding performance specifications in the frequency domain are $B_U(\omega)$ and $B_L(\omega)$ are:

$$B_{L\,ii}(\omega) \leq \left| t_{ii}^{\ y/r}(j\omega) \right| \leq B_{U\,ii}(\omega), \qquad \omega < 10^{-2}\ \text{rad/s}, \ \text{for i} = 1, 2, 3$$

$$B_{L_{ii}}(\omega)=\left|\frac{1.25\times10^{-7}}{0.2\,(j\omega)^3+(j\omega)^2+0.0006\,(j\omega)+1.25\times10^{-7}}\right|$$

for $i=1,2,3$

$$B_{U_{ii}}(\omega)=\left|\frac{10^{-6}}{(j\omega)^2+0.018\,(j\omega)+10^{-6}}\right| \quad \text{for } i=1,2,3$$

iii.- Minimizing the coupling effects c_{ij} as much as possible and for $\omega < 10^{-2}$ rad/sec.

10.6. Repeat Prob. 10.5, designing a non-diagonal compensator $\boldsymbol{G}(s)$ compensator to satisfy the following performance specifications:

i.- Robust stability

$|t_{ii}(j\omega)| \leq 1.2$ for i = 1, 2, 3, $\forall\omega$, where the terms $t_{ii}(j\omega)$ are the diagonal elements of the matrix $T_{y/r}$. This condition implies at least 50° lower phase margin and at least 1.833 (5.26 dB) lower gain margin.

ii.- Rejection of disturbances at plant output

As the disturbance at the plant (furnace) output (due to the changes in external temperature, presence of couplings or heat losses, heat dissipation, etc.) the control system has to reject the effect of that disturbance in the inner temperature. The following equation defines the required disturbance rejection specification:

$$\left|t_{ii}^{\,y/do}(j\omega)\right| =$$

$$\left|\frac{0.5}{(j\omega)^2+0.3(j\omega)+0.6}\right|, \quad \omega<10^{-2}\ \text{rad/sec}, \quad \text{for } i=1,2,3$$

where $t_{ii}^{\,y/do}$ is the transfer function between the inner temperature $T_i(t)$ and the disturbance $d_o(t)$.

iii.- Minimizing the coupling effects c_{ij} as much as possible for $\omega < 10^{-2}$ rad/sec.

10.7. Consider the popular highly interacting 2x2 distillation column, described by Skogestad and Postlethwaite as a benchmark problem.[130] In the last decade several researchers have use that model to validate new MIMO compensator designs. The system presents the next transfer function matrix is:

$$\begin{bmatrix} y_1(s) \\ y_2(s) \end{bmatrix} = \frac{1}{1+75s} \begin{bmatrix} 0.878 & -0.864 \\ 1.082 & -1.096 \end{bmatrix} \begin{bmatrix} k_1 e^{-sT_1} & 0 \\ 0 & k_2 e^{-sT_2} \end{bmatrix} \begin{bmatrix} u_1(s) \\ u_2(s) \end{bmatrix}$$

where the parameters are: $k_1, k_2 \in [0.8, 1.2]$; $T_1, T_2 \in [0, 60]$ sec.

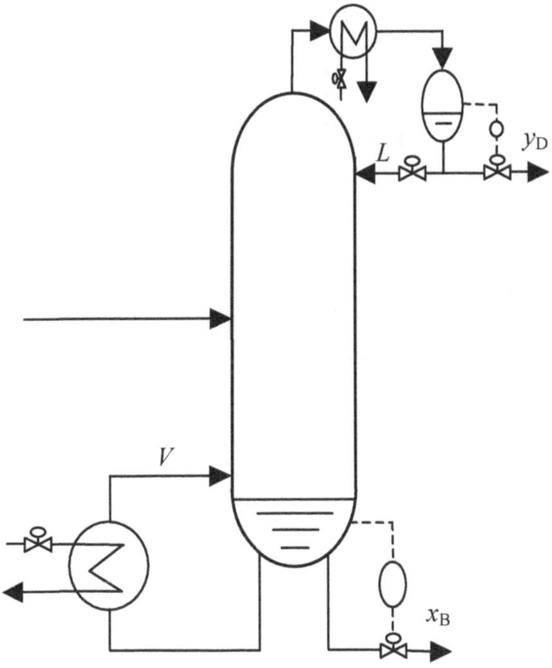

The inputs are the reflux (u_1 = L) and the boil up (u_2 = V), and the controlled outputs are the top and bottom product compositions ($y_1 = y_D$ and $y_2 = x_B$).

Calculate the RGA matrix and design a non-diagonal MIMO compensator so that the required specifications are:

- Closed-loop stability.

- For a unit step demand in channel 1 at $t = 0$ the plant output y_1 (tracking) and y_2 (interaction) should satisfy:

  - Tracking: $y_1(t) \geq 0.9$ for all $t \geq 30$ min.
    $y_1(t) \leq 1.1$ for all t.
    $0.99 \leq y_1(\infty) \leq 1.01$

  - Interaction: $y_2(t) \leq 0.5$ for all t.
    $-0.01 \leq y_2(\infty) \leq 0.01$

10.8. Consider a 2x2 linear multivariable system with uncertainty whose transfer function matrix is:

$$\boldsymbol{P}(s) = \begin{bmatrix} \dfrac{k_{11}}{s/p_{11}+1} & \dfrac{k_{12}}{s/p_{12}+1} \\ \dfrac{k_{21}}{s/p_{21}+1} & \dfrac{k_{22}}{s/p_{22}+1} \end{bmatrix}$$

where,

$$\begin{array}{ll} k_{11} \in [5,8], & p_{11} \in [100,150] \\ k_{12} \in [1,2], & p_{12} \in [5,6] \\ k_{21} \in [-2,-1], & p_{21} \in [5,6] \\ k_{22} \in [5,8], & p_{22} \in [10,15] \end{array}$$

Calculate the RGA matrix for the whole set of parameter uncertainty and for every ω, and design a non-diagonal MIMO compensator to fulfill the required specifications, so that:

i.- <u>Robust stability</u>
Stability of at least 50° lower phase margin and at least 1.8333 (5.26 dB) lower gain margin.

ii.- <u>Rejection of disturbances at plant output</u>
The error should be lower than 0.5 with disturbances of 1.0 at plant output for frequencies lower than 120 rad/s for loop 1, and error lower than 0.5 with disturbances of 1.0 at plant output for frequencies lower than 8 rad/s for loop 2.

iii.- Control effort restriction:
The control signal should be lower than 0.2 when rejecting disturbances at plant output of 0.2 considering a bandwidth of 120 rad/s for loop 1 and of 8 rad/s for loop 2.

10.9. Consider a 2x2 linear multivariable system with uncertainty whose transfer function matrix is:

$$P(s)=\begin{bmatrix} \dfrac{k_{11}}{\tau_{11}\,s+1} & \dfrac{k_{12}}{\tau_{12}\,s+1} \\ \dfrac{k_{21}}{\tau_{21}\,s+1} & \dfrac{k_{22}}{\tau_{22}\,s+1} \end{bmatrix}$$

where,

$$\begin{array}{ll} k_{11}\in[0.5,3], & \tau_{11}\in[0.5,3] \\ k_{12}\in[-2.2,-1.8], & \tau_{12}\in[8,12] \\ k_{21}\in[11,15], & \tau_{21}\in[3,8] \\ k_{22}\in[2,7], & \tau_{22}\in[5,10] \end{array}$$

Calculate the RGA matrix for the whole set of parameter uncertainty and for every ω, and design a non-diagonal MIMO compensator to fulfill the required specifications, so that:

i.- Robust stability
$|\,t_{ii}(j\omega)\,|\leq 1.2$ for i = 1, 2, $\forall\omega$, where the terms $t_{ii}(j\omega)$ are the diagonal elements of the matrix $T_{y/r}$. This condition implies at least 50° lower phase margin and at least 1.833 (5.26 dB) lower gain margin.

ii.- Rejection of disturbances at plant output

$$\left|\,t_{ii}^{\;y/do}(j\omega)\,\right| = \left|\frac{(j\omega)}{(j\omega)+10}\right|, \quad \omega<50\,\text{rad/sec}, \quad \text{for } i=1,2$$

iii.- Minimizing the coupling effects c_{ij} as much as possible and for $\omega < 10^{-2}$ rad/sec.

CHAPTER 12

12.1. Consider a central heating system of a three floor building (see the figure).[107] The transfer function that describes the inner room temperature $T_r(t)$ in terms of the desired mixed water temperature $T_{md}(t)$, that comes from the mixing valve is linearized around the equilibrium point, is:

$$\frac{T_r(s)}{T_{md}(s)} = P(s) = \frac{k}{\tau s + 1} e^{-Ls}$$

Using system identification techniques from experimental data, the parameter variations obtained are: $k \in [40, 60]$; $\tau \in [800, 1200]$ sec.; $L \in [1100, 1300]$ sec.

Design a loop compensator $G_1(s)$ and tune a Smith Predictor compensator $G_2(s)$ so that the system reaches a good level of stability and disturbance rejection for the whole set of parameter uncertainty (see the figures).

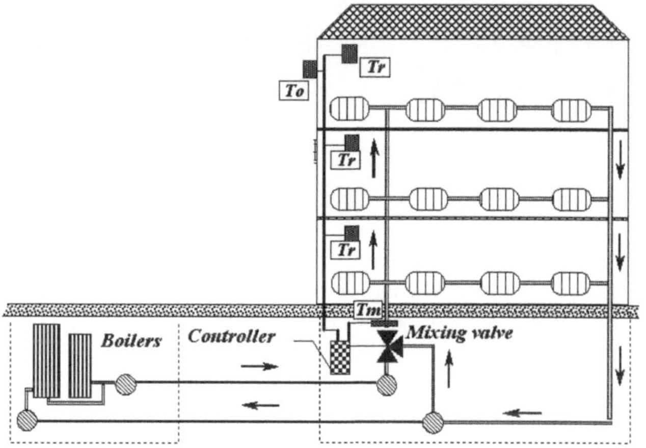

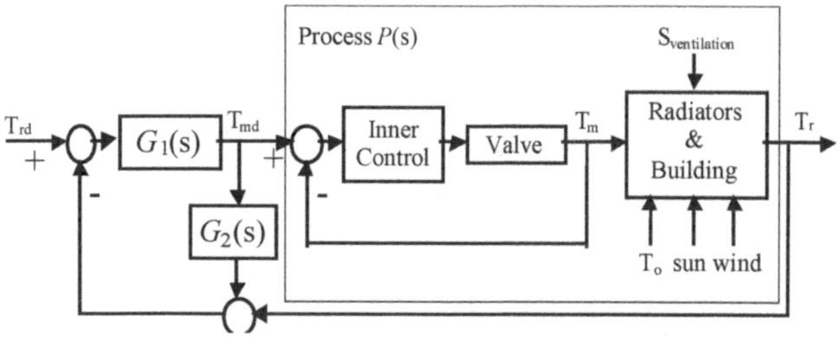

12.2. Let us consider a second order process with dead time,

$$\frac{y(s)}{u(s)} = P(s) = \frac{k}{(as+1)\,(bs+1)}\, e^{-Ls}$$

where the parameter variations are: $k \in [3, 7]$; $a \in [0.1, 0.3]$; $b \in [18, 22]$; $L \in [20, 40]$ sec.

(a) Design a loop compensator $G_1(s)$ and tune a Smith Predictor compensator $G_2(s)$ so that the system reaches a good level of stability and disturbance rejection for the whole set of parameter uncertainty (see figure).

(b) For the nominal case, compare the closed-loop response of the system to a unit step in set point with and without the Smith Predictor Compensator.

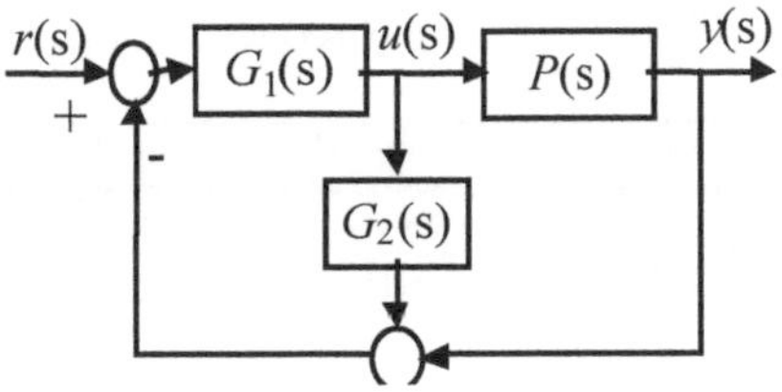

Answers to Selected Problems

CHAPTER 2

2.1. The following figure shows the plots ($Lm\ P(j\omega)$ and $\angle P(j\omega)$ vs ω) for nine different plants selected according to several sets of parameters within the uncertainty (see Table), and for the frequency range of $0.05 < \omega < 500$ rad/s.

Plant	k	a	b
P1	610	1	150
P2	610	15	150
P3	610	1	170
P4	820	1	150
P5	820	15	150
P6	820	1	170
P7	1050	1	150
P8	1050	15	150
P9	1050	1	170

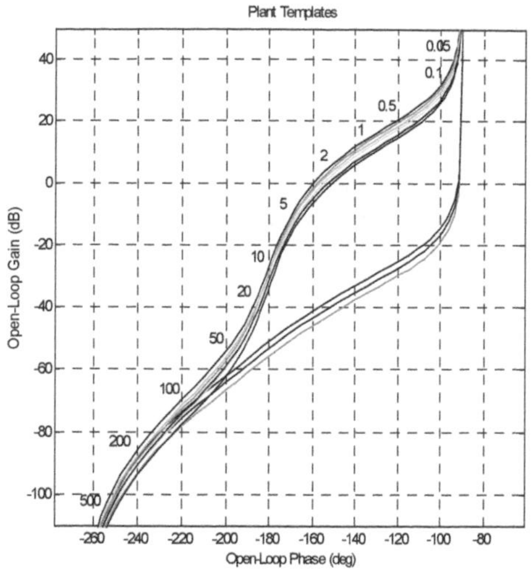

In order to introduce the meaning of the QFT templates, as it is explained in Chapter 3, the next figure superimposes the variation of the plants due to the uncertainty at every frequency.

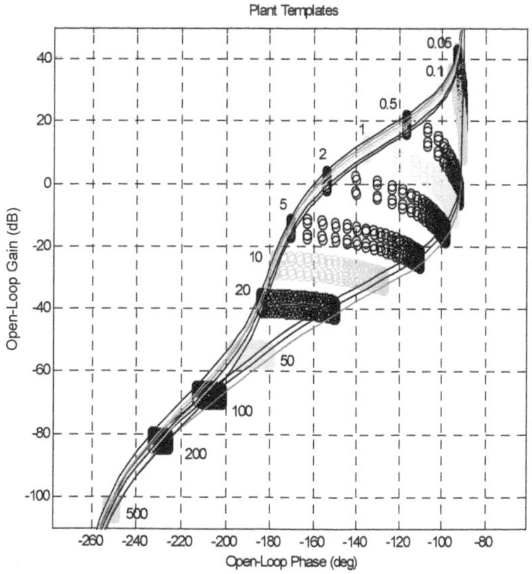

CHAPTER 3

3.1.

a)

$$B_U(s)=\frac{2.222s+19.75}{s^2+5.333s+19.75} \quad B_L(s)=\frac{84.31}{s^3+13.16s^2+57.69s+84.31}$$

d)

$$B_U(s)=\frac{2.193s+30.07}{s^2+5.818s+30.07} \quad B_L(s)=\frac{88.66}{s^3+13.38s^2+59.66s+88.66}$$

3.2.

a) $T_D(s)=\frac{3.423s}{s^2+18.7s+349.7}$ b) $T_D(s)=\frac{3.559s}{s^2+24.3s+590.5}$

3.5.

$$G(s)=\frac{23.6\ (0.8s+1)}{\left(1.81\times10^{-5}s^2+0.006s+1\right)} \qquad F(s)=\frac{1}{(0.2\ s+1)}$$

3.10.

$$G(s)=2708.36\ \frac{(s+10.32)(s+1.61)}{s(s+300)} \qquad F(s)=\frac{16}{(s+4)^2}$$

3.11.

$$G(s)=7030.75\ \frac{(s+75.13)(s+26.08)}{(s+114.8)(s+400)} \qquad F(s)=\frac{17.5}{(s+3.5)(s+5)}$$

3.13.

$$G(s) = 5913.95\ \frac{(s+5.15)(s+1.97)}{s(s+400)}$$

3.15.

$$G(s) = 3138\frac{(s+6)}{s(s+104.6)} \qquad F(s) = \frac{21}{(s+7)(s+3)}$$

3.16.

$$G(s) = 2237.94\frac{(s+6.22)}{s(s+87)} \qquad F(s) = \frac{28.42}{(s+9.8)(s+2.9)}$$

3.20.

$$G(s) = 561.14\frac{(s+7.05)(s+1.82)}{s(s+240)} \qquad F(s) = \frac{10.5}{(s+3)(s+3.5)}$$

3.21.

$$G(s) = 473.24\frac{(s+5.65)(s+1.87)}{s(s+200)} \qquad F(s) = \frac{15}{(s+3)(s+5)}$$

3.23.

a) The complete template is:

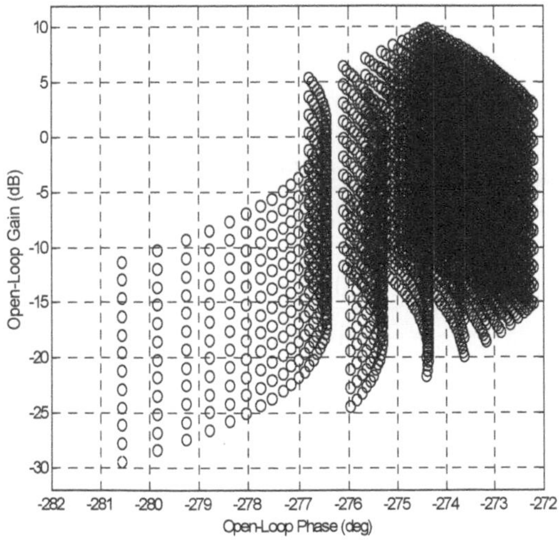

b) The template of the contour of the parameter uncertainty space is:

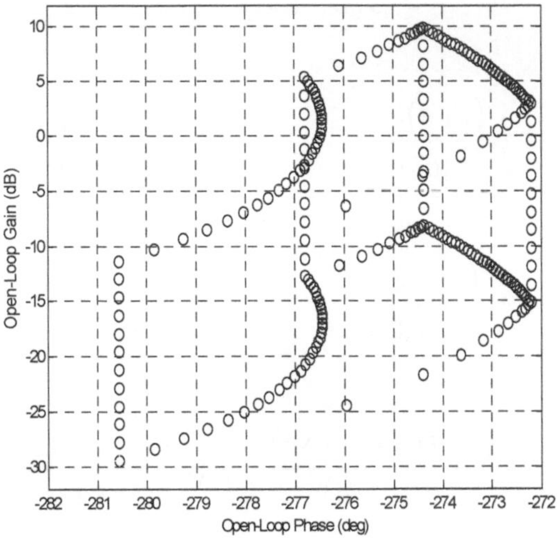

3.28. According to the stability $|1/(1+L)| \leq 1.3$ specification and some other classical performance specifications (overshoot $\leq$ 22%, etc.), and taking into account the parameter uncertainty, a solution for the compensator is:

$$G(s) = \frac{20\left(\frac{1}{0.05}s+1\right)\left(\frac{1}{1.5}s+1\right)}{s\left(\frac{1}{15}s+1\right)\left(\frac{1}{50}s+1\right)}$$

3.30. According to the defined stability and performance specifications, and taking into account the parameter uncertainty, a solution for the compensator is:

$$G(s) = \frac{0.0537s^3 + 5.0905s^2 + 110.8s + 294.3}{-0.144\times10^{-3}s^3 + -0.0289s^2 + 1.4415s + 1}$$

3.31. According to the defined stability and performance specifications, and taking into account the parameter uncertainty, a solution for the compensator and the pre-filter is:

$$G(s) = \frac{37.71\left(\dfrac{s}{0.1}+1\right)}{s\left(\dfrac{s}{175.5}+1\right)} \qquad F(s) = \frac{28}{s^2+11s+28}$$

CHAPTER 4

4.1. According to the defined stability specification and taking into account the parameter uncertainty, a solution for the compensator G(s) and the corresponding controller G(z) is:

$$G(s) = \frac{0.8(s+1)\left(\dfrac{s}{0.5}+1\right)^2\left(\dfrac{s}{0.1}+1\right)}{s\left(\dfrac{s}{150}+1\right)\left(\dfrac{s}{112}+1\right)\left(\dfrac{s}{90}+1\right)}$$

$$G(z) = \frac{(1.2350z^4 - 4.9150z^3 + 7.3340z^2 - 4.8630z + 1.2090)\times 10^7}{z^4 - 1.8040z^3 + 1.0060z^2 - 0.2167z + 0.01528}$$

Sampling time: $T = 0.01$ seconds for the Tustin discretization approach.

CHAPTER 5

5.1.

(a)

$$\boldsymbol{P}^{-1} = \begin{bmatrix} \dfrac{s(s+1)}{s-0.2} & \dfrac{-0.4(s+1)}{s-0.2} \\ \dfrac{-(s+1)(s\ 2)}{s-0.2} & \dfrac{2(s+1)(s+2)}{s-0.2} \end{bmatrix}$$

(b) Since the desired $T(s)$ is specified then from

$$T = [I + PG]^{-1} PG = [I + L]^{-1} L$$

solving for L yields

$$L = \frac{2}{s}\begin{bmatrix} 1 & 0 \\ 0 & 1 \end{bmatrix} = PG$$

Thus:

$$\mathrm{L} = \mathrm{PG} \rightarrow \mathrm{G} = \mathrm{P}^{-1}\mathrm{L} = \frac{2(s+1)}{s(s-0.1)}\begin{bmatrix} 1 & -0.4 \\ -0.5(s+2) & 2(s+2) \end{bmatrix}$$

(c) With the specified G the following is obtained:

$$L = PG = \begin{bmatrix} \dfrac{(0.5+\varepsilon_1)(s-0.2)s - 0.2(0.5+\varepsilon_2)}{(s+1)(s+2)(s-0.2)} & \dfrac{(s+2)g_{12} + 0.2g_{22}}{(s+1)(s+2)} \\ \dfrac{0.5s(\varepsilon_1 - 0.5\varepsilon_2)}{(s+1)(s+2)} & \dfrac{0.5(s+2)g_{12} + 0.5sg_{12}}{(s+1)(s+2)} \end{bmatrix}$$

Analyzing the elements of the first column it is noted that they can be n.m.p. which in turn impose restrictions on the elements of $T(s)$ in order to avoid in designing an unstable system. The problem can be foreseen by examining *det* $P(s)$ which has a RHP zero thus resulting in an n.m.p. system. This imposes restrictions on the elemenets of T because

$$G = \frac{(Adj\,P)L}{\det P} \quad \textit{is required.}$$

(d) Since $G = P^{-1}L = P^{-1}T[I - T]^{-1}$, one can assign the r.h.p. zeros of *det* P to each element of T. Thus, substituting the specified T and P^{-1} of part (a) yields

$$G(s) = \frac{(s+1)(s+2)}{\tau s^2 + 2\tau s + 0.2}\begin{bmatrix} \dfrac{s}{s+2} & \dfrac{-0.4}{s+2} \\ -1 & 2 \end{bmatrix}$$

which does not have any RHP poles. The pole at the origin for the **G** of part (b) is replaced by the LHP poles $\tau s^2 + 2\tau s + 0.2$.

5.4.

(a) For Fig. A: $T_A = G_I P / (1 + G_I P)$; for Fig. B: $T_B = G_2 P / [1 + P(H + G_2)]$. Setting these two equal to one another and substituting in the known transfer functions yields $G_2(s) = 1/(s + 1)$.

(b) The sensitivity functions for the structures are, respectively,

$$\left(S_P^T\right)_A = \frac{s(s+2)}{(s+1)^2} \qquad \left(S_P^T\right)_B = \frac{s}{s+1}$$

The ratio of these two functions is:

$$\left|\frac{(S_P^T)_B}{(S_P^T)_A}\right| = \left|\frac{s+1}{s+2}\right| < 1 \textit{ for all finite } \omega$$

Thus, since $|S_P^T|_B < |S_P^T|_A$ then the structure of Fig. B, which has a *2* degree of freedom structure, yields a better design and can be made even better in terms of the system sensitivity to *P*.

CHAPTER 6

6.2.

(a) From Eq. (5.44):

$$\tau_{c_{11}} = \frac{b_{21}(q_{11}/q_{12})}{1 + g_1 q_{11}} \approx \frac{b_{21}}{g_1 q_{12}} \quad \text{at low frequency where} \quad |g_1 q_{11}| >> 1$$

$$\approx \left(\frac{0.01}{0.5}\right)_{\max}\left(\frac{1}{g_1}\right) = \frac{0.02}{g_1}$$

where the maximum value of $|b_{21}/q_{12}|$ is used. Thus,

$$g_1 q_{11_o} \geq \frac{0.02 q_{11_o}}{\tau_{c_{11}}} = \frac{0.004}{\tau_{c_{11}}}$$

Therefore the q_{11} template is *14 dB* in height and the $\tau_{c_{11}}$ bound can be ignored resulting in the optimal bound for the 1,1 loop being essentially the same as the tracking bound due to τ_{11}. Note that

$$\tau_{c_{11}} \approx \frac{0.004}{\ell_{1_o}} \quad for\ |\ \ell_{1_o}\ | > 0$$

For the frequency range of 0.1 to 2 *rad/sec* the bounds on L_{1_o} are obtained on the basis of τ_{11}. For values of frequency greater than 2 rad/sec the bounds are obtained based on the following:

$$|\,1+\ell_1\,| \geq \frac{b_{21}}{b_{11}}\left|\frac{q_{11}}{q_{12}}\right|; \quad |\,1+\ell_2\,| \geq \frac{b_{11}}{b_{21}}\left|\frac{q_{22}}{q_{21}}\right|$$

(b) For the bounds on L_{2_o} use

$$\left|\frac{1}{1+\boldsymbol{L}_2}\right| \leq \frac{b_{21}}{b_{11}}\left|\frac{q_{21}}{q_{22}}\right|$$

for the "low frequency range." For e.g., $b_{21}/b_{11} = 0.01/0.65 \rightarrow -36$ dB at $\omega = 2$. This gives the bound on $\boldsymbol{L}_{2_o}$ $(j2)$ at approximately *+10* dB at *–100°* which is much tougher than that on $\boldsymbol{L}_{1_o}(j2)$. Thus, a "trade-off" is in order to improve (lower) the bounds on $\boldsymbol{L}_{2_o}$ at the expense of raising the bounds on $\boldsymbol{L}_{1_o}$. It is left to the reader to determine the bounds with b_{11} = *0.18* and compare them with those determined for the original value of b_{11}.

6.3.

$$L_{1_0}(s) = \frac{8}{2s+8}$$

$$L_{1_0}(s) = \frac{12}{0.5s + 5} \qquad g_1(s) = \frac{6.5s + 11}{s^2 + 20s}$$

$$f_1(s) = \frac{3.583s^2 + 2.583s + 1.755}{s^3 + 4.242s^2 + 2.851s + 1.755}$$

$$g_2(s) = \frac{104s + 2095}{s^2 + 75s} \qquad f_2(s) = \frac{9}{s+9}$$

6.4.

$$L_{1_0}(s) = \frac{8}{2s + 8} \qquad L_{2_0}(s) = \frac{12}{0.5s + 5}$$

$$g_1(s) = \frac{55.56s + 50}{s^2 + 20s} \qquad f_1(s) = \frac{2.5}{s + 2.5}$$

$$g_2(s) = \frac{2205s + 44100}{s^2 + 315s} \qquad f_2(s) = \frac{9}{s + 9}$$

CHAPTER 7

7.1.

$$\gamma_{12} = \frac{p_{12}\, p_{21}}{p_{11}\, p_{22}}$$

$$q_{11} = \frac{\det \boldsymbol{P}}{p_{22}}, \quad q_{22} = \frac{\det \boldsymbol{P}}{p_{11}}, \quad q_{12} = \frac{-\det \boldsymbol{P}}{p_{12}}, \quad q_{21} = \frac{-\det \boldsymbol{P}}{p_{21}}$$

$$p_{22} = \frac{\det \boldsymbol{P}}{q_{11}}, \quad p_{11} = \frac{\det \boldsymbol{P}}{q_{22}}, \quad p_{12} = \frac{-\det \boldsymbol{P}}{q_{12}}, \quad p_{21} = \frac{-\det \boldsymbol{P}}{q_{21}}$$

$$\gamma_{12} = \frac{q_{11} q_{22}}{q_{12} q_{21}} \qquad L_{1_0}(s) = \frac{8}{2s+8} \qquad L^e_{2_0}(s) = \frac{q_{11}\left(1+q_{11}g_{11}\right)}{1-\gamma_{12}+q_{11}g_{11}}$$

$$g_1(s) = \frac{2.12\,s+2.75}{s^2+5\,s} \qquad f_1(s) = \frac{3.583s^2+2.583s+1.755}{s^3+4.242s^2+2.851s+1.755}$$

$$g_2(s) = 10^5\,\frac{5.753s^2+5s+6}{s^3+1618s^2+1.3\times10^4} \qquad f_2(s) = \frac{9}{s+9}$$

7.6.

$$L_{1_0}(s) = \frac{8}{2s+8} \qquad L^e_{2_0}(s) = \frac{q_{11}\left(1+q_{11}g_{11}\right)}{1-\gamma_{12}+q_{11}g_{11}}$$

$$g_1(s) = \frac{55.56s+50}{s^2+20s} \qquad f_1(s) = \frac{2.5}{s+2.5}$$

$$g_2(s) = 10^5\,\frac{5.753s^2+5s+6}{s^3+1618s^2+1.3\times10^4 s} \qquad f_2(s) = \frac{8.7}{s+8.7}$$

CHAPTER 8

8.7. According to the defined stability and performance specifications, and taking into account the parameter uncertainty, a solution for the diagonal MIMO compensator is:

$$g_{22} = \frac{7.23s^2+56.96s+111.88}{s\left(s^2+21.31s+111.88\right)} \qquad g_{11} = \frac{0.306s+14.176}{s}$$

CHAPTER 10

10.4. The RGA matrix of the plant is,

$$\Lambda = \begin{bmatrix} \mu & 1-\mu \\ 1-\mu & \mu \end{bmatrix}, \quad \text{where } \mu \in [0.75;0.818] \text{ due to the uncertainty.}$$

Thus, the input 1 controls the output 1, and the input 2 controls the output 2. The p_{11} is chosen to be the first loop to be closed because it has the lowest bandwidth compared to p_{22}. According to the defined stability and performance specifications, and taking into account the parameter uncertainty, a solution for the compensator is:

$$\boldsymbol{G}(\mathrm{s}) = \begin{bmatrix} \dfrac{0.6496}{(s/0.8556+1)} & \dfrac{0.48}{(s/47+1)} \\ \dfrac{-0.43}{(s/0.8556+1)} & \dfrac{1.197}{(s/47.66+1)} \end{bmatrix}$$

10.5. According to the defined stability and performance specifications, and taking into account the parameter uncertainty, a solution for the non-diagonal MIMO compensator is:

$$g_{11}(s) = \frac{3(s+0.0006)}{(s+0.8)(s+0.05)(s+0.001)}$$

$$g_{21}(s) = \frac{-(s+0.0007)}{(s+0.8)(s+0.05)(s+0.002)(s+0.0015)}$$

$$g_{22}(s) = \frac{4(s+0.00042)}{(s+0.3)(s+0.04)(s+0.0006)}$$

$$g_{12}(s) = \frac{-(s+0.0004)}{(s+2)(s+0.001)(s+0.02)}$$

$$g_{32}(s) = \frac{-0.01(s+0.1)(s+0.0004)}{(s+0.3)(s+0.04)(s+0.02)(s+0.0006)}$$

$$g_{33}(s) = \frac{0.4\ (s+0.05)}{(s+0.4)\ (s+0.07)\ (s+0.008)}$$

$$g_{23}(s) = \frac{-(s+0.05)}{(s+0.008)}$$

The pre-filter matrix is diagonal, so that:

$$f_{11}(s) = \frac{0.0004}{s+0.0004} \qquad f_{22}(s) = \frac{0.000265}{s+0.000265}$$

$$f_{33}(s) = \frac{0.00035}{s+0.00035}$$

10.9. According to the defined stability and performance specifications, and taking into account the parameter uncertainty, a solution for the non-diagonal MIMO compensator is:

$$g_{11}(s) = \frac{7550\ s+2718}{s^2+302\ s} \qquad g_{21}(s) = \frac{-18875s^2-10570s-1359}{3s^3+907s^2+302s}$$

$$g_{22}(s) = \frac{26745s^2+1207\times10^3 s+168498}{s^3+277s^2+17020s}$$

$$g_{12}(s) = \frac{7430s^3+3.4\times10^5 s^2+1.8\times10^5 s+18725}{s^4+277.1s^3+17047.7s^2+1418s}$$

Index

MIMO QFT technique, sequentially non-diagonal, some practical issues, 310-311